Data Analysis for Scientists and Engineers

Data Analysis for Scientists and Engineers

Edward L. Robinson

PRINCETON UNIVERSITY PRESS

PRINCETON AND OXFORD

Published by Princeton University Press, 41 William Street, Princeton, New Jersey 08540
In the United Kingdom: Princeton University Press, 6 Oxford Street, Woodstock,
Oxfordshire OX20 1TW

press.princeton.edu

Jacket art courtesy of Shutterstock

ISBN 978-0-691-16992-7

Library of Congress Control Number: 2016930361

British Library Cataloging-in-Publication Data is available

This book has been composed by S4Carlisle Publishing Limited
Printed on acid-free paper.
Printed in the United States of America
10 9 8 7 6 5 4 3 2 1

Contents

Preface

Therefore reasoning does not suffice, but experience does.

Mathematics is the door and key to the sciences.[1]

—Roger Bacon (ca. 1214–1294)

Modern computers have revolutionized statistics. Techniques now routinely employed to analyze data were impractical and even unthinkable just a few years ago. Large data sets and exhaustive calculations can be handled comfortably, often by a typical laptop computer. Techniques once thought abstruse have become standard tools: principle component analysis, Markov chain Monte Carlo sampling, nonlinear model fitting, Bayesian statistics, Lomb-Scargle periodograms, and so on. Scientists and engineers must be conversant with more ways and more sophisticated ways to analyze data than ever before.

For many years I have taught a graduate course on data analysis for astronomers, physicists, and the occasional engineer. The purpose of the course has been to equip experimenters with the skills needed to interpret their data and theoreticians with enough knowledge to understand (and sometimes to question!) those interpretations. I was unable to find a book—or even a selection of books—that could serve as a textbook for the course. Much of the material in the course is not elementary and is not usually included in the many introductory books on data analysis. Books that did cover the material were highly specialized and written in a style and language opaque to most of my students. Books covering specific algorithms in specific computer languages were more appropriate as supplementary sources.

Driven by need, I wrote my own notes for the course, and the notes eventually grew to become this book. It is meant to be a book on advanced data analysis, not an introduction to statistics. Indeed, one may doubt whether yet another elementary introduction to, say, linear regression is really needed. At the same time, the book must be self-contained and must be understandable to readers with a wide variety of backgrounds, so it does cover basic concepts and tools. It includes many specific examples, but it is not a cookbook of statistical methods and contains not a line of computer code. Instead, the course and the

---

[1] "Ergo argumentum non sufficit, sed experientia." *Opus Maius*, Part 6, Chapter 1; "Et harum scientiarum porta et clavis est Mathematica." *Opus Maius*, Part 4, Chapter 1.

book emphasize the principles behind the various techniques, so that practitioners can apply the techniques to their own problems and develop new techniques when necessary. While the target audience is graduate students, the book should be accessible to advanced undergraduates and, of course, to working professionals.

The book is focused on the needs of people working in the physical sciences and engineering. Many of the statistical tools commonly used in other areas of research do not play a large role in the physical sciences and receive only minimal coverage. Thus, the book gives little space to hypothesis testing and omits ANOVA techniques altogether, even though these tools are widely used in the life sciences. In contrast, fits of models to data and analyses of sequences of data are common in the physical sciences, and Bayesian statistics is spreading ever more widely. These topics are covered more thoroughly.

Even so restricted, the subject matter must be heavily pruned to fit in a single book. The criterion I have used for inclusion is utility. The book covers data analysis tools commonly used by working physical scientists and engineers. It is divided into three main sections:

- The first consists of three chapters on probability: Chapter 1 covers basic concepts in probability, then Chapter 2 is on useful probability distributions, and finally Chapter 3 discusses random numbers and Monte Carlo methods, including Markov chain Monte Carlo sampling.
- The next section begins with Chapter 4 introducing basic concepts in statistics and then moves on to model fitting, first from a frequentist point of view (maximum likelihood, and linear and nonlinear χ^2 minimization; Chapters 5 and 6) and then from a Bayesian point of view (Chapter 7).
- The final section is devoted to sequences of data. After reviewing Fourier analysis (Chapter 8), it discusses power spectra and periodograms (Chapter 9), then convolution and image reconstruction, and ends with autocorrelation and cross correlation (Chapter 10).

An emphasis on error analysis pervades the book. This reflects my deep conviction that data analysis should yield not just a result but also an assessment of the reliability of the result. This might be a number plus a variance, but it could also be confidence limits, or, when dealing with likelihood functions or a Bayesian analysis, it could be plots of one- or two-dimensional marginal distributions.

Committed Bayesians may initially be unhappy that only one chapter is devoted to Bayesian statistics. In fact, though, the first two chapters on probability provide the necessary foundations for Bayesian statistics; and the third chapter, which includes a lengthy discussion of Markov chain Monte Carlo sampling, is motivated almost entirely by Bayesian statistics. Likelihood functions are covered extensively, albeit often tacitly, in the two chapters on least squares estimation. The book could easily serve as a textbook for a course devoted solely to Bayesian statistics. Because the book discusses both Bayesian and frequentist approaches to data analysis, it allows a direct comparison of the two. I have found that this comparison greatly improves students' comprehension of Bayesian statistics.

Nearly all the material in the book has already been published in other places by other people, but the presentation is my own. My goal has been to write about the material in a way that is understandable to my students and my colleagues. Much of the book is a translation from the elegant and precise language of mathematicians to the looser, workaday language of scientists and engineers who wrestle directly with data. The book nowhere mentions heteroskedastic data; it does discuss data with variable, sometimes correlated measurement errors!

The presentation is, nevertheless, mathematical, but the style is that of the physical sciences, not of mathematics. I have aimed for clarity and accuracy, not rigor, and the reader will find neither proofs nor lemmas. The book assumes that the reader is conversant with the calculus of several variables and is acquainted with complex numbers. It makes heavy use of linear algebra. It is my experience that most graduate students have taken at least one course on linear algebra, but their knowledge has rusted from lack of use, especially when it comes to eigenvalues and eigenvectors. A lengthy review of linear algebra is provided in appendix E. Specialized topics that would distract from the flow of the book have also been relegated to the appendices. Because of its crucial importance for the analysis of sequences, an entire chapter is devoted to Fourier analysis.

Finally, if you plan to learn or teach from this book, here is an observation, as valid today as 2400 years ago: "for the things we have to learn before we can do them, we learn by doing them, as builders by building and lyre players by playing the lyre."[2] To learn how to analyze data, analyze data—real data if available, artificial data if not. Pre-existing computer programs can readily be found for the analysis techniques discussed in the book, but, especially when first encountering a technique, use the programs only if you must and never without testing them extensively. It is better by far to write your own code.

Few people, certainly not I, can write a book like this without extensive help from colleagues, staff, and students. To all my students, current and former, especially to the students who have taken my course and given me feedback on earlier versions of the book; to my former postdocs, especially Allen Shafter, Janet (née) Wood, Coel Hellier, William Welsh, and Robert Hynes; and to my colleagues in the Department of Astronomy at the University of Texas, especially Terrence Deeming, William Jefferys, Pawan Kumar, Edward Nather, and Donald Winget, thank you.

---

[2] "ἃ γὰρ δεῖ μαθόντας ποιεῖν, ταῦτα ποιοῦντες μανθάνομεν, οἷον οἰκοδομοῦντες οἰκοδόμοι γίνονται καὶ κιθαρίζοντες κιθαρισταί." Aristotle, *Nicomachean Ethics*, Book II.

Probability

1.1 The Laws of Probability

Figure 1.1 shows a standard six sided die, each side distinguished by a unique number of dots from 1 to 6. Consider these three questions:

1. The die is thrown and lands with one side up. Before we look at the die and know the answer, we ask "What is the probability that the side with three dots is up?"
2. The die is weighed many times, each time by a different scale, and the weights are recorded. "How much does the die weigh, and how reliable is the measurement of the weight?"
3. The die is thrown many times, and the up side is recorded each time. Let H be the hypothesis that all faces are equally likely to be on the up side. "What is the probability that H is true?"

Probability is a deductive discipline relying on pre-existing information. Statistics deals with observations of the real world and inferences made from them. The first question can be answered before actually throwing the die if enough is known about the die. It lies in the province of probability. The answers to the second and third questions depend on the measurements. They lie in the province of statistics. The subject of this book is data analysis, which lies firmly in the province of statistics; but the language and mathematical tools of statistics rely heavily on probability. We begin, therefore, with a discussion of probability.

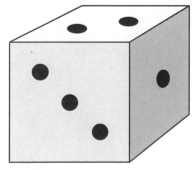

Figure 1.1: A typical six-sided die.

The classical definition of probability is based on frequency. Suppose that an event A can happen in k ways out of a total of n possible, equally likely, ways. The probability that A will occur is

$$P(A) = \frac{k}{n}. \tag{1.1}$$

This definition presupposes that enough is known to permit the calculation of k and n in advance, so it is sometimes called the *a priori* probability. Although the definition may seem reasonable, it has several limitations. First, both k and n must be finite. The answer to the question "If one picks out a star in the universe at random, what is the probability that it is a neutron star?" cannot be calculated from the classical definition if the universe is infinite in size. If k and n are not finite, it is better to define probability as the limit

$$P(A) = \lim_{n \to \infty} \frac{k}{n}. \tag{1.2}$$

Next, k and n must both be countable. Consider an archer shooting an arrow at a target. The answer to the question "What is the probability that an arrow will hit within 1 inch of the center of a target?" cannot be calculated from equations 1.1 and 1.2, because the possible places the arrow might hit form a continuum and are not countable. This problem can be fixed by replacing k and n with integrals over regions. The probability that A occurs is the fraction of the regions in which event A occurs, producing yet a third definition of probability. These three definitions are, of course, closely related. We will need to use all three.

Another problem with the classical definition of probability is that the words "equally likely" actually mean "equally probable," so the definition is circular! One must add an independent set of rules for calculating k and n, or at least for assessing whether probabilities for two events are equal. If the rules are chosen cleverly, the calculations will yield results in accord with our intuition about probability; but there is no guarantee the rules have been chosen correctly and certainly no guarantee that the calculations will apply to our universe. In practice one breaks the circularity by invoking external arguments or information. One might, for example, claim that a six-sided die is symmetric, so each side has the same probability of landing face up.

Finally, we shall see that Bayesian statistics allows calculation of probabilities for unique events. Practitioners of Bayesian statistics can and do calculate the probability that, for example, a specific person will be elected president of the United States in a specific election year. Frequency is meaningless for unique events, so Bayesian statistics requires a reassessment of the meaning of probability. We will delay a discussion of nonfrequentist interpretations of probability until Chapter 7.

One typically calculates probabilities for a complicated problem by breaking it into simpler parts whose probabilities are more easily calculated and then using the laws of probability to combine the probabilities for the simple parts into probabilities for the more complicated problem. The laws of probability can be derived with the help of the Venn diagrams shown in Figure 1.2. Let S be the set of all possible outcomes of an experiment and let A be a subset of S, denoted by

$$A \subset S \tag{1.3}$$

(see the top Venn diagram in Figure 1.2). If the outcomes can be counted and the number of outcomes is finite, then

$$P(A) = \frac{n_A}{n_S}, \tag{1.4}$$

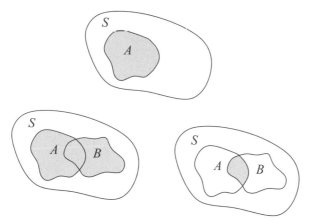

Figure 1.2: The upper figure shows the subset A of the set S. The two lower figures include a second subset B. The shaded region in the left-hand figure is $A \cup B$, and the shaded region in the right-hand figure is $A \cap B$.

where n_S is the total number of outcomes in S, and n_A is the number of outcomes in the subset A. If outcomes in A never occur, so that A is an empty set, then $P(A) = 0/n_S = 0$. If A includes all of S, so that all outcomes are in S, then $n_A = n_S$, and $P(A) = n_S/n_S = 1$. The probability must, then, lie in the range

$$0 \leq P(A) \leq 1. \tag{1.5}$$

The intersection and union of two sets are denoted by the symbols $\cap$ and $\cup$:

$A \cup B = B \cup A =$ the union of sets A and B with the overlap counted just once,
$A \cap B = B \cap A =$ the intersection of A and B.

The meanings of these operations are shown by the two lower Venn diagrams in Figure 1.2. Suppose we have two subsets of S, A and B. The probability that an outcome lies in A or B or both is given by

$$P(A \cup B) = P(A) + P(B) - P(A \cap B). \tag{1.6}$$

The first two terms on the right-hand side are simply the probabilities that the outcome lies in either A or B. We cannot just add the two probabilities together to get $P(A \cup B)$, because the two sets might have outcomes in common—the two sets might overlap. If they do, the overlap region is counted twice by the first two terms. The third term is the overlap of the two sets. Subtracting it removes one of the double-counted overlap regions. If the two sets do not intersect, so that $P(A \cap B) = 0$, then equation 1.6 reduces to

$$P(A \cup B) = P(A) + P(B). \tag{1.7}$$

The complement of A, denoted by $\bar{A}$, is the subset of S consisting of all members of S not in A (see Figure 1.3). Together A and $\bar{A}$ make up all of S, so

$$P(A \cup \bar{A}) = P(S). \tag{1.8}$$

They do not overlap, so

$$P(A \cap \bar{A}) = 0. \tag{1.9}$$

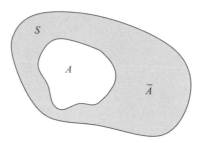

Figure 1.3: The subset $\bar{A}$ of the set S consists of all members of S not in A.

Therefore,

$$1 = P(S) = P(A \cup \bar{A}) = P(A) + P(\bar{A}), \tag{1.10}$$

from which we find

$$P(\bar{A}) = 1 - P(A). \tag{1.11}$$

The conditional probability, denoted by $P(A|B)$, is the probability that A occurs if B occurs, often stated as the probability of A "given" B to avoid implications of causality. It is defined by

$$P(A|B) = \frac{P(A \cap B)}{P(B)}. \tag{1.12}$$

Example: Suppose we were to throw two dice and the sum of their faces is 7. What is the probability that one of the faces is a 5? In other words, what is the conditional probability $P(5|\text{sum} = 7)$?

There are 36 possible ways the faces of the two dice can land up. The faces can add to 7 in the following 6 ways:

1-6	4-3
2-5 X	5-2 X
3-4	6-1

Therefore $P(\text{sum} = 7) = 6/36 = 1/6$. Two of the ways of adding to 7, the ones marked with an X, contain a 5. So, the probability that the sum is 7 and also that one die is a 5 is $P(5 \cap \text{sum} = 7) = 2/36 = 1/18$. The conditional probability is, therefore,

$$P(5|\text{sum} = 7) = \frac{P(5 \cap \text{sum} = 7)}{P(\text{sum} = 7)} = \frac{1/18}{1/6} = 1/3$$

We could, of course, have calculated the conditional probability directly by noting that 1/3 of the entries in the list of combinations are marked with an X.

Events A and B are said to be independent of each other if the occurrence of one event does not affect the probability that the other occurs. This is expressed by

$$P(A|B) = P(A). \tag{1.13}$$

Plugging equation 1.13 into equation 1.12, we derive another way of expressing independence:

$$P(A \cap B) = P(A)P(B). \tag{1.14}$$

So, if events A and B are independent of each other, the probability that both occur is given by the products of the probabilities that each occurs.

Because of the symmetry between A and B, there are two ways to write $P(A \cap B)$:

$$P(A \cap B) = P(A|B)\,P(B) \tag{1.15}$$

and

$$P(A \cap B) = P(B|A)\,P(A). \tag{1.16}$$

Together, these two equations mean

$$P(A|B)\,P(B) = P(B|A)\,P(A) \tag{1.17}$$

and, therefore,

$$P(B|A) = \frac{P(A|B)\,P(B)}{P(A)}. \tag{1.18}$$

This important result is the basis of Bayesian statistics.

1.2 Probability Distributions

1.2.1 Discrete and Continuous Probability Distributions

Up to now we have assumed that an experiment has two possible outcomes, A or B. Now let there be many possible outcomes. We denote the outcomes by A_j, where j is an integer index, and the probability of outcome A_j by $P(A_j)$. To be a valid probability distribution, $P(A_j)$ must be single valued and satisfy

$$P(A_j) \geq 0 \tag{1.19}$$

$$\sum_j P(A_j) = 1. \tag{1.20}$$

These mild constraints leave wide latitude for possible discrete probability distributions. The other laws of probability generalize in obvious ways. One is worth mentioning explicitly since it occurs so often: If n events A_j are independent of one another, the probability that all will occur is equal to the product of the individual probabilities of occurrence:

$$P(A_1 \cap A_2 \cdots \cap A_n) = P(A_1)P(A_2) \cdots P(A_n). \tag{1.21}$$

It is also worth working out one nontrivial example for overlapping probabilities: The probability that any of three possible outcomes occur is

$$
\begin{aligned}
P(A_1 \cup A_2 \cup A_3) &= P((A_1 \cup A_2) \cup A_3) \\
&= P(A_1 \cup A_2) + P(A_3) - P((A_1 \cup A_2) \cap A_3) \\
&= P(A_1) + P(A_2) - P(A_1 \cap A_2) + P(A_3) \\
&\quad - [P(A_1 \cap A_3) + P(A_2 \cap A_3) - P(A_1 \cap A_2 \cap A_3)] \\
&= P(A_1) + P(A_2) + P(A_3) \\
&\quad - P(A_1 \cap A_2) - P(A_1 \cap A_3) - P(A_2 \cap A_3) \\
&\quad + P(A_1 \cap A_2 \cap A_3).
\end{aligned}
\tag{1.22}
$$

The following example shows how these rules can be used to calculate a priori probabilities for a symmetric six-sided die.

Example: Calculate from first principles the probability that any one face of a six-sided die will land face up.

From the physical properties of dice, one or another face must land face up. Therefore probability that the die lands with a 1 or 2 or 3 or 4 or 5 or 6 on the top face must equal 1. We express this by the equation

$$P(1 \cup 2 \cup 3 \cup 4 \cup 5 \cup 6) = 1.$$

Again from physical properties of dice, only one of the faces can land face up. This requires that all the intersections are equal to zero:

$$P(1 \cap 2) = P(1 \cap 3) = \cdots = P(5 \cap 6) = 0.$$

Extending equation 1.22 to six possible outcomes, we have

$$P(1) + P(2) + P(3) + P(4) + P(5) + P(6) = 1.$$

If the die is nearly symmetric, the probabilities must all be equal:

$$P(1) = P(2) = P(3) = P(4) = P(5) = P(6).$$

We have, then

$$6P(1) = 1$$

$$P(1) = \frac{1}{6},$$

and the other faces have the same probability of landing face up.

The outcomes of an experiment can also form a continuum. Let each outcome be any real number x. A function $f(x)$ can be a probability distribution function if it is single-valued and if

$$f(x) \geq 0 \tag{1.23}$$

$$\int_{-\infty}^{\infty} f(x)\, dx = 1 \tag{1.24}$$

$$P(a \leq x \leq b) = \int_{a}^{b} f(x)\, dx, \tag{1.25}$$

where $P(a \leq x \leq b)$ is the probability that x lies in the range $a \leq x \leq b$. Equation 1.23 ensures that all probabilities are positive, and equation 1.24 ensures that the probability of all possible outcomes is equal to 1. A function that satisfies equation 1.24 is said to be *normalized*. Together these two equations replace equations 1.19 and 1.20 for discrete probabilities. The third requirement defines the relation between $f(x)$ and probability and replaces equation 1.4.

The rectangle function is a simple example of a valid continuous probability distribution function:

$$f(x) = \begin{cases} 0, & x < 0 \\ 1/a, & 0 \le x \le a \\ 0, & x > a \end{cases} . \tag{1.26}$$

Note that $f(x) > 1$ if $a < 1$. Since probabilities must be less than 1, $f(x)$ is clearly not itself a probability. The probability is the integral of $f(x)$ between two limits (equation 1.25), so one must always discuss the probability that x lies in a given range. Because of this, $f(x)$ is sometimes called the *probability density distribution function*, not the probability distribution function. For example, one should properly speak of the rectangular probability density distribution function, not the rectangular probability distribution function. In practice the distinction is rarely emphasized, and the word "density" remains unverbalized.

Continuous probability distribution functions need not be continuous everywhere, and they need not even be finite everywhere! The Dirac delta function $\delta(x)$ is a legitimate probability distribution function but has the unusual properties

$$\int_{-\infty}^{\infty} \delta(x)dx = 1 \tag{1.27}$$

$$\int_{-\infty}^{\infty} g(x)\delta(x)dx = g(0), \tag{1.28}$$

where $g(x)$ is any reasonable continuous function. It can loosely be thought of as a function with properties

$$\delta(x) = \begin{cases} \infty, & x = 0 \\ 0, & x \ne 0 \end{cases} . \tag{1.29}$$

Example: Consider the exponential function

$$f(x) = \begin{cases} 0, & x < 0 \\ a\exp[-ax], & x \ge 0 \end{cases} \tag{1.30}$$

with $a > 0$. This function satisfies the requirements for it to be a probability distribution function:

- It is single valued.
- It is greater than or equal to 0 everywhere.
- It is normalized:

$$\int_{-\infty}^{\infty} f(x)\, dx = \int_{0}^{\infty} a\exp[-ax]\, dx = -\exp[-ax]\Big|_{0}^{\infty} = 1.$$

- Its integral exists for all intervals:

$$P(x_a \le x \le x_b) = \int_{x_a}^{x_b} f(x)\, dx = \exp[-x_a] - \exp[-x_b] \ \text{ for } 0 \le x_a \le x_b,$$

with analogous results if the limits are less than 0.

Continued on page 8

The exponential probability distribution is widely applicable. For example, the distribution of power in the power spectrum of white noise and the distribution of intervals between decays of a radioactive source are given by an exponential probability distribution.

1.2.2 Cumulative Probability Distribution Function

The cumulative distribution function, $F(x)$, is defined to be

$$F(x) = \int_{-\infty}^{x} f(y)\,dy. \tag{1.31}$$

The relation between $f(x)$ and $F(x)$ is shown graphically in Figure 1.4. Since $F(x)$ is the integral over a probability density function, it is a true probability—the probability that x lies between $-\infty$ and x. The cumulative distribution function for the Dirac delta function is a step function, sometimes called the *Heaviside function*:

$$H(x) = \int_{-\infty}^{x} \delta(y)\,dy = \left\{ \begin{array}{ll} 0, & x < 0 \\ 1, & x > 0 \end{array} \right. . \tag{1.32}$$

In fact, the delta function can be defined as the derivative of the Heaviside function. The cumulative distribution function is useful for working with sparse or noisy data.

1.2.3 Change of Variables

Suppose we wish to change the independent variable in a probability distribution function $f(x)$ to a different variable y, where the coordinate transformation is given by $x(y)$. We require that the probability in an interval dx be the same as the probability in the

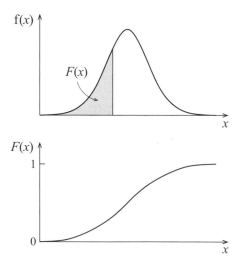

Figure 1.4: The relation between the probability distribution function f(x) and the cumulative distribution function $F(x)$ derived from it.

corresponding dy. The probability distribution for y is then given by

$$f(x)dx = f(x(y)) \left| \frac{dx}{dy} \right| dy = g(y)dy, \tag{1.33}$$

from which we find

$$g(y) = f(x(y)) \left| \frac{dx}{dy} \right|. \tag{1.34}$$

The factor $|dx/dy|$ accounts for the difference between the lengths of dy and dx.

1.3 Characterizations of Probability Distributions

The full description of a probability distribution is the entire distribution itself, given either by its functional form or by an equivalent, such as a table, a graph, or the complete set of moments of the distribution. It is often more convenient to deal with a small set of quantities that succinctly describe the most important properties of a distribution. The descriptors could be a single number that summarizes the entire distribution, such as the mean, mode, or median; a measure of the width of the distribution, such as its variance or full width at half maximum; a measure of the asymmetry, such as skewness; or, more generally, a few of the lower moments of the distribution.

1.3.1 Medians, Modes, and Full Width at Half Maximum

A simple way to summarize a probability distribution $f(x)$ with a single number is to give the value of x at which the distribution reaches its maximum value. This is called the mode of a distribution, x_{mode}. For a continuous function with a single peak, the mode can be calculated from

$$\left. \frac{df(x)}{dx} \right|_{x_{mode}} = 0. \tag{1.35}$$

The mode is most useful for probability distributions with a single maximum or just one dominant maximum.

Another useful single-number descriptor is the median x_{median}, defined by

$$\frac{1}{2} = \int_{-\infty}^{x_{median}} f(x) \, dx. \tag{1.36}$$

The median is the "middle" of a distribution in the sense that a sample from $f(x)$ has equal probabilities of lying above and below x_{median}. The median is particularly useful when one wants to reduce the influence of distant outliers or long tails of a distribution.

The full width at half maximum (often abbreviated by FWHM) is a useful and easily measured descriptor of the width of a probability distribution. The half maximum of a distribution is $f(x_{mode})/2$. To calculate the FWHM, find the values a and b such that

$$f(a) = f(b) = \frac{1}{2} f(x_{mode}); \tag{1.37}$$

then FWHM $= b - a$. Like the mode from which it is derived, the FWHM is most useful for probability distributions with a single maximum or just one dominant maximum; it may not be a sensible way to describe more complicated distributions.

1.3.2 Moments, Means, and Variances

The mth moment M_m of a continuous probability distribution function $f(x)$ is defined to be

$$M_m = \int_{-\infty}^{\infty} x^m f(x) \, dx. \tag{1.38}$$

If $P(A_j)$ is a discrete probability distribution and the outcomes A_j are real numbers, the moments of $P(A_j)$ are

$$M_m = \sum_j A_j^m P(A_j), \tag{1.39}$$

where the sum is taken over all possible values of j. Moment M_m is called the mean value of x^m, symbolically, $M_m = \langle x^m \rangle$. It can be thought of as the average of x^m weighted by the probability that x will occur.

The zeroth moment is equal to 1 for a correctly normalized distribution function:

$$M_0 = \int_{-\infty}^{\infty} x^0 f(x) \, dx = \int_{-\infty}^{\infty} f(x) \, dx = 1. \tag{1.40}$$

The first moment is

$$M_1 = \langle x \rangle = \int_{-\infty}^{\infty} x f(x) \, dx \tag{1.41}$$

or, for a discrete probability distribution,

$$M_1 = \langle A \rangle = \sum_j A_j P(A_j). \tag{1.42}$$

The first moment is usually called the *mean*. To gain a feel for the significance of the mean, suppose that n samples a_i have been generated from the discrete probability distribution $P(A_j)$ and that the number of times each value A_j has been generated is k_j. By definition

$$P(A_j) = \lim_{n \to \infty} \frac{k_j}{n}, \tag{1.43}$$

so the mean is

$$\langle A \rangle = \lim_{n \to \infty} \frac{1}{n} \sum_j A_j k_j. \tag{1.44}$$

Since the quantity k_j means that A_j appears k_j times, we can list all the a_i individually, producing a list of n values of a_i, $i = 1, \ldots, n$, with each value A_j appearing k_j times in the list. The weighted sum in equation 1.44 can therefore be replaced by the unweighted sum over all the individual values of a_i, and the mean value becomes

$$\langle x \rangle = \lim_{n \to \infty} \frac{1}{n} \sum_{i=1}^{n} a_i. \tag{1.45}$$

This corresponds to our intuitive understanding of a mean value.

The mean is yet another way to describe a probability distribution function with a single number. If $f(x)$ is symmetric about x_{median}, then $M_1 = x_{median}$. There is no guarantee that

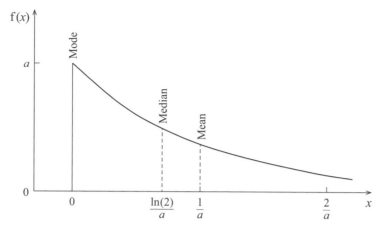

Figure 1.5: The mean, median, and mode of the exponential distribution function $f(x) = a \exp[-ax]$.

a probability distribution has a mean, though. The *Lorentzian distribution* (also called the *Cauchy distribution*)

$$f(x) = \frac{1}{\pi} \frac{b}{b^2 + (x-a)^2} \tag{1.46}$$

satisfies equations 1.23–1.25 and is a perfectly valid probability distribution function, but it does not have a mean. As x becomes large, $xf(x)$ becomes proportional to $1/x$, and the integral of $xf(x)$ approaches a logarithmic function, which increases without limit. The integral in equation 1.41 is therefore undefined, and the mean does not exist.[1] The mode and median of the Lorentzian function do exist and are $x_{mode} = x_{median} = a$. They provide alternative single-number descriptions of the Lorentzian distribution.

Example: Find the mean, median, and mode of the exponential probability distribution function

$$f(x) = a \exp[-ax], \quad x \geq 0.$$

By inspection, the maximum of the function occurs at $x = 0$, so the mode is $x_{mode} = 0$. The median is given by

$$\frac{1}{2} = \int_{-\infty}^{x_{median}} f(x)dx = \int_{0}^{x_{median}} a \exp[-ax] \, dx = 1 - \exp[-ax_{median}],$$

which yields

$$x_{median} = \frac{1}{a} \ln 2.$$

The mean value of x is

$$\langle x \rangle = \int_{-\infty}^{\infty} x a \exp[-ax] \, dx = \left. -\frac{1}{a}(1+x)\exp[-x] \right|_{0}^{\infty} = \frac{1}{a}.$$

The mean, median, and mode for the exponential function are shown in Figure 1.5.

---

[1] If equation 1.5 is taken to be an improper integral, $\langle x \rangle$ does exist for a Lorentzian distribution function, and $\langle x \rangle = a$. The higher moments of the distribution remain undefined, though.

The concept of mean value can be generalized to the mean value of a function. Suppose that A has possible values A_j and that the probability distribution function for the A_j is $P(A_j)$. Also suppose that $g(A)$ is a function of A. The mean value of g is

$$\langle g \rangle = \sum_j g(A_j)P(A_j). \tag{1.47}$$

If $f(x)$ is a continuous probability distribution function and $g(x)$ is also a continuous function, the mean value of $g(x)$, denoted by $\langle g(x) \rangle$, is

$$\langle g(x) \rangle = \int_{-\infty}^{\infty} g(x)f(x)\,dx. \tag{1.48}$$

The quantity $\langle g(x) \rangle$ is a weighted average of $g(x)$, where the weight is the probability that x occurs. For example, $f(x)$ might be the probability distribution for the radii of raindrops in a storm and $g(x)$ the mass of the raindrop as a function of radius. Then $\langle g(x) \rangle$ is the average mass of raindrops. For completeness we note that

$$\langle a_1 g_1(x) + a_2 g_2(x) \rangle = \int_{-\infty}^{\infty} [a_1 g_1(x) + a_2 g_2(x)]f(x)\,dx = a_1 \langle g_1(x) \rangle + a_2 \langle g_2(x) \rangle, \tag{1.49}$$

so calculating a mean is a linear operation.

The *variance*, σ^2, of a distribution is defined to be the mean value of $(x - \langle x \rangle)^2$:

$$\sigma^2 = \langle (x - \mu)^2 \rangle, \tag{1.50}$$

where for notational convenience we have set $\langle x \rangle = \mu$. The positive square root of the variance, denoted by σ, is called the *standard deviation*. The variance and the standard deviation are measures of the width or spread of a distribution function. Their precise meaning depends on the specific distribution, but roughly half of the probability lies between $\mu - \sigma$ and $\mu + \sigma$. The variance of a distribution is related to the second moment of the distribution. Expanding the expression for σ^2, we find

$$\sigma^2 = \langle x^2 - 2x\mu + \mu^2 \rangle = \langle x^2 \rangle - 2\mu \langle x \rangle + \mu^2$$
$$= M_2 - \mu^2 \tag{1.51}$$
$$= M_2 - M_1^2. \tag{1.52}$$

The variance does not exist for all functions—it does not exist for the Lorentzian function because M_2 diverges. But the variance is not the only way to characterize the width of a distribution function, and even when it exists, it is not necessarily the best way to characterize the width. For example, the quantity $\langle |x - \mu| \rangle$ also measures the width and since it is linear in $x - \mu$, not quadratic, it is less sensitive to parts of the distribution far from μ than is the variance. Or one may want to use the FWHM, which does exist for the Lorentzian. The variance may not even be the most informative way to describe the width of a distribution. The variance of a rectangular distribution function is less useful than the full width of the rectangle.

The asymmetry of a probability distribution can be characterized by its *skewness*. There are three common definitions of skewness:

$$\text{skewness} = \frac{\text{mean} - \text{mode}}{\text{standard deviation}} = \frac{\langle x \rangle - x_{mode}}{\sigma}, \tag{1.53}$$

$$\text{skewness} = \frac{\text{mean} - \text{median}}{\text{standard deviation}} = \frac{\langle x \rangle - x_{median}}{\sigma}, \tag{1.54}$$

$$\text{skewness} = \frac{\langle (x - \mu)^3 \rangle}{\sigma^3}. \tag{1.55}$$

One must always specify which definition is being used.

Example: Calculate the second moment, the variance, and the skewness of the exponential distribution function

$$f(x) = a \exp[-ax], \qquad x \geq 0.$$

The second moment is

$$M_2 = \int_{-\infty}^{\infty} x^2 f(x)\, dx$$

$$= \int_0^{\infty} ax^2 \exp[-ax]\, dx = \left. -\frac{1}{a^2}[(ax)^2 + 2ax + 2] \exp[-ax] \right|_0^{\infty}$$

$$= \frac{2}{a^2}.$$

Calculate σ from equation 1.52 and the value of μ just found:

$$\sigma^2 = M_2 - \mu^2 = \frac{2}{a^2} - \left(\frac{1}{a}\right)^2 = \frac{1}{a^2}.$$

The standard deviation is shown in Figure 1.6. For comparison, the probability distribution drops to half its maximum value at $x = 0$ and at

$$a \exp[-ax_{1/2}] = \frac{1}{2},$$

so its full width at half maximum is

$$\text{FWHM} = \frac{1}{a} \ln(2a).$$

We have already shown that $\langle x \rangle = \mu = 1/a$ and $x_{median} = \ln(2)/a$ for the exponential distribution. Then the skewness as defined by equation 1.54 is

$$\text{skewness} = \frac{\langle x \rangle - x_{median}}{\sigma} = \frac{1/a - \ln(2)/a}{1/a^2} = a(1 - \ln 2).$$

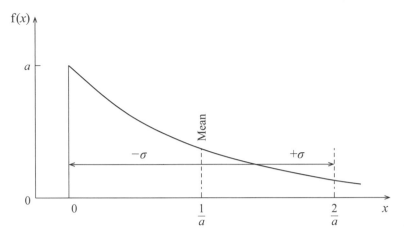

Figure 1.6: The mean $<x>$ and standard deviation σ of the exponential distribution function $f(x) = a\exp[-ax]$.

1.3.3 Moment Generating Function and the Characteristic Function

The moment generating function provides a convenient way to calculate moments for some probability distributions. For a continuous probability distribution $f(x)$, the moment generating function is defined to be

$$M(\zeta) = \int_{-\infty}^{\infty} \exp[\zeta x]\, f(x)dx, \tag{1.56}$$

where the integral needs to exist only in the neighborhood of $\zeta = 0$. To understand the moment generating function, expand $\exp[\zeta x]$ in a Taylor series and carry out the integration term by term. The moment generating function becomes

$$M(\zeta) = \int_{-\infty}^{\infty} \left(1 + \zeta x + \frac{\zeta^2 x^2}{2!} + \cdots\right) f(x)dx$$

$$= M_0 + \zeta M_1 + \frac{\zeta^2}{2!} M_2 + \cdots. \tag{1.57}$$

For a correctly normalized probability distribution, $M_0 = M(0) = 1$. The higher moments of $f(x)$ can be calculated from the derivatives of the moment generating function:

$$M_m = \left.\frac{\partial^m M(\zeta)}{\partial \zeta^m}\right|_{\zeta=0}, \qquad m \geq 1. \tag{1.58}$$

Example: The moment generating function for the exponential probability distribution

$$f(x) = a\exp[-ax], \quad x \geq 0,$$

is

$$M(\zeta) = a \int_0^\infty \exp[\zeta x] \exp[-ax] \, dx = a \int_0^\infty \exp[(\zeta - a)x] \, dx = \frac{a}{a - \zeta}.$$

The first and second moments of the probability distribution are

$$M_1 = \left. \frac{\partial M(\zeta)}{\partial \zeta} \right|_{\zeta = 0} = \left. \frac{a}{(a - \zeta)^2} \right|_{\zeta = 0} = \frac{1}{a}$$

$$M_2 = \left. \frac{\partial^2 M(\zeta)}{\partial \zeta^2} \right|_{\zeta = 0} = \left. \frac{2a}{(a - \zeta)^3} \right|_{\zeta = 0} = \frac{2}{a^2}.$$

The *characteristic function* $\phi(v)$ of a continuous probability distribution function $f(x)$ is defined to be

$$\phi(v) = \int_{-\infty}^\infty \exp[ivx] \, f(x) \, dx, \tag{1.59}$$

where $i = \sqrt{-1}$, and v is a real number. Looking ahead to Chapter 8 on Fourier analysis, we recognize the characteristic function as the inverse Fourier transform of $f(x)$ (see equation 8.61). Expanding the exponential, we have

$$\phi(v) = \int_{-\infty}^\infty \left(1 + ivx + \frac{1}{2!}(iv)^2 x^2 + \frac{1}{3!}(iv)^3 x^3 + \cdots \right) f(x) \, dx$$

$$= 1 + iv M_1 + \frac{(iv)^2}{2!} M_2 + \cdots + \frac{(iv)^n}{n!} M_n + \cdots. \tag{1.60}$$

The Fourier transform of $\phi(v)$ is

$$\frac{1}{2\pi} \int_v \phi(v) \exp[-ivx'] dv = \frac{1}{2\pi} \int_v \left\{ \int_x \exp[ivx] \, f(x) \, dx \right\} \exp[-ivx'] dv$$

$$= \int_x f(x) \left\{ \frac{1}{2\pi} \int_v \exp[iv(x - x')] dv \right\} dx$$

$$= \int_x f(x) \, \delta(x - x') \, dx$$

$$= f(x'), \tag{1.61}$$

where $\delta(x - x')$ is the Dirac delta function (see Table 8.1, which lists some Fourier transform pairs). As expected, the original probability distribution function is the Fourier transform of $\phi(v)$.

Combining equations 1.60 and 1.61 we find

$$f(x') = \frac{1}{2\pi} \int_v \left[1 + iv M_1 + \frac{(iv)^2}{2!} M_2 + \cdots + \frac{(iv)^n}{n!} M_n + \cdots \right] \exp[-ivx'] dv. \tag{1.62}$$

This shows that a probability distribution function can be reconstructed from the values of its moments, a remarkable and unexpected relation between the Fourier transform and the moment equation (equation 1.38). We will need this result when deriving the Gaussian probability distribution by way of the central limit theorem.

1.4 Multivariate Probability Distributions

1.4.1 Distributions with Two Independent Variables

Probability distribution functions can have more than one independent variable (see Figure 1.7). A function of two independent variables x_1 and x_2 is a valid probability distribution function if it is single-valued and satisfies the requirements:

$$\text{f}(x_1, x_2) \geq 0 \tag{1.63}$$

$$\int_{-\infty}^{\infty} \int_{-\infty}^{\infty} \text{f}(x_1, x_2) \, dx_1 dx_2 = 1 \tag{1.64}$$

$$P(a_1 < x_1 < b_1, a_2 < x_2 < b_2) = \int_{a_1}^{b_1} \int_{a_2}^{b_2} \text{f}(x_1, x_2) \, dx_1 dx_2, \tag{1.65}$$

where $P(a_1 < x_1 < b_1, a_2 < x_2 < b_2)$ is the probability that x_1 lies in the range $a_1 < x_1 < b_1$ and x_2 lies in the range $a_2 < x_2 < b_2$. Equation 1.65 can be generalized to the probability that x_1 and x_2 lie in an arbitrary area A in the (x_1, x_2) plane:

$$P(x_1, x_2 \subset A) = \int_A \text{f}(x_1, x_2) \, dx_1 dx_2. \tag{1.66}$$

There are two *marginal probability distributions*:

$$\text{g}_1(x_1) = \int_{x_2 = -\infty}^{\infty} \text{f}(x_1, x_2) \, dx_2 \tag{1.67}$$

$$\text{g}_2(x_2) = \int_{x_1 = -\infty}^{\infty} \text{f}(x_1, x_2) \, dx_1. \tag{1.68}$$

Thus, $\text{g}_1(x_1)$ is the integral of all the probability that is spread out in x_2 (see Figure 1.7).

Given a two-parameter distribution $\text{f}(x_1, x_2)$ and its marginal probability $\text{g}_1(x_1)$, the *conditional distribution* $\text{h}(x_2|x_1)$ is defined to be

$$\text{h}(x_2|x_1) = \frac{\text{f}(x_1, x_2)}{\text{g}_1(x_1)}. \tag{1.69}$$

The conditional distribution is a cut through $\text{f}(x_1, x_2)$ at some constant value of x_1. Compare this to the marginal distribution, which is an integral, not a cut. The conditional distribution is already properly normalized, since

$$\int_{-\infty}^{\infty} \text{h}(x_2|x_1) dx_2 = \int_{-\infty}^{\infty} \frac{\text{f}(x_1, x_2)}{\text{g}_1(x_1)} \, dx_2 = \frac{1}{\text{g}_1(x_1)} \int_{-\infty}^{\infty} \text{f}(x_1, x_2) \, dx_2 = \frac{\text{g}_1(x_1)}{\text{g}_1(x_1)} = 1. \tag{1.70}$$

We say that x_2 is independent of x_1 if the probability that x_2 occurs is independent of the probability that x_1 occurs. The conditional probability then becomes

$$\text{h}(x_2|x_1) = \text{h}(x_2). \tag{1.71}$$

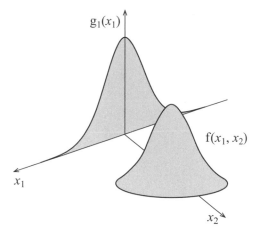

Figure 1.7: A two-dimensional distribution function, $f(x_1, x_2)$, and the marginal distribution, $g_1(x_1)$, derived from it.

If x_1 and x_2 are independent, equation 1.69 can be rewritten as

$$f(x_1, x_2) = g_1(x_1)h(x_2|x_1) = g_1(x_1)h(x_2), \tag{1.72}$$

so the probability that both x_1 and x_2 will occur is the product of the individual probabilities of their occurrence.

Moments can be calculated for each of the variables individually:

$$\langle x_1^m \rangle = \int_{x_1=-\infty}^{\infty} \int_{x_2=-\infty}^{\infty} x_1^m f(x_1, x_2)\,dx_1\,dx_2 \tag{1.73}$$

$$\langle x_2^n \rangle = \int_{x_1=-\infty}^{\infty} \int_{x_2=-\infty}^{\infty} x_2^n f(x_1, x_2)\,dx_1\,dx_2, \tag{1.74}$$

but it is also possible to calculate joint moments:

$$\langle x_1^m x_2^n \rangle = \int_{x_1=-\infty}^{\infty} \int_{x_2=-\infty}^{\infty} x_1^m x_2^n f(x_1, x_2)\,dx_1\,dx_2. \tag{1.75}$$

If x_1 and x_2 are independent, the integrals over x_1 and x_2 in equation 1.75 separate, and $\langle x_1^m x_2^n \rangle = \langle x_1^m \rangle \langle x_2^n \rangle$.

1.4.2 Covariance

If a probability distribution function has two independent variables x_1 and x_2, the *covariance* between x_1 and x_2 is defined to be

$$\sigma_{12} = \sigma_{21} = \langle (x_1 - \langle x_1 \rangle)(x_2 - \langle x_2 \rangle) \rangle \tag{1.76}$$

$$= \int_{x_1=-\infty}^{\infty} \int_{x_2=-\infty}^{\infty} (x_1 - \langle x_1 \rangle)(x_2 - \langle x_2 \rangle)f(x_1, x_2)\,dx_1\,dx_2. \tag{1.77}$$

The covariance measures the extent to which the value of a sample from x_1 depends on the value of a sample from x_2. If x_1 is independent of x_2, then $f(x_1, x_2) = f_1(x_1)f_2(x_2)$, and the

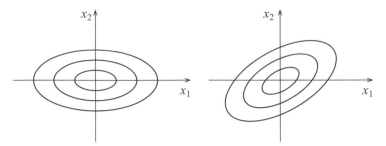

Figure 1.8: Contour plots of two two-parameter probability distributions $f(x_1, x_2)$ for which the contours of constant probability are ellipses centered on the origin. The covariance between x_1 and x_2 is zero for the distribution on the left, but it is nonzero for the distribution on the right.

covariance becomes

$$
\begin{aligned}
\sigma_{12} &= \int_{x_1} \int_{x_2} (x_1 - \langle x_1 \rangle)(x_2 - \langle x_2 \rangle) f_1(x_1) f_2(x_2) dx_1 dx_2 \\
&= \left[\int (x_1 - \langle x_1 \rangle) f_1(x_1) dx_1 \right] \left[\int (x_2 - \langle x_2 \rangle) f_2(x_2) dx_2 \right] \\
&= (\langle x_1 \rangle - \langle x_1 \rangle)(\langle x_2 \rangle - \langle x_2 \rangle) \\
&= 0.
\end{aligned}
\tag{1.78}
$$

To gain some insight into the meaning of covariance, consider a two-parameter distribution for which the contours of constant probability are ellipses centered on the origin. Two possible contour plots of the distribution are shown in Figure 1.8. Ellipses centered on the origin have the general functional form $a_1 x_1^2 + a_{12} x_1 x_2 + a_2 x_2^2 = \text{constant}$, so the probability distribution must have the form $f(a_1 x_1^2 + a_{12} x_1 x_2 + a_2 x_2^2)$. If $a_{12} = 0$, the major and minor axes of the ellipses are horizontal and vertical, and they coincide with the coordinate axes as shown in the left panel of the figure. If $a_{12} > 0$, the axes are tilted as shown in the right panel. Consider the case

$$
f(x_1, x_2) = a_1 x_1^2 + a_{12} x_1 x_2 + a_2 x_2^2
\tag{1.79}
$$

with $-\alpha \le x_1 \le \alpha$ and $-\beta \le x_2 \le \beta$. Since the ellipses are symmetric about the origin, the mean values of x_1 and x_2 are zero. The covariance becomes

$$
\begin{aligned}
\sigma_{12} &= \int_{x_1=-\alpha}^{\alpha} \int_{x_2=-\beta}^{\beta} x_1 x_2 (a_1 x_1^2 + a_{12} x_1 x_2 + a_2 x_2^2) dx_1 dx_2 \\
&= \int_{x_1=-\alpha}^{\alpha} \int_{x_2=-\beta}^{\beta} (a_1 x_1^3 x_2 + a_{12} x_1^2 x_2^2 + a_2 x_1 x_2^3) dx_1 dx_2 \\
&= \frac{4 a_{12}}{9} \alpha^3 \beta^3.
\end{aligned}
\tag{1.80}
$$

The covariance is nonzero if a_{12} is nonzero, so the covariance is nonzero if the ellipses are tilted.

1.4.3 Distributions with Many Independent Variables

These results for two independent variables generalize easily to many independent variables. The multivariate function $f(x_1, x_2, \ldots, x_n)$ can be a probability distribution function if it is single valued and if

- $f(x_1, x_2, \ldots, x_n) \geq 0$ (1.81)

- $\displaystyle\int_{x_1=-\infty}^{\infty} \int_{x_2=-\infty}^{\infty} \cdots \int_{x_n=-\infty}^{\infty} f(x_1, x_2, \ldots, x_n) \, dx_1 dx_2 \cdots dx_n = 1$ (1.82)

- $P(a_1 < x_1 < b_1, a_2 < x_2 < b_2, \ldots, a_n < x_n < b_n)$

$$= \int_{a_1}^{b_1} \int_{a_2}^{b_2} \cdots \int_{a_n}^{b_n} f(x_1, x_2, \ldots, x_n) \, dx_1 dx_2 \cdots dx_n, \qquad (1.83)$$

where P is the probability that x_1 lies in the range $a_1 < x_1 < b_1$, that x_2 lies in the range $a_2 < x_2 < b_2$, and so on. If V is an arbitrary volume in the n-dimensional space spanned by $(x_1, x_2, \ldots, x_n)$, the probability that a point lies within the volume is

$$P(x_1, x_2, \ldots, x_n \subset V) = \int_V f(x_1, x_2, \ldots, x_n) \, dx_1 dx_2 \cdots dx_n. \qquad (1.84)$$

Marginal and Conditional Distributions: Many different marginal probability distributions can be constructed from $f(x_1, x_2, \ldots, x_n)$. For example, there is a marginal distribution given by integrating over x_3 and x_5,

$$g(x_1, x_2, x_4, x_6, \ldots, x_n) = \int_{x_3=-\infty}^{\infty} \int_{x_5=-\infty}^{\infty} f(x_1, x_2, \ldots, x_n) \, dx_3 dx_5; \qquad (1.85)$$

and there is a conditional distribution for each marginal distribution. For example, the conditional distribution $h(x_3, x_5 | x_1, x_2, x_4, x_6, \ldots, x_n)$ is given by

$$h(x_3, x_5 | x_1, x_2, x_4, x_6, \ldots, x_n) = \frac{f(x_1, x_2, \ldots, x_n)}{g(x_1, x_2, x_4, x_6, \ldots, x_n)}. \qquad (1.86)$$

In fact, $f(x_1, x_2, \ldots, x_n)$ can be marginalized in any direction, not just along the $(x_1, x_2, \ldots, x_n)$ coordinate axes. To marginalize in any other desired direction, simply rotate the coordinate system so that one of the rotated coordinate axes points in the desired direction. Then integrate the distribution along that coordinate.

If variables x_i and x_j are independent of each other and their probability distribution functions are $f_i(x_i)$ and $f_j(x_j)$, the joint probability distribution function can be written

$$f(x_1, x_2, \ldots, x_n) = f_i(x_i) f_j(x_j) \bar{f}_{ij}, \qquad (1.87)$$

where $\bar{f}_{ij}$ is short-hand notation for the probability distribution function of all the other variables. If all the x_i are all independent of one another, their joint probability distribution function is the product of their individual probability distribution functions:

$$f(x_1, x_2, \ldots, x_n) = f_1(x_1) f_2(x_2) \cdots f_n(x_n). \qquad (1.88)$$

Covariances: The covariance between parameters x_i and x_j, defined to be

$$\sigma_{ij} = \sigma_{ji} = \langle (x_i - \langle x_i \rangle)\,(x_j - \langle x_j \rangle) \rangle \tag{1.89}$$

$$= \int \cdots \int (x_i - \langle x_i \rangle)\,(x_j - \langle x_j \rangle)\, f(x_1, x_2, \ldots, x_n)\,dx_1\,dx_2 \cdots dx_n, \tag{1.90}$$

measures the extent to which the value of x_i depends on the value of x_j. If $i = j$, equation 1.90 becomes

$$\sigma_{ii} = \langle (x_i - \langle x_i \rangle)^2 \rangle, \tag{1.91}$$

which is the variance of x_i, so it is common to use the notation $\sigma_{ii} = \sigma_i^2$.

If any pair of parameters x_i and x_j are independent of each other, then, using the notation of equation 1.87, their covariance is

$$\sigma_{ij} = \int \cdots \int (x_i - \langle x_i \rangle)\,(x_j - \langle x_j \rangle)\; f_i(x_i) f_j(x_j) \bar{f}_{ij}\; dx_n$$

$$= \left[\int (x_i - \langle x_i \rangle) f_i(x_i) dx_i \right] \left[\int (x_j - \langle x_j \rangle)\, f_j(x_j) dx_j \right]$$

$$= (\langle x_i \rangle - \langle x_i \rangle)\,(\langle x_j \rangle - \langle x_j \rangle)$$

$$= 0. \tag{1.92}$$

Moment Generating Function: The moment generating function for a multivariate probability distribution is

$$M(\zeta_1, \zeta_2, \ldots, \zeta_n) = \int_{x_1 = -\infty}^{\infty} \cdots \int_{x_n = -\infty}^{\infty} \exp[\zeta_1 x_1 + \zeta_2 x_2 + \cdots + \zeta_n x_n]$$

$$\times f(x_1, x_2, \ldots, x_n) dx_1 dx_2 \cdots dx_n. \tag{1.93}$$

The first and second moments of the probability distribution can be calculated by taking first and second derivatives, respectively, of the moment generating function:

$$\langle x_j \rangle = \left. \frac{\partial M(\zeta_1, \zeta_2, \ldots, \zeta_n)}{\partial \zeta_j} \right|_{\zeta_1 = \zeta_2 = \cdots = \zeta_n = 0} \tag{1.94}$$

$$\langle x_j x_k \rangle = \left. \frac{\partial^2 M(\zeta_1, \zeta_2, \ldots, \zeta_n)}{\partial \zeta_j \partial \zeta_k} \right|_{\zeta_1 = \zeta_2 = \cdots = \zeta_n = 0} \tag{1.95}$$

For those who collect inscrutable formulas, the general case is

$$\langle x_1^{m_1} x_2^{m_2} \cdots x_n^{m_n} \rangle = \left. \frac{\partial^{m_1 + m_2 + \cdots + m_n} M(\zeta_1, \zeta_2, \ldots, \zeta_n)}{\partial \zeta_1^{m_1} \partial \zeta_2^{m_2} \cdots \partial \zeta_n^{m_n}} \right|_{\zeta_1 = \zeta_2 = \cdots = \zeta_n = 0}. \tag{1.96}$$

Transformation of Variables: Finally, it is sometimes necessary to transform a probability distribution function $f(x_1, x_2, \ldots, x_n)$ to a different set of variables $(y_1, y_2, \ldots, y_n)$. Let the equations for the coordinate transformation be

$$x_1 = x_1(y_1, y_2, \ldots, y_n)$$

$$\vdots \quad \vdots \tag{1.97}$$

$$x_n = x_n(y_1, y_2, \ldots, y_n).$$

The transformation must be invertible for the probability distribution in the new coordinates to be meaningful. Following standard methods for changing variables in multiple integrals,[2] we have

$$f(x_1, x_2, \ldots, x_n) dx_1 dx_2 \cdots dx_n = f(y_1, y_2, \ldots, y_n) \left| \frac{\partial(x_1, x_2, \ldots, x_n)}{\partial(y_1, y_2, \ldots, y_n)} \right| dy_1 dy_2 \cdots dy_n$$

$$= g(y_1, y_2, \ldots, y_n) dy_1 dy_2 \cdots dy_n, \qquad (1.98)$$

from which we find

$$g(y_1, y_2, \ldots, y_n) = f(y_1, y_2, \ldots, y_n) \left| \frac{\partial(x_1, x_2, \ldots, x_n)}{\partial(y_1, y_2, \ldots, y_n)} \right|. \qquad (1.99)$$

The Jacobian determinant in equation 1.99 accounts for the difference between the sizes of the volume elements in the two coordinate systems.

---

[2] See, for example, Riley et al. (2002).

2

Some Useful Probability Distribution Functions

Any function that satisfies the requirements listed in Section 1.2.1 is a valid probability distribution function, but some probability distributions are more useful than others. In this section we discuss some of the most useful distributions: the binomial distribution, the Poisson distribution, the gaussian (or normal) distribution, the χ^2 distribution, and the beta distribution. We have already given passing attention to several other useful distributions, including the rectangular distribution, the delta function, the exponential distribution, and the Lorentzian distribution. Student's t distribution will be discussed in Chapter 4, and the gamma distribution in Chapter 7. An index to the probability distributions discussed in this book is given in Table 2.1.

2.1 Combinations and Permutations

Calculating probabilities usually requires counting possible outcomes, but counting can be tedious or difficult for any but the simplest problems. Combinatorial analysis—the study of combinations and permutations—eases the labor. We need a few simple results from this extensive subject.

Permutations: Consider a set of n distinguishable objects. Choose r of the objects, and arrange them in a row in the order they were chosen. The ordered arrangement is called a *permutation*. Exchanging the positions of any of the objects produces a different permutation.

Table 2.2 shows how to count the number of permutations of r objects selected from n objects. The first of the r objects can be chosen from any of n different objects, but the second must be chosen from the remaining $n-1$ objects. Thus there are n ways to choose the first object but only $n-1$ ways to choose the second. The total number of permutations of the first two object is then $n(n-1)$. After three choices there are $n(n-1)(n-2)$ permutations; and after all r objects have been chosen, there are $n(n-1)(n-2)\cdots(n-r+1)$ permutations. This can be written more compactly as

$$\frac{n!}{(n-r)!} \tag{2.1}$$

Table 2.1: Index to probability distribution functions

Distribution	Definition	Discussion
Rectangle	$f(x) = \begin{cases} 1/a, & \|x\| \le a/2 \\ 0, & \|x\| > a/2 \end{cases}$	Equation 1.26
Delta function	$f(x) = \delta(x-a)$	Equations 1.27, 1.28
Exponential	$f(x) = \begin{cases} 0, & x < 0 \\ a\exp[-ax], & x \ge 0 \end{cases}$	Equation 1.30, Section 1.3.2
Lorentzian	$f(x) = \dfrac{1}{\pi}\dfrac{b}{b^2 + (x-a)^2}$	Equation 1.46
Binomial	$P(k) = \dfrac{n!}{k!(n-k)!}p^k q^{n-k}$	Section 2.2
Poisson	$P(k) = \dfrac{\mu^k}{k!}\exp[-\mu]$	Section 2.3
Gaussian	$f(x) = \dfrac{1}{\sqrt{2\pi\sigma^2}}\exp\left[-\dfrac{1}{2}\dfrac{(x-\mu)^2}{\sigma^2}\right]$	Section 2.4, Appendix C
Multivariate Gaussian	$f(\mathbf{x}) = \dfrac{1}{(2\pi)^{n/2}\|\mathbf{C}\|^{1/2}}\exp\left[-\dfrac{1}{2}(\mathbf{x}-\boldsymbol{\mu})^{\mathrm{T}}\mathbf{C}^{-1}(\mathbf{x}-\boldsymbol{\mu})\right]$	Section 2.5, Appendix C
χ^2	$f_n(\chi^2) = \dfrac{\left(\chi^2\right)^{(n-2)/2}}{2^{n/2}(n/2-1)!}\exp\left[-\dfrac{1}{2}\chi^2\right]$	Section 2.6
Beta	$\beta(x) = \dfrac{(a+b-1)!}{(a-1)!(b-1)!}x^{a-1}(1-x)^{b-1}$	Section 2.7
Student's t	$s_n(t) = \dfrac{[(n-1)/2]!}{[(n-2)/2]!\sqrt{n\pi}}\left[1+\dfrac{t^2}{n}\right]^{-(n+1)/2}$	Section 4.5
Gamma	$f(x) = \dfrac{1}{\Gamma(k)\theta^k}x^{k-1}\exp\left[-\dfrac{x}{\theta}\right]$	Equation 7.31

Table 2.2: Permutations of r objects selected from n objects

Choice number	Number of objects from which to choose	Cumulative number of permutations
1	n	n
2	$n-1$	$n(n-1)$
3	$n-2$	$n(n-1)(n-2)$
$\vdots$	$\vdots$	$\vdots$
r	$n-r+1$	$n(n-1)(n-2)\cdots(n-r+1)$

Combinations: Now choose r objects out of n distinguishable objects, but toss the r objects into a bowl, so the order in which the objects were chosen is irrelevant. A selection of objects without regard to the order in which they were selected is called a *combination*. To count the number of unique combinations, first count all the permutations, and then divide by the number of permutations that yield the same combination. For n objects taken r at a time, the number of permutations is given by equation 2.1, but all these permutations count as just one combination. Again from equation 2.1 the number of permutations of r objects taken r at a time is $r!$. We have, then,

$$\frac{\text{number of permutations of } n \text{ taken } r \text{ at a time}}{\text{number of permutations of } r} = \frac{\frac{n!}{(n-r)!}}{r!} = \frac{n!}{r!(n-r)!}. \tag{2.2}$$

We will denote this expression by the special symbol

$$\binom{n}{r} = \frac{n!}{r!(n-r)!}, \tag{2.3}$$

which is called the binomial coefficient and typically is verbalized as "n choose r."

2.2 Binomial Distribution

Suppose an experiment can have two possible outcomes, A and B, and the probabilities of the outcomes are

$$P(A) = p \tag{2.4}$$

$$P(B) = q = 1 - p. \tag{2.5}$$

Perform the experiment n independent times. What is the probability $P(k)$ that outcome A occurs k times and outcome B occurs $n - k$ times without regard to the order in which the results occur?

To calculate the probability, first consider the specific result A occurs k times in succession and then B occurs $(n - k)$ times in succession:

$$\underbrace{A\,A\cdots A}_{k \text{ times}}\ \underbrace{B\,B\cdots B}_{n-k \text{ times}}.$$

Since the successive experiments are independent, the probability of getting this result is the product of the individual probabilities:

$$\underbrace{P(A)\,P(A)\cdots P(A)}_{k \text{ times}}\ \underbrace{P(B)\,P(B)\cdots P(B)}_{n-k \text{ times}} = p^k\, q^{n-k}. \tag{2.6}$$

Any other order that has the same number of As and Bs has the same probability of occurrence, so to find the total probability of k occurrences of A, we multiply the probability of getting this specific order by the number of orderings that also have k occurrences of A and $n - k$ occurrences of B:

$$P(k) = \left(\begin{array}{c} \text{Number of ways the experiments} \\ \text{can yield } k \text{ occurrences of } A \\ \text{and } n - k \text{ occurrences of } B \end{array} \right) \times p^k q^{n-k}. \tag{2.7}$$

Table 2.3: The binomial coefficients for $n = 4$

Number of times A occurs	4	3	2	1	0
Number of times B occurs	0	1	2	3	4
Possible orderings	$AAAA$	$AAAB$ $AABA$ $ABAA$ $BAAA$	$AABB$ $ABAB$ $ABBA$ $BAAB$ $BABA$ $BBAA$	$ABBB$ $BABB$ $BBAB$ $BBBA$	$BBBB$
Number of orderings	$\binom{4}{0} = 1$	$\binom{4}{1} = 4$	$\binom{4}{2} = 6$	$\binom{4}{3} = 4$	$\binom{4}{4} = 1$

Table 2.3 shows the orderings of A and B that yield the same probabilities for an experiment performed four times. One can think of the experiment as four flips of a coin. There are, for example, six ways that heads and tails can both occur two times.

More generally, from equation 2.1 with $r = n$, there are $n!$ permutations of sequence 2.6. But since all the A outcomes are identical, all the permutations of the A among themselves count as just one ordering. As there are $k!$ permutations of the A, we must divide $n!$ by $k!$. Likewise, there are $n - k$ permutations of B, so we must also divide by $(n - k)!$. The number of orderings with k occurrences of A and $n - k$ occurrences of B is then

$$\binom{n}{k} = \frac{n!}{k!(n-k)!}.$$
(2.8)

This is evaluated for $n = 4$ in the last row of Table 2.3. Expressions 2.3 and 2.8 have the same form, and both are called the *binomial coefficient*. They were, however, derived in different ways, and their meanings are not the same. Even though equation 2.3 was used to derive equation 2.8, it was not at all obvious beforehand that equation 2.8 would wind up having the same form as equation 2.3.

Combining equation 2.7 and equation 2.8, we arrive at the probability that an experiment will yield k occurrences of A and $n - k$ occurrences of B:

$$P(k) = \frac{n!}{k!\,(n-k)!}\,p^k q^{n-k} = \binom{n}{k} p^k q^{n-k}.$$
(2.9)

$P(k)$ is called the *binomial probability distribution*, because $P(k)$ is identical with the terms in the binomial series:

$$(p + q)^n = p^n + np^{n-1}q + \frac{n(n-1)}{2}p^{n-2}q^2 + \cdots + npq^{n-1} + q^n$$
(2.10)

$$= \sum_{k=0}^{n} \frac{n!}{k!\,(n-k)!}\,p^k q^{n-k}$$
(2.11)

$$= \sum_{k=0}^{n} P(k).$$
(2.12)

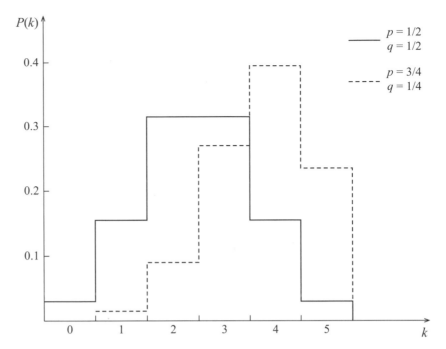

Figure 2.1: Two examples of the binomial distribution (equation 2.9). The solid line is the distribution for $n = 5$ and $p = q = 1/2$, where p is the probability of getting event A in a single experiment. The dashed line is the distribution for $n = 5$, $p = 3/4$, and $q = 1/4$.

The binomial distribution is already correctly normalized, since

$$\sum_{k=0}^{n} P(k) = (p+q)^n = (p+1-p)^n = 1^n = 1. \tag{2.13}$$

Plots of the binomial distribution for $n = 5$ and two different values of p are shown in Figure 2.1.

Example: Suppose one flips a coin 10 times. What is the probability that heads will *not* occur 5 times?

According to the binomial distribution, the probability that heads will not occur 5 times is

$$P(\bar{5}) = 1 - P(5) = 1 - \binom{10}{5} \left(\frac{1}{2}\right)^5 \left(\frac{1}{2}\right)^{10-5}$$

$$= 1 - \frac{10!}{5! \,(10-5)!} \left(\frac{1}{2}\right)^{10}$$

$$= 0.754$$

To find the mean and variance of the binomial distribution, we need its moments, where the mth moment is

$$M_m = \langle k^m \rangle = \sum_{k=0}^{n} k^m \binom{n}{k} p^k q^{n-k}. \tag{2.14}$$

The easiest way to calculate the moments is from a recurrence relation that gives higher moments in terms of lower moments. Take the derivative of M_m with respect to p (remember that $p + q = 1$):

$$\frac{\partial M_m}{\partial p} = \frac{\partial}{\partial p} \left\{ \sum_{k=0}^{n} k^m \binom{n}{k} p^k (1-p)^{n-k} \right\}$$

$$= \sum_{k=0}^{n} k^{m+1} \binom{n}{k} p^{k-1} (1-p)^{n-k} - \sum_{k=0}^{n} k^m (n-k) \binom{n}{k} p^k (1-p)^{n-k-1}$$

$$= \frac{1}{p} \sum_{k=0}^{n} k^{m+1} \binom{n}{k} p^k q^{n-k} - \frac{n}{q} \sum_{k=0}^{n} k^m \binom{n}{k} p^k q^{n-k} + \frac{1}{q} \sum_{k=0}^{n} k^{m+1} \binom{n}{k} p^k q^{n-k}$$

$$= \frac{1}{p} M_{m+1} - \frac{n}{q} M_m + \frac{1}{q} M_{m+1}. \tag{2.15}$$

Rearranging and simplifying, we obtain the desired recurrence relation:

$$M_{m+1} = np M_m + pq \frac{\partial M_m}{\partial p}. \tag{2.16}$$

The first three moments of the binomial distribution are then

$$M_0 = 1 \quad \text{(the distribution is normalized)} \tag{2.17}$$

$$M_1 = \langle k \rangle = np M_0 + pq \frac{\partial M_0}{\partial p} = np \tag{2.18}$$

$$M_2 = np M_1 + pq \frac{\partial M_1}{\partial p} = np(np) + pq \frac{\partial}{\partial p}(np)$$

$$= n^2 p^2 + npq. \tag{2.19}$$

The variance follows from

$$\sigma^2 = M_2 - M_1^2 = n^2 p^2 + npq - (np)^2$$

$$= npq. \tag{2.20}$$

2.3 Poisson Distribution

We derive the Poisson distribution as a limiting case of the binomial distribution. It is the probability of getting k events as the number of trials becomes large but the probability of an event occurring in any individual trial becomes small. Specifically, it is the probability of k occurrences of an event as n becomes much larger than k, and as p goes to zero in such a way that np remains constant. An example from nature is the probability of k

decays of a radioactive element per unit time. Suppose we expect a mean of 10 decays per second. Imagine dividing 1 second into n extremely small subdivisions. The probability of a decay within a subdivision is also small, but after sampling n subdivisions, we have sampled a whole second, and the mean is back up to 10 decays per second. An essential difference between the Poisson distribution and the binomial distribution with small n is that the number of occurrences in 1 second can potentially be extremely large, although the probability of this happening is small.

To derive the Poisson distribution, we start with the binomial distribution and let n become much larger than k. Let $P(k)$ denote the Poisson distribution. Then

$$P(k) = \lim_{n \gg k} \left[\binom{n}{k} p^k (1-p)^{n-k} \right] = \lim_{n \gg k} \left[\frac{n!}{k!\,(n-k)!} p^k (1-p)^{n-k} \right]$$
$$= \frac{n^k}{k!} p^k (1-p)^n. \tag{2.21}$$

Setting $\mu = np$ and rearranging equation 2.21 a bit, we get

$$P(k) = \frac{\mu^k}{k!} \left[(1-p)^{1/p} \right]^{np} = \frac{\mu^k}{k!} \left[(1-p)^{1/p} \right]^{\mu}. \tag{2.22}$$

Now let $p \to 0$. From equation F.8 in Appendix F,

$$\lim_{p \to 0} (1-p)^{1/p} = e^{-1}, \tag{2.23}$$

so we arrive at the Poisson distribution

$$P(k) = \frac{\mu^k}{k!} \exp[-\mu]. \tag{2.24}$$

$P(k)$ is already correctly normalized, because

$$\sum_{k=0}^{\infty} P(k) = \sum_{k=0}^{\infty} \frac{\mu^k}{k!} \exp[-\mu] = \exp[-\mu] \sum_{k=0}^{\infty} \frac{\mu^k}{k!}$$
$$= \exp[-\mu] \left\{ 1 + \frac{\mu}{1} + \frac{\mu^2}{2!} + \frac{\mu^3}{3!} + \cdots \right\}$$
$$= \exp[-\mu] \exp[+\mu] = 1. \tag{2.25}$$

Plots of the Poisson distribution for two different values of μ are shown in Figure 2.2.

The moments of the Poisson distribution can be calculated from a recurrence relation. We derive it in much the same way as we derived the recurrence relation for the moments of the binomial distribution. Begin with the expression for the mth moment,

$$M_m = \langle k^m \rangle = \sum_{k=0}^{\infty} k^m \frac{\mu^k}{k!} \exp[-\mu]. \tag{2.26}$$

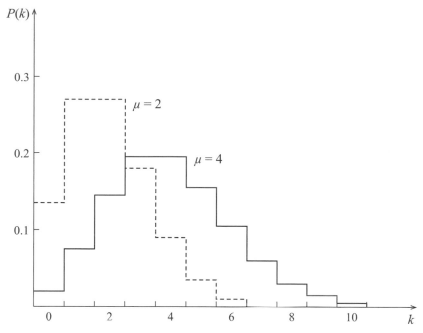

Figure 2.2: Two examples of the Poisson distribution (equation 2.24). The dashed line is the distribution for $\mu = 2$, and the solid line is the distribution for $\mu = 4$.

Taking the derivative with respect to μ, we have

$$\frac{\partial M_m}{\partial \mu} = \sum_{k=0}^{\infty} k^{m+1} \frac{\mu^{k-1}}{k!} \exp[-\mu] - \sum_{k=0}^{\infty} k^m \frac{\mu^k}{k!} \exp[-\mu]$$

$$= \frac{1}{\mu} M_{m+1} - M_m. \tag{2.27}$$

Rearranging equation 2.27, we find the desired recurrence relation:

$$M_{m+1} = \mu M_m + \mu \frac{\partial M_m}{\partial \mu}. \tag{2.28}$$

Since the distribution is correctly normalized, we already know that $M_0 = 1$. The first moment is

$$M_1 = \mu M_0 + \mu \frac{\partial M_0}{\partial \mu} = \mu \times 1 + \mu \frac{\partial}{\partial \mu}(1) = \mu. \tag{2.29}$$

Since $M_1 = \langle k \rangle$, we see that μ is the mean value of k. The second moment is

$$M_2 = \mu M_1 + \mu \frac{\partial}{\partial \mu}(M_1) = \mu^2 + \mu \frac{\partial}{\partial \mu}(\mu) = \mu^2 + \mu, \tag{2.30}$$

so the variance is

$$\sigma^2 = M_2 - M_1^2 = \mu^2 + \mu - \mu^2$$
$$\sigma^2 = \mu. \tag{2.31}$$

The Poisson distribution often applies when one simply counts something. For the distribution to apply, the result of a measurement (1) must be an integer, and (2) the integer can potentially be large, but (3) it must have a definable mean value. There is also a tacit assumption that the things being counted do not interact among themselves. Some examples: the number of photons detected from a star in 1 second; the number of defective light bulbs produced by a factory in 1 day; the number of active telephones in one of the cells of a cell phone network.

Example: Under many conditions the photons arriving from an astronomical source obey Poisson statistics. Let N be the mean number of photons detected in an interval. Then, since $\mu = N$ for that interval, the Poisson distribution becomes

$$P(k) = \frac{N^k}{k!} \exp[-N],$$

where k is the number of photons actually detected in the interval. From equation 2.31 the variance and standard deviation are

$$\sigma^2 = N$$
$$\sigma = \sqrt{N},$$

which is the famous $\sqrt{N}$ of photon-counting statistics. The signal-to-noise ratio is often taken to be the mean divided by the variance. For the Poisson distribution this is

$$\frac{\text{Signal}}{\text{Noise}} = \frac{\mu}{\sigma} = \frac{N}{\sqrt{N}} = \sqrt{N}.$$

If the mean photon counting rate is constant, so that $N = Rt$, where R is the photon counting rate and t is the sampling interval, then

$$\frac{\text{Signal}}{\text{Noise}} \propto \sqrt{t}.$$

The signal-to-noise ratio improves as the square root of the integration time.

Example: Instruments for detecting and counting individual photons sometimes have a "dead time" after the detection of a photon, during which no additional photons can be detected. A typical value for the dead time of photomultiplier tubes is

$$\tau = 50 \times 10^{-9} \text{ seconds.}$$

The dead time causes a systematic error in measured count rates, because photons arriving during the dead time are not counted. If the error is unacceptably large, one must correct for the uncounted photons. The correction is called the *dead time correction*.

If the mean number of photons N that arrive in an interval t is

$$N = Rt,$$

the probability that one or more photons arrive in the dead time interval τ is given by

$$P(k > 0) = 1 - P(0) = 1 - \frac{N^0}{0!} \exp[-N]$$

$$= 1 - \exp[-N] = 1 - \exp[-R\tau].$$

Since the photons arriving during the dead time are not detected, the measured number of photons N_m and the measured photon counting rate R_m are less than the true values N and R by a factor $1 - P(k > 0)$:

$$N_m = N[1 - P(k > 0)]$$
$$R_m t = Rt \exp[-R\tau]$$
$$R_m = R \exp[-R\tau].$$

We measure R_m and τ and then, in principle, we can determine R from the previous equation. The equation is transcendental and cannot be inverted algebraically.

In practice, one tries to design experiments to make the dead time correction small. If it is, a first-order algebraic solution may be acceptable:

$$R = R_m \exp[R\tau] \approx R_m \exp[R_m \tau] \approx R_m(1 + R_m \tau).$$

To keep the dead time correction less than 1% when the dead time is 50×10^{-9} seconds, one must have

$$R_m \tau < 0.01,$$

so

$$R_m < \frac{0.01}{50 \times 10^{-9}} = 2 \times 10^5 \text{ counts/second.}$$

2.4 Gaussian or Normal Distribution

2.4.1 Derivation of the Gaussian Distribution—Central Limit Theorem

The Gaussian or normal probability distribution function can be derived in a variety of ways. We derive it here by way of the central limit theorem, because this derivation gives some insight into why the Gaussian distribution is so widely applicable. Several other derivations are given in Section C1 of Appendix C, including the important derivation as a limit of the Poisson distribution.

Suppose we have n independent samples z_j from an unknown probability distribution function $g(z)$. We take the sum of the samples,

$$s = \sum_{j=1}^{n} z_j, \tag{2.32}$$

and wish to find f(s), the probability distribution for s. We will derive f(s) by constructing the characteristic function for f(s) from its moments and then taking the Fourier transform of the characteristic function.

Let $\phi_s(v)$ be the characteristic function for f(s). From equation 1.60 the characteristic function can be written in terms of the moments of s as

$$\phi_s(v) = \int \exp[ivs]f(s)ds = 1 + iv\langle s \rangle + \frac{(iv)^2}{2!}\langle s^2 \rangle + \cdots. \tag{2.33}$$

Let $h(z_1, z_2, \ldots, z_n)$ be the joint probability of getting $z_1, z_2, \ldots, z_n$. Since the z_j are independent of one another, $h(z_1, z_2, \ldots, z_n)$ is just the product of the individual probabilities:

$$h(z_1, z_2, \ldots, z_n) = g(z_1)g(z_2)\cdots g(z_n). \tag{2.34}$$

The mth moment of s is, therefore, given by

$$\langle s^m \rangle = \int_{z_1}\int_{z_2}\cdots\int_{z_n} s^m h(z_1, z_2, \ldots, z_n)\, dz_1 dz_2 \cdots dz_n$$
$$= \int_{z_1}\int_{z_2}\cdots\int_{z_n} s^m g(z_1)g(z_2)\cdots g(z_n)\, dz_1 dz_2 \cdots dz_n. \tag{2.35}$$

Inserting these expressions for the moments into equation 2.33, we get

$$\phi_s(v) = 1 + iv\langle s \rangle + \frac{(iv)^2}{2!}\langle s^2 \rangle + \cdots$$
$$= \int_{z_1}\int_{z_2}\cdots\int_{z_n}(1 + ivs + \frac{(iv)^2}{2!}s^2 + \cdots)g(z_1)g(z_2)\cdots g(z_n)\, dz_1 dz_2 \cdots dz_n$$
$$= \int_{z_1}\int_{z_2}\cdots\int_{z_n}\exp[ivs]\,g(z_1)g(z_2)\cdots g(z_n)\, dz_1 dz_2 \cdots dz_n$$
$$= \int_{z_1}\int_{z_2}\cdots\int_{z_n}\exp[iv\sum_{j=1}^{n} z_j]\,g(z_1)g(z_2)\cdots g(z_n)\, dz_1 dz_2 \cdots dz_n$$
$$= \int_{z_1}\exp[ivz_1]\,g(z_1)\,dz_1 \int_{z_2}\exp[ivz_2]\,g(z_2)\,dz_2 \cdots \int_{z_n}\exp[ivz_n]\,g(z_n)\,dz_n. \tag{2.36}$$

Referring again to equation 1.60, we recognize that each of the integrals is $\phi_z(v)$, the characteristic function of $g(z)$, so

$$\phi_s(v) = \phi_z(v)\phi_z(v)\cdots\phi_z(v) = [\phi_z(v)]^n. \tag{2.37}$$

Equation 2.37 says that the characteristic function for the distribution of a sum of n independent samples is equal to the characteristic function of the distribution for a single sample raised to the nth power, an interesting result in itself.

For convenience let us shift the origin of the z-coordinate to give $g(z)$ a zero mean. Then $\langle z \rangle = 0$ and $\langle z^2 \rangle = \sigma_z^2$. Referring again to equation 1.60, we can write the characteristic function for g(z) as

$$\phi_z(v) = 1 + iv\langle z\rangle + \frac{(iv)^2}{2!}\langle z^2\rangle + \cdots$$
$$= 1 - \frac{v^2}{2}\sigma_z^2 + \cdots, \tag{2.38}$$

and equation 2.37 becomes

$$\phi_s(v) = \left[1 - \frac{v^2}{2}\sigma_z^2 + \cdots\right]^n = 1 - n\left(\frac{v^2}{2}\sigma_z^2\right) + \frac{n(n-1)}{2}\left(\frac{v^2}{2}\sigma_z^2\right)^2 - \cdots. \tag{2.39}$$

Up to this point we have not made any approximations.

The terms on the right-hand side of equation 2.39 are similar—but not identical—to the terms in a Taylor series expansion for $\exp[-nv^2\sigma_z^2/2]$. As n becomes large, however, the series approaches

$$1 - \left(\frac{nv^2}{2}\sigma_z^2\right) + \frac{1}{2}\left(\frac{nv^2}{2}\sigma_z^2\right)^2 - \cdots, \tag{2.40}$$

which *is* the Taylor series expansion for the exponential. So in the limit of large n we find

$$\lim_{n\to\infty}\phi_s(v) = \exp\left[-n\frac{v^2}{2}\sigma_z^2\right]. \tag{2.41}$$

The probability distribution for s is the the Fourier transform of $\phi_s(v)$ (equation 1.61), so

$$f(s) = \frac{1}{2\pi}\int_{-\infty}^{\infty}\phi_s(v)\exp[-ivs]\,dv = \frac{1}{2\pi}\int_{-\infty}^{\infty}\exp\left[-n\frac{v^2}{2}\sigma_z^2\right]\exp[-ivs]\,dv. \tag{2.42}$$

This is a standard integral in Fourier analysis. Looking forward to equations 8.71–8.74 in Chapter 8 on Fourier Analysis, or referring to tables of integrals, we find

$$f(s) \propto \exp\left[-\frac{1}{2}\frac{s^2}{n\sigma_z^2}\right]. \tag{2.43}$$

Short of the normalization constant, equation 2.43 is the sought-for expression for $f(s)$. It is, though, more useful when converted from a distribution for s to a distribution for $x = s/n$:

$$f(x) \propto \exp\left[-\frac{1}{2}\frac{x^2}{(\sigma_z^2/n)}\right]. \tag{2.44}$$

One normally writes the Gaussian distribution in the form

$$f(x) \propto \exp\left[-\frac{(x-\mu)^2}{2\sigma_x^2}\right], \tag{2.45}$$

where μ is a constant that allows the probability distribution function to be centered on an arbitrary x. We will see that the variance $\sigma_x^2 = \sigma_z^2/n$ is the variance of x. It is smaller than the variance of z by the factor $1/n$. Finally, to normalize the distribution we make use of the result derived in Section A.2 of Appendix A:

$$\int_{-\infty}^{\infty}\exp[-bx^2]\,dx = \sqrt{\frac{\pi}{b}}, \tag{2.46}$$

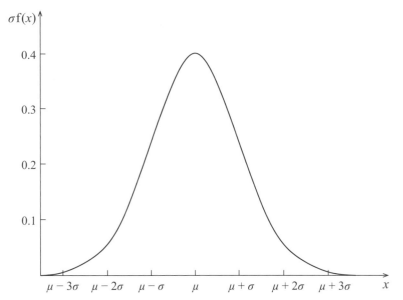

Figure 2.3: A plot of $\sigma f(x)$, where $f(x)$ is the Gaussian probability distribution function $f(x) = (2\pi\sigma^2)^{-1/2} \exp\left[-(x-\mu)^2/2\sigma^2\right]$.

where b is an arbitrary constant. The normalized Gaussian distribution becomes

$$f(x) = \frac{1}{\sqrt{2\pi\sigma_x^2}} \exp\left[-\frac{(x-\mu)^2}{2\sigma_x^2}\right]. \tag{2.47}$$

A plot of the normalized Gaussian distribution function is shown in Figure 2.3.

2.4.2 Summary and Comments on the Central Limit Theorem

Here, in summary, is our version of the central limit theorem and its relation to the Gaussian probability distribution. Let x be the mean value of n samples z_j chosen from an arbitrary probability distribution function whose mean is μ and variance is σ_z^2:

$$x = \frac{1}{n}\sum_{j=1}^{n} z_j. \tag{2.48}$$

As n becomes large, the probability distribution function for x approaches the Gaussian distribution

$$f(x) = (2\pi\sigma_x^2)^{-1/2} \exp\left[-\frac{(x-\mu)^2}{2\sigma_x^2}\right], \tag{2.49}$$

where $\sigma_x^2 = \sigma_z^2/n$.

Thus, the average of many independent samples taken from an arbitrary, unknown distribution always approaches a Gaussian distribution. The only constraint is that moments of the original distribution must exist. For clarity of presentation, we sampled all the z_j from the same probability distribution $g(z)$. In fact, this restriction is not needed; each z_j can be drawn from a different distribution, $g_j(z)$. Thereafter the derivation proceeds in much the same way. This is truly remarkable, because we never needed to specify the form of the original probability distribution, and, moreover, the form does not even need to be the same for all the samples.

The central limit theorem is one reason—a fundamental reason—the Gaussian distribution is so widely applicable in nature. The Gaussian distribution is, for example, likely to be a good description of measurement errors when the errors result from small contributions from many sources. Indeed, the central limit theorem is a reasonable justification for assuming that an unknown probability distribution function is Gaussian until and unless the assumption is contradicted by additional information.

Example: Let us start with a distribution function that is far from Gaussian:

$$g(z) = \frac{1}{2} [\delta(z-0) + \delta(z-1)].$$

This probability distribution function is two Dirac delta functions, one located at $z = 0$ and the other at $z = 1$. The distribution actually occurs in nature. It could describe the flip of a coin, for example, if we associate $z = 0$ with tails and $z = 1$ with heads. The first and second moments of the distribution exist and are given by

$$M_1 = \int_{\infty}^{\infty} \frac{z}{2} \delta(z-0) \, dz + \int_{-\infty}^{\infty} \frac{z}{2} \delta(z-1) \, dz = 0 + \frac{1}{2} = \frac{1}{2}$$

$$M_2 = \int_{-\infty}^{\infty} \frac{z^2}{2} \delta(z-0) \, dz + \int_{-\infty}^{\infty} \frac{z^2}{2} \delta(z-1) \, dz = 0 + \frac{1}{2} = \frac{1}{2}.$$

The mean of the distribution is $\mu = M_1 = 1/2$, and the variance is

$$\sigma_z^2 = M_2 - \mu^2 = \frac{1}{2} - \left(\frac{1}{2}\right)^2 = \frac{1}{4}.$$

Now suppose we sample the distribution $n = 10$ times and form the mean value of the samples:

$$x = \frac{1}{n} \sum_{j=1}^{n} z_j = \frac{1}{10} \sum_{j=1}^{10} z_j.$$

The probability distribution for x is the binomial distribution with 10 trials, equal probabilities for the two outcomes, and $k = 10x$:

$$P(k = 10x) = \binom{n}{k} p^k q^{n-k} = \binom{10}{k} \left(\frac{1}{2}\right)^{10}. \tag{2.50}$$

According to the central limit theorem, the distribution is also approximately a Gaussian with $\mu = 1/2$ and standard deviation $\sigma_x^2 = \sigma_z^2 / 10 = 1/40$:

$$f(x) \approx (\pi/20)^{1/2} \exp\left[-\frac{1}{2} \frac{(x-1/2)^2}{(1/40)}\right]. \tag{2.51}$$

Figure 2.4.2 compares the Gaussian distribution for x to the binomial distribution. Even though x is calculated from only 10 samples, the pathological nature of the original distribution has disappeared, and x is approaching a Gaussian distribution.

Continued on page 36

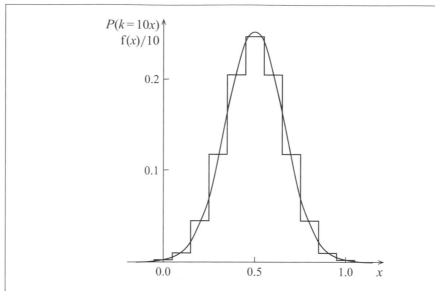

Figure 2.4: The central limit theorem at work. The histogram is the binomial distribution $P(k = 10x)$ for 10 samples and equal probabilities for the two possible results (equation 2.50). The limiting case is the scaled Gaussian $f(x)/10$, shown as the smooth curve (equation 2.51).

2.4.3 Mean, Moments, and Variance of the Gaussian Distribution

We will work with the Gaussian probability distribution in the form

$$f(x) = \frac{1}{\sqrt{2\pi\sigma^2}} \exp\left[-\frac{(x-\mu)^2}{2\sigma^2}\right]. \tag{2.52}$$

To calculate the moments of the distribution, we perform a change of variables, setting $z = x - \mu$ to get

$$f(z) = \frac{1}{\sqrt{2\pi\sigma^2}} \exp\left[-\frac{z^2}{2\sigma^2}\right]. \tag{2.53}$$

This equation is symmetric about $z = 0$, so it is immediately obvious that all odd moments of the distribution are zero. Specifically, we have

$$M_1 = \langle z \rangle = \langle x - \mu \rangle = \langle x \rangle - \mu = 0 \tag{2.54}$$

and thus we have found

$$\langle x \rangle = \mu. \tag{2.55}$$

This justifies a tacit assumption made when writing down equation 2.45.

We derive the variance from the moment generating function. For the Gaussian distribution this is

$$M(\zeta) = \frac{1}{\sqrt{2\pi\sigma^2}} \int_{-\infty}^{\infty} \exp[z\zeta] \exp\left[-\frac{z^2}{2\sigma^2}\right] dz$$

$$= \frac{1}{\sqrt{2\pi\sigma^2}} \int_{-\infty}^{\infty} \exp\left[-\frac{1}{2}\left(-2z\zeta + \frac{z^2}{\sigma^2}\right)\right] dz. \tag{2.56}$$

Completing the square and pulling the constant term outside the integral, we get

$$M(\zeta) = \exp\left[\frac{1}{2}\sigma^2\zeta^2\right]\frac{1}{\sqrt{2\pi\sigma^2}}\int_{-\infty}^{\infty}\exp\left[-\frac{1}{2}\left(\sigma^2\zeta^2 - 2z\zeta + \frac{z^2}{\sigma^2}\right)\right]dz$$

$$= \exp\left[\frac{1}{2}\sigma^2\zeta^2\right]\frac{1}{\sqrt{2\pi\sigma^2}}\int_{-\infty}^{\infty}\exp\left[-\frac{1}{2}\frac{(z-\sigma^2\zeta)^2}{\sigma^2}\right]dz. \tag{2.57}$$

The integral is simply the integral over a normalized Gaussian distribution with $\mu = \sigma^2\zeta$, so by inspection, the moment generating function is

$$M(\zeta) = \exp\left[\frac{1}{2}\sigma^2\zeta^2\right]. \tag{2.58}$$

The variance of x can be calculated from

$$\langle(x-\mu)^2\rangle = \langle z^2\rangle = M_2 = \left.\frac{\partial^2 M(\zeta)}{\partial\zeta^2}\right|_{\zeta=0}$$

$$= \left[\sigma^2\exp[\frac{1}{2}\sigma^2\zeta^2] + \sigma^4\zeta^2\exp[\frac{1}{2}\sigma^2\zeta^2]\right]\Big|_{\zeta=0}$$

$$= \sigma^2, \tag{2.59}$$

as expected.

2.5 Multivariate Gaussian Distribution

Suppose a Gaussian distribution has two independent variables with different variances:

$$f(x_1, x_2) \propto \exp\left[-\frac{1}{2}\left(\frac{x_1^2}{\sigma_1^2} + \frac{x_2^2}{\sigma_2^2}\right)\right]. \tag{2.60}$$

The contours of constant probability for this distribution are ellipses centered on the origin with their axes aligned with the x_1- and x_2-axes. Now rotate the Gaussian (or rotate the coordinate axes) so that the axes of the ellipses are tilted with respect to the coordinate axes (refer back to Figure 1.8). The rotation introduces a cross term $x_1 x_2$, so that the function takes the form

$$f(x_1, x_2) \propto \exp\left[-\frac{1}{2}\left(a_{11}x_1^2 + 2a_{12}x_1 x_2 + a_{22}x_2^2\right)\right], \tag{2.61}$$

where a_{11}, a_{12}, and a_{22} are constants that depend on σ_1^2, σ_2^2, and the angle of rotation. If the Gaussian is now moved away from the origin, its form becomes

$$f(x_1, x_2) \propto \exp\left[-\frac{1}{2}\left\{a_{11}(x_1 - \mu_1)^2 + 2a_{12}(x_1 - \mu_1)(x_2 - \mu_2) + a_{22}(x_2 - \mu_2)^2\right\}\right], \tag{2.62}$$

where μ_1 and μ_2 are constants. Generalized to n variables, the multivariate Gaussian has the form

$$f(x_1, x_2, \ldots, x_n) = B \exp\left[-\frac{1}{2}\sum_{j=1}^{n}\sum_{k=1}^{n}(x_j - \mu_j)a_{jk}(x_k - \mu_k)\right], \tag{2.63}$$

where B is the normalization constant, and $a_{jk} = a_{kj}$. The following discussion becomes more compact if we switch to vector notation. Let $\mathbf{x}$ and $\boldsymbol{\mu}$ be the column vectors

$$\mathbf{x} = \begin{pmatrix} x_1 \\ x_2 \\ \vdots \\ x_n \end{pmatrix} \quad \text{and} \quad \boldsymbol{\mu} = \begin{pmatrix} \mu_1 \\ \mu_2 \\ \vdots \\ \mu_n \end{pmatrix}, \tag{2.64}$$

and let $\mathbf{A}$ be the symmetric matrix

$$\mathbf{A} = \begin{pmatrix} a_{11} & a_{12} & \cdots & a_{1n} \\ a_{21} & a_{22} & \cdots & a_{2n} \\ \vdots & \vdots & & \vdots \\ a_{n1} & a_{n2} & \cdots & a_{nn} \end{pmatrix}. \tag{2.65}$$

With these identities, the multivariate Gaussian takes the simple form

$$f(\mathbf{x}) = B \exp\left[-\frac{1}{2}(\mathbf{x} - \boldsymbol{\mu})^{\mathrm{T}}\mathbf{A}(\mathbf{x} - \boldsymbol{\mu})\right], \tag{2.66}$$

where $(\mathbf{x} - \boldsymbol{\mu})^{\mathrm{T}}$ is the transpose of $(\mathbf{x} - \boldsymbol{\mu})$.

The matrix $\mathbf{A}$ is the inverse of the covariance matrix. To see this, we first need to calculate the moment generating function for $f(\mathbf{x})$. Let $\boldsymbol{\zeta}$ be the vector

$$\boldsymbol{\zeta} = \begin{pmatrix} \zeta_1 \\ \zeta_2 \\ \vdots \\ \zeta_n \end{pmatrix}, \tag{2.67}$$

so that we can write

$$\exp[\zeta_1(x_1 - \mu_1) + \zeta_2(x_2 - \mu_2) + \cdots + \zeta_n(x_n - \mu_n)] = \exp\left[\boldsymbol{\zeta}^{\mathrm{T}}(\mathbf{x} - \boldsymbol{\mu})\right]. \tag{2.68}$$

The moment generating function becomes

$$M(\boldsymbol{\zeta}) = B \int_{x_1} \cdots \int_{x_n} \exp\left[\boldsymbol{\zeta}^{\mathrm{T}}(\mathbf{x} - \boldsymbol{\mu}) - \frac{1}{2}(\mathbf{x} - \boldsymbol{\mu})^{\mathrm{T}}\mathbf{A}(\mathbf{x} - \boldsymbol{\mu})\right] dx_1 \cdots dx_n. \tag{2.69}$$

It is easier to evaluate this integral than one might expect. To do so, first convert the coordinates from $\mathbf{x}$ to $\mathbf{z}$, where $\mathbf{z} = \mathbf{x} - \boldsymbol{\mu}$, so that the moment generating function becomes

$$M(\boldsymbol{\zeta}) = B \int_{z_1} \cdots \int_{z_n} \exp\left[\boldsymbol{\zeta}^{\mathrm{T}}\mathbf{z} - \frac{1}{2}\mathbf{z}^{\mathrm{T}}\mathbf{A}\mathbf{z}\right] dz_1 \cdots dz_n. \tag{2.70}$$

Now, by completing the square, write the exponent as

$$\mathbf{z}^{\mathrm{T}}\mathbf{A}\mathbf{z} - 2\boldsymbol{\zeta}^{\mathrm{T}}\mathbf{z} = (\mathbf{z} - \mathbf{A}^{-1}\boldsymbol{\zeta})^{\mathrm{T}}\mathbf{A}(\mathbf{z} - \mathbf{A}^{-1}\boldsymbol{\zeta}) - \boldsymbol{\zeta}^{\mathrm{T}}\mathbf{A}^{-1}\boldsymbol{\zeta}. \tag{2.71}$$

(To verify this identity, expand the right-hand side of the equation, remembering that $\mathbf{A}$ and $\mathbf{A}^{-1}$ are symmetric, so that $(\mathbf{A}^{-1}\boldsymbol{\zeta})^{\mathrm{T}} = \boldsymbol{\zeta}^{\mathrm{T}}(\mathbf{A}^{-1})^{\mathrm{T}} = \boldsymbol{\zeta}^{\mathrm{T}}\mathbf{A}^{-1}$.) Making use of this expression, we can write the moment generating function as

$$M(\boldsymbol{\zeta}) = \exp\left[\frac{1}{2}\boldsymbol{\zeta}^{\mathrm{T}}\mathbf{A}^{-1}\boldsymbol{\zeta}\right] B \int_{z_1} \cdots \int_{z_n} \exp\left[-\frac{1}{2}(\mathbf{z} - \mathbf{A}^{-1}\boldsymbol{\zeta})^{\mathrm{T}}\mathbf{A}(\mathbf{z} - \mathbf{A}^{-1}\boldsymbol{\zeta})\right] dz_1 \cdots dz_n.$$

(2.72)

The integral on the right-hand side of this equation is just the integral over the Gaussian distribution, which is normalized because of the factor B. The moment generating function simplifies to

$$M(\boldsymbol{\zeta}) = \exp\left[\frac{1}{2}\boldsymbol{\zeta}^{\mathrm{T}}\mathbf{A}^{-1}\boldsymbol{\zeta}\right] = \exp\left[\frac{1}{2}\sum_q\sum_r \zeta_q\zeta_r\left(\mathbf{A}^{-1}\right)_{qr}\right],$$

(2.73)

where the notation $(\mathbf{A}^{-1})_{qr}$ means the q, r component of the inverse of $\mathbf{A}$.

We are now equipped to calculate the means, variances, and covariances of the multivariate Gaussian distribution. The first moments are

$$\langle z_i \rangle = \left.\frac{\partial M(\boldsymbol{\zeta})}{\partial \zeta_i}\right|_{\boldsymbol{\zeta}=0}$$

$$= \left.\sum_k \zeta_k\left(\mathbf{\Lambda}^{-1}\right)_{ik}\exp\left[\frac{1}{2}\sum_q\sum_r \zeta_q\zeta_r\left(\mathbf{A}^{-1}\right)_{qr}\right]\right|_{\boldsymbol{\zeta}=0}$$

$$= 0.$$

(2.74)

Since $z_i = x_i - \mu_i$, we arrive at

$$\langle x_i \rangle = \langle z_i + \mu_i \rangle = \mu_i,$$

(2.75)

or in vector notation

$$\langle \mathbf{x} \rangle = \boldsymbol{\mu}.$$

(2.76)

The variances and covariances are given by

$$\sigma_{ij} = \langle (x_i - \mu_i)(x_j - \mu_j) \rangle = \langle z_i z_j \rangle$$

$$= \left.\frac{\partial^2 M(\boldsymbol{\zeta})}{\partial \zeta_i \partial \zeta_j}\right|_{\boldsymbol{\zeta}=0}$$

$$= \left.\frac{\partial}{\partial \zeta_j}\left[\sum_k \zeta_k\left(\mathbf{A}^{-1}\right)_{ik}\exp\left\{\frac{1}{2}\sum_q\sum_r \zeta_q\zeta_r\left(\mathbf{A}^{-1}\right)_{qr}\right\}\right]\right|_{\boldsymbol{\zeta}=0}$$

$$= (\mathbf{A}^{-1})_{ij}.$$

(2.77)

Since $\mathbf{A}$ is symmetric, $\mathbf{A}^{-1}$ is symmetric and $\sigma_{ij} = \sigma_{ji}$.

One normally prefers to work with the covariance matrix $\mathbf{C}$, where

$$\mathbf{C} = \mathbf{A}^{-1} = \begin{pmatrix} \sigma_{11} & \sigma_{12} & \cdots & \sigma_{1n} \\ \sigma_{21} & \sigma_{22} & \cdots & \sigma_{2n} \\ \vdots & \vdots & & \vdots \\ \sigma_{n1} & \sigma_{n2} & \cdots & \sigma_{nn} \end{pmatrix}. \tag{2.78}$$

The multivariate Gaussian distribution then takes the form

$$f(\mathbf{z}) = \frac{1}{(2\pi)^{n/2} |\mathbf{C}|^{1/2}} \exp\left[-\frac{1}{2}\mathbf{z}^{\mathrm{T}}\mathbf{C}^{-1}\mathbf{z}\right], \tag{2.79}$$

or alternatively,

$$f(\mathbf{x}) = \frac{1}{(2\pi)^{n/2} |\mathbf{C}|^{1/2}} \exp\left[-\frac{1}{2}(\mathbf{x}-\boldsymbol{\mu})^{\mathrm{T}}\mathbf{C}^{-1}(\mathbf{x}-\boldsymbol{\mu})\right], \tag{2.80}$$

where $|\mathbf{C}|$ is the determinant of $\mathbf{C}$. This is the form of the multivariate Gaussian distribution most typically encountered. The surfaces defined by

$$G = (\mathbf{x}-\boldsymbol{\mu})^{\mathrm{T}}\mathbf{C}^{-1}(\mathbf{x}-\boldsymbol{\mu}) = \text{constant} \tag{2.81}$$

are ellipsoids centered on $\mathbf{x} = \boldsymbol{\mu}$. These ellipsoids are the surfaces of constant probability density.

It is relatively easy to show that this form of the multivariate Gaussian is properly normalized. In outline the calculation runs as follows. Suppose the Gaussian has n independent variables. Rotate the coordinate system so that $\mathbf{C}$ transforms into a diagonal matrix $\mathbf{D}$ whose diagonal elements $(\mathbf{D})_{ii}$ are all real. Since the covariance matrix is symmetric and all its components are real, this can always be done (see Section E.11 in Appendix E). Because the off-diagonal terms in $\mathbf{D}$ are all zero, its inverse is also a diagonal matrix, and its elements are $1/(\mathbf{D})_{ii}$. Therefore, the multivariate Gaussian separates into the product of n independent distributions whose variances are $\sigma_i^2 = (\mathbf{D})_{ii}$. For each independent distribution the normalization constant is $[2\pi(\mathbf{D})_{ii}]^{1/2}$. The normalization constant for the entire distribution is the product of the individual normalization constants, $(2\pi)^{n/2}\left[\prod_i(\mathbf{D})_{ii}\right]^{1/2}$. Because $\mathbf{D}$ is diagonal, the product of its diagonal terms is equal to its determinant, $\prod_i(\mathbf{D})_{ii} = |\mathbf{D}|$, and the normalization constant can be written as $(2\pi)^{n/2}|\mathbf{D}|^{1/2}$. Now rotate back to the original coordinate system. The value of the determinant in the original coordinate system is the same as in the rotated coordinate system, $|\mathbf{C}| = |\mathbf{D}|$. Thus, the normalization factor is $(2\pi)^{n/2}|\mathbf{C}|^{1/2}$.

For the specific case $n = 2$ the inverse of $\mathbf{C}$ is

$$\mathbf{C}^{-1} = \frac{1}{\sigma_{11}\sigma_{22} - \sigma_{12}^2} \begin{pmatrix} \sigma_{22} & -\sigma_{12} \\ -\sigma_{12} & \sigma_{11} \end{pmatrix}, \tag{2.82}$$

and the exponent of the Gaussian distribution is

$$\frac{1}{2}(\mathbf{x}-\boldsymbol{\mu})^{\mathrm{T}}\mathbf{C}^{-1}(\mathbf{x}-\boldsymbol{\mu}) = \frac{1}{2}\left(\frac{\sigma_{22}(x_1-\mu_1)^2}{\sigma_{11}\sigma_{22}-\sigma_{12}^2} + \frac{2\sigma_{12}(x_1-\mu_1)(x_2-\mu_2)}{\sigma_{11}\sigma_{22}-\sigma_{12}^2} + \frac{\sigma_{11}(x_2-\mu_2)^2}{\sigma_{11}\sigma_{22}-\sigma_{12}^2}\right). \tag{2.83}$$

If x_1 and x_2 are uncorrelated, which means $\sigma_{12} = 0$, equation 2.83 simplifies to

$$\frac{1}{2}(\mathbf{x} - \boldsymbol{\mu})^{\mathrm{T}} \mathbf{C}^{-1} (\mathbf{x} - \boldsymbol{\mu}) = \frac{1}{2} \left(\frac{(x_1 - \mu_1)^2}{\sigma_{11}} + \frac{(x_2 - \mu_2)^2}{\sigma_{22}} \right). \tag{2.84}$$

In this case the distribution function can be separated into the product of a distribution that depends only on x_1 and one that depends only on x_2, so x_1 and x_2 are independent.

2.6 χ^2 Distribution

2.6.1 Derivation of the χ^2 Distribution

Suppose we have two uncorrelated samples, ϵ_1 and ϵ_2, with means equal to zero and variances of σ_1^2 and σ_2^2, both drawn from Gaussian distributions. The probability of finding the samples in the range ϵ_1 to $\epsilon_1 + d\epsilon_1$ and ϵ_2 to $\epsilon_2 + d\epsilon_2$ is

$$f(\epsilon_1, \epsilon_2) d\epsilon_1 d\epsilon_1 = \frac{1}{\sqrt{2\pi}\,\sigma_1} \exp\left[-\frac{\epsilon_1^2}{2\sigma_1^2} \right] \frac{1}{\sqrt{2\pi}\,\sigma_2} \exp\left[-\frac{\epsilon_2^2}{2\sigma_2^2} \right] d\epsilon_1 d\epsilon_2. \tag{2.85}$$

Defining the new variables

$$x_1 = \frac{\epsilon_1}{\sigma_1}, \qquad x_2 = \frac{\epsilon_2}{\sigma_2} \tag{2.86}$$

the probability becomes

$$f(x_1, x_2) dx_1 dx_2 = \frac{1}{2\pi} \exp\left[-\frac{1}{2}(x_1^2 + x_2^2) \right] dx_1 dx_2. \tag{2.87}$$

Let S be the sum of the squares of the variables:

$$S = x_1^2 + x_2^2. \tag{2.88}$$

What is the probability of finding S in the interval S to $S + dS$? The probability of finding S in a certain range is equivalent to the probability that S lies in an annulus centered on the origin, as shown in Figure 2.5.

To calculate this probability we change variables from (x_1, x_2) to the polar coordinates (χ, θ), where

$$x_1 = \chi \cos\theta \tag{2.89}$$

$$x_2 = \chi \sin\theta \tag{2.90}$$

$$\chi^2 = S = x_1^2 + x_2^2. \tag{2.91}$$

The probability distribution becomes

$$f(x_1, x_2) dx_1 dx_2 = f(\chi, \theta) \chi \, d\chi \, d\theta = \frac{1}{2\pi} \exp\left[-\frac{1}{2}\chi^2 \right] \chi \, d\chi \, d\theta. \tag{2.92}$$

Integrating over θ, we find

$$f(\chi) \chi \, d\chi = \exp\left[-\frac{1}{2}\chi^2 \right] \chi \, d\chi \tag{2.93}$$

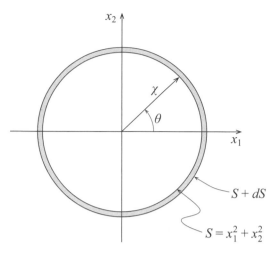

Figure 2.5: Curves on which $S = x_1^2 + x_2^2$ is a constant are circles around the origin. The annulus is the region in which $x_1^2 + x_2^2$ lies between S and $S + dS$. The radius of the circle is χ, where $\chi^2 = S$.

or

$$f(\chi^2)d(\chi^2) = \frac{1}{2}\exp\left[-\frac{1}{2}\chi^2\right]d(\chi^2) \tag{2.94}$$

and, finally, dropping the differentials, we arrive at

$$f(\chi^2) = \frac{1}{2}\exp\left[-\frac{1}{2}\chi^2\right]. \tag{2.95}$$

This is called the χ^2 *distribution for two degrees of freedom.*

The general case of n points (n degrees of freedom), proceeds in exactly the same way, although there is a complication, because we must evaluate the volume of a spherical shell in n-dimensional space. Set

$$\chi^2 = S = \sum_{i=1}^{n}\frac{\epsilon_i^2}{\sigma_i^2} = \sum_{i=1}^{n}x_i^2 \tag{2.96}$$

and the probability distribution becomes

$$f(x_1, x_2, \ldots, x_n)dx_1 dx_2 \cdots dx_n = \left(\frac{1}{\sqrt{2\pi}}\right)^n\exp\left[-\frac{1}{2}\chi^2\right]dx_1 dx_2 \cdots dx_n. \tag{2.97}$$

Now convert this to a distribution of the form $f(\chi^2)d(\chi^2)$. In Appendix D we show that the volume of a spherical shell with radius r and width dr in n-dimensional space is

$$dV = \frac{2\pi^{n/2}}{(n/2 - 1)!}r^{n-1}dr. \tag{2.98}$$

If n is even, $(n/2 - 1)!$ is the standard factorial. If it is odd, then $(n/2 - 1)! = (n/2 - 1)$ $(n/2 - 2)!$, ending at $(1/2)! = \sqrt{\pi}/2$ (see Section A.3 in Appendix A). Recognizing that χ

plays the role of r, we have, after integrating over $n - 1$ angles,

$$dx_1 dx_2 \cdots dx_n \Rightarrow \frac{2\pi^{n/2}}{(n/2 - 1)!} \chi^{n-1} d\chi$$

$$= \frac{\pi^{n/2}}{(n/2 - 1)!} \left(\chi^2\right)^{(n-2)/2} d(\chi^2), \qquad (2.99)$$

so the distribution becomes

$$f(\chi^2) d(\chi^2) = \left(\frac{1}{\sqrt{2\pi}}\right)^n \exp\left[-\frac{1}{2}\chi^2\right] \frac{\pi^{n/2}}{(n/2 - 1)!} \left(\chi^2\right)^{(n-2)/2} d(\chi^2). \qquad (2.100)$$

Simplifying and dropping the differentials, we arrive at the standard form for the χ^2 probability distribution:

$$f_n(\chi^2) = \frac{\left(\chi^2\right)^{(n-2)/2}}{2^{n/2}(n/2 - 1)!} \exp\left[-\frac{1}{2}\chi^2\right]. \qquad (2.101)$$

Example: The explicit forms for 4 and 9 degrees of freedom are

$$f_4(\chi^2) = \frac{1}{4}\chi^2 \exp\left[-\frac{1}{2}\chi^2\right]$$

$$f_9(\chi^2) = \frac{\left(\chi^2\right)^{7/2}}{105\sqrt{2\pi}} \exp\left[-\frac{1}{2}\chi^2\right].$$

These two distributions are plotted in Figure 2.6.1. The χ^2 distribution is equal to zero at $\chi^2 = 0$ for $n > 2$, and has an exponential tail for large χ^2, so the distribution always looks somewhat like the distributions shown in the figure.

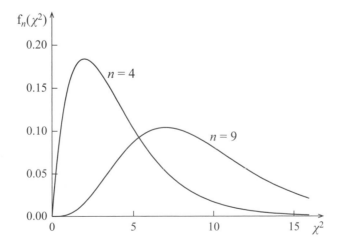

Figure 2.6: The χ^2 distributions for 4 and 9 degrees of freedom.

2.6.2 Mean, Mode, and Variance of the χ^2 Distribution

The mode is the peak value of the distribution. Taking the derivative of the distribution and setting it equal to zero, we find

$$
0 = \frac{d}{d\chi^2} f_n\left(\chi^2\right) = \frac{d}{d\chi^2}\left\{ \left(\chi^2\right)^{(n-2)/2} \exp\left[-\frac{1}{2}\chi^2 \right] \right\}
$$

$$
= \frac{n-2}{2} \left(\chi^2\right)^{(n-4)/2} \exp\left[-\frac{1}{2}\chi^2 \right] - \frac{1}{2}\left(\chi^2\right)^{(n-2)/2} \exp\left[-\frac{1}{2}\chi^2 \right]
$$

$$
= \frac{n-2}{2} - \frac{1}{2}\chi^2, \tag{2.102}
$$

so

$$
\chi^2_{mode} = n-2. \tag{2.103}
$$

The mean of χ^2 is given by

$$
\langle \chi^2 \rangle = \int_0^\infty \chi^2 f_n(\chi^2) d\chi^2. \tag{2.104}
$$

After changing variables to

$$
t = \frac{1}{2}\chi^2 \tag{2.105}
$$

the mean becomes

$$
\langle \chi^2 \rangle = \int_0^\infty 2t \, \frac{(2t)^{(n-2)/2}}{2^{n/2} \, (n/2-1)!} \exp[-t] \, 2dt = \frac{2}{(n/2-1)!} \int_0^\infty t^{n/2} \exp[-t] \, dt. \tag{2.106}
$$

The integral is a Γ function (see Section A.3 in Appendix A), so

$$
\langle \chi^2 \rangle = \frac{2}{(n/2-1)!} \Gamma\left(\frac{n}{2} + 1 \right)
$$

$$
= \frac{2}{(n/2-1)!} \left(\frac{n}{2} \right)! = \frac{2}{(n/2-1)!} \frac{n}{2} \left(\frac{n}{2} - 1 \right)!
$$

$$
= n. \tag{2.107}
$$

This is the expected result because, by definition,

$$
\chi^2 = \sum_i^n \frac{\epsilon_i^2}{\sigma_i^2}, \tag{2.108}
$$

so the mean value of χ^2 is

$$
\langle \chi^2 \rangle = \sum_i^n \frac{\langle \epsilon_i^2 \rangle}{\sigma_i^2} = \sum_i^n \frac{\sigma_i^2}{\sigma_i^2} = n. \tag{2.109}
$$

To calculate the variance of χ^2, we begin by calculating the mean value of $\left(\chi^2\right)^2$:

$$
\left\langle \left(\chi^2\right)^2 \right\rangle = \int_0^\infty \left(\chi^2\right)^2 f_n(\chi^2) d\chi^2. \tag{2.110}
$$

As before, change variables to

$$t = \frac{1}{2}\chi^2, \tag{2.111}$$

and then

$$\left\langle (\chi^2)^2 \right\rangle = \int_0^\infty (2t)^2 \frac{(2t)^{(n-2)/2}}{2^{n/2}(n/2-1)!} \exp[-t]\, 2dt$$

$$= \frac{4}{(n/2-1)!} \int_0^\infty t^{(1+n/2)} \exp[-t]\, dt. \tag{2.112}$$

We again recognize the integral on the right-hand side as a Γ function and find

$$\left\langle (\chi^2)^2 \right\rangle = \frac{4}{(n/2-1)!} \Gamma\left(\frac{n}{2}+2\right)$$

$$= \frac{4}{(n/2-1)!} \left(\frac{n}{2}+1\right)! = \frac{4}{(n/2-1)!} \left(\frac{n}{2}+1\right)\left(\frac{n}{2}\right)\left(\frac{n}{2}-1\right)!$$

$$= n^2 + 2n. \tag{2.113}$$

We can now calculate the variance of χ^2:

$$\sigma_{\chi^2}^2 = \left\langle \left(\chi^2 - \langle \chi^2 \rangle\right)^2 \right\rangle = \left\langle (\chi^2)^2 \right\rangle - \langle \chi^2 \rangle^2 = n^2 + 2n - n^2$$

$$= 2n. \tag{2.114}$$

2.6.3 χ^2 *Distribution in the Limit of Large* n

The central limit theorem guarantees that the χ^2 distribution approaches a Gaussian distribution as n becomes large. The mean and variance of the χ^2 distribution become the mean and variance of the Gaussian, so we have:

$$\lim_{n\to\infty} f_n(\chi^2) = \frac{1}{\sqrt{2n}\sqrt{2\pi}} \exp\left[-\frac{1}{2}\frac{(\chi^2-n)^2}{2n}\right]. \tag{2.115}$$

This approximation becomes quite good for $n > 30$.

Example: What is the probability that χ^2 is greater than 6 if $n = 4$? Greater than 60 if $n = 50$?

We evaluate the probability in the usual way:

$$P_n(\chi^2 > a) = \int_a^\infty f_n(\chi^2)d\chi^2.$$

For $n = 4$ we simply look up the result in tables for the χ^2 distribution, finding $P_4(\chi^2 > 6) = 0.20$. For $n = 50$, we assume that the χ^2 distribution is close to a Gaussian distribution with a mean of 50 and a variance of $\sigma^2 = 2n = 100$. Again using tables we find $P_{50}(\chi^2 > 60) = 0.16$

2.6.4 Reduced χ^2

Instead of the usual χ^2, one sometimes uses the reduced χ^2, defined to be

$$\chi^2_{red} = \frac{1}{n}\chi^2 = \frac{1}{n}\sum_i \frac{\epsilon_i^2}{\sigma_i^2}. \tag{2.116}$$

The point of using χ^2_{red} instead of χ^2 is that

$$\langle\chi^2_{red}\rangle = 1, \tag{2.117}$$

which seems easier to interpret than χ^2, since the number of degrees of freedom does not appear explicitly. The simplification is not real, though, because the number of degrees of freedom does appear explicitly in the expression for the variance of χ^2_{red}:

$$\sigma^2_{\chi^2_{red}} = \frac{2}{n}. \tag{2.118}$$

Whether to use χ^2 or χ^2_{red} is a matter of taste.

2.6.5 χ^2 for Correlated Variables

Up to now we have, for clarity, defined χ^2 to be

$$\chi^2 = \sum_{i=1}^{n}\frac{\epsilon_i^2}{\sigma_i^2}. \tag{2.119}$$

This is too restrictive. Consider the general n-parameter Gaussian (equation 2.80)

$$\begin{aligned} f(\mathbf{x}) &= \frac{1}{(2\pi)^{n/2}|\mathbf{C}|^{1/2}}\exp\left[-\frac{1}{2}(\mathbf{x}-\boldsymbol{\mu})^{\mathrm{T}}\mathbf{C}^{-1}(\mathbf{x}-\boldsymbol{\mu})\right] \\ &= \frac{1}{(2\pi)^{n/2}|\mathbf{C}|^{1/2}}\exp\left[-\frac{1}{2}\boldsymbol{\epsilon}^{\mathrm{T}}\mathbf{C}^{-1}\boldsymbol{\epsilon}\right]. \end{aligned} \tag{2.120}$$

where $\boldsymbol{\epsilon}^{\mathrm{T}} = (\epsilon_1, \epsilon_2, \ldots, \epsilon_n)$ is the vector of residuals, the components of the covariance matrix are

$$(\mathbf{C})_{ij} = \sigma_{ij}, \tag{2.121}$$

and σ_{ij} is the covariance between ϵ_i and ϵ_j.

We now show that the quantity

$$S = \boldsymbol{\epsilon}^{\mathrm{T}}\mathbf{C}^{-1}\boldsymbol{\epsilon}. \tag{2.122}$$

is a χ^2 variable with n degrees of freedom. To do this, we rotate to a new coordinate system in which the all the correlations become zero, making $\mathbf{C}$ diagonal. It is always possible to do this if $\mathbf{C}$ is real and symmetric (see Section E.12 in Appendix E for a discussion of diagonalization and Section 4.6 for the physical meaning of diagonalizing $\mathbf{C}$). For us the only constraint is that $\sigma_{ij} = \sigma_{ji}$. Let the residuals in the rotated coordinate system be $(\epsilon')^{\mathrm{T}} = (\epsilon'_1, \epsilon'_2, \cdots, \epsilon'_n)$ and the diagonal components of the covariance matrix be

$$\left(\mathbf{C}'\right)_{ii} = \sigma'_{ii} = \sigma'^2_i. \tag{2.123}$$

Since all the off-diagonal components of the covariance matrix are zero in the rotated coordinate system, all the off-diagonal components of its inverse are zero, and its diagonal components are

$$(\mathbf{C}'^{-1})_{ii} = \frac{1}{\sigma_i'^2}. \tag{2.124}$$

Therefore, in the rotated coordinate system, equation 2.122 becomes

$$S = \epsilon'^{\mathrm{T}} \mathbf{C}'^{-1} \epsilon' = \sum_{i=1}^{n} \frac{\epsilon_i^2}{\sigma_i'^2}. \tag{2.125}$$

This has the same form as equation 2.119, so S is a χ^2 variable with n degrees of freedom, and it is correct to write

$$\chi^2 = \epsilon^{\mathrm{T}} \mathbf{C}^{-1} \epsilon. \tag{2.126}$$

The only complication that might arise is that some of the diagonal components of $\mathbf{C}'$ might be zero. If so, each zero reduces the number of degrees of freedom by one.

2.7 Beta Distribution

The beta probability distribution function is useful when the probability of getting x is zero outside a finite range of x. It is defined to be

$$\beta(x) \propto x^{a-1}(1-x)^{b-1}, \tag{2.127}$$

where x lies in the range $0 \leq x \leq 1$, and a and b are positive integers. The functional form of the beta distribution looks somewhat like that of the binomial probability distribution, but the independent variable is x, not a or b. The beta distribution can be normalized using the identity (see equations A.32 and A.40 in Appendix A)

$$\frac{\Gamma(a)\Gamma(b)}{\Gamma(a+b)} = \int_0^1 x^{a-1}(1-x)^{b-1}dx, \tag{2.128}$$

where $\Gamma(a)$ is the gamma function. Since a and b are positive integers, the gamma function reduces to the factorial, and the integral becomes

$$\frac{(a-1)!(b-1)!}{(a+b-1)!} = \int_0^1 x^{a-1}(1-x)^{b-1}dx. \tag{2.129}$$

The normalized beta distribution is, then,

$$\beta(x) = \frac{(a+b-1)!}{(a-1)!(b-1)!} x^{a-1}(1-x)^{b-1}. \tag{2.130}$$

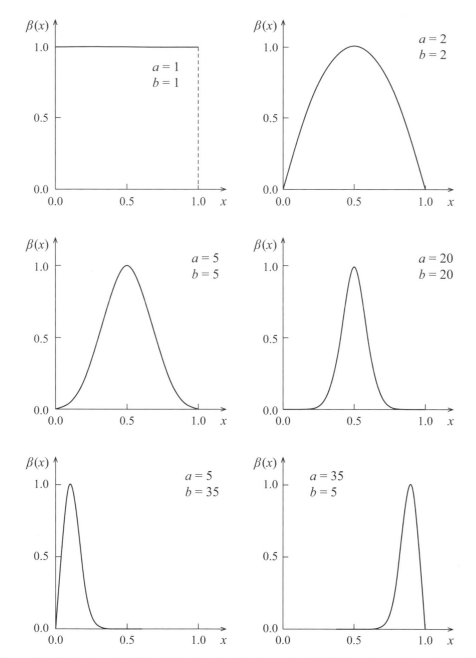

Figure 2.7: The beta probability distribution function (equation 2.130) for various values of a and b. The plots have been normalized to 1 at the peak of the distribution for convenience in plotting.

Plots of the beta distribution for various values of a and b are shown in Figure 2.7. It is easy to show that the mean and variance of x are

$$\langle x \rangle = \frac{a}{a+b} \tag{2.131}$$

$$\sigma_x^2 = \frac{ab}{(a+b)^2(a+b+1)}. \tag{2.132}$$

Thus, for the case $a = 35$ and $b = 5$ shown in the bottom right panel of Figure 2.7, the mean and variance are

$$\langle x \rangle = 35/40 = 0.875$$

$$\sigma_x^2 = 175/(40 \times 40 \times 41) = 0.00267$$

$$\sigma_x = 0.058.$$

A bit of algebra yields the inverse relations

$$a = \langle x \rangle \left(\frac{\langle x \rangle \, [1 - \langle x \rangle]}{\sigma_x^2} - 1 \right) \tag{2.133}$$

$$b = a \frac{1 - \langle x \rangle}{\langle x \rangle}. \tag{2.134}$$

3

Random Numbers and Monte Carlo Methods

3.1 Introduction

Uses of random numbers in statistics tend to fall into two broad categories. The first is estimation of the reliability of a statistical analysis. Artificial data with noise properties that mimic those of real data are generated and then analyzed in the same way as the real data. The reliability of results for the real data can be estimated by generating and analyzing many different sets of artificial data. The second is sampling or integrating complicated functions. The posterior probability distribution functions of Bayesian analyses are notable examples (see Chapter 7). Posterior distributions are often multidimensional functions from which it is difficult to extract meaningful information. They can defy analytic integration, and standard numerical techniques may be too computationally expensive to be practical. Information can often be extracted from the distributions rapidly and efficiently using Monte Carlo techniques.

This chapter is devoted to random numbers and Monte Carlo methods. This first section continues with a brief discussion of the meaning of random numbers generated by mathematical algorithms. Section 3.2 discusses methods for converting random numbers with uniform distributions to random numbers with other distributions. Section 3.3 is a short introduction to Monte Carlo integration. Then, motivated by the needs of Bayesian statistics, Section 3.4 discusses Markov chains, and Section 3.5 discusses Markov chain Monte Carlo sampling.

It is easy to find random number generators in nature, not so easy in mathematics. The intervals between the arrivals of photons from a faint star and the intervals between the decays of a radioactive element are random numbers, both with exponential distributions. More generally, all quantum mechanical processes have some aspect that is fundamentally and irreducibly random. The numbers produced by a mathematical algorithm are never truly random, because the algorithm will always produce the same numbers given the same starting conditions. A string of numbers produced by an algorithm can, however, act like random numbers if no information about the next number in a string can be determined from the previous numbers without knowing the algorithm. Numbers that are random in this sense are sometimes called pseudo-random numbers. Although pseudo-randomness is sometimes called a defect, in practice it is often the opposite. Strings of random numbers that can be reproduced at will are highly useful. Computer programs that make use of random numbers are, for example, much easier to debug and verify if the same sequence of random numbers can be regenerated.

It is difficult to invent good, practical algorithms for producing pseudo-random numbers. One can prove that nearly all irrational numbers expressed in ordinary decimal form consist of a sequence of digits that are random integers between 0 and 9. Unfortunately, there is no such proof for any specific irrational number. Thus, while it is widely suspected and even supposed that the digits of π, e, and such algebraic irrationals as $\sqrt{2}$ are random numbers, this has never been shown. Furthermore, algorithms for producing random numbers must be fast and use little computer memory if they are to be useful. Suppose we accept that the digits of $\sqrt{2}$ are random numbers and compute $\sqrt{2}$ using a Newton's iteration,

$$x_{k+1} = \frac{1}{2}\left(x_k + \frac{2}{x_k}\right). \tag{3.1}$$

This simple algorithm converges quadratically and yields 10^6 or even 10^9 digits with just 20 or 30 iterations. Unfortunately, to produce the billionth digit, one must retain and divide by a number with 10^9 digits (or more), so the algorithm is an exceedingly slow and memory-intensive way to produce random numbers.

Fortunately, many good random number generators are now available. Your author is currently using a version of the Mersenne twister.[1] For other random number generators see, for example, Press et al. (2007). Note in passing that the requirements on random number generators for statistical applications are not the same as those for cryptography. A random number generator that, for example, relies on a seed that is known to be a small integer, as many do, would be a poor choice for cryptography.

The numbers produced by a random number generator are called *random variates* or *random deviates*. From now on we will assume that we have a source of random deviates u_x drawn from a flat probability distribution between 0 and 1:

$$p(x) = \begin{cases} 1, & 0 \leq x \leq 1 \\ 0, & \text{otherwise} \end{cases}. \tag{3.2}$$

The u_x are called *uniform random deviates* or just *uniform deviates*.

3.2 Nonuniform Random Deviates

It is often necessary to generate random deviates with nonuniform distributions. One usually does this by generating uniform random deviates and then converting them to the desired nonuniform distribution. There is as much art as science to these conversion methods, and whole books have been written on the subject.[2] The rest of this section and much of the rest of the chapter is a brief introduction to techniques for generating nonuniform deviates.

---

[1] M. Matsumoto and T. Nishimura. 1998. "Mersenne Twister: A 623-Dimensionally Equidistributed Uniform Pseudo-random Number Generator." *ACM Transactions on Modeling and Computer Simulations* vol. 8, p. 3.

[2] For example, J.E. Gentle. 2004. *Random Number Generation and Monte Carlo Methods*, second edition. New York: Springer.

3.2.1 Inverse Cumulative Distribution Function Method

A conceptually simple way to convert a uniform random deviate u_x to a random deviate u_s with a probability distribution g(s) is to choose u_s so that (see Figure 3.1)

$$\int_0^{u_x} dx = \int_{-\infty}^{u_s} g(s)ds \tag{3.3}$$

$$u_x = G(u_s), \tag{3.4}$$

where $G(u_s)$ is the cumulative distribution function (equation 1.31). Equation 3.4 is sometimes written

$$u_s = G^{-1}(u_x), \tag{3.5}$$

so the method is sometimes called the inverse cumulative distribution function method or inverse CDF method. If G is an elementary function and its inverse can be calculated rapidly, the inverse CDF method is a good way to convert random deviates. Thus, if g(s) is the triangular distribution

$$g(s) = \begin{cases} 2s, & 0 \le s \le 1 \\ 0, & \text{otherwise} \end{cases}, \tag{3.6}$$

then

$$u_x = \int_0^{u_s} 2s\,ds = u_s^2 \tag{3.7}$$

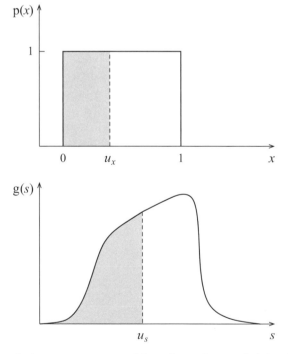

Figure 3.1: A random deviate u_x was generated from the uniform probability distribution p(x). We wish to convert it to a random deviate u_s that has a probability distribution g(s). In the inverse cumulative distribution function method, u_s is chosen so that the shaded area in the lower graph is equal to the shaded area in the upper one.

and the conversion is just

$$u_s = u_x^{1/2}. \tag{3.8}$$

The inverse CDF method has discouragingly few applications, though it fails when g(s) is even so simple as a Gaussian distribution.

Example: Suppose we wish to generate deviates from the exponential probability distribution function

$$g(s) = \begin{cases} 0, & s < 0 \\ a\exp[-as], & s \geq 0 \end{cases} .$$

Equation 3.4 becomes

$$u_x = \int_{s=0}^{u_s} a\exp[-as]ds = 1 - \exp[-au_s],$$

so the exponential random deviates are given by

$$u_s = -\frac{1}{a}\ln(1 - u_x),$$

where the u_x are uniform deviates.

3.2.2 Multidimensional Deviates

If a multidimensional probability distribution can be separated into independent probability distributions, generating deviates for the multidimensional distribution is straightforward. To distribute points uniformly over the surface of a sphere, for example, take

$$\int_0^{u_x}\int_0^{u_y} dxdy = u_x u_y = \frac{1}{4\pi}\int_0^{u_\phi}\int_0^{u_\theta}\sin\theta d\theta d\phi = \left[\frac{1}{2}\int_0^{u_\theta}\sin\theta d\theta\right]\left[\frac{1}{2\pi}\int_0^{u_\phi}d\phi\right] \tag{3.9}$$

and use the inverse CDF method separately for the deviates in the θ and ϕ coordinates

$$u_\theta = \cos^{-1}(1 - 2u_y) \tag{3.10}$$

$$u_\phi = 2\pi u_x. \tag{3.11}$$

3.2.3 Box-Müller Method for Generating Gaussian Deviates

If the desired multidimensional distribution cannot be separated into independent distributions, one must use equation 1.99 to transform the deviates. This is illustrated by the two-dimensional Box-Müller method, which provides a convenient way to generate Gaussian random deviates.

Let $f(x_1, x_2)$ be a two-dimensional Gaussian distribution whose independent variables have means equal to 0, variances equal to 1, and are uncorrelated:

$$f(x_1, x_2) = \frac{1}{\sqrt{2\pi}}\exp\left[-\frac{x_1^2}{2}\right]\frac{1}{\sqrt{2\pi}}\exp\left[-\frac{x_2^2}{2}\right] = \frac{1}{2\pi}\exp\left[-\frac{x_1^2 + x_2^2}{2}\right]. \tag{3.12}$$

Consider the coordinate transformation

$$x_1 = \sqrt{-2\ln s_1} \, \cos 2\pi s_2 \tag{3.13}$$

$$x_2 = \sqrt{-2\ln s_1} \, \sin 2\pi s_2, \tag{3.14}$$

where the new variables s_1 and s_2 range from 0 to 1. With this transformation the two-dimensional Gaussian becomes (see equation 1.98)

$$f(x_1,x_2)dx_1dx_2 = f(s_1,s_2)\left|\frac{\partial(x_1,x_2)}{\partial(s_1,s_2)}\right| ds_1 ds_2 = g(s_1,s_2)ds_1 ds_2, \tag{3.15}$$

where

$$f(s_1,s_2) = f(x_1(s_1,s_2),x_2(s_1,s_2))$$

$$= \frac{1}{2\pi}\exp\left[-\frac{(\sqrt{-2\ln s_1}\,\cos 2\pi s_2)^2 + (\sqrt{-2\ln s_1}\,\sin 2\pi s_2)^2}{2}\right]$$

$$= \frac{1}{2\pi}\exp[\ln s_1]$$

$$= \frac{s_1}{2\pi}. \tag{3.16}$$

After some work, the Jacobian determinant reduces to

$$\left|\frac{\partial(x_1,x_2)}{\partial(s_1,s_2)}\right| = \left|\frac{\partial x_1}{\partial s_1}\frac{\partial x_2}{\partial s_2} - \frac{\partial x_2}{\partial s_1}\frac{\partial x_1}{\partial s_2}\right| = \frac{2\pi}{s_1}, \tag{3.17}$$

so $g(s_1,s_2) = 1$. Thus, the coordinate transformation converts the two-dimensional Gaussian distribution with both variables ranging between $-\infty$ and $+\infty$ to a two-dimensional uniform distribution with both variables ranging between 0 and 1. The procedure for producing random deviates with a Gaussian distribution is then as follows:

1. Generate two uniform random deviates u_{s_1} and u_{s_2} between 0 and 1.
2. Transform u_{s_1} and u_{s_2} to new deviates u_{x_1} and u_{x_2} using equations 3.13 and 3.14. The two new deviates have a Gaussian distribution with mean equal to 0 and variance equal to 1.

Although the Box-Müller method is convenient, it is not efficient. Each random deviate requires calculating a trigonometric function, a logarithm, and a square root.

3.2.4 Acceptance-Rejection Algorithm

The acceptance-rejection algorithm lies at the heart of many methods for producing nonuniform random deviates. A simple example highlights the essence of the method. Suppose one wants to generate a set of random deviates that uniformly cover the two-dimensional area inside the first quadrant of the unit circle $x^2 + y^2 = 1$ (see Figure 3.2). Generate two uniform deviates u_x and u_y between 0 and 1, one for the x-coordinate and one for the y-coordinate. The points (u_x, u_y) cover the unit square in the (x,y) plane uniformly. Accept the pair if $u_x^2 + u_y^2 \leq 1$; otherwise throw the pair away.

Figure 3.3 shows how the standard acceptance-rejection algorithm works for a one-dimensional probability distribution. We wish to generate deviates drawn from the probability distribution function $f(x)$, where $0 \leq x \leq x_{max}$, and the maximum value of

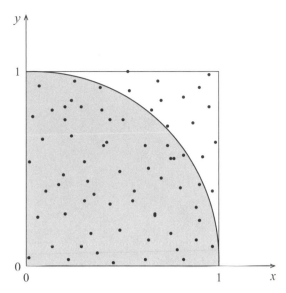

Figure 3.2: An example of the acceptance-rejection technique. The coordinates of the dots are (u_x, u_y), where u_x and u_y are independent uniform deviates between 0 and 1. To generate a set of random points that uniformly cover the two-dimensional area inside the first quadrant of the unit circle, just reject the points outside the shaded area bounded by $x^2 + y^2 = 1$.

$f(x)$ is f_{max}. Define a new, two-dimensional probability distribution function $g(x, y)$ that is constant within the rectangle bounded by x_{max} and $y_{max} = f_{max}$:

$$g(x, y) = \begin{cases} \text{constant,} & \begin{array}{l} 0 \leq x \leq x_{max} \\ 0 \leq y \leq y_{max} \end{array} \\ 0, & \text{otherwise} \end{cases} . \qquad (3.18)$$

To generate a two-dimensional random deviate from $g(x, y)$, first generate two independent one-dimensional uniform deviates, u_x over the range $0 \leq u_x \leq x_{max}$ and u_y over the range $0 \leq u_y \leq y_{max}$, and assign them to x and y to give the two-dimensional deviate (u_x, u_y). The deviates generated in this way are shown as dots in Figure 3.3 and cover the rectangle uniformly.

Now draw the curve $y = f(x)$. This divides $g(x, y)$ into two areas, one in which y is less than or equal to $f(x)$, shown by the shaded area in Figure 3.3, and the other in which it is greater. The two-dimensional random deviates for which $u_y \leq f(u_x)$ are those lying within the shaded area. If we accept the deviates for which $u_y \leq f(u_x)$, rejecting the rest, the u_x part of the accepted (u_x, u_y) deviates has a distribution proportional to $f(x)$. To see this, note that the marginal distribution of the part of $g(x, y)$ that is within the shaded area is

$$\int_0^{y=f(x)} g(x, y) dy \propto \int_0^{y=f(x)} dy \propto f(x). \qquad (3.19)$$

We can now state the standard version of the acceptance-rejection algorithm:

1. Generate a uniform random deviate u_x between 0 and x_{max}.
2. Generate a second random deviate u_y between 0 and $y_{max} = f_{max}$.
3. If u_y is less than or equal to $f(u_x)$, keep u_x as the random deviate from $f(x)$.
4. Otherwise throw away u_x and u_y and try again.

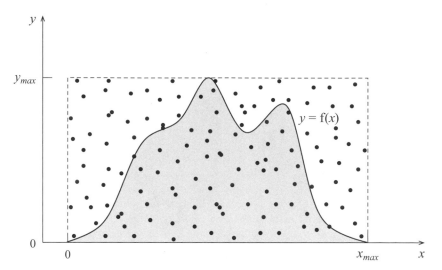

Figure 3.3: The acceptance-rejection method for a one-dimensional probability distribution. The dots are deviates (u_x, u_y) drawn from the two-dimensional probability distribution function g(x,y), where g(x,y) is constant inside the dashed rectangle and is zero outside. The one-dimensional probability distribution function f(x) is plotted as $y = f(x)$ and divides the rectangle into two areas. The deviates for which $u_y \leq f(u_x)$ lie within the shaded area. The u_x part of the two-dimensional deviates within the shaded area are one-dimensional deviates drawn from f(x).

In principle the acceptance-rejection algorithm works for any f(x), but it is inefficient at values of x where f(x) is much less than f_{max}. Thus, the algorithm is an efficient generator of Gaussian deviates near the center of the Gaussian, but for $x > 3\sigma$ it throws away 99 out of every 100 trial points. The algorithm is also a poor way to find deviates for probability distributions in which x can become infinite and the probability density decreases only slowly as x becomes large. The Lorentzian distribution is an outstanding example.

These problems can be ameliorated, at least in part, by choosing the first random number from a probability distribution h(x) that roughly approximates f(x). Specifically, choose a distribution g(x, y) such that

$$g(x, y) = \begin{cases} \text{constant}, & 0 \leq x \leq x_{max} \\ & 0 \leq y \leq ah(x) \\ 0, & \text{otherwise} \end{cases} , \qquad (3.20)$$

where $ah(x) \geq f(x)$ for all x. If h(x) is to be useful, $ah(x)$ should not be much larger than f(x), and it must be possible to generate deviates from h(x) efficiently. The modified acceptance-rejection algorithm is

1. Generate a uniform random deviate u_x between 0 and x_{max}.
2. Generate a second deviate u_y that is uniform between 0 and $ah(u_x)$.
3. If $u_y \leq f(u_x)$, keep u_x as the random deviate from f(x).
4. Otherwise throw away u_x and u_y and try again.

This modified version of the standard acceptance-rejection algorithm is less useful than one might hope, because it is often difficult to devise a good h(x).

3.2.5 Ratio of Uniforms Method

The ratio of uniforms method is a variant of the acceptance-rejection algorithm that avoids many of the difficulties encountered with the standard algorithm. The method is deceptively simple.[3] Suppose one wishes to generate random deviates from the probability distribution function $y = f(x)$. Transform coordinates from (x, y) to (s, t), where

$$y = s \tag{3.21}$$

$$x = t/s, \tag{3.22}$$

and define a curve in the (s, t) plane by the equation $s = h(s, t) = [f(x)]^{1/2} = [f(t/s)]^{1/2}$. Find a rectangle in the (s, t) plane that encloses the curve and has sides parallel to the s and t axes. Then:

1. Generate pairs of uniform random deviates (u_s, u_t) that populate the rectangle.
2. If the pair satisfies the requirement $u_s \leq [f(u_t/u_s)]^{1/2}$, accept the pair. The ratio $u_x = u_t/u_s$ is a random deviate drawn from $f(x)$.

The following example shows how the ratio of uniforms method works for the Lorentzian probability distribution.

Example: The Lorentzian probability distribution function is

$$y = f(x) = \frac{1}{\pi} \frac{b}{b^2 + x^2}.$$

The curve in the (s, t) plane is

$$s = \left[f(t/s) \right]^{1/2} = \left[\frac{1}{\pi} \frac{b}{b^2 + (t/s)^2} \right]^{1/2}.$$

Expanding and rearranging, we find

$$b^2 s^2 + t^2 = \frac{b}{\pi}.$$

This is the ellipse shown in Figure 3.4 Generate pairs of independent uniform deviates (u_s, u_t) covering the ranges $0 \leq u_s < (\pi b)^{-1/2}$ and $0 \leq u_t \leq (b/\pi)^{1/2}$. Each point in Figure 3.4 is one such pair. If the pair satisfies

$$u_s \leq \left[\frac{1}{\pi} \frac{b}{b^2 + (u_t/u_s)^2} \right]^{1/2},$$

accept the pair and set $u_x = u_t/u_s$. This ratio is a Lorentzian deviate. Note that the inequality is equivalent to

$$b^2 u_s^2 + u_t^2 \leq \frac{b}{\pi},$$

so the acceptance criterion accepts all points in the shaded region of the figure.

Continued on page 58

---

[3] A.J. Kinderman, and J. F. Monahan. 1977. "Computer Generation of Random Variables Using the Ratio of Uniform Deviates." *ACM Transactions on Mathematical Software* vol. 3, p. 257.

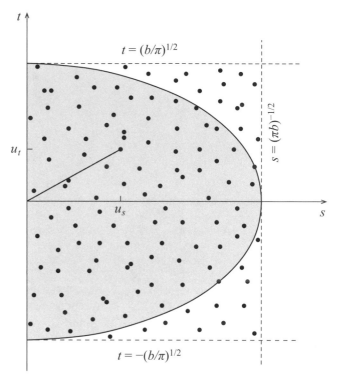

Figure 3.4: The ratio of uniforms method for generating random deviates for a Lorentzian probability distribution function. In the (x, y) plane, the distribution is $f(x) = (b/\pi)/(b^2 + x^2)$. The distribution transforms to the ellipse $b^2 s^2 + t^2 = b/\pi$ in the (s, t) plane. The dots are uniform random deviates in the rectangle that bounds the ellipse. The ratio of uniforms method retains the points in the shaded region bounded by the ellipse. If point has coordinates (u_s, u_t), the ratio $u_x = u_t/u_s$ is a random deviate for the Lorentzian distribution.

To see why the ratio of uniforms algorithm works, we begin with equation 3.19, which describes the acceptance-rejection algorithm for the probability distribution function $f(x)$ in the (x, y) coordinate system:

$$f(x)dx \propto \int_{y=0}^{f(x)} g(x, y)dy dx = \int_{y=0}^{f(x)} dy dx, \tag{3.23}$$

where $g(x, y)$ is a uniform probability distribution. When transformed to the (s, t) coordinate system, this becomes

$$f(x)dx \propto \int_{s=0}^{h(s,t)} g(s, t)ds dt. \tag{3.24}$$

Neither $h(s, t)$ nor $g(s, t)$ has been defined yet. Since we wish to use the acceptance-rejection algorithm in the (s, t) coordinate system, we force $g(s, t)$ to be a constant and then carry out the coordinate transformation explicitly to determine what $h(s, t)$ must be:

$$f(x)dx \propto \int_{s=0}^{h(s,t)} ds dt = \int_{y=0}^{h(s,t)} \left| \frac{\partial(s, t)}{\partial(x, y)} \right| dy dx. \tag{3.25}$$

After a small amount of work, we find the Jacobian to be

$$\left| \frac{\partial (s,t)}{\partial (x,y)} \right| = y, \tag{3.26}$$

so equation 3.25 becomes

$$f(x)dx \propto \int_{y=0}^{h(s,t)} y\,dy\,dx \propto h^2(s,t)dx. \tag{3.27}$$

Thus we find

$$h(s,t) \propto [f(x)]^{1/2} = [f(t/s)]^{1/2}, \tag{3.28}$$

and

$$f(x)dx \propto \int_{s=0}^{[f(t/s)]^{1/2}} ds\,dt. \tag{3.29}$$

Equation 3.29 says that one produces random deviates u_x with a distribution proportional to $f(x)$ by generating uniform deviates in the (s,t)-plane, accepting all those for which $0 \le s \le [f(t/s)]^{1/2}$, and then setting $u_x = u_t/u_s$.

Why does the ratio of uniforms method work so well? The standard acceptance-rejection method cannot be applied to probability distributions for which x can extend to infinity. The ratio of uniforms method implicitly replaces the x-coordinate by an angle coordinate θ, where $\tan\theta = t/s = x$. As x goes from 0 to infinity, θ sweeps from $0°$ to $90°$. Large values of x are transformed into a few degrees near $\theta = 90°$, making them tractable for sampling by the acceptance-rejection algorithm.

3.2.6 Generating Random Deviates from More Complicated Probability Distributions

One often needs to generate vectors of random deviates from multiparameter probability distributions. This requires little extra work if the parameters are uncorrelated, since the deviates can be generated independently for each parameter. Substantial extra work may be needed to generate the deviates if the parameters are correlated. An algorithm for generating vectors of correlated random deviates from multiparameter Gaussian distributions is described at the end of Section 5.4. The algorithm is reasonably efficient when the number of parameters in the distribution is not large. In general, however, simple methods for generating random deviates become inefficient as the number of parameters in the probability distribution increases and unworkable if the number of independent parameters is large. Random deviates from more difficult distributions can always be generated using the Markov chain Monte Carlo techniques that are discussed in Sections 3.4 and 3.5.

3.3 Monte Carlo Integration

Suppose I is the definite integral of the function $y(x)$ over the interval $a \le x \le b$,

$$I = \int_a^b y(x)dx, \tag{3.30}$$

and suppose we wish to find the value of I by numerical integration. One typically evaluates $y(x)$ at prespecified points x_i and approximates I by the weighted sum

$$I \approx \sum_{i=1}^{N} w_i y(x_i), \tag{3.31}$$

where the w_i are weights. The points are often evenly spaced but, as in Gaussian quadratures, need not be. In Monte Carlo integration $y(x)$ is evaluated at random values of x, not prespecified values. For example, if the u_i are random deviates uniformly distributed between a and b, intuition suggests that the integral can be approximated by

$$I \approx (b-a)\frac{1}{N}\sum_{i=1}^{N} y(u_i). \tag{3.32}$$

As N becomes large, this approaches the mean value of $y(x)$ over the interval times the width of the interval.

We need to sharpen this intuition. Let $p(x)$ be a probability distribution function. For the moment its functional form is arbitrary, except that it is defined over the interval $a \leq x \leq b$ and is greater than zero everywhere in the interval.[4] Let u_i be random deviates generated from $p(x)$. Form new random deviates $v_i = y(u_i)/p(u_i)$, and take the sum

$$\xi = \frac{1}{N}\sum_{i=1}^{N} v_i. \tag{3.33}$$

We first see that

$$\langle \xi \rangle = \frac{1}{N}\sum_{i=1}^{N}\langle v_i \rangle = \langle v \rangle. \tag{3.34}$$

Since v is a function of u and since the u_i are drawn from $p(x)$, the mean value of v is

$$\langle v \rangle = \lim_{N\to\infty}\frac{1}{N}\sum_{i=1}^{N} v(u_i)p(u_i) = \lim_{N\to\infty}\frac{1}{N}\sum_{i=1}^{N}\frac{y(u_i)}{p(u_i)}p(u_i) = \lim_{N\to\infty}\frac{1}{N}\sum_{i=1}^{N} y(u_i) = \langle y \rangle. \tag{3.35}$$

We can, therefore, approximate the integral of $y(x)$ by

$$I \approx (b-a)\xi = \frac{(b-a)}{N}\sum_{i=1}^{N} v_i = \frac{(b-a)}{N}\sum_{i=1}^{N}\frac{y(u_i)}{p(u_i)}, \tag{3.36}$$

since this approaches $I = (b-a)\langle y \rangle$ as N becomes large, which is the correct answer. Furthermore, from the central limit theorem (Section 2.4), the probability distribution for ξ approaches the Gaussian distribution

$$f(\xi) \propto \exp\left[-\frac{(\xi - I)^2}{2\sigma_\xi^2}\right], \tag{3.37}$$

---

[4] Much of this section follows the discussion in I. M. Sobol. 1975. *Monte Carlo Method. Popular Lectures in Mathematics.* Chicago: University of Chicago Press.

where

$$\sigma_\xi^2 = \frac{\sigma_v^2}{N}, \tag{3.38}$$

and σ_v^2 is the variance of the individual deviates v. Since σ_ξ^2 is a measure of the expected difference between ξ and I, it measures the accuracy of the Monte Carlo integration. Thus the accuracy of equation 3.36 improves as $N^{-1/2}$.

Perhaps surprisingly, the functional form of $p(x)$ is not specified in equation 3.36—any probability distribution will do. This traces back to the insensitivity of the central limit theorem to the original distribution of the sample points. The simplest choice for $p(x)$ is the flat distribution

$$p(x) = \frac{1}{b-a}, \quad a \le x \le b, \tag{3.39}$$

so that the u_i are uniform deviates between a and b. In this case equation 3.36 reduces to

$$I \approx \frac{b-a}{N} \sum_{i=1}^N y(u_i), \tag{3.40}$$

and we have retrieved equation 3.32.

The distribution need not be flat, though; indeed, a flat distribution is not usually the best choice. We now show that the best choice is $p(x) \propto |y(x)|$. For convenience, define a new function $h(x)$ by

$$h(x) = \frac{y(x)}{p(x)}. \tag{3.41}$$

The mean value of h is

$$\langle h \rangle = \int_a^b h(x)p(x)dx = \int_a^b \frac{y(x)}{p(x)}p(x)dx = \int_a^b y(x)dx = I. \tag{3.42}$$

Define the variance of h by

$$\sigma_h^2 = \langle (h - I)^2 \rangle = \langle h^2 \rangle - I^2. \tag{3.43}$$

Since $v_i = h_i = y(u_i)/p(u_i)$, we have $\sigma_v^2 = \sigma_h^2$. We now search for the $p(x)$ that minimizes σ_h^2 and, therefore (through equation 3.38), minimizes σ_ξ^2. Since

$$\langle h^2 \rangle = \int_a^b h^2(x)p(x)dx = \int_a^b \left[\frac{y(x)}{p(x)}\right]^2 p(x)dx = \int_a^b \frac{y^2(x)}{p(x)}dx, \tag{3.44}$$

we have

$$\sigma_h^2 = \int_a^b \frac{y^2(x)}{p(x)}dx - I^2. \tag{3.45}$$

One form of the Cauchy-Schwarz inequality states that if $f(x)$ and $g(x)$ are two functions, then

$$\left[\int_a^b |f(x)g(x)|\, dx\right]^2 \le \left[\int_a^b f^2(x)dx\right]\left[\int_a^b g^2(x)dx\right]. \tag{3.46}$$

Set $f^2(x) = y^2(x)/p(x)$ and $g^2(x) = p(x)$. Then

$$\left[\int_a^b |y(x)|\, dx\right]^2 \le \left[\int_a^b \frac{y^2(x)}{p(x)}dx\right]\left[\int_a^b p(x)dx\right] = \int_a^b \frac{y^2(x)}{p(x)}dx, \tag{3.47}$$

where the last equality holds because p is normalized. Using equation 3.47 in equation 3.45, we find

$$\sigma_h^2 \geq \left[\int_a^b |y(x)| \, dx \right]^2 - I^2. \tag{3.48}$$

The minimum value of σ_h^2 occurs at the equality. Let us guess that we achieve the equality by choosing

$$p(x) = \frac{|y(x)|}{\int_a^b |y(x)| \, dx}. \tag{3.49}$$

For this choice, we have

$$\int_a^b \frac{y^2(x)}{p(x)} \, dx = \left[\int_a^b |y(x)| \, dx \right]^2, \tag{3.50}$$

and substituting this result into equation 3.43, we do retrieve the equality

$$\sigma_h^2 = \left[\int_a^b |y(x)| \, dx \right]^2 - I^2. \tag{3.51}$$

Therefore, the error in the estimated value of I is minimized by choosing $p(x) \propto |y(x)|$, and the most efficient way to distribute the sample points for a Monte-Carlo integration of $y(x)$ is to draw the sample points from a probability distribution that is proportional to $|y(x)|$.

It makes sense that the accuracy of a Monte Carlo integration would be improved by placing more sample points in regions where $y(x)$ is large and contributes more to the integral, and fewer points where it is smaller and contributes less. It is not so obvious that the distribution should be exactly proportional to $|y(x)|$. However, it is generally not feasible to set $p(x) \propto |y(x)|$, because normalizing $p(x)$ would require integrating $|y(x)|$, which is almost the same integral as we were attempting to integrate in the first place. One may, however, be able to devise a $p(x)$ that has roughly the same shape as $|y(x)|$ but is easier to normalize. While the simpler distribution will not be optimal, it will still give better results than a uniform distribution.

Likewise, equations 3.45 and 3.51 are not useful for calculating σ_ξ^2. Instead, return to equations 3.38 and 3.43:

$$\sigma_\xi^2 = \frac{1}{N}\sigma_h^2 = \frac{1}{N}\left[\langle h^2 \rangle - I^2 \right] = \frac{1}{N}\left[\langle h^2 \rangle - \langle h \rangle^2 \right]. \tag{3.52}$$

Set $h_i = y(u_i)/p(u_i)$, and use the approximations

$$\langle h^2 \rangle \approx \frac{1}{N}\sum_{i=1}^N h_i^2, \qquad \langle h \rangle \approx \frac{1}{N}\sum_{i=1}^N h_i \tag{3.53}$$

to get

$$\sigma_\xi^2 \approx \frac{1}{N}\left[\frac{1}{N}\sum_{i=1}^N h_i^2 - \left(\frac{1}{N}\sum_{i=1}^N h_i \right)^2 \right]. \tag{3.54}$$

If $p(x)$ is a uniform distribution, then equation 3.41 reduces to $h(x) = (b-a)y(x)$ or $h_i = (b-a)y(u_i)$; and equation 3.54 reduces to the more familiar

$$\sigma_\xi^2 \approx \frac{1}{N} \left[\frac{(b-a)^2}{N} \sum_{i=1}^N y(u_i)^2 - \left(\frac{(b-a)^2}{N} \sum_{i=1}^N y(u_i) \right)^2 \right]. \tag{3.55}$$

To summarize the results of this section: An approximate value for the definite integral of $y(x)$ over the interval $a \le x \le b$ can be calculated from

$$I \approx \frac{(b-a)}{N} \sum_{i=1}^N \frac{y(u_i)}{p(u_i)}, \tag{3.56}$$

where $p(x)$ is an arbitrary probability distribution function over the same range, and u_i is a random deviate generated from $p(x)$. The standard deviation of the approximation is $\sigma_I = (b-a)\sigma_\xi$, where σ_ξ is given by equation 3.54. The standard deviation decreases as $N^{-1/2}$. The best choice for the probability distribution function is $p(x) \propto |y(x)|$, but this choice is generally not feasible, because it is difficult to normalize $p(x)$. One may instead use a probability distribution that has roughly the same shape as $|y(x)|$ but is easier to normalize. Monte Carlo integration is rarely useful for such simple one-dimensional integrals. As the dimensionality increases, however, the space over which the integral must be evaluated can become too large to be easily integrated by standard techniques, and Monte Carlo integration becomes more useful. The choice of $p(x)$ also becomes more important as the dimensionality increases.

3.4 Markov Chains

Markov chain Monte Carlo (MCMC) techniques provide a way to improve the efficiency of some kinds of Monte Carlo calculations. This section describes Markov chains, and Section 3.5 shows how to use them in some typical applications.

Let u be a random deviate generated from a probability distribution, and let

$$u^{(1)}, u^{(2)}, u^{(3)}, \ldots, u^{(n-1)}$$

be a sequence of $n-1$ values of u. The superscript in parentheses denotes the position in the sequence. The sequence is a Markov chain if the probability distribution for the next member of the sequence, $u^{(n)}$, depends on $u^{(n-1)}$ but not on values of u before $u^{(n-1)}$. Unlike, say, successive flips of a coin, which are independent of each other, a Markov chain has a history, and the probability of generating a particular value of u depends on that history. The dependence is limited, though, extending only to the immediately previous value of u. Random walks are an example of Markov chains. Suppose a pointer picks out an integer. Create a series of integers by moving the pointer randomly to the next higher or next lower integer. Since the position of the pointer after any step depends on its previous position but not on any earlier position, the sequence of integers is a Markov chain.

3.4.1 Stationary, Finite Markov Chains

The dependence of a Markov chain on its history can be represented by a conditional probability distribution,

$$T^{(n-1)}(x^{(n)}|x^{(n-1)})$$

that gives the probability of selecting a particular value of x at step n given a value of x at the previous step. There is a superscript in parentheses on T because the conditional probabilities might change from step to step. Thus $T^{(n-1)}(x^{(n)}|x^{(n-1)})$ is the conditional probability distribution at step $n-1$. For a symmetrical random walk among the integers, the conditional probability has the simple form

$$T(x^{(n)}|x^{(n-1)}) = \begin{cases} 1/2, & x^{(n)} - x^{(n-1)} = 1 \\ 1/2, & x^{(n)} - x^{(n-1)} = -1 \\ 0, & \text{otherwise} \end{cases}. \tag{3.57}$$

For our purposes we need to consider only stationary Markov chains, also called *homogeneous Markov chains*. These are Markov chains whose conditional probabilities do not change from step to step. Since the conditional probabilities do not change, the superscript on T can be dropped, and the conditional probabilities can be written more simply as $T(x^{(n)}|x^{(n-1)})$.

Suppose that x can have only a finite number of discrete values x_k, $k = 1, \ldots, m$. Markov chains for which this is true are called *finite Markov chains*. With the understanding that successive values of x form a sequence, the conditional probabilities can now be written even more simply as $T(x_k|x_j)$. Since $T(x_k|x_j)$ is a conditional probability, it must be normalized, so

$$\sum_{k=1}^{m} T(x_k|x_j) = 1. \tag{3.58}$$

The conditional probabilities can be thought of as transition probabilities. That is, $T(x_k|x_j)$ is the probability of a transition to x_j from x_k.

Let the probability that value x_k will occur at step n be $P^{(n)}(x_k)$. One must carefully distinguish $P^{(n)}(x_k)$ from $T(x_k|x_j)$. The value of $u^{(n)}$ is a random number generated from the conditional distribution $T(x_k|x_j)$, *not* from the probability distribution $P^{(n)}(x_k)$. The random numbers generated at step n from $T(x_k|x_j)$ do, nevertheless, have a distribution, and that distribution is $P^{(n)}(x_k)$. The probability distribution generally changes from step to step. The relation between the probability distribution at step n and the distribution at step $n-1$ is

$$P^{(n)}(x_k) = \sum_{j=1}^{m} P^{(n-1)}(x_j) T(x_k|x_j). \tag{3.59}$$

Note that although $P^{(n)}(x_k)$ changes, the transition probabilities do not, so equation 3.59 does describe a stationary Markov chain.

Stationary, finite Markov chains can be described with a matrix formalism. Equation 3.59 is equivalent to the matrix equation

$$\boldsymbol{\pi}^{(n)} = \boldsymbol{\pi}^{(n-1)}\mathbf{T}, \tag{3.60}$$

where $\boldsymbol{\pi}^{(n)}$ is the row vector

$$\boldsymbol{\pi}^{(n)} = \big(P^{(n)}(x_1), P^{(n)}(x_2), \ldots, P^{(n)}(x_m)\big), \tag{3.61}$$

and **T** is the transition matrix

$$\mathbf{T} = \begin{pmatrix} T(x_1|x_1) & T(x_2|x_1) & \cdots & T(x_m|x_1) \\ T(x_1|x_2) & T(x_2|x_2) & \cdots & T(x_m|x_2) \\ \vdots & \vdots & & \vdots \\ T(x_1|x_m) & T(x_2|x_m) & \cdots & T(x_m|x_m) \end{pmatrix}. \tag{3.62}$$

Note that **T** multiplies $\boldsymbol{\pi}^{(n-1)}$ from the right. It is convenient to introduce a more compact notation for the components of $\boldsymbol{\pi}^{(n)}$ and **T**. Set

$$\pi_j^{(n)} = P^{(n)}(x_j) \tag{3.63}$$

$$T_{jk} = T(x_k|x_j). \tag{3.64}$$

(Note the order of the indices on T_{jk}.) In component form equation 3.60 is

$$\pi_k^{(n)} = \sum_j \pi_j^{(n-1)} T_{jk}, \tag{3.65}$$

and equation 3.58 is

$$\sum_k T_{jk} = 1. \tag{3.66}$$

The evolution of the probability distribution function is particularly simple for finite, stationary Markov chains. From equation 3.60 we have

$$\boldsymbol{\pi}^{(n)} = \boldsymbol{\pi}^{(n-1)}\mathbf{T} = \left[\boldsymbol{\pi}^{(n-2)}\mathbf{T}\right]\mathbf{T} = \boldsymbol{\pi}^{(n-2)}\mathbf{T}^2 = \cdots = \boldsymbol{\pi}^{(n-k)}\mathbf{T}^k, \tag{3.67}$$

where the notation $\mathbf{T}^2$ means that the matrix **T** has been multiplied by itself. Suppose the chain begins with an initial probability distribution $\boldsymbol{\pi}^{(0)}$. Then

$$\boldsymbol{\pi}^{(n)} = \boldsymbol{\pi}^{(0)}\mathbf{T}^n. \tag{3.68}$$

Equations 3.60 and 3.68 show that the probability distribution of a stationary Markov chain evolves deterministically. Again, this deterministic behavior must be carefully distinguished from the properties of specific realizations $u^{(1)}, u^{(2)}, u^{(3)}, \ldots, u^{(n)}$, which is a sequence of random deviates generated from $T(x_k|x_j)$.

3.4.2 Invariant Probability Distributions

An *invariant probability distribution*, also called an *equilibrium distribution*, remains the same at each step in the chain:

$$\boldsymbol{\pi}^{(n)} = \boldsymbol{\pi}^{(n-1)}\mathbf{T} = \boldsymbol{\pi}^{(n-1)}. \tag{3.69}$$

In the language of linear algebra, a distribution is invariant if it is a left eigenvector of **T** with an eigenvalue equal to 1. Do not confuse invariance with stationarity: If **T** does not change, the Markov chain is stationary; if $\boldsymbol{\pi}$ does not change, $\boldsymbol{\pi}$ is invariant.

Later we will be given a distribution $P(x_k)$ and will need to generate a transition matrix for which $P(x_k)$ is the invariant distribution. One way to do this is to require detailed balance. In detailed balance the rate of transitions from x_j to x_k is the same as the reverse rate of transitions:

$$P^{(n-1)}(x_j)T(x_k|x_j) = P^{(n-1)}(x_k)T(x_j|x_k), \tag{3.70}$$

or in component form

$$\pi_j^{(n-1)} T_{jk} = \pi_k^{(n-1)} T_{kj}. \tag{3.71}$$

To see that this guarantees invariance, sum both sides of this equation over j:

$$\sum_j \pi_j^{(n-1)} T_{jk} = \sum_j \pi_k^{(n-1)} T_{kj}. \tag{3.72}$$

From equation 3.65, the left-hand side of the equation is $\pi_k^{(n)}$. From equation 3.66 the right-hand side of the equation is

$$\sum_j \pi_k^{(n-1)} T_{kj} = \pi_k^{(n-1)} \sum_j T_{kj} = \pi_k^{(n-1)}. \tag{3.73}$$

Thus,

$$\pi_k^{(n)} = \pi_k^{(n-1)}, \tag{3.74}$$

so the probability distribution is invariant.

Detailed balance (equation 3.70 or 3.71) does not fully specify the components of the transition matrix. One way to completely specify them is to set

$$T(x_k|x_j) = P(x_k) \tag{3.75}$$

or

$$T_{jk} = \pi_k, \tag{3.76}$$

so that the conditional probabilities for the x_k become independent of the x_j. Substitution of equation 3.75 into equation 3.70 shows that this choice does satisfy detailed balance. Equation 3.75 seems strange, because the Markov chain to which it corresponds has no history. It is a Markov chain nevertheless, and a generalization of it is at the heart of the Gibbs sampler described in Section 3.5.

A stationary Markov chain is *ergodic* if $\pi^{(n)}$ converges to the invariant distribution for **T** as n increases. Consider the (left) eigenvalue equation

$$\pi \mathbf{T} = \lambda \pi. \tag{3.77}$$

Let the solutions be the eigenvectors π_i with eigenvalues λ_i. If any $\lambda_i = 1$, the corresponding eigenvector is an invariant distribution, because

$$\pi_i \mathbf{T} = \lambda_i \pi_i = \pi_i. \tag{3.78}$$

We can distinguish three cases. First, if more than one eigenvalue is equal to 1, there is more than one invariant distribution. Once a chain converges to any of these invariant distributions, it is trapped there and cannot get to the other invariant distributions. These Markov chains are not invariant.

Second, if any of the eigenvalues is equal to -1, the chain can become trapped in a repeating cycle, never reaching the invariant distribution. These Markov chains are also not invariant. For example, consider the Markov chain for which the transition matrix is

$$\mathbf{T} = \begin{pmatrix} 0 & 1 \\ 1 & 0 \end{pmatrix}. \tag{3.79}$$

The reader can easily verify that this matrix has two eigenvalues, 1 and -1, and that the normalized invariant distribution is $(1/\sqrt{2}, 1/\sqrt{2})$. The chain can become trapped in an endless cycle between $(1, 0)$ and $(0, 1)$, never reaching the invariant distribution.

The third case is those chains for which only one eigenvalue is equal to 1, so there is just one invariant distribution; and $|\lambda_i| < 1$ for all other eigenvalues, so the chain neither cycles endlessly nor produces distributions with probabilities greater than 1. Order the eigenvectors so that $\lambda_0 = 1$ and $\boldsymbol{\pi}_0 = \boldsymbol{\pi}_s$ is the corresponding invariant distribution. Since any arbitrary vector can be decomposed into a sum of the eigenvectors, the probability distribution $\boldsymbol{\pi}^{(0)}$ at the beginning of a Markov chain can always be written

$$\boldsymbol{\pi}^{(0)} = \alpha_0 \boldsymbol{\pi}_s + \alpha_1 \boldsymbol{\pi}_1 + \cdots + \alpha_m \boldsymbol{\pi}_m. \tag{3.80}$$

Now apply $\mathbf{T}^n$ to $\boldsymbol{\pi}^{(0)}$ to get $\boldsymbol{\pi}^{(n)}$:

$$\boldsymbol{\pi}^{(n)} = \boldsymbol{\pi}^{(0)} \mathbf{T}^n = \alpha_0 \boldsymbol{\pi}_s + \alpha_1 \lambda_1^n \boldsymbol{\pi}_1 + \cdots + \alpha_m \lambda_m^n \boldsymbol{\pi}_m. \tag{3.81}$$

Since $|\lambda_i|^n \to 0$ as n becomes large for all $i \geq 1$,

$$\boldsymbol{\pi}^n \to \alpha_0 \boldsymbol{\pi}_s. \tag{3.82}$$

These Markov chains are ergodic.

The following example shows how to construct a Markov chain that converges to a simple two-value invariant probability distribution.

Example: Consider a probability distribution function $P(x_j)$ for which x_j has just two values, 1 or 2, that occur with probabilities

$$P(1) = 1/4 \qquad \text{and} \qquad P(2) = 3/4.$$

An example might be a coin that lands heads 3/4 of the time and tails 1/4 of the time.

We now construct the transition matrix for which $P(x_j)$ is an invariant distribution. The components of the invariant distribution $\boldsymbol{\pi}$ are

$$\pi_1 = 1/4 \qquad \text{and} \qquad \pi_2 = 3/4.$$

Since $\boldsymbol{\pi}$ has two states, the transition matrix $\mathbf{T}$ is a 2×2 matrix. From equation 3.66, the components of $\mathbf{T}$ satisfy

$$T_{11} + T_{12} = 1$$
$$T_{21} + T_{22} = 1,$$

so the matrix has the form

$$\mathbf{T} = \begin{pmatrix} a & 1-a \\ 1-b & b \end{pmatrix},$$

Continued on page 68

---

where a and b are constants. It is easily verified that at least one eigenvalue of this matrix is equal to 1. Now apply detailed balance:

$$\pi_1 T_{12} = \pi_2 T_{21}$$

$$\frac{1}{4}(1-a) = \frac{3}{4}(1-b),$$

which yields $(1-b) = (1-a)/3$. The transition matrix is, therefore,

$$\mathbf{T} = \begin{pmatrix} a & 1-a \\ (1-a)/3 & (a+2)/3 \end{pmatrix}.$$

This transition matrix has π for an invariant distribution for all values of a, so we are free to choose a at our convenience. One choice could be the value of a that makes the second eigenvalue λ_2 equal to 0, since the resulting Markov chain would converge to π immediately. For $\lambda_2 = 0$, the determinant equation for the eigenvalues becomes

$$\begin{vmatrix} a - \lambda_2 & 1-a \\ (1-a)/3 & (a+2)/3 - \lambda_2 \end{vmatrix} = \begin{vmatrix} a & 1-a \\ (1-a)/3 & (a+2)/3 \end{vmatrix} = 0.$$

After a bit of algebra, this yields $a = 1/4$, and the transition matrix becomes

$$\mathbf{T} = \begin{pmatrix} 1/4 & 3/4 \\ 1/4 & 3/4 \end{pmatrix}.$$

This transition matrix is the same as that produced by imposing equation 3.75 and gives some insight into the meaning of that equation: Equation 3.75 yields a transition matrix for which only one eigenvalue is nonzero. The nonzero eigenvalue is equal to 1, and the corresponding eigenvector is proportional to the desired probability distribution. Then, from equation 3.81, the resulting Markov chain converges to the desired distribution immediately.

Suppose that we had set $\lambda_2 = 2/3$. After some more algebra, the transition matrix becomes

$$\mathbf{T} = \begin{pmatrix} 3/4 & 1/4 \\ 1/12 & 11/12 \end{pmatrix}.$$

This also is a perfectly acceptable transition matrix. Because the second eigenvalue is not much smaller than 1, the second eigenvector dies out slowly, and the Markov chain will approach its invariant distribution only slowly.

3.4.3 Continuous Parameter and Multiparameter Markov Chains

Most of the results in the previous sections carry over to Markov chains with continuous probability distribution functions. The chain is still defined to be a sequence of random numbers

$$u^{(1)}, u^{(2)}, u^{(3)}, \ldots, u^{(n-1)}$$

in which the value of $u^{(n)}$ depends on $u^{(n-1)}$ but not values of u before $u^{(n-1)}$; but now the u are continuous, not discrete. The $u^{(n)}$ are random deviates generated from continuous conditional probability density distribution functions $t^{(n-1)}(x^{(n)}|x^{(n-1)})$. For stationary Markov chains, the conditional probability density distribution function is independent of n and can be written $t(x^{(n)}|x^{(n-1)})$, or even more simply as $t(x|x')$, where x' is the previous value of x. By analogy to the discrete case, $t(x|x')$ can be thought of as a transition probability, or more accurately as a transition probability density. Because $t(x|x')$ is a probability distribution, it must satisfy the normalization constraint

$$\int t(x|x')dx = 1. \tag{3.83}$$

The conditional distribution

$$t(x|x') = \frac{1}{\sqrt{2\pi\sigma^2}}\exp\left[-\frac{1}{2}\frac{(x-x')^2}{\sigma^2}\right] \tag{3.84}$$

would, for example, produce a Markov chain with random walk behavior but with steps whose lengths have a Gaussian distribution—a Gaussian random walk.

The probability density distribution function for x at step n is $f^{(n)}(x)$. Because $t(x|x')$ is a continuous function, the transition probabilities can no longer be written as a matrix, and $f^{(n)}(x)$ is calculated by an integral instead of a matrix multiplication:

$$f^{(n)}(x) = \int f^{(n-1)}(x')t(x|x')dx'. \tag{3.85}$$

A probability distribution $\pi(x)$ is invariant if

$$\pi^{(n)}(x) = \int \pi^{(n-1)}(x')t(x|x')dx' = \pi^{(n-1)}(x). \tag{3.86}$$

If $\pi(x)$ is the invariant probability distribution for $t(x|x')$ and if $f^{(n)}(x)$ approaches $\pi(x)$ as n increases, that is, if

$$\lim_{n\to\infty} f^{(n)}(x) = \pi(x), \tag{3.87}$$

then the Markov chain is ergodic. Determining ergodicity is more complicated for continuous distributions than for discrete distributions. We state without proof that a chain is ergodic if it can reach any value of x from any value of x', which is true if $t(x|x')$ is nowhere equal to 0.

Finally, one can find conditional probability distributions for which $\pi(x)$ is the invariant distribution by requiring detailed balance:

$$\pi^{(n-1)}(x)t(x'|x) = \pi^{(n-1)}(x')t(x|x'). \tag{3.88}$$

To see that this is true, integrate equation 3.88 over x':

$$\int \pi^{(n-1)}(x)t(x'|x)dx' = \int \pi^{(n-1)}(x')t(x|x')dx' \tag{3.89}$$

From equation 3.83, the left-hand side of this equation is just $\pi^{(n-1)}(x)$, and from equation 3.85 the right-hand side is $\pi^{(n)}(x)$. We find

$$\pi^{(n-1)}(x) = \pi^{(n)}(x), \tag{3.90}$$

so $\pi(x)$ is invariant. As in the discrete case, detailed balance does not fully specify $t(x|x')$. The equivalent of equation 3.75 is

$$t(x|x') = \pi(x), \tag{3.91}$$

which does fully specify $t(x|x')$ and satisfies equation 3.88, although with the penalty that the resulting Markov chain has no memory.

Up to now, we have assumed that a Markov chain is a sequence of numbers. In fact, the concept is much broader: A Markov chain can be a series of states

$$s^{(1)}, s^{(2)}, s^{(3)}, \ldots, s^{(n-1)}$$

in which the probability for the occurrence of state $s^{(n)}$ depends on state $s^{(n-1)}$ but not states before $s^{(n-1)}$. The conditional probabilities $t(s|s')$ are the probabilities for transitions from state s' to state s, and the probability $f^{(n)}(s)$ is the probability that the chain will be in state s after n steps. The distinction between a state and a number is that it may require several numbers to specify a state, not just one. It may be useful to think of an electron in a hydrogen atom: Several quantum numbers are needed to fully specify the possible states of the electron. The conditional probabilities are the probabilities for the transitions of the electron from one state to another.

Suppose that a state is specified by k continuous variables $(x_1, x_2, \ldots, x_k)$, and that state $s^{(j)}$ in a Markov chain is specified by

$$s^{(j)} = (u_1^{(j)}, u_2^{(j)}, \ldots, u_k^{(j)}), \tag{3.92}$$

where the $u_i^{(j)}$ are random deviates. The transition probabilities for a stationary Markov chain are the conditional probabilities $t(x_1, \ldots, x_k | x_1', \ldots, x_k')$. The probability distribution for the x_i evolves from step to step according to

$$f^{(n)}(x_1, \ldots, x_k) = \int_{x_1'} \cdots \int_{x_k'} f^{(n-1)}(x_1', \ldots, x_k') t(x_1, \ldots, x_k | x_1', \ldots, x_k') dx_1' \cdots dx_k', \tag{3.93}$$

and a distribution $\pi(x_1, \ldots, x_k)$ is invariant if

$$\pi^{(n)}(x_1, \ldots, x_k) = \int_{x_1'} \cdots \int_{x_k'} \pi^{(n-1)}(x_1', \ldots, x_k') t(x_1, \ldots, x_k | x_1', \ldots, x_k') dx_1' \cdots dx_k'$$
$$= \pi^{(n-1)}(x_1, \ldots, x_k). \tag{3.94}$$

If $f^{(n)}(x_1, \ldots, x_k)$ approaches $\pi(x_1, \ldots, x_k)$ as n increases, the Markov chain is ergodic. As before, a chain is ergodic if any state can be reached from any other state, which is true if the transition probability is nowhere equal to 0.

Transition probabilities for which $\pi(x_1, \ldots, x_k)$ is the invariant distribution can be found by imposing detailed balance:

$$\pi(x_1, \ldots, x_k) t(x_1', \ldots, x_k' | x_1, \ldots, x_k) = \pi(x_1', \ldots, x_k') t(x_1, \ldots, x_k | x_1', \ldots, x_k'). \tag{3.95}$$

As before, detailed balance does not fully specify the transition probabilities. The generalization of equation 3.91 to multiparameter distributions is

$$t(x_1, \ldots, x_k | x_1', \ldots, x_k') = \pi(x_1, \ldots, x_k). \tag{3.96}$$

In practice it is not feasible to apply equation 3.95 as it stands, even when simplified by adopting equation 3.96. Instead one breaks the single equation for detailed balance into a set of k equations for detailed balance, a separate equation for each parameter. This will be covered more fully in the discussion of the Metropolis-Hastings algorithm and the Gibbs sampler in Section 3.5.

3.5 Markov Chain Monte Carlo Sampling

The techniques for generating random deviates discussed in Section 3.2 are ineffective for complicated, multiparameter probability distributions. Coming to the rescue, Markov chains can generate random deviates efficiently from complicated distributions and can generate them even if the distributions are not normalized. This section begins with a pair of simple examples—calculation of a mean value and calculation of a marginal distribution—to illustrate the main ideas Markov chain Monte Carlo (MCMC) sampling. The section then outlines the two most important techniques for generating Markov chains: the Metropolis-Hastings algorithm and the Gibbs sampler.

3.5.1 Examples of Markov Chain Monte Carlo Calculations

Calculation of a Mean Value: Suppose that x can have m discrete values x_i and that the probability distribution function for the x_i is $P(x_i)$. The mean value of a quantity $g(x_i)$ is

$$\langle g \rangle = \sum_{i=1}^{m} g(x_i) P(x_i) \tag{3.97}$$

(see equation 1.47). Suppose that n samples of x have been generated from its probability distribution and that the number of times value x_i has been generated is $k(x_i)$. From equation 1.2

$$P(x_i) = \lim_{n \to \infty} \frac{k(x_i)}{n}, \tag{3.98}$$

so the mean value of g is

$$\langle g \rangle = \lim_{n \to \infty} \frac{1}{n} \sum_{i=1}^{m} g(x_i) k(x_i). \tag{3.99}$$

The $g(x_i)$ can be listed individually, the resulting list having n values of g, with each value $g(x_i)$ appearing $k(x_i)$ times. If each individual value is denoted by g_ℓ, the sum in equation 3.99 can be replaced by the sum over all n individual values of g_ℓ, so that the mean value becomes

$$\langle g \rangle = \lim_{n \to \infty} \frac{1}{n} \sum_{\ell=1}^{n} g_\ell. \tag{3.100}$$

This is the basis for an MCMC evaluation of equation 3.97:

1. Devise a Markov chain whose invariant distribution is $P(x_i)$.
2. Generate numbers $u^{(j)}$ in the Markov chain. If the chain is ergodic, the distribution of the $u^{(j)}$ will converge to $P(x_i)$. After the distribution is sufficiently close to $P(x_i)$, take n values of $u^{(j)}$ from the sequence, and use them as random deviates u_ℓ generated from $P(x_i)$.
3. Calculate a mean value for $g(x_i)$ from $\langle g \rangle \approx (1/n) \sum_{\ell=1}^{n} g(u_\ell)$.

There is no standard way to decide how long a Markov chain should be before beginning to accumulate the u_ℓ. A common way is to monitor the properties of the $u^{(j)}$ as they are being generated, perhaps by calculating a running mean and standard deviation, and assume the chain has converged to the invariant distribution when the properties are no longer changing rapidly. The chain will generally converge rapidly if $u^{(0)}$ is chosen from a region where $P(x_i)$ is large.

Calculation of a Marginal Distribution: Let f(x, y) be a two-parameter probability distribution function. One of the marginal distributions of f(x, y) is

$$g(x) = \int_y f(x, y) dy. \tag{3.101}$$

To calculate g(x) using MCMC sampling,

1. Devise a two-parameter Markov chain whose invariant distribution is f(x, y).
2. Generate a sequence of states $(u^{(j)}, v^{(j)})$ in the Markov chain. If the chain is ergodic, the distribution of the $(u^{(j)}, v^{(j)})$ will converge to f(x, y). After the distribution has converged, take n states from the sequence, and use them as random deviates (u_ℓ, v_ℓ) generated from f(x, y).
3. Discard the v_ℓ and tabulate the u_ℓ. The tabulated u_ℓ have a distribution that is proportional to g(x).

The result is a set of random deviates drawn from the marginal distribution, not the functional form of the distribution. If desired, there are a variety of ways to convert the deviates into something akin to a function. The simplest is to divide x into bins and then count the number of u_ℓ in each bin. If the middle value of x in bin k is x_k and the number of u_ℓ in bin k is n_k, then g$(x_k) \approx n_k/n$. Note that g(x_k) determined this way is normalized even if f(x, y) was not. There are many minor variants of MCMC sampling. One might, for example, choose the (u_ℓ, v_ℓ) from several different Markov chains, each chain beginning from a different first state $(u^{(0)}, v^{(0)})$.

3.5.2 Metropolis-Hastings Algorithm

The Metropolis-Hastings algorithm is a way to produce a Markov chain for an arbitrary probability distribution function. To show how and why the algorithm works, we will first describe the algorithm for one-parameter discrete probability distributions, then generalize to one-parameter continuous distributions, and finally generalize to multiparameter continuous distributions.

One-Parameter Discrete Probability Distributions: Suppose that x can have discrete values x_i, and let $\pi_i = P(x_i)$ be the probability that value x_i will occur. To generate a Markov chain $u^{(0)}, u^{(1)}, u^{(2)}, \ldots, u^{(n)}$ whose invariant distribution is equal to $P(x_i)$, we must devise transition probabilities $T_{ij} = T(x_j|x_i)$ that satisfy detailed balance (see equations 3.60–3.66 and 3.71),

$$\pi_i T_{ij} = \pi_j T_{ji}. \tag{3.102}$$

The Metropolis-Hastings algorithm does this by way of a conditional probability distribution $q_{ij} = Q(x_j|x_i)$ called the *candidate distribution*. Candidate values $u_c^{(k+1)}$ for the next member of the Markov chain are generated from $Q(u_c^{(k+1)}|u^{(k)})$. The candidate distribution does not generally satisfy detailed balance, and because it does not, the algorithm uses a second conditional distribution $\alpha_{ij} = A(x_j|x_i)$ to restore detailed balance. A candidate drawn from $Q(u_c^{(k+1)}|u^{(k)})$ is accepted or rejected with probability $A(u_c^{(k+1)}, u^{(k)})$. The Metropolis-Hastings algorithm is, therefore, a version of the acceptance-rejection algorithm (Section 3.2).

There are relatively few constraints on $Q(x_j|x_i)$. The most important is that it must allow the Markov chain to reach all x_j for which $P(x_j) > 0$ and to do so from any x_i. One way to guarantee this is to make $Q(x_j|x_i) > 0$ for all i and j. Ideally, it should be possible to generate deviates from $Q(x_j|x_i)$ rapidly and efficiently, and the distribution of the deviates should not differ greatly from $P(x_i)$. Once $Q(x_j|x_i)$ has been chosen, $A(x_j|x_i)$ is constrained by setting $T_{ij} = q_{ij}\alpha_{ij}$ and then substituting $q_{ij}\alpha_{ij}$ for T_{ij} in equation 3.102 for detailed balance:

$$\pi_i q_{ij} \alpha_{ij} = \pi_j q_{ji} \alpha_{ji}. \tag{3.103}$$

This constraint does not, however, fully determine the functional form of α_{ij}. Hastings[5] noted that

$$\alpha_{ij} = \frac{s_{ij}}{1 + \dfrac{\pi_i}{\pi_j} \dfrac{q_{ij}}{q_{ji}}} \tag{3.104}$$

satisfies detailed balance for any symmetric s_{ij}. This is easily verified by substituting equation 3.104 into equation 3.103 and making use of $s_{ij} = s_{ji}$.

While equation 3.104 and other alternative forms for $A(x_j|x_i)$ are sometimes useful, the form that has gained almost universal favor and has come to be known as *the* Metropolis-Hastings algorithm is

$$\alpha_{ij} = \min\left[1, \frac{\pi_j}{\pi_i} \frac{q_{ji}}{q_{ij}}\right], \tag{3.105}$$

which is understood to mean that α_{ij} is set equal to whichever is smaller, 1 or $\pi_j q_{ji}/\pi_i q_{ij}$. For the special case $(\pi_j q_{ji})/(\pi_i q_{ij}) = 1$, detailed balance is already satisfied by q_{ij}, so $\alpha_{ij} = 1$, and all the candidates are accepted. To see that equation 3.105 preserves detailed balance when $(\pi_j q_{ji})/(\pi_i q_{ij}) \neq 1$, first note that $(\pi_j q_{ji})/(\pi_i q_{ij})$ is either greater than or less than 1. If it is greater than 1, then $\alpha_{ij} = 1$, and the left-hand side of equation 3.103 becomes

$$\pi_i q_{ij} \alpha_{ij} = \pi_i q_{ij}. \tag{3.106}$$

But, this also means that $(\pi_i q_{ij})/(\pi_j q_{ji})$ is less than 1, so $\alpha_{ji} = (\pi_i q_{ij})/(\pi_j q_{ji})$, and the right-hand side equation 3.103 becomes

$$\pi_j q_{ji} \alpha_{ji} = \pi_j q_{ji} \left[\frac{\pi_i}{\pi_j} \frac{q_{ij}}{q_{ji}}\right] = \pi_i q_{ij}, \tag{3.107}$$

and the equality is satisfied. If $(\pi_j q_{ji})/(\pi_i q_{ij})$ is less than 1, then $\alpha_{ij} = (\pi_j q_{ji})/(\pi_i q_{ij})$, and $\alpha_{ji} = 1$, and again the equality is satisfied.

Here, then, is the Metropolis-Hastings algorithm for generating a Markov chain that converges to $P(x_i)$. Choose a candidate distribution $Q(x_j|x_i)$ and a starting value $u^{(0)}$. To generate additional members of the chain:

1. Generate a candidate value $u_c^{(k+1)}$ from the conditional distribution $Q(u_c^{(k+1)}|u^{(k)})$.
2. Calculate α_{ij} from

$$\alpha_{ij} = \min\left[1, \frac{P(u_c^{(k+1)})}{P(u^{(k)})} \frac{Q(u^{(k)}|u_c^{(k+1)})}{Q(u_c^{(k+1)}|u^{(k)})}\right]. \tag{3.108}$$

3. Generate a uniform random deviate v between 0 and 1.
4. If $v < \alpha_{ij}$, accept the candidate, and set $u^{(k+1)} = u_c^{(k+1)}$. Otherwise reject the candidate, and set $u^{(k+1)} = u^{(k)}$.

---

[5] W. K. Hastings. 1970. "Monte Carlo Sampling Methods Using Markov Chains and Their Applications." *Biometrika* vol. 57, p. 97.

At first sight it looks strange to set $u^{(k+1)}$ equal to $u^{(k)}$ when the candidate is rejected. Why not keep generating candidates until one is finally accepted? Generating multiple candidates creates an excess flow away from $u^{(k)}$ that is no longer necessarily matched by reverse flows back to $u^{(k)}$, which violates detailed balance. It may look like $u^{(k)}$ is being double counted when a candidate is rejected. If, however, the Markov chain has not converged to the invariant distribution, the extra point will not be counted as a sample from $P(x_i)$ and merely delays convergence. If the Markov chain has converged to the invariant distribution, the extra point is a valid sample from $P(x_i)$. After many steps in the Markov chain, all such extra points will collectively have the correct distribution.

Finally, note that the only place $P(x_i)$ appears in the Metropolis-Hastings algorithm is in equation 3.108, and there it appears only in the division $P(u_c^{(k+1)})/P(u^{(k)})$. The normalization constant in $P(x_i)$ divides out, so the distribution does not need to be normalized. The same is true for the conditional distribution, since it, too, only occurs in the ratio $Q(u^{(k)}|u_c^{(k+1)})/Q(u_c^{(k+1)}|u^{(k)})$.

One-Parameter Continuous Probability Distributions: The generalization to one-parameter continuous probability distributions is straightforward. The probability distribution function is now $\pi(x)$, where x is a continuous variable. To generate a Markov chain $u^{(0)}, u^{(1)}, u^{(2)}, \ldots, u^{(n)}$ whose invariant distribution is $\pi(x)$, we must devise a conditional probability distribution function $t(x|x')$ that satisfies detailed balance (see equation 3.88)

$$\pi(x')t(x|x') = \pi(x)t(x'|x). \tag{3.109}$$

The conditional probability distribution $q(x|x')$, again called the candidate distribution, is now continuous, as is the probability of acceptance, $\alpha(x|x')$. The product $t(x|x') = q(x|x')\alpha(x|x')$ must satisfy detailed balance

$$\pi(x')q(x|x')\alpha(x|x') = \pi(x)q(x'|x)\alpha(x'|x). \tag{3.110}$$

This equation is satisfied if $\alpha(x|x')$ is chosen to be

$$\alpha(x|x') = \min\left[1, \frac{\pi(x)}{\pi(x')}\frac{q(x'|x)}{q(x|x')}\right]. \tag{3.111}$$

To generate a Markov chain that converges to $\pi(x)$ using the Metropolis-Hastings algorithm, choose a candidate distribution $q(x'|x)$ and a starting value $u^{(0)}$. Then,

1. Generate a candidate value $u_c^{(k+1)}$ from the conditional distribution $q(u_c^{(k+1)}|u^{(k)})$.
2. Calculate $\alpha(u_c^{(k+1)}|u^{(k)})$ from

$$\alpha(u_c^{(k+1)}|u^{(k)}) = \min\left[1, \frac{\pi(u_c^{(k+1)})}{\pi(u^{(k)})}\frac{q(u^{(k)}|u_c^{(k+1)})}{q(u_c^{(k+1)}|u^{(k)})}\right]. \tag{3.112}$$

3. Generate a uniform random deviate v between 0 and 1.
4. If $v < \alpha(u_c^{(k+1)}u^{(k)})$, accept the candidate, and set $u^{(k+1)} = u_c^{(k+1)}$. Otherwise, reject the candidate, and set $u^{(k+1)} = u^{(k)}$.

As for the discrete case, the only places where $\pi(x)$ and $q(x|x')$ appear in the algorithm are in the ratios in equation 3.112. Because the normalization factor divides out in the ratios, neither $\pi(x)$ nor $q(x|x')$ needs to be normalized.

One is free to choose whatever $q(x|x')$ serves the problem at hand, but it should be easy to generate deviates from the candidate distribution, and a high proportion of the candidates should be accepted. A common choice for q$(x|x')$ is the Gaussian random-walk distribution (see equation 3.84)

$$q(x|x') = \frac{1}{\sqrt{2\pi}\sigma} \exp\left[-\frac{1}{2}\frac{(x-x')^2}{\sigma^2}\right]. \qquad (3.113)$$

Note that this distribution is adaptive, recentering itself to $x' = u^{(k)}$ at every step. The efficiency of the Metropolis-Hastings algorithm depends sensitively on the typical size of the steps from x' to x, which is set by the value of σ. The steps need to be large enough that all regions where $\pi(x)$ is large can be reached without an inordinately long Markov chain; but they should not be so large that most candidate points fall where $\pi(x)$ is small and are rejected. Rejections rates near 50% often give good results, but choosing the step size can be a challenge if $\pi(x)$ is large in narrow regions separated by wide gaps where $\pi(x)$ is small.

Equation 3.113 is symmetric in x and x'. This means q$(u^{(k)}|u_c^{(k+1)})/$q$(u_c^{(k+1)}|u^{(k)})$ is equal to 1, so the expression for $\alpha(u_c^{(k+1)}|u^{(k)})$ simplifies to

$$\alpha(u_c^{(k+1)}|u^{(k)}) = \min\left[1, \frac{\pi(u_c^{(k+1)})}{\pi(u^{(k)})}\right], \qquad (3.114)$$

materially speeding the acceptance-rejection step in the algorithm. This simplification occurs for any symmetric candidate distribution.

Multiparameter Continuous Probability Distributions: These are the distributions for which MCMC sampling displays its true power. We wish to generate a Markov chain for which the invariant distribution is the k-dimensional multivariate probability distribution function $\pi(x_1,\ldots,x_k)$, where the x_i are continuous variables. The array of values $(x_1,\ldots,x_k)$ is a state. The Markov chain is the sequence of states $s^{(0)}, s^{(1)}, s^{(2)}, \ldots, s^{(n)}$, where state $s^{(j)}$ is

$$s^{(j)} = (u_1^{(j)},\ldots,u_k^{(j)}), \qquad (3.115)$$

and the $u_i^{(j)}$ are random deviates corresponding to the x_i. The conditional probability for a transition from state $(x_1',\ldots,x_k')$ to state $(x_1,\ldots,x_k)$ is t$(x_1,\ldots,x_k|x_1',\ldots,x_k')$. The method for generating the Markov chain must explicitly or implicitly satisfy the detailed balance equation

$$\pi(x_1',\ldots,x_k')\text{t}(x_1,\ldots,x_k|x_1',\ldots,x_k') = \pi(x_1,\ldots,x_k)\text{t}(x_1',\ldots,x_k'|x_1,\ldots,x_k). \qquad (3.116)$$

It is tempting to generalize the Metropolis-Hastings algorithm from the one-parameter case to the multiparameter case by devising a conditional distribution q$(x_1',\ldots,x_k'|x_1,\ldots,x_k)$ that produces entire candidate states in one step. The acceptance probability $\alpha(x_1',\ldots,x_k'|x_1,\ldots,x_k)$ would be calculated by replacing x by $(x_1,\ldots,x_k)$ and x' by $(x_1',\ldots,x_k')$ in equation 3.111. While this may work when k is small, it does not work if the number of parameters is large. As k increases, the volume of the multidimensional space becomes so large that most of the candidate states will fall where $\pi(x_1,\ldots,x_k)$ is small and will be rejected, making the algorithm inefficient.

The usual way to extend the Metropolis-Hastings algorithm to multiparameter distributions is to generate values for the $u_i^{(j)}$ one at a time, cycling through the k parameters to obtain the new state. There is a separate conditional probability for each parameter. The conditional probability for the first parameter, for example, is (keep track of which parameters are primed and which are not!)

$$\text{t}_1(x_1, x_2',\ldots,x_k'|x_1', x_2',\ldots,x_k'),$$

which must separately satisfy detailed balance:

$$\pi(x'_1, x'_2, \ldots, x'_k) t_1(x_1 | x'_1, x'_2, \ldots, x'_k) = \pi(x_1, x'_2, \ldots, x'_k) t_1(x'_1 | x_1, x'_2, \ldots, x'_k). \quad (3.117)$$

There is a separate candidate distribution for each parameter. The candidate distribution for the first parameter is

$$q_1(x_1, x'_2, \ldots, x'_k | x'_1, x'_2, \ldots, x'_k).$$

A candidate deviate for the first parameter is generated from $q_1(x_1, x'_2, \ldots, x'_k | x'_1, x'_2, \ldots, x'_k)$, and the candidate deviate is accepted with probability

$$\alpha_1(x_1, x'_2, \ldots, x'_k | x'_1, x'_2, \ldots, x'_k) = \min\left[1, \frac{\pi(x_1, x'_2, \ldots, x'_k)}{\pi(x'_1, x'_2, \ldots, x'_k)} \frac{q_1(x'_1, x'_2, \ldots, x'_k | x_1, x'_2, \ldots, x'_k)}{q_1(x_1, x'_2, \ldots, x'_k | x'_1, x'_2, \ldots, x'_k)}\right].$$

$$(3.118)$$

There are equivalent candidate distributions and acceptance probabilities for the other parameters. The distributions can be and often are different for each parameter. One might, for example, use Gaussian candidate distributions with different variances for different parameters.

It is not correct to generate values for all parameters before updating the state, because a state made by simply conjoining the individually generated values will not satisfy detailed balance, even though the individual values do. The Markov chain generated this way is not guaranteed to converge to $\pi(x_1, \ldots, x_k)$. The proper way to proceed is to replace the old value of a parameter by the new value immediately, before generating deviates for the other parameters. Thus, the procedure for generating the next state after $s^{(j)} = (u_1^{(j)}, u_2^{(j)}, \ldots, u_k^{(j)})$ in a Markov chain is as follows.

1. Generate a candidate $u_{1c}^{(j+1)}$ for the first dimension from $q_1(u_{1c}^{(j+1)}, u_2^{(j)}, \ldots, u_k^{(j)} | u_1^{(j)}, u_2^{(j)}, \ldots, u_k^{(j)})$, and calculate the acceptance probability $\alpha_1(u_{1c}^{(j+1)}, u_2^{(j)}, \ldots, u_k^{(j)} | u_1^{(j)}, u_2^{(j)}, \ldots, u_k^{(j)})$ from equation 3.118. Generate a uniform random deviate v between 0 and 1. If $v < \alpha_1$, accept the candidate, and set $u_1^{(j+1)} = u_{1c}^{(j+1)}$. Otherwise, set $u_1^{(j+1)} = u_1^{(j)}$.

2. State $(u_1^{(j+1)}, u_2^{(j)}, \ldots, u_k^{(j)})$ is a valid state in the Markov chain, because the way it was generated preserved detailed balance. This intermediate state can, therefore, be used to generate a valid deviate for the second parameter. Generate a candidate $u_{2c}^{(j+1)}$ from $q_2(u_1^{(j+1)}, u_{2c}^{(j+1)}, \ldots, u_k^{(j)} | u_1^{(j+1)}, u_2^{(j)}, \ldots, u_k^{(j)})$ and accept it with a probability

$$\alpha_2 = \min\left[1, \frac{\pi(u_1^{(j+1)}, u_{2c}^{(j+1)}, \ldots, u_k^{(j)})}{\pi(u_1^{(j+1)}, u_2^{(j)}, \ldots, u_k^{(j)})} \frac{q_2(u_1^{(j+1)}, u_2^{(j)}, \ldots, u_k^{(j)} | u_1^{(j+1)}, u_{2c}^{(j+1)}, \ldots, u_k^{(j)})}{q_2(u_1^{(j+1)}, u_{2c}^{(j+1)}, \ldots, u_k^{(j)} | u_1^{(j+1)}, u_2^{(j)}, \ldots, u_k^{(j)})}\right].$$

$$(3.119)$$

Generate a uniform random deviate v between 0 and 1. If $v < \alpha_2$, set $u_2^{(j+1)} = u_{2c}^{(j+1)}$. Otherwise, set $u_2^{(j+1)} = u_2^{(j)}$.

3. State $(u_1^{(j+1)}, u_2^{(j+1)}, \cdots, u_k^{(j)})$ is a valid state, because the way it was generated satisfies detailed balance. Generate $u_3^{(j+1)}$ from this state. Continue in this way for the remaining parameters, always immediately using the new deviates for generating deviates for the other parameters.

At the end of this procedure, one has $s^{(j+1)} = (u_1^{(j+1)}, u_2^{(j+1)}, \ldots, u_k^{(j+1)})$. Since detailed balance was enforced at each step in the generation of $s^{(j+1)}$, the entire state satisfies detailed balance. It can be shown that the Markov chain produced this way is ergodic under a wide variety of conditions. In particular, it is ergodic if the marginal distributions are everywhere greater than 0. The Markov chain will, therefore, converge to $\pi(x_1,\ldots,x_k)$. One must step through enough states to ensure that the chain has converged before using the states as samples from the distribution. If $s^{(0)}$ is chosen from a region where $\pi(x_1,\ldots,x_k)$ is large, convergence is likely to be rapid.

One must calculate enough states in the chain to fully sample the distribution. It can take many samples to fully sample the distribution if it has widely separated peaks. In addition, the procedure described produces a multiparameter Markov chain by generating deviates for the parameters one at a time from a series of one-parameter probability distributions. If any of the parameters in $\pi(x_1,\ldots,x_k)$ are strongly correlated, the Markov chain may need to be extremely long to fully sample $\pi(x_1,\ldots,x_k)$. Consider, for example, a distribution that is large on a long narrow ridge aligned along $x_1 \approx x_2$. The Metropolis-Hastings algorithm samples first along x_1 then along x_2, and it can take many steps to fully sample the ridge even after the Markov chain has converged. It may be useful in situations like this to rotate the coordinate system so that one coordinates axis lies in the direction of the ridge line, removing the correlation.

3.5.3 Gibbs Sampler

The Gibbs sampler is a special case of the Metropolis-Hastings algorithm. It has two distinguishing features. First, the transition probabilities and the candidate distributions are set equal to the target probability distribution (see equation 3.96):

$$q(x_1,\ldots,x_k|x_1',\ldots,x_k') = t(x_1,\ldots,x_k|x_1',\ldots,x_k') = \pi(x_1,\ldots,x_k). \tag{3.120}$$

Since every candidate is now drawn directly from $\pi(x_1,\ldots,x_k)$, every candidate can be accepted, and there is no need to calculate $\alpha(x_1,\cdots,x_k|x_1',\ldots,x_k')$ or to go through the acceptance-rejection steps. Furthermore, since the Markov chain has no memory, the chain starts off already converged to $\pi(x_1,\ldots,x_k)$. Indeed, the chain is just a series of samples from $\pi(x_1,\ldots,x_k)$, which is, of course, goal of MCMC sampling.

As in the standard Metropolis-Hastings algorithm, the next state in the Markov chain is generated one parameter at a time. The second distinguishing feature of the Gibbs sampler is that the conditional distributions for the individual parameters,

$$t_1 = t_1(x_1|x_2',x_3',\ldots,x_k') \tag{3.121}$$

$$t_2 = t_2(x_2|x_1',x_3',\ldots,x_k') \tag{3.122}$$

$$\vdots$$

$$t_k = t_k(x_k|x_1',x_2',x_3',\ldots,x_{k-1}'), \tag{3.123}$$

are constructed directly from $\pi(x_1,\ldots,x_k)$. The conditional distribution for the first parameter, for example, is

$$t_1(x_1|x_2',\ldots,x_k') = \frac{\pi(x_1,x_2',\ldots,x_k')}{\int_{x_1} \pi(x_1,x_2',\ldots,x_k')dx_1}. \tag{3.124}$$

The meaning of equation 3.124 is that all parameters except x_1 are kept fixed, turning $\pi(x_1,\ldots,x_k)$ into a one-dimensional probability distribution for x_1. The next value for x_1 is generated from this distribution. The integral in the denominator normalizes t_1.

Expressions similar to equation 3.124 hold for the remaining parameters. Thus, the conditional distributions are just $\pi(x_1,\ldots,x_k)$ itself but with all parameters except the one to be generated fixed at their previous values. Note that $\pi(x_1,\ldots,x_k)$ itself need not be normalized.

Detailed balance is automatically preserved when using t_1 to generate the next value of x_1. To see this, note that

$$\pi(x_1',x_2',\ldots,x_k')t_1(x_1|x_2',\ldots,x_k') = \pi(x_1',x_2',\ldots,x_k')\frac{\pi(x_1,x_2',\ldots,x_k')}{\int_{x_1}\pi(x_1,x_2',\ldots,x_k')dx_1} \tag{3.125}$$

and

$$\pi(x_1,x_2',\ldots,x_k')t_1(x_1'|x_2',\ldots,x_k') = \pi(x_1,x_2',\ldots,x_k')\frac{\pi(x_1',x_2',\ldots,x_k')}{\int_{x_1'}\pi(x_1',x_2',\ldots,x_k')dx_1'}. \tag{3.126}$$

The integrals in equations 3.125 and 3.126 are the same, so we have

$$\pi(x_1',x_2',\ldots,x_k')t_1(x_1|x_2',\ldots,x_k') = \pi(x_1,x_2',\ldots,x_k')t_1(x_1'|x_2',\ldots,x_k'), \tag{3.127}$$

and detailed balance holds. In fact, since the normalization factors are identical in equations 3.125 and 3.126, they are not really needed, and detailed balance holds without them. Thus, the conditional distributions do not need to be normalized, and we can simply take

$$t_1(x_1|x_2',\ldots,x_k') = \pi(x_1,x_2',\ldots,x_k'). \tag{3.128}$$

As in the standard Metropolis-Hastings algorithm, one generates the next state in a Markov chain after $s^{(j)} = (u_1^{(j)},u_2^{(j)},\ldots,u_k^{(j)})$ by generating $u_1^{(j+1)}$ from $t_1(u_1^{(j+1)}|u_2^{(j)},x_u^{(j)},\ldots,u_k^{(j)})$, then immediately using $u_1^{(j+1)}$ when generating $u_2^{(j+1)}$ from $t_2(u_2^{(j+1)}|u_1^{(j+1)},u_3^{(j)},\ldots,u_k^{(j)})$, and so on. A complete cycle through all parameters yields $s^{(j+1)}$. Because detailed balance was preserved for each intermediate state, each intermediate state is a valid sample from $\pi(x_1,\ldots,x_k)$, so the final result $s^{(j+1)}$ is also a valid sample. In one sense the choice of the initial state $s^{(0)}$ is irrelevant, since all the states are generated from $\pi(x_1,\ldots,x_k)$ and there is no need to "burn in" the Markov chain. If, however, $s^{(0)}$ has a low probability of occurrence, the Markov chain may need to be inconveniently long to generate enough compensating states with high probabilities. Thus, the initial state should be chosen to have a high probability of occurrence.

The efficiency of the Gibbs sampler depends crucially on the speed with which deviates can be generated from the conditional distributions for the individual parameters. If deviates can be generated from all the t_i efficiently, the Gibbs sampler is an excellent way to generate the corresponding Markov chain. If it is time consuming to generate deviates from the t_i, the considerable advantages of the Gibbs sampler can be lost.

The following example shows how the Gibbs sampler might be used to generate samples from a two-parameter distribution of the kind that often arises in Bayesian analyses.

Example: We often think of the Gaussian distribution

$$f(x) \propto \sigma^{-1}\exp\left[-\frac{1}{2}\frac{(x-\mu)^2}{\sigma^2}\right]$$

as being the probability of getting x. The variable parameter is x, while the mean μ and variance σ^2 are constants. If instead we take x to be constant and let μ and σ^2 be variable parameters, this becomes a two-parameter distribution for μ and σ^2:

$$\mathrm{f}(\mu, \sigma^2) \propto \sigma^{-1} \exp\left[-\frac{1}{2} \frac{(x-\mu)^2}{\sigma^2}\right].$$

Now assume that n of these two-parameter distributions have been multiplied together. To avoid unnecessary complications, we assume that the distributions have the same μ and σ^2 but different values of x. The product distribution is

$$\pi(\mu, \sigma^2) \propto \sigma^{-n} \exp\left[-\frac{1}{2} \frac{\sum(x_i-\mu)^2}{\sigma^2}\right],$$

where the x_i are the n values of x.

Suppose we wish to generate deviates from this distribution using the Gibbs sampler. The deviates are (u_μ, u_{σ^2}) pairs, so the Gibbs sampler requires two conditional distributions, one for μ and one for σ^2. The conditional distribution for σ^2 is

$$\mathrm{t}_1(\sigma^2|\mu) \propto \sigma^{-n} \exp\left[-\frac{1}{2} \frac{\sum(x_i-\mu)^2}{\sigma^2}\right].$$

It is, however, difficult to generate random deviates from this distribution, so we change the independent parameter from σ^2 to $\chi^2 = \sum(x_i-\mu)^2/\sigma^2$. With μ held constant, $\sigma^2 \propto 1/\chi^2$ and $d\sigma^2 \propto d\chi^2/(\chi^2)^2$. The conditional distribution for χ^2 becomes

$$\mathrm{t}_1(\chi^2|\mu) \propto (\chi^2)^{n/2-2} \exp[-\chi^2/2],$$

which is the standard χ^2 distribution for $n-1$ degrees of freedom. Efficient random number generators for χ^2 distributions are readily available.

The conditional probability distribution for μ is

$$\mathrm{t}_2(\mu|\sigma^2) \propto \exp\left[-\frac{1}{2} \frac{\sum(x_i-\mu)^2}{\sigma^2}\right].$$

Once again it is inconvenient to generate deviates for this form of the distribution. From Section C.2 in Appendix C the product of n Gaussian distributions can be rewritten as a single Gaussian distribution, so t_2 can be converted to

$$\mathrm{t}_2(\mu|\sigma^2) \propto \exp\left[-\frac{1}{2} \frac{(\bar{x}-\mu)^2}{\sigma^2/n}\right],$$

where

$$\bar{x} = \frac{1}{n}\sum x_i.$$

Efficient random number generators for Gaussian distributions are also readily available.

Continued on page 80

We are now prepared to use the Gibbs sampler. Given a state $s^{(j)} = (u_\mu^{(j)}, u_{\sigma^2}^{(j)})$ in a Markov chain, we wish to generate $s^{(j+1)}$. To do this,

1. Generate $u_{\chi^2}^{(j+1)}$ from $t_1(u_{\chi^2}^{(j+1)}|u_\mu^{(j)})$, where t_1 is a χ^2 distribution for $n - 1$ degrees of freedom. Calculate $u_{\sigma^2}^{(j+1)})$ from

$$u_{\sigma^2}^{(j+1)} = \frac{\sum(x_i - u_\mu^{(j)})^2}{u_{\chi^2}^{(j+1)}}.$$

2. Generate $u_\mu^{(j+1)}$ from $t_2(u_\mu^{(j+1)}|u_{\sigma^2}^{(j+1)})$, where the value of $u_{\sigma^2}^{(j+1)}$ generated in step 1 is used immediately when generating $u_\mu^{(j+1)}$.

The utility of the Gibbs sampler for this problem depended explicitly on finding conditional probability distributions from which it was easy to generate random deviates for μ and χ^2.

4

Elementary Frequentist Statistics

4.1 Introduction to Frequentist Statistics

This chapter begins the discussion of standard statistics, sometimes called *frequentist statistics* to distinguish it from Bayesian statistics. To understand the basic assumptions of frequentist statistics, return to the definition of probability given at the beginning of Chapter 1 (equation 1.1). Suppose that an event A can happen in k ways out of a total of n possible, equally likely, ways. The probability that A will occur is defined to be

$$P(A) = \frac{k}{n}. \tag{4.1}$$

Since this probability is calculated without reference to experiments or data, it is sometimes called a priori probability. It is tempting to adopt a similar definition for statistical probability: If an event A is *observed* to occur k times out of n trials, the probability of A occurring is k/n. This definition is, however, inadequate. If we perform a series of independent experiments, each experiment consisting of n trials, we will generally observe a different value of k in each experiment and would infer a different probability. We need a definition that recognizes that individual experiments can yield probabilities that are different from the true probability.

Instead, define the probability as the limit of k/n as n becomes large:

$$P(A) = \lim_{n \to \infty} \frac{k}{n}. \tag{4.2}$$

The definition has two crucial features. It identifies probability with a frequency of occurrence (k/n is a frequency), and the true probability only emerges in the limit as n becomes large. Since the probability is calculated after the experiment and does not require that anything be known in advance, it is sometimes called the a posteriori probability.

Equation 4.2 gives intuitively reasonable results. Suppose, for example, that we make n error-free measurements of a discrete quantity x, yielding the values x_i, $i = 1, \ldots, n$. Suppose another quantity g is a function of x, and set $g_i = g(x_i)$. Let us evaluate

$$\lim_{n \to \infty} \frac{1}{n} \sum_{i=1}^{n} g_i. \tag{4.3}$$

Because x is discrete, there will be many duplications among the x_i, so the set of samples can be written

$$(k_j, x_j), \quad j = 1, \ldots, m, \tag{4.4}$$

where k_j is the number of times x_j occurs, and m is the number of unique values of x_j. Taking advantage of the duplications and using equation 4.2, we have

$$\lim_{n \to \infty} \frac{1}{n} \sum_{i=1}^{n} g_i = \lim_{n \to \infty} \frac{1}{n} \sum_{j=1}^{m} k_j g_j = \sum_{j=1}^{m} \left[\lim_{n \to \infty} \frac{k_j}{n} \right] g_j = \sum_{j=1}^{m} g_j P(x_j) = \langle g \rangle, \tag{4.5}$$

where we have recognized the penultimate term as the mean value of g (see equation 1.69). Succinctly we have,

$$\lim_{n \to \infty} \frac{1}{n} \sum_{i=1}^{n} g_i = \langle g \rangle. \tag{4.6}$$

With somewhat more work, this generalizes to continuous quantities:

$$\lim_{n \to \infty} \frac{1}{n} \sum_{i=1}^{n} g_i = \int_{-\infty}^{\infty} g(x) f(x) dx = \langle g \rangle, \tag{4.7}$$

where $f(x)$ is the continuous probability density distribution for x. Equations 4.6 and 4.7 state that in the limit as n becomes large, the quantity $(1/n) \sum_{i=1}^{n} g_i$ does, indeed, equal the true mean value of g. If, for example, the g_i are a simple set of numbers $g_i = x_i$, we have the pleasing (and desired!) result

$$\lim_{n \to \infty} \frac{1}{n} \sum_{i=1}^{n} x_i = \langle x \rangle. \tag{4.8}$$

Equation 4.2 is far from obviously correct. A mathematician might call it an axiom, while an experimental scientist might call it an observed property of nature; but neither can demonstrate its validity. Indeed, it cannot even be applied in practice, since one can never carry out the limit (n never goes to infinity in a real experiment). This is a deep weakness in the foundations of frequentist statistics. It also leads to the question: What is the meaning of k/n and of quantities like $(1/n) \sum_{i=1}^{n} g_i$ when n remains finite? Much of frequentist statistics is devoted to this question.

Sections 4.2–4.4 derive the standard expressions for the means and variances of sets of data, first for unweighted data, then for data with uncorrelated measurement errors, and finally for data with correlated measurement errors. The estimated variances can themselves have large errors. Section 4.5 shows how to estimate those errors. Finally, Sections 4.6 and 4.7 introduce principle component analysis and one aspect of hypothesis testing, the Kolmogorov-Smirnov test.

4.2 Means and Variances for Unweighted Data

Suppose that x has a probability distribution $f(x)$ as shown in the upper panel of Figure 4.1, and that n samples (measurements!) of x_i were taken from the distribution. We would like to estimate the mean $\mu = \langle x \rangle$ and variance $\sigma_x^2 = \langle (x - \mu)^2 \rangle$ of the distribution from the measurements. These two quantities are usually called the population mean and variance. To be concrete, suppose a farmer with a cherry orchard wishes to know μ and σ_x^2, the mean

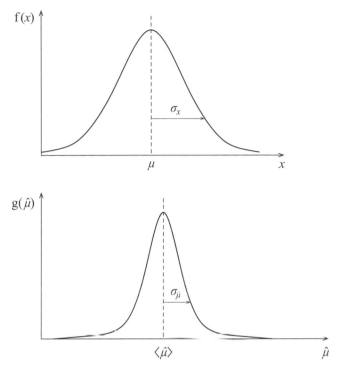

Figure 4.1: f(x) is the probability distribution for x. The mean value and variance of the distribution are μ and σ_x^2, respectively, usually called the population mean and variance. Let $\hat{\mu}$ be an estimate of μ determined from measurements. The estimates have a distribution g($\hat{\mu}$) with variance $\sigma_{\hat{\mu}}^2$, usually called the variance of the sample mean. If $\langle\hat{\mu}\rangle = \mu$, then $\hat{\mu}$ is called an unbiased estimator of μ. The variances of the two distributions are different. Furthermore, σ_x^2 is constant, because the original population is fixed, but $\sigma_{\hat{\mu}}^2$ decreases (one hopes!) as the number of measurements increases.

diameter and the range of diameters of the cherries from one of his trees. The farmer samples some cherries from one branch of the tree and measures their diameters, x_i. The farmer would like to determine μ and σ_x^2 from the x_i.

In real life we never have the infinite number of measurements required by equation 4.8 and, indeed, n is often small (the farmer picks just a handful of cherries). We cannot calculate—can never calculate—μ and σ_x^2! We are forced instead to calculate such quantities as

$$\hat{\mu} = \frac{1}{n}\sum_{i=1}^{n} x_i, \tag{4.9}$$

called the *sample mean*, and use them as estimates of the quantities we really want.[1] In general, $\hat{\mu} \neq \mu$. Does equation 4.9 at least give the right answer on average? An estimate $\hat{a}$ of the quantity a is said to be *unbiased* if $\langle\hat{a}\rangle = a$. Is $\hat{\mu}$ calculated from equation 4.9 unbiased?

---

[1] We reserve the hat accent for estimated quantities.

Taking the mean of both sides of equation 4.9, we have

$$\langle\hat{\mu}\rangle = \left\langle \frac{1}{n}\sum_{i=1}^{n} x_i \right\rangle = \frac{1}{n}\sum_{i=1}^{n}\langle x_i\rangle = \frac{1}{n}\sum_{i=1}^{n}\langle x\rangle = \frac{1}{n}n\langle x\rangle = \langle x\rangle = \mu, \qquad (4.10)$$

so $\hat{\mu}$ is unbiased. This is a more interesting issue than it might appear at first sight. When discussing maximum likelihood methods and, to a lesser extent, Bayesian statistics, we will run across often-used estimates that are biased, albeit by only a small amount.

We also need to know how much $\hat{\mu}$ is likely to differ from μ. Suppose we measure $\hat{\mu}$ several times. The farmer might, for example, measure the diameters of cherries first from one branch of the tree, then from other branches, each branch yielding a separate mean diameter. The values of $\hat{\mu}$ will differ from one another, and together they form a distribution of measured values, $g(\hat{\mu})$, as shown in the lower panel of Figure 4.1. The scatter in the measurements is given by the width of the distribution, which can be characterized by the variance of $\hat{\mu}$,

$$\sigma_{\hat{\mu}}^2 = \langle(\hat{\mu}-\mu)^2\rangle, \qquad (4.11)$$

usually called the variance of the sample mean. Note that $\sigma_{\hat{\mu}}^2$ is *not* σ_x^2. As shown in Figure 4.1, $\sigma_{\hat{\mu}}^2$ characterizes the width of $g(\hat{\mu})$, the distribution of the measured values of μ; while σ_x^2 characterizes the width of $f(x)$, the original distribution. In terms of our cherry tree example, σ_x^2 characterizes the range of cherry diameters, while $\sigma_{\hat{\mu}}^2$ characterizes the accuracy of their measured mean diameter. Using equation 4.9 to eliminate $\hat{\mu}$ from equation 4.11, we have

$$\sigma_{\hat{\mu}}^2 = \langle(\hat{\mu}-\mu)^2\rangle = \left\langle\left(\frac{1}{n}\sum_{i=1}^{n} x_i - \mu\right)^2\right\rangle = \left\langle\left(\frac{1}{n}\sum_{i=1}^{n}[x_i - \mu]\right)^2\right\rangle$$

$$= \left\langle\frac{1}{n^2}\sum_{i=1}^{n}\sum_{j=1}^{n}(x_i-\mu)(x_j-\mu)\right\rangle = \frac{1}{n^2}\sum_{i=1}^{n}\sum_{j=1}^{n}\langle(x_i-\mu)(x_j-\mu)\rangle$$

$$= \frac{1}{n^2}\sum_{i=1}^{n}\sum_{j=1}^{n}\sigma_{ij}, \qquad (4.12)$$

where we have recognized that $\langle(x_i-\mu)(x_j-\mu)\rangle$ is the covariance between x_i and x_j (see equation 1.76). In this section we assume that the measurements are independent, so that there are no correlations among them. Then, for $i\neq j$,

$$\sigma_{ij} = \langle(x_i-\mu)(x_j-\mu)\rangle = \langle(x_i-\mu)\rangle\langle(x_j-\mu)\rangle = 0. \qquad (4.13)$$

We also assume that all the x_i have the same variance, so

$$\sigma_{ii} = \langle(x_i-\mu)^2\rangle = \sigma_x^2, \qquad (4.14)$$

where σ_x^2 is the variance of x. Equations 4.13 and 4.14 can be combined into the single equation

$$\sigma_{ij} = \sigma_x^2\delta_{ij}, \qquad (4.15)$$

where δ_{ij} is the Kronecker delta. If the x_i are uncorrelated and all have the same variance, equation 4.12 becomes

$$\sigma_{\hat{\mu}}^2 = \frac{1}{n^2} \sum_{i=1}^{n} \sum_{j=1}^{n} \sigma_x^2 \delta_{ij} = \frac{1}{n^2} \sum_{i=1}^{n} \sigma_x^2$$

$$= \frac{1}{n} \sigma_x^2. \tag{4.16}$$

This important result is good news. We see that $\sigma_{\hat{\mu}}^2$ is no worse than σ_x^2 and decreases as $1/n$. In other words, the accuracy with which we measure μ improves as $1/\sqrt{n}$.

However, we do not know σ_x^2! It too must be estimated from the data sample. Consider the quantity $\hat{S}$ (note that $\hat{S}$ is defined in terms of the measured value $\hat{\mu}$, not the true but unknown value μ), where

$$\hat{S} = \sum_i (x_i - \hat{\mu})^2 = \sum_i \left\{ x_i - \frac{1}{n} \sum_j x_j \right\}^2$$

$$= \sum_i \left\{ (x_i - \mu) - \frac{1}{n} \sum_j (x_j - \mu) \right\}^2$$

$$= \sum_i \left\{ (x_i - \mu)^2 - \frac{2}{n} \sum_j (x_i - \mu)(x_j - \mu) + \frac{1}{n^2} \sum_j \sum_k (x_j - \mu)(x_k - \mu) \right\}. \tag{4.17}$$

Now take the mean value of $\hat{S}$. Remembering that taking a mean value is a linear operation, we find

$$\langle \hat{S} \rangle = \sum_i \left\{ \langle (x_i - \mu)^2 \rangle - \frac{2}{n} \sum_j \langle (x_i - \mu)(x_j - \mu) \rangle + \frac{1}{n^2} \sum_j \sum_k \langle (x_j - \mu)(x_k - \mu) \rangle \right\}$$

$$= \sum_i \left\{ \sigma_x^2 - \frac{2}{n} \sum_j \sigma_{ij} + \frac{1}{n^2} \sum_j \sum_k \sigma_{jk} \right\}. \tag{4.18}$$

As before, we assume that the x_i are uncorrelated samples, so $\sigma_{ij} = \sigma_x^2 \delta_{ij}$. Equation 4.18 collapses to

$$\langle \hat{S} \rangle = \sum_i \left\{ \sigma_x^2 - \frac{2}{n} \sigma_x^2 + \frac{1}{n^2} \sum_j \sigma_x^2 \right\} = \sum_i \left\{ \sigma_x^2 - \frac{1}{n} \sigma_x^2 \right\}$$

$$= (n-1)\sigma_x^2. \tag{4.19}$$

Let $\hat{\sigma}_x^2$ be an estimate of σ_x^2. Guided by equation 4.19, we take

$$\hat{\sigma}_x^2 = \frac{1}{n-1} \hat{S} = \frac{1}{n-1} \sum_{i=1}^{n} (x_i - \hat{\mu})^2. \tag{4.20}$$

Note that $\hat{\sigma}_x^2$ is an estimate of σ_x^2, equaling σ_x^2 only in the limit as n goes to infinity. As will be shown in Section 4.5, the difference between $\hat{\sigma}_x^2$ and σ_x^2 can be substantial. It is, though, an unbiased estimate, because

$$\langle \hat{\sigma}_x^2 \rangle = \frac{1}{n-1} \langle \hat{S} \rangle = \frac{1}{n-1}(n-1)\sigma_x^2 = \sigma_x^2. \tag{4.21}$$

Why is the divisor $n-1$, not n? The individual quantities $(x_i - \hat{\mu})^2$ in the definition of $\hat{S}$ are not independent of one another, because $\hat{\mu}$ depends on all the x_i. The use of $\hat{\mu}$ instead of μ makes the differences slightly smaller on average, which in turn makes $\hat{S}$ smaller than S on average. Dividing $\hat{S}$ by $n-1$ instead of n exactly compensates for this bias.

We also need an estimate of $\hat{\sigma}_{\hat{\mu}}^2$. Equipped with an estimate for σ_x^2 and guided by equation 4.16, we adopt

$$\hat{\sigma}_{\hat{\mu}}^2 = \frac{1}{n}\hat{\sigma}_x^2 = \frac{1}{n(n-1)}\sum_{i=1}^{n}(x_i - \hat{\mu})^2. \tag{4.22}$$

As before, this is an estimate of $\hat{\sigma}_{\hat{\mu}}^2$, not $\hat{\sigma}_{\hat{\mu}}^2$ itself, and the difference between the two can be substantial. It is, though, an unbiased estimate of $\sigma_{\hat{\mu}}^2$, because from equations 4.16 and 4.21,

$$\langle \hat{\sigma}_{\hat{\mu}}^2 \rangle = \frac{1}{n}\langle \hat{\sigma}_x^2 \rangle = \frac{1}{n}\sigma_x^2 = \sigma_{\hat{\mu}}^2. \tag{4.23}$$

Summary: If $\{x_i\}$ is a set of n measurements, all with equal weights, then

$$\hat{\mu} = \frac{1}{n}\sum_{i=1}^{n}x_i \tag{4.24}$$

$$\hat{\sigma}_x^2 = \frac{\hat{S}}{n-1} = \frac{\sum_i(x_i - \hat{\mu})^2}{n-1} \tag{4.25}$$

$$\hat{\sigma}_{\hat{\mu}}^2 = \frac{1}{n}\hat{\sigma}_x^2 = \frac{\hat{S}}{n(n-1)} = \frac{\sum_i(x_i - \hat{\mu})^2}{n(n-1)}. \tag{4.26}$$

Equations 4.24–4.26 are the standard equations found in any elementary book on statistics. Since they were derived without explicitly specifying $f(x)$ and $g(\hat{\mu})$, the equations are valid for any probability distribution. The usefulness of $\hat{\mu}$ and $\hat{\sigma}_x^2$ for characterizing $f(x)$ depends on the properties of the distribution, though. Neither are good characterizations of, for example, a strongly bimodal distribution.

4.3 Data with Uncorrelated Measurement Errors

Let us now suppose we have n measurements (x_i, σ_i) of a quantity x, where σ_i is the measurement error for measurement x_i. By measurement error, I mean that each x_i is drawn from a different probability distribution and that the distributions all have the same mean μ but different variances σ_i^2. In this section, we assume that the measurement errors are uncorrelated: $\sigma_{ij} = \sigma_i^2 \delta_{ij}$. Note that we use σ_i^2, not σ_x^2 as in equation 4.15, because each x_i has a different variance. We would like to include the measurement errors when calculating the estimates $\hat{\mu}$ and $\hat{\sigma}_{\hat{\mu}}^2$.

This is a common situation in data analysis. For example, the Large Magellanic Cloud is a small, nearby galaxy in orbit around our own galaxy, the Milky Way. The distance to the Large Magellanic Cloud has been measured in dozens of different ways, each way with a different reliability. The individual measurements are the x_i, the reliability of the measurements is given by their standard deviations σ_i, and the true distance to the Large Magellanic Cloud is μ. We would like to estimate μ from the measurements. Intuitively, the x_i with smaller errors should have greater effect on the estimated distance, but exactly how should the measurements be combined to give the best estimate? And how much should the estimate be trusted? This section provides answers to these questions.

For a estimated mean value to make any sense, it must be linear in the measurements, so we take

$$\hat{\mu} = \sum_{i=1}^{n} a_i x_i, \tag{4.27}$$

where the a_i are constants. We also require that $\hat{\mu}$ be unbiased, so

$$\langle \hat{\mu} \rangle = \mu = \left\langle \sum_{i=1}^{n} a_i x_i \right\rangle = \sum_{i=1}^{n} a_i \langle x_i \rangle = \mu \sum_{i=1}^{n} a_i. \tag{4.28}$$

The a_i must, therefore, satisfy the constraint equation

$$\sum_{i=1}^{n} a_i = 1. \tag{4.29}$$

We further require that the a_i give the best possible estimate of μ in the sense that the difference between $\hat{\mu}$ and μ as measured by the variance is minimized. The variance of $\hat{\mu}$ is

$$\sigma_{\hat{\mu}}^2 = \langle (\hat{\mu} - \mu)^2 \rangle$$

$$= \left\langle \left\{ \sum_i a_i x_i - \mu \right\}^2 \right\rangle = \left\langle \left\{ \sum_i a_i x_i - \mu \sum_i a_i \right\}^2 \right\rangle = \left\langle \left\{ \sum_i a_i (x_i - \mu) \right\}^2 \right\rangle$$

$$= \sum_i \sum_j a_i a_j \langle (x_i - \mu)(x_j - \mu) \rangle$$

$$= \sum_i \sum_j a_i a_j \sigma_{ij} = \sum_i \sum_j a_i a_j \sigma_i^2 \delta_{ij}$$

$$= \sum_i a_i^2 \sigma_i^2. \tag{4.30}$$

The goal is to choose values for the a_i that minimize $\sigma_{\hat{\mu}}^2$ subject to the constraint $\sum_i a_i = 1$.

This is a classic problem in constrained minimization. It can be solved using the method of Lagrange multipliers (see Appendix B for a review of Lagrange multipliers). Following the notation of Appendix B, we wish to choose values of a_i that minimize

$$f = \sigma_{\hat{\mu}}^2 = \sum_i a_i^2 \sigma_i^2 \tag{4.31}$$

subject to the constraint

$$g = \sum_i a_i - 1 = 0. \tag{4.32}$$

The constrained minimum is located where the gradients of f and g are parallel to each other:

$$\nabla f = \lambda \nabla g \tag{4.33}$$

or, in component form,

$$\frac{\partial f}{\partial a_i} - \lambda \frac{\partial g}{\partial a_i} = 0. \tag{4.34}$$

The partial derivatives are $\partial f / \partial a_i = 2 a_i \sigma_i^2$ and $\partial g / \partial a_i = 1$, so equations 4.34 become

$$2 a_i \sigma_i^2 - \lambda = 0, \tag{4.35}$$

which have solutions

$$a_i = \frac{\lambda}{2 \sigma_i^2}. \tag{4.36}$$

To determine λ, we insert these expressions for a_i back into the constraint equation:

$$1 = \sum_i a_i = \sum_i \frac{\lambda}{2 \sigma_i^2} = \frac{\lambda}{2} \sum_i \frac{1}{\sigma_i^2}, \tag{4.37}$$

so λ is

$$\lambda = \frac{2}{\sum_i (1/\sigma_i^2)}. \tag{4.38}$$

The final result is

$$a_i = \frac{1/\sigma_i^2}{\sum_i (1/\sigma_i^2)}. \tag{4.39}$$

For convenience, define the weight w_i of data point i to be

$$w_i = \frac{1}{\sigma_i^2}. \tag{4.40}$$

The sample mean of x then takes the simple form

$$\hat{\mu} = \frac{\sum_i w_i x_i}{\sum_i w_i}. \tag{4.41}$$

Returning to equation 4.30, we find the variance of the sample mean to be

$$\sigma_{\hat{\mu}}^2 = \left\langle (\hat{\mu} - \mu)^2 \right\rangle = \sum_i a_i^2 \sigma_i^2$$

$$= \sum_i \left[\frac{1/\sigma_i^2}{\sum_j (1/\sigma_j^2)} \right]^2 \sigma_i^2 = \frac{1}{\left[\sum_j (1/\sigma_j^2) \right]^2} \sum_i (1/\sigma_i^2) = \frac{1}{\sum_j (1/\sigma_j^2)}$$

$$= \frac{1}{\sum_j w_j}. \tag{4.42}$$

We need one further generalization. The measurement errors σ_i are often not quite correct and, therefore, the weights determined from them are not correct. This situation happens so often in real experiments that we need to know how to handle it. If the w_i are seriously in error, all is lost, and one should redo the experiment and measure the σ_i correctly. If, however, the σ_i are at least proportional to the real measurement errors, we can still calculate meaningful means and variances. Let us assume that the true errors of measurement are given by

$$\langle (x_i - \mu)^2 \rangle = \sigma^2 \sigma_i^2, \tag{4.43}$$

so that the true weights are

$$\alpha_i = \frac{1}{\sigma^2 \sigma_i^2} = \frac{w_i}{\sigma^2}, \tag{4.44}$$

where σ^2 is a proportionality constant sometimes called the *variance of unit weight*. Because we have assumed the measurement errors to be uncorrelated, we also have $\langle (x_i - \mu)(x_j - \mu) \rangle = 0$ for $i \neq j$. This can be combined with equation 4.43 into the single equation

$$\langle (x_i - \mu)(x_j - \mu) \rangle = \sigma^2 \sigma_i^2 \delta_{ij}. \tag{4.45}$$

The estimated mean is unaffected by the incorrect weights, because

$$\hat{\mu} = \frac{\sum_i \alpha_i x_i}{\sum_i \alpha_i} = \frac{\sum_i (w_i/\sigma^2) x_i}{\sum_i (w_i/\sigma^2)} = \frac{\sum_i w_i x_i}{\sum_i w_i}, \tag{4.46}$$

which is the same as equation 4.41. However, the variance of the sample mean becomes

$$\sigma_{\hat{\mu}}^2 = \frac{1}{\sum_i \alpha_i} = \frac{1}{\sum_i (w_i/\sigma^2)} = \frac{\sigma^2}{\sum_i w_i}, \tag{4.47}$$

which is not the same as equation 4.42, so the variance *is* affected by the incorrect weights. We need a way to estimate σ^2 and $\sigma_{\hat{\mu}}^2$. Define the quantity $\hat{S}$ by

$$\hat{S} = \frac{\sum_i w_i (x_i - \hat{\mu})^2}{\sum_i w_i}. \tag{4.48}$$

Note that $\hat{S}$ can be calculated without knowing σ^2. Expanding $\hat{S}$ and using equation 4.46 to eliminate $\hat{\mu}$, we get

$$\hat{S} = \frac{1}{\sum_i w_i} \sum_i w_i \left\{ (x_i - \mu) - (\hat{\mu} - \mu) \right\}^2$$

$$= \frac{1}{\sum_i w_i} \sum_i w_i \left\{ (x_i - \mu)^2 - 2(x_i - \mu)(\hat{\mu} - \mu) + (\hat{\mu} - \mu)^2 \right\}$$

$$= \frac{1}{\sum_i w_i} \sum_i w_i \left\{ (x_i - \mu)^2 - 2(x_i - \mu) \left[\frac{1}{\sum_i w_i} \sum_j w_j (x_j - \mu) \right] \right.$$

$$\left. + \left[\frac{1}{\sum_i w_i} \sum_j w_j (x_j - \mu) \right]^2 \right\}. \tag{4.49}$$

Rearranging this equation and then taking the mean value of $\hat{S}$, we have

$$\langle \hat{S} \rangle = \frac{1}{\sum_i w_i} \sum_i w_i \langle (x_i - \mu)^2 \rangle - 2 \left(\frac{1}{\sum_i w_i} \right)^2 \sum_i \sum_j w_i w_j \langle (x_i - \mu)(x_j - \mu) \rangle$$

$$+ \left(\frac{1}{\sum_i w_i} \right)^3 \sum_i \sum_j \sum_k w_i w_j w_k \langle (x_j - \mu)(x_k - \mu) \rangle. \quad (4.50)$$

With the use of equation 4.45, this imposing equation collapses to

$$\langle \hat{S} \rangle = \frac{1}{\sum_i w_i} \sum_i w_i \sigma^2 \sigma_i^2 - 2 \left(\frac{1}{\sum_i w_i} \right)^2 \sum_i \sum_j w_i w_j \sigma^2 \sigma_i^2 \delta_{ij}$$

$$+ \left(\frac{1}{\sum_i w_i} \right)^3 \sum_i \sum_j \sum_k w_i w_j w_k \sigma^2 \sigma_k^2 \delta_{jk}$$

$$= \frac{1}{\sum_i w_i} \sum_i \sigma^2 - 2 \left(\frac{1}{\sum_i w_i} \right)^2 \sum_i w_i \sigma^2 + \left(\frac{1}{\sum_i w_i} \right)^3 \sum_i \sum_j w_i w_j \sigma^2$$

$$= \frac{\sigma^2}{\sum_i w_i} \sum_i 1 - 2\sigma^2 \left(\frac{1}{\sum_i w_i} \right)^2 \left(\sum_i w_i \right) + \sigma^2 \left(\frac{1}{\sum_i w_i} \right)^3 \left(\sum_j w_i \right)^2$$

$$= \frac{1}{\sum_i w_i} (n-1)\sigma^2. \quad (4.51)$$

Guided by this result, we can estimate the variance of unit weight from

$$\hat{\sigma}^2 = \left(\frac{1}{n-1} \sum_i w_i \right) \hat{S} = \frac{1}{n-1} \sum_i w_i (x_i - \hat{\mu})^2. \quad (4.52)$$

Finally, guided by equation 4.47, we can estimate the variance of $\hat{\mu}$ from

$$\hat{\sigma}_{\hat{\mu}}^2 = \frac{\hat{\sigma}^2}{\sum_i w_i} = \frac{1}{n-1} \frac{\sum_i w_i (x_i - \hat{\mu})^2}{\sum_j w_j}. \quad (4.53)$$

This is an unbiased estimate of $\sigma_{\hat{\mu}}^2$, because from equations 4.51 and 4.47 we have

$$\langle \hat{\sigma}_{\hat{\mu}}^2 \rangle = \frac{1}{n-1} \left\langle \frac{\sum_i w_i (x_i - \hat{\mu})^2}{\sum_j w_j} \right\rangle = \frac{1}{n-1} \langle \hat{S} \rangle = \frac{\sigma^2}{\sum_i w_i} = \sigma_{\hat{\mu}}^2. \quad (4.54)$$

Summary: Given a set of measurements $\{x_i\}$, $i = 1, \ldots, n$, with uncorrelated measurement errors σ_i, the best estimate of $\langle x \rangle$ is

$$\hat{\mu} = \frac{\sum_i w_i x_i}{\sum_i w_i}, \quad (4.55)$$

where the weights w_i are

$$w_i = \frac{1}{\sigma_i^2}. \quad (4.56)$$

If and only if the measurement errors σ_i are correct, the variance of the estimate is

$$\sigma_{\hat{\mu}}^2 = \frac{1}{\sum_j w_j}. \tag{4.57}$$

If, however, the σ_i are only proportional to (or approximately proportional to) the true measurement errors, the variance of the estimate is

$$\hat{\sigma}_{\hat{\mu}}^2 = \frac{1}{n-1} \frac{\sum_i w_i (x_i - \hat{\mu})^2}{\sum_j w_j}. \tag{4.58}$$

The variance of unit weight is usually not needed explicitly in this context, but if it is, it can be estimated from

$$\hat{\sigma}^2 = \frac{1}{n-1} \sum_i w_i (x_i - \hat{\mu})^2. \tag{4.59}$$

Equations 4.55–4.59 are the standard equations found in any elementary book on statistics. Since they do not depend on the probability distributions for the x_i and σ_i, they are true for all probability distributions. Furthermore, we now have a justification for taking the weights to be the inverse of the variances—these are the weights that produce the most reliable value for $\hat{\mu}$ in the sense of minimizing its variance.

4.4 Data with Correlated Measurement Errors

We now generalize to data with correlated measurement errors. Suppose the data consists of n measurements x_i of a quantity x, and an $n \times n$ array of measurement errors σ_{ij}, where the σ_{ij} are the covariances between the errors in measuring x_i and x_j:

$$\sigma_{ij} = \langle (x_i - \mu)(x_j - \mu) \rangle. \tag{4.60}$$

We no longer assume that σ_{ij} is equal to 0, but do assume that $\sigma_{ji} = \sigma_{ij}$. Data with correlated errors are quite common, especially in sequences of data. An astronomer, for example, might want to improve the accuracy of a star's estimated brightness by averaging ten individual measurements of the brightness taken in rapid succession. Some of the error in the individual measurements is introduced by the atmosphere (thin cloud, twinkling) and by the experimental equipment (dust on a lens). These kind of errors change with time but often rather slowly. If, therefore, a measurement yields an incorrectly high brightness for a star, the previous and next measurements are also likely to give an incorrectly high brightness. Correlated errors can greatly increase the difficulty of extracting meaningful information from data. The estimated brightness of a star is not improved by taking the mean of ten measurements if a cloud is passing overhead during all ten measurements!

We wish to estimate μ. We proceed in much the same way as for data with uncorrelated errors (Section 4.3). The mean value must be linear in the measurements if it is to make any sense, so we require

$$\hat{\mu} = \sum_{i=1}^{n} a_i x_i, \tag{4.61}$$

where the a_i are constants. We also require that $\hat{\mu}$ be unbiased:

$$\langle\hat{\mu}\rangle = \mu = \left\langle\sum_{i=1}^{n} a_i x_i\right\rangle = \sum_{i=1}^{n} a_i \langle x_i\rangle = \mu\sum_{i=1}^{n} a_i, \qquad (4.62)$$

so the a_i must satisfy the constraint

$$\sum_{i=1}^{n} a_i = 1. \qquad (4.63)$$

As before, the best estimate of μ is the one that minimizes the variance of $\hat{\mu}$. The expression for the variance

$$\sigma_{\hat{\mu}}^2 = \langle(\hat{\mu}-\mu)^2\rangle$$

$$= \left\langle\left\{\sum_i a_i x_i - \mu\right\}^2\right\rangle = \left\langle\left\{\sum_i a_i x_i - \mu\sum_i a_i\right\}^2\right\rangle = \left\langle\left\{\sum_i a_i(x_i-\mu)\right\}^2\right\rangle$$

$$= \sum_i\sum_j a_i a_j\langle(x_i-\mu)(x_j-\mu)\rangle$$

$$= \sum_i\sum_j a_i a_j\sigma_{ij}, \qquad (4.64)$$

and this cannot be further simplified, because the covariances are nonzero. Our goal is to choose values of a_i that minimize

$$f = \sigma_{\hat{\mu}}^2 = \sum_i\sum_j a_i a_j\sigma_{ij} \qquad (4.65)$$

subject to the constraint

$$g = \sum_{i=1}^{n} a_i - 1 = 0. \qquad (4.66)$$

We solve this problem using the method of Lagrange multipliers. The minimum occurs where the gradient of f is parallel to the gradient of g:

$$\nabla f = \lambda\nabla g, \qquad (4.67)$$

or, in component form,

$$\frac{\partial f}{\partial a_i} = \lambda\frac{\partial g}{\partial a_i}. \qquad (4.68)$$

The partial derivatives are

$$\frac{\partial f}{\partial a_i} = 2\sum_j a_j\sigma_{ij}, \qquad \frac{\partial g}{\partial a_i} = 1, \qquad (4.69)$$

so equation 4.68 becomes

$$2\sum_j a_j\sigma_{ij} = \lambda. \qquad (4.70)$$

We recognize this as a matrix equation

$$2\mathbf{C}\mathbf{a} = \boldsymbol{\lambda}\,, \tag{4.71}$$

where $\mathbf{C}$, $\mathbf{a}$, and $\boldsymbol{\lambda}$ are

$$\mathbf{C} = \begin{pmatrix} \sigma_{11} & \sigma_{12} & \cdots & \sigma_{1n} \\ \sigma_{21} & \sigma_{22} & \cdots & \sigma_{2n} \\ \vdots & \vdots & & \vdots \\ \sigma_{n1} & \sigma_{n2} & \cdots & \sigma_{nn} \end{pmatrix}, \quad \mathbf{a} = \begin{pmatrix} a_1 \\ a_2 \\ \vdots \\ a_n \end{pmatrix}, \quad \boldsymbol{\lambda} = \begin{pmatrix} \lambda \\ \lambda \\ \vdots \\ \lambda \end{pmatrix}. \tag{4.72}$$

The matrix $\mathbf{C}$ is the covariance matrix of the observations. The solution to equation 4.71 is

$$\mathbf{a} = \frac{1}{2}\mathbf{C}^{-1}\boldsymbol{\lambda}\,. \tag{4.73}$$

The inverse of the covariance is called the *weight matrix*:

$$\mathbf{W} = \begin{pmatrix} w_{11} & w_{12} & \cdots & w_{1n} \\ w_{21} & w_{22} & \cdots & w_{2n} \\ \vdots & \vdots & & \vdots \\ w_{n1} & w_{n2} & \cdots & w_{nn} \end{pmatrix} = \mathbf{C}^{-1}. \tag{4.74}$$

Written in terms of the weight matrix, equation 4.73 becomes

$$\mathbf{a} = \frac{1}{2}\mathbf{W}\boldsymbol{\lambda}\,, \tag{4.75}$$

which, in component form, is

$$a_i = \frac{\lambda}{2}(w_{i1} + w_{i2} + \cdots + w_{in}) = \frac{\lambda}{2}\sum_j w_{ij}. \tag{4.76}$$

Plugging these expressions for the a_i back into the constraint equation, we find

$$1 = \sum_i a_i = \frac{\lambda}{2}\sum_i\sum_j w_{ij}, \tag{4.77}$$

which yields an expression for λ in terms of the weights:

$$\lambda = \frac{2}{\sum_i\sum_j w_{ij}}. \tag{4.78}$$

Using this expression to eliminate λ in equation 4.76, we arrive at the solution for the a_i:

$$a_i = \frac{\sum_j w_{ij}}{\sum_i\sum_j w_{ij}}. \tag{4.79}$$

Plugging these a_i back into the original equation for $\hat{\mu}$ (equation 4.61), we obtain the desired expression for the estimated mean:

$$\hat{\mu} = \sum_i a_i x_i = \frac{\sum_i\sum_j w_{ij}x_i}{\sum_i\sum_j w_{ij}}. \tag{4.80}$$

The variance of the estimated mean now becomes

$$\sigma_{\hat{\mu}}^2 = \sum_i \sum_j a_i a_j \sigma_{ij} = \left(\sum_\rho \sum_\sigma w_{\rho\sigma} \right)^{-2} \sum_i \sum_j \left(\sum_k w_{ik} \right) \left(\sum_\ell w_{j\ell} \right) \sigma_{ij}. \quad (4.81)$$

This impressive equation is easy to simplify. Rearranging the order of the summation, we have

$$\sigma_{\hat{\mu}}^2 = \left(\sum_\rho \sum_\sigma w_{\rho\sigma} \right)^{-2} \sum_i \sum_\ell \left(\sum_k w_{ik} \right) \left(\sum_j w_{j\ell} \sigma_{ij} \right). \quad (4.82)$$

Because σ_{ij} and $w_{j\ell}$ are symmetric, $\sum_j w_{j\ell} \sigma_{ij}$ is the product of the matrices $\mathbf{W}$ and $\mathbf{C}$ written in component form. Since they are inverses of each other, $\sum_j w_{j\ell} \sigma_{ij} = \delta_{i\ell}$, and the variance of the estimated mean becomes

$$\sigma_{\hat{\mu}}^2 = \left(\sum_\rho \sum_\sigma w_{\rho\sigma} \right)^{-2} \sum_i \sum_\ell \left(\sum_k w_{ik} \right) \delta_{i\ell}. \quad (4.83)$$

Once again reversing the order of the summation and dropping the unnecessary parentheses, we find the desired expression for the variance of the estimated mean:

$$\sigma_{\hat{\mu}}^2 = \left(\sum_\rho \sum_\sigma w_{\rho\sigma} \right)^{-2} \sum_k \sum_\ell \sum_i w_{ik} \delta_{i\ell} = \left(\sum_\rho \sum_\sigma w_{\rho\sigma} \right)^{-2} \sum_k \sum_\ell w_{\ell k}$$

$$= \left(\sum_\rho \sum_\sigma w_{\rho\sigma} \right)^{-1}. \quad (4.84)$$

Note that equations 4.80 and 4.84 do reduce to our earlier results for uncorrelated measurement errors, because, if $\sigma_{ij} = 0$, the covariance matrix becomes

$$\mathbf{C} = \begin{pmatrix} \sigma_{11} & 0 & \cdots & 0 \\ 0 & \sigma_{22} & \cdots & 0 \\ \vdots & \vdots & & \vdots \\ 0 & 0 & \cdots & \sigma_{nn} \end{pmatrix}, \quad (4.85)$$

and the weight matrix becomes

$$\mathbf{W} = \mathbf{C}^{-1} = \begin{pmatrix} 1/\sigma_{11} & 0 & \cdots & 0 \\ 0 & 1/\sigma_{22} & \cdots & 0 \\ \vdots & \vdots & & \vdots \\ 0 & 0 & \cdots & 1/\sigma_{nn} \end{pmatrix}. \quad (4.86)$$

In component form, the weights are

$$w_{ij} = \frac{1}{\sigma_{ij}} \delta_{ij} = \frac{1}{\sigma_i^2} \delta_{ij}, \quad (4.87)$$

so equation 4.80 becomes

$$\hat{\mu} = \frac{\sum_i \sum_j (1/\sigma_i^2)\delta_{ij} x_i}{\sum_i \sum_j (1/\sigma_i^2)\delta_{ij}} = \frac{\sum_i (1/\sigma_i^2) x_i}{\sum_i (1/\sigma_i^2)} = \frac{\sum_i w_i x_i}{\sum_i w_i}. \tag{4.88}$$

and, likewise, equation 4.84 reduces to $\sigma_{\hat{\mu}}^2 = 1/\sum_i w_i$.

At this stage one could argue that it would make sense to find an equation for $\hat{\sigma}_{\hat{\mu}}^2$, as we did for data with uncorrelated measurement errors. Instead, we will merge that discussion with the discussion of linear least squares estimation in Chapter 5. In that context, calculation of $\hat{\mu}$ can be thought of as a one-parameter linear least squares fit to data with correlated measurement errors. Then $\hat{\sigma}_{\hat{\mu}}^2$ is the sole component of the 1×1 estimated covariance matrix.

Summary: Given n measurements x_i of quantity x with correlated measurement errors σ_{ij}, the estimated mean of x is

$$\hat{\mu} = \frac{\sum_i \sum_j w_{ij} x_i}{\sum_i \sum_j w_{ij}}, \tag{4.89}$$

and the variance of the estimated mean is

$$\sigma_{\hat{\mu}}^2 = \left(\sum_\rho \sum_\sigma w_{\rho\sigma} \right)^{-1}. \tag{4.90}$$

The w_{ij} are the components of the weight matrix $\mathbf{W}$, which is the inverse of the covariance matrix for the measurement errors:

$$\mathbf{W} = \mathbf{C}^{-1} = \begin{pmatrix} \sigma_{11} & \sigma_{12} & \cdots & \sigma_{1n} \\ \sigma_{21} & \sigma_{22} & \cdots & \sigma_{2n} \\ \vdots & \vdots & & \vdots \\ \sigma_{n1} & \sigma_{n2} & \cdots & \sigma_{nn} \end{pmatrix}^{-1}. \tag{4.91}$$

Since we never specified the form of the probability distribution from which the x_i were sampled, these expressions, like the expressions for the estimated mean and variance derived in the previous two sections, are independent of the underlying distributions.

4.5 Variance of the Variance and Student's *t* Distribution

Suppose we have performed an experiment that yields n measured values x_i of a quantity x, from which we have derived estimates $\hat{\mu}$ and $\hat{\sigma}_x^2$ for the mean and standard deviation, respectively, of the unknown probability distribution f(x). We would like to know how much the measured values are likely to differ from the true values μ and σ_x^2. It may also be that we have performed the experiment several times, each experiment yielding a different value for $\hat{\mu}$ and $\hat{\sigma}_x^2$, and we want to know whether the differences among the values are significant.

In this section we first calculate the variance of $\hat{\sigma}_x^2$—the variance of the variance—which we denote by $\sigma_{\hat{\sigma}_x^2}^2$, and show that measured variances are highly uncertain. We will also find that $\hat{\sigma}_x^2$ has a χ^2 distribution with $n-1$ degrees of freedom. If the original distribution f(x) is known to have a Gaussian distribution—or if the central limit theorem is acting so that the difference between the $\hat{\mu}$ and μ has a Gaussian distribution—it is possible to derive an analytic expression for the distribution of the difference in terms of the measured value $\hat{\sigma}^2_{\hat{\mu}}$ instead of the unknown value σ^2_x. This distribution is called Student's t distribution. In the second part of this section we derive Student's t distribution.

4.5.1 Variance of the Variance

We begin the calculation of the variance of the variance by looking more carefully at $\hat{S}$. At first glance the quantity

$$\frac{1}{\sigma^2_x}\hat{S} = \sum_{i=1}^{n} \frac{(x_i - \hat{\mu})^2}{\sigma^2_x} \tag{4.92}$$

looks like a χ^2_n variable and should follow a χ^2_n distribution, because it is a summation over n residuals. However, the individual components of the sum are not independent, because $\hat{\mu}$ depends on the x_i. To be a χ^2 variable, the components of the sum must be independent. To overcome this problem, we transform the quantities $(x_i - \hat{\mu})$ to a different set of variables y_i that are independent of one another.

We require that the transformation from x_i to y_i be a real linear transformation

$$y_i = \sum_{j=1}^{n} M_{ij} x_j \tag{4.93}$$

that preserves lengths:

$$\sum_{i=1}^{n} y_i^2 = \sum_{i=1}^{n} x_i^2. \tag{4.94}$$

In the language of linear algebra, M_{ij} is an *orthogonal* matrix. The inverse of an orthogonal matrix is equal to its transpose (see the discussion following equation E.137 in Appendix E). Thus,

$$\sum_{k=1}^{n} M_{ik} M_{jk} = \delta_{ij}. \tag{4.95}$$

We specify that

$$y_1 = \frac{1}{\sqrt{n}}(x_1 + x_2 + \cdots + x_n) = \frac{1}{\sqrt{n}} \sum_{i=1}^{n} x_1 = \sqrt{n}\hat{\mu}, \tag{4.96}$$

which implies

$$M_{1j} = \frac{1}{\sqrt{n}}; \tag{4.97}$$

and we also specify that the means of the remaining y_i be equal to zero:

$$\langle y_i \rangle = 0 = \left\langle \sum_{j=1}^{n} M_{ij} x_j \right\rangle = \sum_{j=1}^{n} M_{ij} \langle x_j \rangle = \mu \sum_{j=1}^{n} M_{ij}, \quad i \geq 2, \tag{4.98}$$

which implies

$$\sum_{j=1}^{n} M_{ij} = 0, \quad i \geq 2. \tag{4.99}$$

For $n > 2$ it is always possible to find a linear transformation that satisfies equations 4.94–4.99. For our purposes, the transformation does not need to be specified more completely.

We also need the variances of the y_i. We first calculate $\sigma_{y_i}^2$ for $i \geq 2$ and then calculate $\sigma_{y_1}^2$ separately. Beginning with the definition of $\sigma_{y_i}^2$, we have for $i \geq 2$

$$
\begin{aligned}
\sigma_{y_i}^2 = \langle (y_i - \langle y_i \rangle)^2 \rangle = \langle y_i^2 \rangle &= \left\langle \left(\sum_{j=1}^{n} M_{ij} x_j \right)^2 \right\rangle \\
&= \left\langle \left[\sum_{j=1}^{n} M_{ij}(x_j - \mu) + \mu \sum_{j=1}^{n} M_{ij} \right]^2 \right\rangle \\
&= \mu^2 \sum_{j=1}^{n} \sum_{k=1}^{n} M_{ij} M_{ik} + \sum_{j=1}^{n} \sum_{k=1}^{n} M_{ij} M_{ik} \langle (x_j - \mu)(x_k - \mu) \rangle \\
&= \mu^2 \left(\sum_{j=1}^{n} M_{ij} \right)^2 + \sigma_x^2 \sum_{j=1}^{n} M_{ij} M_{ij} \\
&= \sigma_x^2, \quad i \geq 2,
\end{aligned}
\tag{4.100}
$$

where the fourth line was derived using equation 4.15, and the final line using equations 4.99 and 4.95. For $i = 1$, we have (leaving out some steps)

$$
\begin{aligned}
\sigma_{y_1}^2 = \langle (y_1 - \sqrt{n}\mu)^2 \rangle &= \left\langle \left[\frac{1}{\sqrt{n}} \sum_{i=1}^{n} x_i - \sqrt{n}\mu \right]^2 \right\rangle = \left\langle \left[\frac{1}{\sqrt{n}} \sum_{i=1}^{n} (x_i - \mu) \right]^2 \right\rangle \\
&= \frac{1}{n} \sum_{i=1}^{n} \sum_{j=1}^{n} \langle (x_i - \mu)(x_j - \mu) \rangle = \frac{1}{n} \sum_{i=1}^{n} \sum_{j=1}^{n} \sigma_x^2 \delta_{ij} \\
&= \sigma_x^2.
\end{aligned}
\tag{4.101}
$$

Since $\sigma_{y_i}^2 = \sigma_x^2$ for all i, we can drop the subscript and simply set $\sigma_y^2 = \sigma_x^2$.

Now convert $\hat{S}$ from a function of the x_i to a function of the y_i. Expanding $\hat{S}$ and using equations 4.94 and 4.96, we find

$$
\hat{S} = \sum_{i=1}^{n} (x_i - \hat{\mu})^2 = \sum_{i=1}^{n} x_i^2 - n\hat{\mu}^2 = \sum_{i=1}^{n} y_i^2 - y_1^2 = \sum_{i=2}^{n} y_i^2.
\tag{4.102}
$$

Note that the sum now begins at $i = 2$, not $i = 1$, and that $\hat{\mu}$ has disappeared from the sum; so $\hat{S}$ is now the sum of $n - 1$ independent components instead of n dependent components. Finally, dividing equation 4.102 by σ_x^2, we have

$$
\frac{1}{\sigma_x^2} \hat{S} = \frac{1}{\sigma_x^2} \sum_{i=2}^{n} y_i^2 = \sum_{i=2}^{n} \frac{y_i^2}{\sigma_y^2} = \chi_{n-1}^2.
\tag{4.103}
$$

Since the y_i are independent, we now see that $\hat{S}/\sigma_x^2$ is a χ^2 variable with $n-1$ degrees of freedom, not n degrees of freedom.

From equation 4.20 we have

$$\hat{\sigma}_x^2 = \frac{1}{n-1}\hat{S} = \frac{\sigma_x^2}{n-1}\chi_{n-1}^2. \tag{4.104}$$

The variance of σ_x^2 can now be calculated from the variance of the χ^2 distribution (see equation 2.114):

$$\sigma_{\hat{\sigma}_x^2}^2 = \left(\frac{\sigma_x^2}{n-1}\right)^2 \sigma_{\chi_{n-1}^2}^2 = \left(\frac{\sigma_x^2}{n-1}\right)^2 2(n-1) = \frac{2}{n-1}(\sigma_x^2)^2, \tag{4.105}$$

and since $\hat{\sigma}_{\hat{\mu}}^2 = \hat{\sigma}_x^2/n$ (equation 4.22), we also have

$$\sigma_{\hat{\sigma}_{\hat{\mu}}^2}^2 = \frac{2}{n^2(n-1)}(\sigma_x^2)^2. \tag{4.106}$$

To get a feel for the size of $\sigma_{\hat{\sigma}_{\hat{\mu}}^2}^2$ we write the measured variance of $\hat{\mu}$ and its uncertainty as $\hat{\sigma}_{\hat{\mu}}^2 \pm \sigma_{\hat{\sigma}_{\hat{\mu}}^2}$. Since $\hat{\sigma}_{\hat{\mu}}^2 \approx \sigma_x^2/n$, we have

$$\hat{\sigma}_{\hat{\mu}}^2 \pm \sigma_{\hat{\sigma}_{\hat{\mu}}^2} \approx \frac{1}{n}\sigma_x^2 \pm \left(\frac{2}{n-1}\right)^{1/2}\frac{\sigma_x^2}{n} = \frac{1}{n}\sigma_x^2\left[1 \pm \left(\frac{2}{n-1}\right)^{1/2}\right]. \tag{4.107}$$

For 100 samples, the measured variance of $\hat{\mu}$ is uncertain by a fractional amount $(2/(100-1))^{1/2}$ or 14%. Thus $\hat{\sigma}_{\hat{\mu}}^2$ has a large standard deviation unless n is large. It is not a well determined quantity.

4.5.2 Student's t Distribution

Equations 4.105 and 4.106 depend on σ_x^2, which is generally not known. The equations retain utility if one is willing to assume that $\sigma_x^2 \approx \hat{\sigma}_x^2$, which we did when writing down equation 4.107; but a better way to deal with this and other properties of $\hat{\mu}$ is to derive them from Student's t distribution. Student's t distribution is the probability distribution function for the quantity

$$t = \frac{\hat{\mu} - \mu}{\hat{\sigma}_{\hat{\mu}}}. \tag{4.108}$$

The reason for using t is that the difference $\hat{\mu} - \mu$ is compared to $\hat{\sigma}_{\hat{\mu}}$, which is a measured quantity, not to σ_x or $\sigma_{\hat{\mu}}$, which are unknown quantities.

It is inconvenient to derive Student's t distribution using expression 4.108 for t, because $\hat{\mu}$ and $\hat{\sigma}_{\hat{\mu}}$ are not independent of each other. We first write t in terms of independent variables. Using equation 4.104 and remembering that $\hat{\sigma}_{\hat{\mu}}^2 = \hat{\sigma}_x^2/n$, we have

$$t = \frac{\hat{\mu} - \mu}{\hat{\sigma}_x/\sqrt{n}} = \frac{\hat{\mu} - \mu}{\sigma_x/\sqrt{n}} \left/ \left(\frac{\hat{\sigma}_x^2}{\sigma_x^2}\right)^{1/2}\right. = \frac{\hat{\mu} - \mu}{\sigma_x/\sqrt{n}} \left/ \left(\frac{\chi_{n-1}^2}{n-1}\right)^{1/2}\right. . \tag{4.109}$$

Remembering that χ^2_{n-1} was specifically constructed to be independent of $\hat{\mu}$ (equation 4.103), we can define two independent variables z and u to be

$$z = \frac{\hat{\mu} - \mu}{\sigma_x/\sqrt{n}}, \quad u = \frac{\chi^2_{n-1}}{n-1}, \tag{4.110}$$

so that

$$t = z/u^{1/2}. \tag{4.111}$$

Since z and u are independent variables, equation 4.111 is the desired expression for t. Also, because they are independent variables, their joint probability distribution $f_0(z, u)$ can be written as the product of their individual probability distributions:

$$f_0(z, u) = g(z)c_{n-1}(u). \tag{4.112}$$

Relying on the central limit theorem, we assume that z has a Gaussian distribution with mean equal to 0 and variance equal to 1:

$$g(z) = \frac{1}{\sqrt{2\pi}} \exp\left[-z^2/2\right]. \tag{4.113}$$

From equation 4.103, we know that $(n-1)u$ has a χ^2_{n-1} distribution. If $f_n(\chi^2)$ is the χ^2 distribution for n degrees of freedom (see equation 2.101), the distribution for u is

$$c_{n-1}(u)du = f_{n-1}\left(\chi^2(u)\right)\left|\frac{\partial \chi^2}{\partial u}\right| du = (n-1)f_{n-1}\left(\chi^2(u)\right) du \tag{4.114}$$

or

$$c_{n-1}(u) = (n-1)\frac{[(n-1)u]^{(n-3)/2}}{2^{(n-1)/2}[(n-3)/2]!} \exp\left[-(n-1)u/2\right]. \tag{4.115}$$

We now perform a second change of variables from (z, u) to (t, u), where

$$u = u \tag{4.116}$$

$$z = tu^{1/2}, \tag{4.117}$$

returning to $t = z/u^{1/2}$. The joint distribution function for t and u is given by

$$f_1(t, u)dtdu = g(z(t, u))c_{n-1}(u)\left|\frac{\partial(z, u)}{\partial(t, u)}\right| dtdu. \tag{4.118}$$

The Jacobian is

$$\begin{vmatrix} \dfrac{\partial z}{\partial t} & \dfrac{\partial u}{\partial t} \\[2mm] \dfrac{\partial z}{\partial u} & \dfrac{\partial u}{\partial u} \end{vmatrix} = \begin{vmatrix} u^{1/2} & 0 \\[2mm] \frac{1}{2}tu^{-1/2} & 1 \end{vmatrix} = u^{1/2}, \tag{4.119}$$

so we have

$$f_1(t, u) = \frac{1}{\sqrt{2\pi}} \exp\left[-ut^2/2\right](n-1)\frac{[(n-1)u]^{(n-3)/2}}{2^{(n-1)/2}[(n-3)/2]!} \exp\left[-(n-1)u/2\right]u^{1/2}. \tag{4.120}$$

We now find the distribution function for t alone by marginalizing over u:

$$s_{n-1}(t) = \int_0^\infty f_1(t, u)du. \tag{4.121}$$

Although the integral looks threatening, it can be converted to the standard form for the gamma function (see Section A.3 in Appendix A). After some algebra, we obtain

$$s_{n-1}(t) = \frac{[(n-2)/2]!}{[(n-3)/2]!\sqrt{(n-1)\pi}}\left[1 + \frac{t^2}{n-1}\right]^{-n/2} \tag{4.122}$$

with t given explicitly by equation 4.108 or 4.111, depending on the problem at hand. This is called Student's t distribution for $n-1$ degrees of freedom. Student's t distribution for n degrees of freedom is

$$s_n(t) = \frac{[(n-1)/2]!}{[(n-2)/2]!\sqrt{n\pi}}\left[1 + \frac{t^2}{n}\right]^{-(n+1)/2}. \tag{4.123}$$

For small n, Student's t distribution is roughly like a Gaussian distribution but with higher probability density in the wings. The distribution rapidly approaches a Gaussian distribution as n becomes large and is almost indistinguishable from a Gaussian for $n > 30$ (see equation F.9 in Appendix F). Since t appears only as t^2, Student's t distribution is symmetric about $t = 0$. Therefore all of its odd moments are equal to zero. In particular,

$$\langle t \rangle = M_1 = 0. \tag{4.124}$$

We state without proof that the even moments of Student's t distribution are given by

$$M_k = n^{k/2}\frac{\Gamma(\frac{k+1}{2})\Gamma(\frac{n-k}{2})}{\Gamma(\frac{n}{2})\Gamma(\frac{1}{2})}, \tag{4.125}$$

but the moments exist only for $k < n$. The variance of t is

$$\sigma_t^2 = \langle (t - \langle t \rangle)^2 \rangle = \langle t^2 \rangle = M_2$$

$$= \frac{n}{n-2}. \tag{4.126}$$

Referring back to the definition of t in equation 4.108, we see that $\sigma_t^2 \approx \sigma_{\hat{\mu}}^2/\hat{\sigma}_{\hat{\mu}}^2$, so on average $\sigma_{\hat{\mu}}^2$ is too small by a fraction $(n-2)/n$. This bias is the result of calculating the estimated variance with respect to the estimated value $\hat{\mu}$ instead of the true value μ. For $n > 30$, this bias becomes negligible and one may, for example, safely use equations 4.105 and 4.106, substituting $\hat{\sigma}_x^2$ and $\hat{\sigma}_{\hat{\mu}}^2$ for σ_x^2 and $\sigma_{\hat{\mu}}^2$.

4.5.3 Summary

The main purpose of this section was to derive Student's t distribution, equation 4.123, where t is given by equation 4.108. This is the distribution of estimated mean values $\hat{\mu}$ as a function of the observed quantity $\hat{\sigma}_{\hat{\mu}}^2$ instead of the unknown quantities $\sigma_{\hat{\mu}}^2$ or σ_x^2. Student's t distribution is almost indistinguishable from a Gaussian distribution for $n > 30$ but has greater probability in its wings for smaller n.

We also calculated the variances of the estimated variances $\sigma^2_{\hat{\sigma}^2_x}$ and $\sigma^2_{\hat{\sigma}^2_\mu}$,

$$\sigma^2_{\hat{\sigma}^2_x} = \frac{2}{n-1}(\sigma^2_x)^2 \tag{4.127}$$

$$\sigma^2_{\hat{\sigma}^2_\mu} = \frac{2}{n^2(n-1)}(\sigma^2_x)^2, \tag{4.128}$$

showing that variances are poorly determined quantities. Although σ^2_x is unknown, these expressions can still be used if n is large, substituting $\hat{\sigma}^2_x$ for σ^2_x without introducing a large error. If, however, n is not large, Student's t distribution suggests that one should take

$$\sigma^2_x \approx \frac{n}{n-2}\hat{\sigma}^2_x, \tag{4.129}$$

increasing the variance of the variance even more. Alternatively one can calculate the variances of the variances directly from Student's t distribution using standard, albeit laborious, techniques.

4.6 Principal Component Analysis

4.6.1 Correlation Coefficient

Figure 4.2 shows three sets of data. The x and y values of the data points are related in the left and right panels of the figure: once x is specified, y is narrowly constrained. The x and y values of the data points in the center panel are not related: the value of x does not constrain y. The covariance is one way to quantify these relations. Suppose we have n equally weighted data points (x_i, y_i), and that μ_x and μ_y are the mean values of x and y, respectively. The covariance between the x_i and y_i is

$$\sigma_{xy} = \frac{1}{n}\sum_{i=1}^{n}(x_i - \mu_x)(y_i - \mu_y). \tag{4.130}$$

For data like that shown in the left panel of Figure 4.2, $(x_i - \mu_x)$ and $(y_i - \mu_y)$ are generally both positive or both negative, so the product is generally positive, and the covariance is a large positive number. For data like that in the right panel of the figure, $(x_i - \mu_x)$ and $(y_i - \mu_y)$ generally have opposite signs, so their product is generally negative, and the covariance

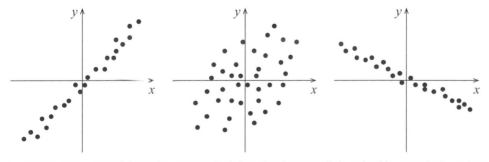

Figure 4.2: Three sets of data. The sets on the left and right are well described by straight lines, but the data set in the center is not. The least squares method will happily find a best-fitting straight line to all three sets.

is a large negative number. For data like that in the middle panel, $(x_i - \mu_x)$ and $(y_i - \mu_y)$ sometimes have the same signs and sometimes opposite signs, so their products tend to cancel each other in the sum, and the covariance is a small number.

The correlation coefficient is defined to be

$$r = \frac{\sigma_{xy}}{\sigma_x \sigma_y}, \tag{4.131}$$

where

$$\sigma_x^2 = \frac{1}{n} \sum_i (x_i - \mu_x)^2 \tag{4.132}$$

$$\sigma_y^2 = \frac{1}{n} \sum_i (y_i - \mu_y)^2. \tag{4.133}$$

Dividing the covariance by the two variances σ_x and σ_y makes the correlation coefficient essentially a scale-free covariance. If the data are uncorrelated, $\sigma_{xy} = 0$ and $r = 0$. For perfect correlation, $y_i = \pm a x_i$ and $\mu_y = \pm a \mu_x$, and the correlation coefficient is ± 1. The negative sign holds for distributions like that in the left panel of Figure 4.2, while the positive sign holds for distributions like that in the right panel. Data for which r is greater than 0 are said to be correlated; while data for which r is less than 0 are said to be anticorrelated.

4.6.2 Principal Component Analysis

Correlations can be powerful tools for understanding data. To be sure, correlation is not causation, and an observed correlation might be nothing but a meaningless random accident. The striking correlation between a country's chocolate consumption per capita and the number of its Nobel laureates per 10^7 population is a likely example.[2] But correlations so often reveal important interrelations in data that the search for correlations is a preoccupation in all fields of science. Correlations can, however, be difficult to find in multidimensional data. *Principal component analysis* (PCA) provides a way to find hidden correlations. We will first show how PCA works for 2-dimensional data and then generalize to n-dimensional data.

Suppose that an experiment produces n pairs of data points (x_{1i}, x_{2i}) with a strong correlation, as shown in Figure 4.3. The true mean values of x_1 and x_2 are unknown, so we must use the estimated means $\hat{\mu}_1 = (\sum_i x_{1i})/n$ and $\hat{\mu}_2 = (\sum_i x_{2i})/n$, and the estimated variances and covariances

$$\hat{\sigma}_1^2 = \frac{1}{n-1} \sum_{i=1}^n (x_{1i} - \hat{\mu}_1)^2 \tag{4.134}$$

$$\hat{\sigma}_2^2 = \frac{1}{n-1} \sum_{i=1}^n (x_{2i} - \hat{\mu}_2)^2 \tag{4.135}$$

$$\hat{\sigma}_{12} = \hat{\sigma}_{21} = \frac{1}{n-1} \sum_{i=1}^n (x_{1i} - \hat{\mu}_1)(x_{2i} - \hat{\mu}_2). \tag{4.136}$$

---

[2] F. H. Messerli. 2012. "Chocolate Consumption, Cognitive Function, and Nobel Laureates." *New England Journal of Medicine* vol. 367, p. 1562.

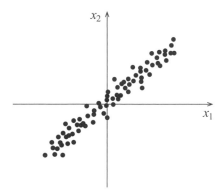

Figure 4.3: A set of data points plotted in the (x_1, x_2) coordinate system. The data points are highly correlated.

These are combined into the covariance matrix of the observations

$$\mathbf{C} = \begin{pmatrix} \hat{\sigma}_1^2 & \hat{\sigma}_{12} \\ \hat{\sigma}_{21} & \hat{\sigma}_2^2 \end{pmatrix}. \tag{4.137}$$

Since the covariances are the off-diagonal components of the matrix, the data are correlated if and only if some off-diagonal components are nonzero.

We now search for a length-preserving coordinate transformation that diagonalizes $\mathbf{C}$. In effect, we rotate into a coordinate system in which all the off-diagonal elements of the matrix are equal to zero. When the off-diagonal components are all zero, there are no correlations, which means that the long axis of the cloud of data points is aligned with one of the rotated coordinate axes. Diagonalize the matrix by finding its eigenvectors. The normalized eigenvectors are the basis vectors of the coordinate system into which the matrix is transformed and the eigenvalues are the diagonal elements. (The reader might find the review of matrix algebra in Sections E.9–E.11 of Appendix E to be useful at this point.) The normalized eigenvector corresponding to the largest eigenvalue is called the first principal component, that corresponding to the second largest eigenvalue is the second principle component, and so on.

Figure 4.4 shows the geometrical meaning of the diagonalization. The left panel of the figure shows the data points in the original (x_1, x_2) coordinate system. The dashed arrows are aligned with the eigenvectors of $\mathbf{C}$ and are the axes of the rotated coordinate system. The first principal component lies on the long axis of the cloud of data points. It is the desired correlation between x_1 and x_2. Call the new coordinates (z_1, z_2), where the z_1 is in the direction of the first principle component, and z_2 is perpendicular to z_1; and call the respective eigenvalues λ_1 and λ_2. The right panel of Figure 4.4 shows the data points replotted in the (z_1, z_2) coordinate system. Most of the scatter is now confined to the z_1 coordinate, so the variance in z_1 is larger than that in z_2. The eigenvalues are the variances along their respective principle components, $\sigma_{\hat{z}_1}^2 = \lambda_1$ and $\sigma_{\hat{z}_2}^2 = \lambda_2$. The strength of the correlation in the z_1 direction is measured by

$$\frac{\sigma_{\hat{z}_1}^2}{\sigma_{\hat{z}_1}^2 + \sigma_{\hat{z}_2}^2} = \frac{\lambda_1}{\lambda_1 + \lambda_2}. \tag{4.138}$$

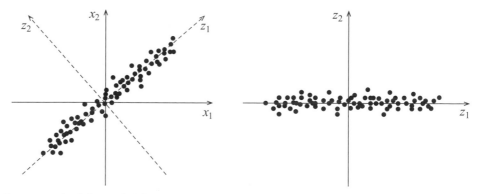

Figure 4.4: The left panel defines a new set of coordinates (z_1, z_2) such that the z_1-axis lies on the long axis of the distribution of data points and z_2 is perpendicular to z_1. Unit vectors along the z_1- and z_2-coordinate axes are the first and second principal components of the distribution. The right panel shows the data points plotted in the (z_1, z_2) coordinate system. The new plot has two important characteristics. First, the data points are no longer correlated. Second, just one coordinate, z_1, gives most of the information about the location of the data point; the z_2-coordinate gives little extra information.

We now generalize to data with m independent parameters that are measured n times to produce the n data points

$$(\mathbf{x}_1)^{\mathbf{T}} = (x_{11}, x_{21}, \ldots, x_{m1})$$
$$(\mathbf{x}_2)^{\mathbf{T}} = (x_{12}, x_{22}, \ldots, x_{m2})$$
$$\vdots \qquad \vdots$$
$$(\mathbf{x}_n)^{\mathbf{T}} = (x_{1n}, x_{2n}, \ldots, x_{mn}).$$

The $\mathbf{x}_i$ are m-dimensional column vectors but have been written in their transposed form to save space. The vector of means is defined to be

$$\hat{\boldsymbol{\mu}}^{\mathbf{T}} = (\hat{\mu}_1, \hat{\mu}_2, \ldots, \hat{\mu}_m), \tag{4.139}$$

where $\hat{\mu}_j$ is the mean value of parameter j, calculated from

$$\hat{\mu}_j = \frac{1}{n} \sum_{i=1}^{n} x_{ji}. \tag{4.140}$$

The vector of means is the vector to the center of the cloud of data points. Subtract the mean vector from the data points to give the mean-subtracted data vectors

$$\mathbf{y}_i = \mathbf{x}_i - \hat{\boldsymbol{\mu}}, \tag{4.141}$$

and then calculate the components the $m \times m$ covariance matrix $\hat{\mathbf{C}}$ from

$$\hat{C}_{ij} = \hat{\sigma}_{ij} = \frac{1}{n-1} \sum_{k=1}^{n} (x_{ik} - \hat{\mu}_i)(x_{jk} - \hat{\mu}_j) = \frac{1}{n-1} \sum_{k=1}^{n} y_{ik} y_{jk}. \tag{4.142}$$

If any off-diagonal elements of $\hat{\mathbf{C}}$ are significantly greater than 0, the cloud of data points is elongated in at least one direction. We wish to find a transformation that rotates the coordinate system so that all the elongated axes of the cloud lie along the axes of an orthogonal coordinate system. After the rotation, all covariances will be 0, so the covariance matrix will be diagonal in the rotated coordinate system. The transformation must preserve the shape of the data cloud, which means the transformation must preserve distances. Succinctly, we wish to find an orthogonal transformation that diagonalizes $\hat{\mathbf{C}}$.

The covariance matrix is diagonalized by finding its eigenvalues and eigenvectors. Let $\mathbf{v}$ be a vector for which

$$\hat{\mathbf{C}}\mathbf{v} = \lambda\mathbf{v}, \tag{4.143}$$

where λ is a scalar constant. The m vectors $\mathbf{v}_j$ and corresponding scalars λ_j that satisfy equation 4.143 are the eigenvectors and eigenvalues of matrix $\hat{\mathbf{C}}$. Eigenvectors have arbitrary length and need to be normalized. Denote the normalized eigenvector corresponding to eigenvalue λ_j by $\mathbf{e}_j = \mathbf{v}_j/|\mathbf{v}_j|$. The $\mathbf{e}_j$ together make up a complete set of orthonormal basis vectors. The matrix that diagonalizes $\hat{\mathbf{C}}$ is given by setting the rows of $\mathbf{U}$ equal to the transposed basis vectors (see Section E.11 in Appendix E)

$$\mathbf{U} = \begin{bmatrix} \mathbf{e}_1^{\mathrm{T}} \\ \mathbf{e}_2^{\mathrm{T}} \\ \vdots \\ \mathbf{e}_m^{\mathrm{T}} \end{bmatrix}. \tag{4.144}$$

The diagonal matrix corresponding to $\hat{\mathbf{C}}$ is calculated from the similarity transformation

$$\hat{\mathbf{C}}' = \mathbf{U}\mathbf{C}\mathbf{U}^{-1}, \tag{4.145}$$

and the diagonal elements of $\hat{\mathbf{C}}'$ are the eigenvalues

$$(\hat{\mathbf{C}}')_{jj} = \lambda_j. \tag{4.146}$$

We are now equipped to write down an algorithm for PCA:

1. Collect the data $\mathbf{x}_i^{\mathrm{T}} = (x_{1i}, x_{2i}, \ldots, x_{mi})$, $i = 1, \ldots, n$, where m is the number of parameters measured in each experiment, and n is the number of experiments,
2. Calculate the vector of mean values $\hat{\boldsymbol{\mu}}$ using equation 4.140 and the mean-subtracted data vectors $\mathbf{y}_i$ using equation 4.141.
3. Calculate the covariance matrix $\hat{\mathbf{C}}$ from equation 4.142.
4. Find the eigenvalues and eigenvectors of $\hat{\mathbf{C}}$. This is generally done using a "canned" routine from a package of routines for matrix operations. Normalize the eigenvectors to give the orthonormal basis vectors $\mathbf{e}_j$. The orthonormal basis vectors are the principle components! The principle component with the largest eigenvalue is the first principle component, the one with the second largest eigenvalue is the second principle component, and so on.
5. Arrange the eigenvalues and their corresponding basis vectors in order of decreasing size of the eigenvectors, so that $\lambda_1 \geq \lambda_2 \geq \cdots \geq \lambda_m$. Form the unitary transformation matrix $\mathbf{U}$ from equation 4.144.

6. Transform the covariance matrix and the data vectors to the rotated coordinate system using

$$\hat{\mathbf{C}}' = \mathbf{U}\hat{\mathbf{C}}\mathbf{U}^{-1} \tag{4.147}$$

$$\mathbf{y}'_i = \mathbf{U}\mathbf{y}_i. \tag{4.148}$$

The diagonal elements of the matrix $\mathbf{C}'$ are the λ_j and its off-diagonal elements are all 0. The $\hat{\mathbf{C}}'$ matrix is the covariance matrix in the rotated coordinate system, so the λ_j are the variances $(\sigma'_j)^2$ along the axes defined by the $\mathbf{e}_j$. The distance of data point $\mathbf{y}'_i$ from the origin in the new coordinate system is given by the square root of

$$d_i^2 = \sum_{j=1}^{m} (\mathbf{y}'_i)_j^2, \tag{4.149}$$

so the mean distance of all the points from the origin is given by

$$\langle d^2 \rangle = \sum_{j=1}^{m} \left\langle (\mathbf{y}'_i)_j^2 \right\rangle = \sum_{j=1}^{m} (\sigma'_j)^2 = \sum_{j=1}^{m} \lambda_j. \tag{4.150}$$

The fraction of the total variance due to each of the various eigenvectors is given by

$$\frac{\lambda_i}{\sum_j \lambda_j}. \tag{4.151}$$

The projection of a mean-subtracted data point i onto an eigenvector j is the dot product

$$a_{ij} = \mathbf{y}_i \cdot \mathbf{e}_j. \tag{4.152}$$

If desired, $\mathbf{y}_i$ can be decomposed into the sum of its projections onto the eigenvectors,

$$\mathbf{y}_i = a_{i1}\mathbf{e}_1 + a_{i2}\mathbf{e}_2 + \cdots + a_{im}\mathbf{e}_m = \sum_{j=1}^{m} a_{ij}\mathbf{e}_j, \tag{4.153}$$

and the original data points can be reconstructed by adding back the mean vector

$$\mathbf{x}_i = \hat{\boldsymbol{\mu}} + \sum_{j=1}^{m} a_{ij}\mathbf{e}_j. \tag{4.154}$$

If some of the λ_i are much smaller than the others, one can dispense with the corresponding eigenvectors and reconstruct most of the properties of the original data with fewer independent parameters.

Any symmetric square matrix can be diagonalized and has real eigenvalues, so the formalism of PCA always works. If, however, the data are not well described as a multidimensional ellipsoidal cloud, the results will be biased, meaningless, or even misleading. The data sets shown in Figure 4.5, for example, have strong patterns but are not well described by a covariance matrix and are not amenable to PCA. Nonlinear patterns in data are so common that one must always be on guard against them when applying PCA. One way to do this is to examine the distribution of the $\mathbf{y}'_i$ on various pairs of axes.

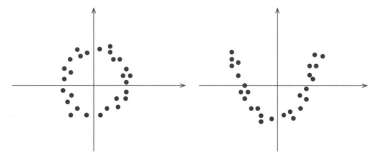

Figure 4.5: Principal component analysis (PCA) will always succeed in the mathematical sense that it will always return a diagonalized matrix. If, however, the properties of the data are not well represented by the correlation matrix, the result will be an incomplete or even misleading characterization of the data. These two sets of data have strong patterns but are poor candidates for PCA.

4.7 Kolmogorov-Smirnov Test

The Kolmogorov-Smirnov[3] (K-S) test is widely used to to decide whether an observed distribution is consistent with a theoretical probability distribution (the one-sample K-S test) or with another observed distribution (the two-sample K-S test).

4.7.1 One-Sample K-S Test

Suppose that we have n occurrences $A_i(x_i)$ that are functions of an independent variable x, and we wish to know whether the A_i are consistent with a theoretical probability distribution $f(x)$. The K-S test compares the cumulative distribution of the A_i to the cumulative distribution of $f(x)$:

$$F(x) = \int_{-\infty}^{x} f(y)dy. \tag{4.155}$$

If the A_i are, for example, a list of stars along with their apparent brightness, we might wish to know whether the number of stars at each brightness is the same as some theoretical distribution of their brightness.

To make the comparison, first place the $A_i(x_i)$ in order of increasing x. For the case of the stars, rearrange the list so the stars are in order of increasing (or decreasing) brightness. Next, find the cumulative probability distribution $S_n(x)$ for the reordered $A_i(x_i)$. $S_n(x)$ is a step function rising from 0 at $x < x_1$ to 1 at $x > x_n$ in n steps. The individual steps are located at the x_i where the the A_i are located (see Figure 4.6). This can be represented by

$$S_n(x) = \frac{k(x)}{n}, \tag{4.156}$$

where $k(x)$ is the number of the x_i that are less than or equal to x. To compare $F(x)$ to $S_n(x)$, plot the two together on the same graph (as in Figure 4.6), and then find the maximum vertical distance between the two functions:

$$D_n = \max |S_n(x) - F(x)|. \tag{4.157}$$

---

[3] See R. von Mises. 1964. *Mathematical Theory of Probability and Statistics*, second edition. Waltham, MA: Academic Press, chapter. IX.E.

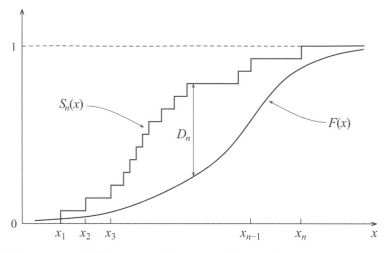

Figure 4.6: The one-sample Kolmogorov-Smirnov test. The step function $S_n(x)$ is the cumulative distribution function for the n data points $A_i(x_i)$. It rises from 0 at $x < x_1$ to 1 at $x \geq x_n$ in n steps, the steps located at the x_i where the A_i are located. $F(x)$ is the cumulative distribution function to which $S_n(x)$ is being compared. D_n is the maximum vertical difference between the two.

Note that D_n uses the absolute value, so it is the maximum distance of $S_n(x)$ above or below $F(x)$, leading to the two-tailed K-S test. A pleasant feature of D_n is that it does not change for arbitrary stretching or contraction of x.

Since $S_n(x)$ is constructed from observational data, D_n will vary from experiment to experiment, and its distribution will be described by a probability distribution. Calculation of the distribution is beyond the scope of this course, but the result is easily understood. Let β be the probability that the difference is greater than ϵ:

$$\beta = P(D_n > \epsilon). \tag{4.158}$$

For n greater than about 50, one may use the approximation

$$\beta \approx 2 \exp[-2n\epsilon^2], \tag{4.159}$$

but for smaller n one must refer to tables. Table 4.1 gives ϵ as a function of β for several values of n between 5 and 50.

To apply the K-S test, first measure D_n. Set $\epsilon = D_n$ in equation 4.158. Then β is the probability that a value greater than D_n will happen by chance. As β becomes smaller, the observed data become less consistent with the theoretical distribution. If n is large enough that equation 4.159 can be used, the calculation is transparent. If n is small and it is necessary to use a table like Table 4.1, choose the line corresponding most closely to the observed value of n. Read across the table to find the value of ϵ closest to D_n. The column head gives the value of β.

As implied by the values in Table 4.1, one usually needs large n and large D_n to convincingly reject the hypothesis that the observations were drawn from a particular theoretical distribution. In addition, the K-S test is invalid if any property of the $A_i(x_i)$ has been adjusted to agree with that of the theoretical distribution. For example, referring to Figure 4.6, we see that offsetting the zero point of the x-coordinate to make the x_{median} of the $A_i(x_i)$ the same as x_{median} for f(x) is equivalent to sliding $S_n(x)$ to the right or left until

Table 4.1: Values of ϵ as a function of β and n for use in the Kolmogorov-Smirnov test

n	β 0.20	0.10	0.02
5	0.447	0.509	0.627
10	0.326	0.369	0.457
20	0.232	0.265	0.328
40	0.165	0.189	0.235
50	0.148	0.170	0.211

it crosses $F(x)$ at $S_n(x_{median}) = F(x_{median}) = 1/2$. This will generally reduce the magnitude of D_n substantially.

4.7.2 Two-Sample K-S Test

The K-S test can also be used to test whether two different data sets come from the same distribution. Suppose the data points are $A_i(x_i)$, $i = 1,\ldots,n$ and $B_j(x_j)$, $j = 1,\ldots,m$. Construct the two step functions $S_{A,n}(x)$ and $S_{B,m}(x)$ from the data sets and then find the maximum distance between the two:

$$D_{n,m} = \max |S_{A,n}(x) - S_{B,m}(x)|, \tag{4.160}$$

where, as before, the absolute value leads to the two-tailed test (see Figure 4.7). If n or m is small, one must use published tables to determine the probability that a value as large or larger than $D_{n,m}$ will arise by chance if the distributions are the same. If both are large,

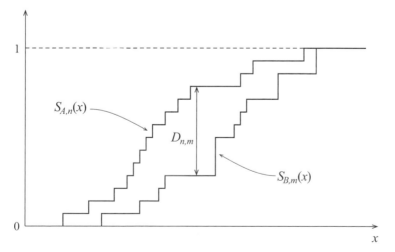

Figure 4.7: The two-sample Kolmogorov-Smirnov test. The step function $S_{A,n}(x)$ is the cumulative distribution function made from n data points $A_i(x_i)$. It rises from 0 at $x < x_1$ to 1 at $x \geq x_n$ in n steps, the steps located at the x_i where the A_i are located. The step function $S_{B,m}(x)$ is made in the same way from m data points $B_j(x_j)$. $D_{n,m}$ is the maximum vertical difference between the two.

merely replace the value of n in the one-sample K-S test with

$$\frac{nm}{n+m},$$ (4.161)

and set $\epsilon = D_n = D_{n,m}$. Then use equation 4.159 to calculate the probability that a value greater than $D_{n,m}$ will happen by chance. Note that the true distributions need not be known to apply the two-sample K-S test.

Linear Least Squares Estimation

5.1 Introduction

One of the most common and important tasks in data analysis is fitting a model to data. A model can be an explicit function or it can be a complicated numerical algorithm—a numerical model. In either case, we will presume the model has parameters that can be adjusted to make it a better representation of the data. Suppose, for example, the data consists of n points (x_i, y_i), where the x_i are known accurately but the y_i have measurement errors, and suppose the model is the straight line $y = a_0 + a_1 x$. To "fit the model to the data" means to adjust the values of the two parameters a_0 and a_1 so that the straight line becomes the best representation of the data consistent with the errors. The adjustable parameters are called the *parameters of fit* or the *fitting parameters*, and their fitted values are estimates of their true values. It is not enough just to estimate the parameters, though. The reliability of the estimates is also needed, and this requires an error analysis.

There is a deep divide between the frequentist and Bayesian approaches to model fitting. A discussion of the Bayesian approach is delayed to Chapter 7. Here and in Chapter 6 we will discuss frequentist techniques with occasional forays into likelihood statistics, the much under-appreciated cousin of frequentist statistics. If the data points have negligible errors in at least one coordinate and if the model depends linearly on the adjustable parameters, it is comparatively easy to fit the model to the data. These elementary cases are discussed in the present chapter. If, however, the model is nonlinear in any of the parameters or if the data points have errors in all their coordinates, fits can become computationally difficult and nonunique. Nonlinear model fitting are discussed in Chapter 6.

The present chapter begins with a brief discussion of likelihood statistics, the likelihood function, and the maximum likelihood principle (Section 5.2). These supply a logical foundation for model fitting when the probability distributions for the errors in the data are known, and they equip us with tools for fitting models to data with non-Gaussian as well as Gaussian error distributions. If the errors do have Gaussian distributions, the maximum likelihood method reduces to χ^2 minimization, also called *least squares optimization* or *least squares estimation*. If the distribution functions for the errors are not known, one commonly simply adopts least squares estimates for the parameters without reference to the likelihood function. These two cases—the error distribution is known and is Gaussian, and the error distribution is unknown—make up the large majority of practical applications. Most of this chapter concentrates, therefore, on techniques for least squares estimation. Beginning with polynomial fits (Section 5.3), we move on to general linear least squares (Section 5.5) and then end with a short section on fitting models with more than one dependent variable (Section 5.6). Much of the discussion is devoted to techniques for error estimation. The chapter is, perforce, rather long, since it introduces concepts that will be used in later chapters.

5.2 Likelihood Statistics

5.2.1 Likelihood Function

The likelihood function plays a fundamental role in both frequentist and Bayesian statistics. For a Bayesian, the final product of statistical inference is the posterior probability distribution. The likelihood function is one of the two required components of the posterior distribution. For a frequentist, the likelihood function incorporates all information produced by an experiment. Given the likelihood function, one can calculate means, medians, and modes of the parameters and estimate their reliabilities from standard deviations or confidence limits. The likelihood function also provides an approach to statistics that is sufficiently distinctive that likelihood statistics has been called a third approach to statistics, one that avoids some of the least appealing aspects of both frequentist and Bayesian statistics. In this section we use the likelihood function and the maximum likelihood principle to generalize least squares estimation, providing the logical foundations for model fitting when the data have non-Gaussian error distributions.

Suppose a probability distribution function for ξ has $m + 1$ parameters a_j,

$$\mathrm{f}(\xi, a_0, a_1, \ldots, a_m) = \mathrm{f}(\xi, \vec{a}), \tag{5.1}$$

where the notation $\vec{a}$ stands for $a_0, a_1, \ldots, a_m$. Extending the notation, let the joint probability distribution for n samples of ξ be

$$\mathrm{f}(\xi_1, \xi_2, \ldots, \xi_n, a_0, a_1, \ldots, a_m) = \mathrm{f}(\vec{\xi}, \vec{a}). \tag{5.2}$$

When discussing probability distributions in previous chapters, we assumed that the $\vec{a}$ are known and fixed, and that $\mathrm{f}(\vec{\xi}, \vec{a})$ gives the probability that specific values of the ξ_i will occur. Now let us assume that the $\vec{a}$ are not known. Instead we are given samples (measurements!) of the ξ_i and wish to infer $\vec{a}$. Let the measured value of ξ_i be x_i. To create the likelihood function $L(\vec{x}, \vec{a})$, simply replace the variables ξ_i in the probability distribution function with their measured values:

$$L(\vec{x}, \vec{a}) \equiv \mathrm{f}(\vec{x}, \vec{a}) = \mathrm{f}(x_1, x_2, \ldots, x_n, \vec{a}). \tag{5.3}$$

In the likelihood function it is the $\vec{x}$ that are known and fixed, and the $\vec{a}$ that are the variables.

Example: Suppose an experiment yields two independent data points with values x_1 and x_2, and suppose we know that both were drawn from the same Gaussian distribution

$$\mathrm{f}(\xi, a_0, a_1) = \frac{1}{\sqrt{2\pi}\, a_1} \exp\left[-\frac{1}{2} \frac{(\xi - a_0)^2}{a_1^2} \right],$$

but we do not know the values of a_0 and a_1. Because they are independent, the joint probability distribution for the two data points is the product of two Gaussians:

$$f(\xi_1, \xi_2, a_0, a_1) = f(\xi_1, a_0, a_1)f(\xi_2, a_0, a_1).$$

The likelihood function is

$$L(x_1, x_2, a_0, a_1) = f(x_1, x_2, a_0, a_1)$$

$$= \frac{1}{\sqrt{2\pi}\, a_1} \exp\left[-\frac{1}{2}\frac{(x_1 - a_0)^2}{a_1^2}\right] \frac{1}{\sqrt{2\pi}\, a_1} \exp\left[-\frac{1}{2}\frac{(x_2 - a_0)^2}{a_1^2}\right]$$

$$= \frac{1}{2\pi a_1^2} \exp\left[-\frac{1}{2}\frac{\sum_{i=1}^{2}(x_i - a_0)^2}{a_1^2}\right].$$

When written out explicitly, $L(\vec{x}, \vec{a})$ looks like it has the same functional form as $f(\vec{\xi}, \vec{a})$, tempting one to identify $L(\vec{x}, \vec{a})$ as a probability distribution function for the parameters. It most emphatically is not. The most obvious reason the likelihood function is not a probability distribution is that it is not normalized. To misidentify it as a probability distribution can lead to disastrous, even iniquitous conclusions. The following example is drawn from the realm of hypothesis testing.

Example: The Centers for Disease Control and Prevention (CDC) summarizes the number of deaths each year in the United States due to various causes. One can use this information to construct the probability that a random person will die from the various causes. Let H be the cause of death, ξ be a random person in the United States, and $P(\xi|H)$ be the probability that person ξ will die because of H. According to the CDC, there were 3,556 deaths due to accidental drowning in 2011. Approximately 3.12×10^8 people lived in the United States in 2011. The probability of death by accidental drowning is, therefore,

$$P(\xi|\text{accidental drowning}) = 3556/3.12 \times 10^8 = 1.14 \times 10^{-5}.$$

Suppose the local newspaper reports that a particular person, John Doe, died by drowning last Sunday. We now have $x =$ "John Doe", and the likelihood function is

$$L(x|H) = L(\text{John Doe}|\text{accidental drowning})$$

$$= P(\text{John Doe}|\text{accidental drowning}) = 1.14 \times 10^{-5}.$$

It is false to interpret this to mean there is an extremely low probability (1.1×10^{-5}) that the cause of the unfortunate John Doe's death was accidental drowning, implying a high probability that the drowning was not an accident. A version of this abuse has been called the prosecutor's fallacy.

Dedicated frequentists may raise a second objection to interpreting the likelihood function as a probability distribution for the parameters. According to this objection, there is only one set of true values for the parameters. For $L(\vec{x}, \vec{a})$ to have meaning as a probability density distribution, the quantity $L(\vec{x}, \vec{a}) \, d\vec{a}$ would have to exist; but $d\vec{a}$ does not have any meaning if there is only one true set of values for the parameters. The use of delta functions does not resolve this problem, since the uniqueness applies to the $d\vec{a}$ factor, not to the likelihood function. Therefore, the likelihood function cannot be a probability distribution for the parameters. It is, instead, a measure of how well the data have allowed us to encompass the true values. We will examine this viewpoint more fully in Section 6.5 when discussing confidence limits for the estimated parameters.

Nevertheless, the likelihood function does encode information about the parameters and can be used to determine which values of the parameters are preferred. While the value of the likelihood function has no meaning in isolation, values that have a higher likelihood are preferred over values that have lower likelihood. The likelihoods can be compared using the likelihood ratio. Suppose one wishes to compare two sets of specific values, $\vec{a}_1$ and $\vec{a}_2$, for the parameters. The likelihood ratio for the two sets of values is

$$ LR = \frac{L(\vec{x}, \vec{a}_1)}{L(\vec{x}, \vec{a}_2)}. \tag{5.4} $$

The likelihood ratio can be understood as the ratio of the probabilities that parameters $\vec{a}_1$ and $\vec{a}_2$ will produce the observed data. However, large likelihood ratio does not necessarily guarantee that one set of parameters is preferred over another, because an independent experiment could yield different values for the $\vec{x}$. We will return to this issue when discussing variances, confidence limits, and tests for goodness of fit.

The following example shows how the use of the likelihood ratio avoids the abuse recounted in the previous example.

Example: According to the FBI, there were 15 homicides by drowning in the United States in 2011. If the hypothesis H is "homicidal drowning," the probability that random person ξ will die by homicidal drowning was

$$ P(\xi|H) = P(\xi|\text{homicidal drowning}) = 15/3.12 \times 10^8 = 4.8 \times 10^{-8}. $$

The likelihood function for the unfortunate John Doe who died last Sunday now has two cases:

$$ L(x|H) = \begin{cases} L(\text{John Doe}|\text{accidental drowning}) = 1.14 \times 10^{-5} \\ L(\text{John Doe}|\text{homicidal drowning}) = 4.8 \times 10^{-8} \end{cases}. $$

The likelihood ratio for the two cases is

$$ LR = \frac{L(\text{John Doe}|\text{accidental drowning})}{L(\text{John Doe}|\text{homicidal drowning})} = \frac{1.14 \times 10^{-5}}{4.8 \times 10^{-8}} = \frac{3556}{15} \approx 240. $$

Thus, John Doe's death by drowning is 240 times more likely to have been caused by an accident than by homicide. Note, however, that the number of homicidal drownings varies considerably from year to year, so using data from different years will yield different likelihood ratios.

The likelihood function can be viewed as a fully frequentist construct. However, its role in Bayesian statistics suggests another way to interpret the likelihood function. In Bayesian statistics the posterior probability distribution is the product of the likelihood function and the prior probability distribution (see Section 7.1). If there is no prior knowledge so that the prior probability distribution is flat, the posterior distribution *is* the likelihood function to within a normalization constant. The odds ratio, which is often used in Bayesian statistics for hypothesis testing, becomes identical to the likelihood ratio.

Frequentists generally object to two aspects of Bayesian statistics. The first is the enforced use of a prior probability distribution, and the second is the interpretation of probability as a degree of belief. Even to a dedicated frequentist much of the rest of Bayesian statistics is attractive. The use of probability distributions to describe parameter values is an example. Thus, the posterior distribution is a measure of our belief that the various values of the parameters should be taken as the true value. Is it possible to salvage the attractive features of Bayesian statistics without resorting to priors or redefining the meaning of probability?

Likelihood statistics proffers a way.[1] Suppose the denominator of the likelihood ratio is the likelihood of an arbitrary but fixed set of fiducial parameter values (perhaps set all the a_j equal to 0). The likelihood ratio becomes the probability of other parameter values with respect to the fiducial values. If one is willing to accept that the likelihood function incorporates our uncertainty about the true values of the parameters, the likelihood ratio can be interpreted as an unnormalized probability that the parameters will yield the observed data. More compact characterizations of the parameters can then be extracted from the likelihood function in ways made familiar by the postprocessing of the posterior probability distribution in Bayesian statistics, perhaps by MCMC sampling or the Laplace approximation (see Section 7.3.2).

In principle, likelihood statistics is a rejection Bayesian statistics, but in practice it can be thought of as Bayesian statistics without a prior, or perhaps with a prior that does not restrict the parameters in any way. It is the author's impression that many practitioners of data analysis are actually using a version of likelihood statistics, even if they do not recognize that they are doing so. The maximum likelihood principle is a prime example.

5.2.2 Maximum Likelihood Principle

As we saw in Section 5.2.1, values of the parameters that have a higher likelihood are preferred over values that have a lower likelihood. The values with the highest likelihood are the most preferred. The maximum likelihood principle codifies this property:

Given data points $\vec{x}$ drawn from a joint probability distribution function whose functional form is known to be $f(\vec{\xi}, \vec{a})$, the best estimate of the parameters $\vec{a}$ are those which maximize the likelihood function $L(\vec{x}, \vec{a}) \propto f(\vec{x}, \vec{a})$.

The following example shows how the maximum likelihood principle works for data points drawn from a binomial distribution.

Example: Suppose we suspect that a coin is unfair, because heads turn up more often that tails when the coin is flipped. We would like to know the probability that a flip gives heads, so we flip the coin 100 times, and it winds up heads 75 times. Use the maximum likelihood principle to estimate the probability that a flip gives heads.

Continued on page 116

---

[1] See, notably, A.W.F. Edwards (1992).

This experiment is described by the binomial distribution,

$$P(k, n, p) = \frac{n!}{k!(n-k)!} p^k (1-p)^{n-k},$$

where p is the probability of heads, n is the number of flips, and k is the number of times heads occurs. The likelihood function is

$$L(75, 100, p) = \frac{100!}{75!25!} p^{75} (1-p)^{25}.$$

To find the maximum of the likelihood function, set its derivative with respect to p equal to zero:

$$0 = \frac{d}{dp} \left\{ L(75, 100, p) \right\} = \frac{100!}{75!25!} \left[75p^{74}(1-p)^{25} - 25p^{75}(1-p)^{24} \right].$$

We denote inferred values of parameters by accenting them with a hat: $\hat{p}$ is the inferred value of p. Dividing this equation by $100!/75!25!$ and by $25p^{74}(1-p)^{24}$, we find

$$0 = 3(1 - \hat{p}) - \hat{p},$$

which yields

$$p = 0.75,$$

a gratifying but unsurprising result. Note that the factor $100!/75!25!$ divides out and does not affect the solution. It is a general property of maximum likelihood solutions that any constant factor multiplying the likelihood function is irrelevant to the solution.

Suppose we perform many experiments, and the experiments produce n independent data points x_i. For now we assume the x_i are free of measurement error. If each sample is drawn from the same distribution, $f(\xi, \vec{a})$, the joint probability distribution for n samples is

$$f(\vec{\xi}, \vec{a}) = f(\xi_1, \vec{a}) f(\xi_2, \vec{a}) \cdots f(\xi_n, \vec{a}) = \prod_{i=1}^{n} f(\xi_i, \vec{a}). \tag{5.5}$$

The likelihood function is therefore

$$L(\vec{x}, \vec{a}) = f(x_1, \vec{a}) f(x_2, \vec{a}) \cdots f(x_n, \vec{a}) = \prod_{i=1}^{n} f(x_i, \vec{a}). \tag{5.6}$$

For computational convenience, one often prefers to work with the log-likelihood function,

$$\ell(\vec{x}, \vec{a}) = \ln L = \ln \left\{ \prod_{i=1}^{n} f(x_i, \vec{a}) \right\} = \sum_{i=1}^{n} \ln \left\{ f(x_i, \vec{a}) \right\}. \tag{5.7}$$

The maximum of the log-likelihood function—which occurs at the same values of $\vec{a}$ as the maximum of $L(\vec{x}, \vec{a})$—can be found by setting its first derivatives with respect to its parameters equal to zero. The maximum likelihood principle thus translates into a set of equations called the *likelihood equations*:

$$\frac{\partial \ell(\vec{x}, \vec{a})}{\partial a_j} = 0, \quad j = 0, \cdots, m. \tag{5.8}$$

The solution to these equations yields the estimates of the a_j. The following example uses the maximum likelihood principle to estimate the parameters of a Gaussian distribution.

Example: Suppose that we have n independent data points x_i and we know they were sampled from the Gaussian distribution

$$f(\xi, a_0, a_1) = \frac{1}{\sqrt{2\pi a_1}} \exp\left[-\frac{1}{2}\frac{(\xi - a_0)^2}{a_1}\right].$$

We wish to estimate a_0 and a_1, the mean and variance of the distribution. Because the data points are independent, their joint likelihood function is the product of their individual likelihood functions:

$$L(\vec{x}, a_0, a_1) = \prod_{i=1}^{n} f(x_i, a_0, a_1) = \left(\frac{1}{2\pi a_1}\right)^{n/2} \exp\left[-\frac{1}{2}\frac{\sum_{i=1}^{n}(x_i - a_0)^2}{a_1}\right].$$

The log-likelihood function is

$$\ell(\vec{x}, a_0, a_1) = -\frac{n}{2}\ln(2\pi a_1) - \frac{1}{2}\frac{\sum_{i=1}^{n}(x_i - a_0)^2}{a_1}.$$

One likelihood equation gives $\hat{a}_0$:

$$\frac{\partial \ell}{\partial a_0} = \frac{\partial}{\partial a_0}\left\{-\frac{n}{2}\ln(2\pi a_1^2) - \frac{1}{2}\frac{\sum_{i=1}^{n}(x_i - a_0)^2}{a_1}\right\}$$

$$0 = \frac{1}{\hat{a}_1}\sum_{i=1}^{n}(x_i - \hat{a}_0) = \frac{1}{\hat{a}_1}\left[\sum_{i=1}^{n}x_i - n\hat{a}_0\right]$$

$$\hat{a}_0 = \frac{1}{n}\sum_{i=1}^{n}x_i.$$

The other likelihood equation gives $\hat{a}_1$:

$$\frac{\partial \ell}{\partial a_1} = \frac{\partial}{\partial a_1}\left\{-\frac{n}{2}\ln(2\pi a_1) - \frac{1}{2}\frac{\sum_{i=1}^{n}(x_i - a_0)^2}{a_1}\right\}$$

$$0 = -\frac{n}{2}\left(\frac{1}{\hat{a}_1}\right) + \frac{1}{2}\frac{\sum_{i=1}^{n}(x_i - \hat{a}_0)^2}{\hat{a}_1^2}$$

$$\hat{a}_1 = \frac{1}{n}\sum_{i=1}^{n}(x_i - \hat{a}_0)^2.$$

The results in the previous example are much as we would expect, but the equation for $\hat{a}_1$ has a factor of $1/n$ instead of $1/(n-1)$ (compare to equation 4.25). This means that the maximum likelihood value of the variance is biased, although the bias is small for large values of n. Maximum likelihood estimators often have a small bias.

While the maximum likelihood principle is easy to apply to Gaussian probability distributions, it is not restricted to them. The following example applies the principle to multiple data points sampled from a Poisson distribution.

Example: Suppose the data consist of n independent integers k_i sampled from a Poisson distribution

$$P(k, a) = \frac{a^k}{k!} \exp[-a].$$

The data might be, for example, the number of photons detected from a faint star in 1 second, measured during several 1-second intervals. The likelihood function is

$$L(\vec{k}, a) = \prod_{i=1}^{n} \frac{a^{k_i}}{k_i!} \exp[-a],$$

and the log-likelihood function is

$$\ell(\vec{k}, a) = \sum_{i=1}^{n} \ln\left[\frac{a^{k_i}}{k_i!} \exp[-a]\right]$$

$$= \sum_{i=1}^{n} \left[k_i \ln a - \ln(k_i!) - a\right].$$

The likelihood equation for $\hat{a}$ is

$$0 = \left.\frac{\partial \ell}{\partial a}\right|_{\hat{a}} = \sum_{i=1}^{n}\left[\frac{k_i}{\hat{a}} - 1\right]_{\hat{a}} = \frac{1}{\hat{a}}\sum_{i=1}^{n} k_i - n$$

$$\hat{a} = \frac{1}{n}\sum_{i=1}^{n} k_i.$$

Since a is also the first moment of the Poisson distribution (equation 2.29), $\hat{a}$ is also maximum likelihood estimate for the mean value of k.

Suppose again we make n independent measurements x_i of a quantity ξ that has a mean value $\langle \xi \rangle = a$, but now let each measurement x_i have a measurement error $\epsilon_i = x_i - a$. Suppose the measurement errors have Gaussian distributions with variance σ_i^2, so that the probability distribution function for data point i is

$$f_i(\xi_i, \sigma_i, a) = \frac{1}{\sqrt{2\pi}\,\sigma_i} \exp\left[-\frac{1}{2}\frac{(\xi_i - a)^2}{\sigma_i^2}\right]. \tag{5.9}$$

We abbreviate this situation by saying we have n data points (x_i, σ_i). The σ_i are almost always called the *errors of the measurements* (instead of the standard deviations of the error distributions).

We wish to estimate the value of a. Because they are independent, the joint likelihood function for all measurements is

$$L(\vec{x},\vec{\sigma},a) = \prod_{i=1}^{n} \frac{1}{\sqrt{2\pi}\,\sigma_i} \exp\left[-\frac{1}{2}\frac{(x_i - a)^2}{\sigma_i^2}\right], \tag{5.10}$$

and the log-likelihood function is

$$\ell(\vec{x},\vec{\sigma},a) = \sum_{i=1}^{n} \ln\left(\frac{1}{\sqrt{2\pi}\,\sigma_i}\right) - \frac{1}{2}\sum_{i=1}^{n}\frac{(x_i - a)^2}{\sigma_i^2}. \tag{5.11}$$

The likelihood equation for $\hat{a}$ is

$$0 = \left.\frac{\partial\ell}{\partial a}\right|_{\hat{a}} = \frac{\partial}{\partial a}\left[-\frac{1}{2}\sum_{i=1}^{n}\frac{(x_i - a)^2}{\sigma_i^2}\right]_{\hat{a}} \tag{5.12}$$

$$0 = \sum_{i=1}^{n}\frac{x_i - \hat{a}}{\sigma_i^2} = \sum_{i=1}^{n}\frac{x_i}{\sigma_i^2} - \hat{a}\sum_{i=1}^{n}\frac{1}{\sigma_i^2}$$

$$\hat{a} - \frac{\sum_{i=1}^{n}\left(1/\sigma_i^2\right)x_i}{\sum_{i=1}^{n}\left(1/\sigma_i^2\right)}. \tag{5.13}$$

Setting $w_i = 1/\sigma_i^2$, we arrive at

$$\hat{a} = \frac{\sum_{i=1}^{n} w_i x_i}{\sum_{i=1}^{n} w_i}. \tag{5.14}$$

This is identical to the expression for the weighted average derived in Chapter 4 (see equation 4.55), although it should be recognized that the present derivation explicitly assumed a Gaussian distribution for the errors, while the earlier derivation did not.

5.2.3 Relation to Least Squares and χ^2 Minimization

Equation 5.12 can be restated as

$$\left.\frac{\partial S}{\partial a}\right|_{\hat{a}} = 0, \tag{5.15}$$

where

$$S = \sum_{i=1}^{n}\frac{\epsilon_i^2}{\sigma_i^2} = \sum_{i=1}^{n}\frac{(x_i - a)^2}{\sigma_i^2} = \sum_{i=1}^{n} w_i(x_i - a)^2. \tag{5.16}$$

The quantity S is the sum of the weighted squared differences between the x_i and a, and equation 5.15 is the requirement that S be minimized. More generally, whenever the error distributions are Gaussian, the likelihood equations (equations 5.8) become derivatives of the weighted sum of the squared residuals. Since the equations specify that the weighted sums are to be minimized, this procedure is called *least squares optimization* or *least squares estimation*. Least squares optimization can thus be derived from the maximum likelihood principle.

We also recognize that S appears to be a χ^2 variable (see equation 2.96). If one identifies S with χ^2, equation 5.15 becomes the requirement that χ^2 be minimized, so the procedure is often called χ^2 minimization. This identification is dangerous, however. Measured values of σ_i^2 are frequently incorrect and, when so, $S \neq \chi^2$. In this case the assumption that S

is equal to χ^2 will lead to incorrect standard deviations and covariances for the estimated parameters. We will, therefore, consistently use the symbol S for the weighted sums of the residuals and use the term "least squares", not "χ^2 minimization", allowing ourselves to be happily surprised if we discover that $S = \chi^2$ for the problem at hand. We will deal with these issues more fully in the discussion of the estimated covariance matrix in Section 5.3.3.

The derivation of least squares optimization from the maximum likelihood principle depends explicitly on the assumption that the residuals have Gaussian distributions. If the residuals have a non-Gaussian distribution and if that distribution is known, least squares optimization is not appropriate, and one should return to the maximum likelihood principle to estimate the parameters. More commonly, though, one does not have any prior information about the distribution. Least squares optimization may still be the method of choice, because the central limit theorem guarantees that many processes are well approximated by a Gaussian distribution. Also, in Chapter 4 (Sections 4.3 and 4.4) we used the variance as the fundamental measure of scatter and then estimated mean values by minimizing the variance subject to the constraint that the estimated parameters are unbiased. The probability distribution function for the errors was never needed. This is, of course, least squares optimization. Thus, least squares optimization has a legitimacy independent of maximum likelihood. As a result, it is usually the default method for fitting models to data when the probability distributions for the residuals are unknown.

Finally, the maximum likelihood values for the parameters are modes, because they are the values corresponding to the maximum of the likelihood function. As always, other single-number characterizations of the likelihood function are possible—notably, the means and the medians of the parameters. For symmetric distributions (e.g., the Gaussian distribution), the mean, median, and mode are the same and the difference is moot. But for strongly asymmetric functions (e.g., the exponential distribution), other characterizations may by more appropriate. We return to this issue when discussing confidence intervals (Section 6.5) and yet again in Chapter 7 on Bayesian statistics.

5.3 Fits of Polynomials to Data

5.3.1 Straight Line Fits

Suppose we have n independent data points $\{x_i, y_i, \sigma_i\}$, where the y_i have errors σ_i but the x_i are known precisely. Suppose further that the data points are distributed as shown in Figure 5.1. To characterize this distribution by just a mean and variance would be inadequate and even misleading. The long, skinny shape of the distribution suggests that it would be more accurate to model it as a straight line with scatter. This section shows how to fit a straight line to data by least squares and then generalizes to fits of polynomials of arbitrary degree.

One can think of each data point (x_i, y_i, σ_i) as being sampled in y from a probability distribution function whose mean value changes with x, as shown in Figure 5.2. Let us assume that the values of y are drawn from Gaussian distributions

$$f(y) \propto \exp\left[-\frac{(y-\mu)^2}{2\sigma^2}\right], \tag{5.17}$$

where both μ and σ depend on x. The values of σ are given explicitly for each x_i. Assume that μ depends linearly on x,

$$\mu = a_0 + a_1 x, \tag{5.18}$$

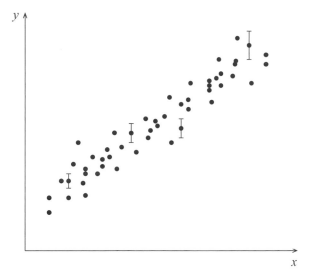

Figure 5.1: The distribution of data points (x_i, y_i, σ_i). The values of x_i are known precisely, but the y_i have errors σ_i. Error bars are shown for only a few representative points. To characterize this distribution by just a mean and variance would be inadequate.

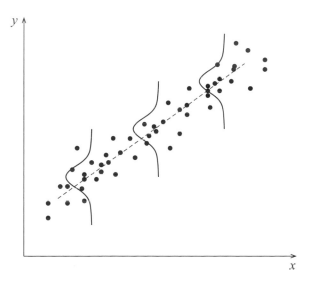

Figure 5.2: The data are distributed in y according to a probability distribution function. The mean of the distribution is a function of x.

so that the probability distributions from which the individual y_i are drawn are

$$f(y) \propto \exp\left[-\frac{(y - a_0 - a_1 x_i)^2}{2\sigma_i^2}\right]. \tag{5.19}$$

The goal is to estimate a_0 and a_1.

We can attack this problem in two ways. Since the probability distributions for the errors are known, we can apply the maximum likelihood principle. The likelihood function for a single data point is

$$L(x_i, y_i, \sigma_i, a_0, a_1) \propto \exp\left[-\frac{(y_i - a_0 - a_1 x_i)^2}{2\sigma_i^2}\right]. \tag{5.20}$$

Because the data points are independent, the joint likelihood function is the product of the individual likelihood functions. The log-likelihood function is then

$$\begin{aligned}
\ell(\vec{x}, \vec{y}, \vec{\sigma}, a_0, a_1) &= -\frac{1}{2}\sum_{i=1}^{n}\frac{(y_i - a_0 - a_1 x_i)^2}{\sigma_i^2} + c \\
&= -\frac{1}{2}\sum_{i=1}^{n} w_i(y_i - a_0 - a_1 x_i)^2 + c,
\end{aligned} \tag{5.21}$$

where c is a constant, and we have set $w_i = 1/\sigma_i^2$. The maximum likelihood estimates for the parameters are found by setting to 0 the partial derivatives of $\ell(\vec{x}, \vec{y}, \vec{\sigma}, a_0, a_1)$ with respect to a_0 and a_1.

Alternatively, if the probability distributions for the y_i are not known, we can bypass the maximum likelihood principle and simply minimize the weighted sum of the squares of the residuals:

$$S = \sum_{i=1}^{n} w_i(y_i - a_0 - a_1 x_i)^2. \tag{5.22}$$

Comparison of equation 5.22 to equation 5.21 shows that S reaches its minimum at the same values of a_0 and a_1 as those where $\ell(\vec{x}, \vec{y}, \vec{\sigma}, a_0, a_1)$ reaches its maximum. Thus the least squares estimates for a_0 and a_1 are the same as the maximum likelihood estimates. As the error distribution functions are usually not known, we will adopt the least squares terminology and formalism, saving an example that requires maximum likelihood estimates for the end of this section.

Proceeding, we minimize S by setting its derivatives with respect to a_0 and a_1 equal to zero:

$$0 = \frac{\partial S}{\partial a_0} = -2\sum_i w_i(y_i - \hat{a}_0 - \hat{a}_1 x_i) \tag{5.23}$$

$$0 = \frac{\partial S}{\partial a_1} = -2\sum_i w_i x_i(y_i - \hat{a}_0 - \hat{a}_1 x_i). \tag{5.24}$$

The hats are placed on $\hat{a}_0$ and $\hat{a}_1$, because they are now fitted values. Expanding and rearranging these equations, we find

$$\hat{a}_0 \sum_i w_i + \hat{a}_1 \sum_i w_i x_i = \sum_i w_i y_i \tag{5.25}$$

$$\hat{a}_0 \sum_i w_i x_i + \hat{a}_1 \sum_i w_i x_i^2 = \sum_i w_i x_i y_i. \tag{5.26}$$

Equations 5.25 and 5.26 are called the *normal equations*. They are linear in $\hat{a}_0$ and $\hat{a}_1$, and are easily solved to give

$$\hat{a}_0 = \frac{\left(\sum_i w_i x_i^2\right)\left(\sum_i w_i y_i\right) - \left(\sum_i w_i x_i\right)\left(\sum_i w_i x_i y_i\right)}{\Delta} \tag{5.27}$$

$$\hat{a}_1 = \frac{\left(\sum_i w_i\right)\left(\sum_i w_i x_i y_i\right) - \left(\sum_i w_i x_i\right)\left(\sum_i w_i y_i\right)}{\Delta}, \tag{5.28}$$

where for compactness, we have set

$$\Delta = \left(\sum_i w_i\right)\left(\sum_i w_i x_i^2\right) - \left(\sum_i w_i x_i\right)^2. \tag{5.29}$$

The following example shows the fit of a straight line to data. We will return to this example repeatedly over the next few chapters.

Example: Fit the straight line

$$y = a_0 + a_1 x$$

to the data points listed in Table 5.1 by least squares. We will perform the fit twice, first without weights, then with weights. The specific form of the probability distributions for the errors is unknown. We initially assume that the variances of the distributions are correctly given by σ^2 but will need to relax this assumption when continuing the example at the end of Section 5.3.3.

Unweighted Fit: We can mimic an unweighted least squares fit by simply setting all standard deviations of the data points equal to 1.0, so that their weights also become 1.0. We find

$$\sum w_i = \sum 1 = 12 \qquad\qquad \sum w_i y_i = \sum y_i = 20.75$$
$$\sum w_i x_i = \sum x_i = 70.56 \qquad\qquad \sum w_i x_i y_i = \sum x_i y_i = 129.57$$
$$\sum w_i x_i^2 = \sum x_i^2 = 425.57 \qquad\qquad \Delta = 128.18$$

Then from equations 5.27 and 5.28, the parameters of the fit are

$$\hat{a}_0 = -2.43$$
$$\hat{a}_1 = 0.707$$

Table 5.1: Data points for x, y, and σ

x	y	σ	x	y	σ	x	y	σ
4.41	0.43	0.08	5.32	1.26	0.32	6.60	2.75	0.18
4.60	0.99	0.15	5.81	0.95	0.40	6.99	2.64	0.08
4.95	0.87	0.22	5.89	1.79	0.35	7.13	3.01	0.05
5.28	2.09	0.32	6.36	2.00	0.25	7.22	1.97	0.99

Continued on page 124

The fit is shown in the left panel of Figure 5.3. Even for this simple example, the calculations are time consuming and tedious if done by hand. Hand calculations are rarely feasible in realistic least squares applications.

Weighted Fit: Setting the weights equal to $w_i = 1/\sigma_1^2$, we find after even more tedious calculations:

$$\sum w_i = 859.43 \qquad \sum w_i y_i = 1917.81$$
$$\sum w_i x_i = 5440.71 \qquad \sum w_i x_i y_i = 13,127.58$$
$$\sum w_i x_i^2 = 35,542.25 \qquad \Delta = 944,946.8$$

The parameters of the fit are

$$\hat{a}_0 = -3.45$$
$$\hat{a}_1 = 0.897.$$

The fit is shown in the right panel of Figure 5.3.

We discuss variances and covariances of the fitted parameters in Section 5.3.3 and will discover that the difference between the weighted and unweighted fits is highly significant. The large difference between the two fits is caused by the large differences among the standard deviations of the data points. The data point at $x = 7.22$, for example, counts as much as the preceding point in the unweighted fit, but its weight is nearly 400 times smaller in the weighted fit.

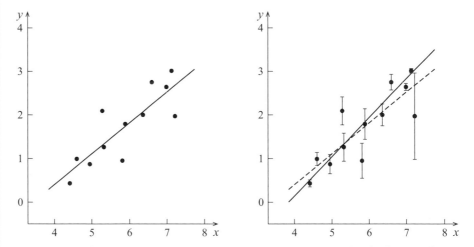

Figure 5.3: Fits of a straight line to the data in the accompanying example. The data are identical in the two panels but are weighted differently. The left panel shows the data points without their error bars, and the fit is unweighted. The right panel shows the data points with their error bars, and the fit is weighted. The dashed line in the right panel repeats the unweighted fit to the data for comparison.

The normal equations can be put in the form of a matrix equation

$$\begin{pmatrix} \sum_i w_i & \sum_i w_i x_i \\ \sum_i w_i x_i & \sum_i w_i x_i^2 \end{pmatrix} \begin{pmatrix} \hat{a}_0 \\ \hat{a}_1 \end{pmatrix} = \begin{pmatrix} \sum_i w_i y_i \\ \sum_i w_i x_i y_i \end{pmatrix}, \tag{5.30}$$

for which the solution is

$$\begin{pmatrix} \hat{a}_0 \\ \hat{a}_1 \end{pmatrix} = \begin{pmatrix} \sum_i w_i & \sum_i w_i x_i \\ \sum_i w_i x_i & \sum_i w_i x_i^2 \end{pmatrix}^{-1} \begin{pmatrix} \sum_i w_i y_i \\ \sum_i w_i x_i y_i \end{pmatrix}. \tag{5.31}$$

If we define the matrix $\mathbf{N}$, called the *normal matrix*, to be

$$\mathbf{N} = \begin{pmatrix} \sum_i w_i & \sum_i w_i x_i \\ \sum_i w_i x_i & \sum_i w_i x_i^2 \end{pmatrix} \tag{5.32}$$

and the two vectors $\hat{\mathbf{a}}$ and $\mathbf{Y}$ to be

$$\hat{\mathbf{a}} - \begin{pmatrix} \hat{a}_0 \\ \hat{a}_1 \end{pmatrix} \quad \text{and} \quad \mathbf{Y} = \begin{pmatrix} \sum_i w_i y_i \\ \sum_i w_i x_i y_i \end{pmatrix}, \tag{5.33}$$

then in matrix notation, the normal equations become

$$\mathbf{N}\hat{\mathbf{a}} = \mathbf{Y}. \tag{5.34}$$

The solution takes the simple form

$$\hat{\mathbf{a}} = \mathbf{N}^{-1}\mathbf{Y}. \tag{5.35}$$

While the matrix formalism does not add much for the fit of a straight line, we will see in the following sections that it greatly simplifies more complicated fits.

Example: For the preceding example the normal matrix and $\mathbf{Y}$ vector for the weighted least squares fit are

$$\mathbf{N} = \begin{pmatrix} 859.43 & 5440.71 \\ 5440.71 & 35,542.25 \end{pmatrix} \quad \text{and} \quad \mathbf{Y} = \begin{pmatrix} 1917.81 \\ 13,127.58 \end{pmatrix}.$$

The inverse of the normal matrix is

$$\mathbf{N}^{-1} = \begin{pmatrix} 3.7612 \times 10^{-2} & -5.7575 \times 10^{-3} \\ -5.7575 \times 10^{-3} & 9.0951 \times 10^{-4} \end{pmatrix},$$

and the solution for the least squares fit is

$$\hat{\mathbf{a}} = \mathbf{N}^{-1}\mathbf{Y} = \begin{pmatrix} .-3.45 \\ 0.897 \end{pmatrix}.$$

This result is, of course, identical to the one in the previous example.

The foregoing discussion explicitly assumes that the errors in the y_i had either an unknown distribution or a Gaussian distribution. The following example uses the maximum likelihood principle to fit a straight line to data with a non-Gaussian distribution.

Example: Suppose we have n independent data points (x_i, k_i), where the k_i are integers with a Poisson distribution

$$P(k, \mu) = \frac{\mu^k}{k!} \exp[-\mu].$$

We wish to fit a straight line to the data using the maximum likelihood principle. Set $\mu = a_0 + a_1 x$. The likelihood function becomes

$$L(\vec{x}, \vec{k}, a_0, a_1) = \prod_{i=1}^{n} \frac{(a_0 + a_1 x_i)^{k_i}}{k_i!} \exp[-a_0 - a_1 x_i]$$

and the log-likelihood function becomes

$$\ell(\vec{x}, \vec{k}, a_0, a_1) = \sum_{i=1}^{n} \ln \left[\frac{(a_0 + a_1 x_i)^{k_i}}{k_i!} \exp[-a_0 - a_1 x_i] \right]$$

$$= \sum_{i=1}^{n} \left[k_i \ln(a_0 + a_1 x_i) - \ln(k_i!) - a_0 - a_1 x_i \right].$$

To find the maximum likelihood estimates of $\hat{a}_0$ and $\hat{a}_1$, set the partial derivatives of ℓ with respect to a_0 and a_1 equal to zero:

$$0 = \frac{\partial \ell}{\partial a_0} = \sum_{i=1}^{n} \left[\frac{k_i}{\hat{a}_0 + \hat{a}_1 x_i} - 1 \right]$$

$$0 = \frac{\partial \ell}{\partial a_1} = \sum_{i=1}^{n} \left[\frac{x_i k_i}{\hat{a}_0 + \hat{a}_1 x_i} - x_i \right],$$

or

$$\sum_{i=1}^{n} \frac{k_i}{\hat{a}_0 + \hat{a}_1 x_i} = n$$

$$\sum_{i=1}^{n} \frac{x_i k_i}{\hat{a}_0 + \hat{a}_1 x_i} = \sum_{i=1}^{n} x_i.$$

These two equations must be solved for $\hat{a}_0$ and $\hat{a}_1$. Since the equations are nonlinear, the values of $\hat{a}_0$ and $\hat{a}_1$ must found using numerical methods.

5.3.2 Fits with Polynomials of Arbitrary Degree

We now generalize to polynomials of arbitrary degree. As before, we are given n independent data points (x_i, y_i, σ_i), where the y_i have errors σ_i, and the x_i are assumed to be known without error. We wish to fit the data points with the function

$$y = a_0 + a_1 x + a_2 x^2 + \cdots + a_m x^m. \tag{5.36}$$

Let us first assume that the errors are independent and have Gaussian distributions. Equation 5.18 generalizes to

$$\mu = a_0 + a_1 x + a_2 x^2 + \cdots + a_m x^m, \tag{5.37}$$

the probability distributions for y become (compare to Equation 5.19)

$$f(y) \propto \exp\left[-\frac{1}{2}\frac{(y - a_0 - a_1 x_i - a_2 x_i^2 - \cdots - a_m x_i^m)^2}{\sigma_i^2}\right], \tag{5.38}$$

and the log-likelihood function becomes

$$\ell = \sum_{i=1}^{n} \ln[f(y_i)] = -\frac{1}{2}\sum_{i=1}^{n} w_i(y_i - a_0 - a_1 x_i - a_2 x_i^2 - \cdots - a_m x_i^m)^2 + c, \tag{5.39}$$

where $w_i = 1/\sigma_i^2$. The maximum likelihood values for the a_j are determined by the solution to the $m + 1$ likelihood equations

$$\frac{\partial \ell}{\partial a_j} = 0. \tag{5.40}$$

In the language of least squares estimation, the sum of the weighted squares of the residuals between the data points and the polynomial model becomes (compare to equation 5.22)

$$S = \sum_{i=1}^{n} w_i(y_i - a_0 - a_1 x_i - a_2 x_i^2 - \cdots - a_m x_i^m)^2. \tag{5.41}$$

The least squares estimates for the a_j are given by the solutions to the $m + 1$ equations

$$\frac{\partial S}{\partial a_j} = 0. \tag{5.42}$$

Comparing equations 5.39 and 5.40 to equations 5.41 and 5.42, we see again that the maximum likelihood and least squares estimates for the a_j are identical if the errors have a Gaussian distribution.

Since equations 5.41 and 5.42 (or equations 5.39 and 5.40, if one desires) are linear in the parameters of fit, they are easy to solve. Generalizing from equation 5.30, we arrive at the normal equations in matrix form:

$$\begin{pmatrix} \sum w_i & \sum w_i x_i & \cdots & \sum w_i x_i^m \\ \sum w_i x_i & \sum w_i x_i^2 & \cdots & \sum w_i x_i^{m+1} \\ \vdots & \vdots & & \vdots \\ \sum w_i x_i^m & \sum w_i x_i^{m+1} & \cdots & \sum w_i x_i^{2m} \end{pmatrix} \begin{pmatrix} \hat{a}_0 \\ \hat{a}_1 \\ \vdots \\ \hat{a}_m \end{pmatrix} = \begin{pmatrix} \sum w_i y_i \\ \sum w_i x_i y_i \\ \vdots \\ \sum w_i x_i^m y_i \end{pmatrix}, \tag{5.43}$$

where all the sums are taken from $i = 1$ to n. The solution is

$$\begin{pmatrix} \hat{a}_0 \\ \hat{a}_1 \\ \vdots \\ \hat{a}_m \end{pmatrix} = \begin{pmatrix} \sum w_i & \sum w_i x_i & \cdots & \sum w_i x_i^m \\ \sum w_i x_i & \sum w_i x_i^2 & \cdots & \sum w_i x_i^{m+1} \\ \vdots & \vdots & & \vdots \\ \sum w_i x_i^m & \sum w_i x_i^{m+1} & \cdots & \sum w_i x_i^{2m} \end{pmatrix}^{-1} \begin{pmatrix} \sum w_i y_i \\ \sum w_i x_i y_i \\ \vdots \\ \sum w_i x_i^m y_i \end{pmatrix}. \tag{5.44}$$

Equations 5.43 and 5.44 can be written more compactly in matrix notation. Set the normal matrix equal to

$$
\mathbf{N} = \begin{pmatrix} \sum w_i & \sum w_i x_i & \cdots & \sum w_i x_i^m \\ \sum w_i x_i & \sum w_i x_i^2 & \cdots & \sum w_i x_i^{m+1} \\ \vdots & \vdots & & \vdots \\ \sum w_i x_i^m & \sum w_i x_i^{m+1} & \cdots & \sum w_i x_i^{2m} \end{pmatrix}, \tag{5.45}
$$

and set

$$
\hat{\mathbf{a}} = \begin{pmatrix} \hat{a}_0 \\ \hat{a}_1 \\ \vdots \\ \hat{a}_m \end{pmatrix} \quad \text{and} \quad \mathbf{Y} = \begin{pmatrix} \sum w_i y_i \\ \sum w_i x_i y_i \\ \vdots \\ \sum w_i x_i^m y_i \end{pmatrix}. \tag{5.46}
$$

The normal equations become

$$
\mathbf{N}\hat{\mathbf{a}} = \mathbf{Y}, \tag{5.47}
$$

for which the solution is

$$
\hat{\mathbf{a}} = \mathbf{N}^{-1}\mathbf{Y}. \tag{5.48}
$$

These are identical to equations 5.34 and 5.35.

5.3.3 Variances, Covariances, and Biases

Extension of the Matrix Notation: While equation 5.44 gives estimates for the $\hat{a}_j$, the estimates by themselves are not enough. We need to know how much to trust them—we need to know their variances, covariances, and biases. First note that because of the measurement errors, a polynomial will not pass through all the individual data points. We can represent this by setting

$$
y_i = a_0 + a_1 x_i + a_2 x_i^2 + \cdots + a_m x_i^m + \epsilon_i, \tag{5.49}
$$

where the ϵ_i is the scatter of the data points around the polynomial. The ϵ_i have zero mean, are uncorrelated, and their individual variances are equal to σ_i^2:

$$
\langle \epsilon_i \rangle = 0 \tag{5.50}
$$

$$
\langle \epsilon_i \epsilon_j \rangle = \sigma_i^2 \delta_{ij}. \tag{5.51}
$$

If it is already known that the errors are independent and have Gaussian distributions, these properties are implied, and equations 5.50 and 5.51 add no new information. If, however, the error distribution is unknown, these are independent assumptions.

To avoid cumbersome and opaque algebra, let us extend the matrix notation. Define the vectors $\mathbf{y}$, $\boldsymbol{\epsilon}$, and $\mathbf{a}$ to be

$$
\mathbf{y} = \begin{pmatrix} y_1 \\ y_2 \\ \vdots \\ y_n \end{pmatrix} \quad \boldsymbol{\epsilon} = \begin{pmatrix} \epsilon_1 \\ \epsilon_2 \\ \vdots \\ \epsilon_n \end{pmatrix} \quad \mathbf{a} = \begin{pmatrix} a_0 \\ a_1 \\ \vdots \\ a_m \end{pmatrix}, \tag{5.52}
$$

and the $n \times (m+1)$ matrix $\mathbf{X}$ and $n \times n$ matrix $\mathbf{W}$ to be

$$\mathbf{X} = \begin{pmatrix} 1 & x_1 & \cdots & x_1^m \\ 1 & x_2 & \cdots & x_2^m \\ \vdots & \vdots & & \vdots \\ 1 & x_n & \cdots & x_n^m \end{pmatrix} \qquad \mathbf{W} = \begin{pmatrix} w_1 & 0 & \cdots & 0 \\ 0 & w_2 & \cdots & 0 \\ \vdots & \vdots & & \vdots \\ 0 & 0 & \cdots & w_n \end{pmatrix}. \qquad (5.53)$$

It is easy to verify that the normal matrix (equation 5.45) is given by

$$\mathbf{N} = \mathbf{X}^{\mathrm{T}}\mathbf{W}\mathbf{X}, \qquad (5.54)$$

and the $\mathbf{Y}$ vector (equation 5.46) is given by

$$\mathbf{Y} = \mathbf{X}^{\mathrm{T}}\mathbf{W}\mathbf{y}, \qquad (5.55)$$

and that equation 5.49 becomes

$$\mathbf{y} = \mathbf{X}\mathbf{a} + \boldsymbol{\epsilon}. \qquad (5.56)$$

Equation 5.41 becomes

$$S = (\mathbf{y} \quad \mathbf{X}\mathbf{a})^{\mathrm{T}}\mathbf{W}(\mathbf{y} - \mathbf{X}\mathbf{a}), \qquad (5.57)$$

the normal equations (equation 5.43) become

$$(\mathbf{X}^{\mathrm{T}}\mathbf{W}\mathbf{X})\hat{\mathbf{a}} = \mathbf{X}^{\mathrm{T}}\mathbf{W}\mathbf{y}, \qquad (5.58)$$

and the solution (equation 5.44) becomes

$$\hat{\mathbf{a}} = (\mathbf{X}^{\mathrm{T}}\mathbf{W}\mathbf{X})^{-1}\mathbf{X}^{\mathrm{T}}\mathbf{W}\mathbf{y}. \qquad (5.59)$$

Estimates of the Parameters Are Unbiased: With this notation, it is now easy to show that the $\hat{\mathbf{a}}$ are unbiased. Use equation 5.56 to eliminate $\mathbf{y}$ in equation 5.59:

$$\begin{aligned} \hat{\mathbf{a}} &= (\mathbf{X}^{\mathrm{T}}\mathbf{W}\mathbf{X})^{-1}\mathbf{X}^{\mathrm{T}}\mathbf{W}(\mathbf{X}\mathbf{a} + \boldsymbol{\epsilon}) \\ &= (\mathbf{X}^{\mathrm{T}}\mathbf{W}\mathbf{X})^{-1}(\mathbf{X}^{\mathrm{T}}\mathbf{W}\mathbf{X})\mathbf{a} + (\mathbf{X}^{\mathrm{T}}\mathbf{W}\mathbf{X})^{-1}\mathbf{X}^{\mathrm{T}}\mathbf{W}\boldsymbol{\epsilon} \\ &= \mathbf{a} + (\mathbf{X}^{\mathrm{T}}\mathbf{W}\mathbf{X})^{-1}\mathbf{X}^{\mathrm{T}}\mathbf{W}\boldsymbol{\epsilon}. \end{aligned} \qquad (5.60)$$

Taking the mean value of $\hat{\mathbf{a}}$, we have

$$\begin{aligned} \langle\hat{\mathbf{a}}\rangle &= \langle\mathbf{a} + (\mathbf{X}^{\mathrm{T}}\mathbf{W}\mathbf{X})^{-1}\mathbf{X}^{\mathrm{T}}\mathbf{W}\boldsymbol{\epsilon}\rangle \\ &= \mathbf{a} + (\mathbf{X}^{\mathrm{T}}\mathbf{W}\mathbf{X})^{-1}\mathbf{X}^{\mathrm{T}}\mathbf{W}\langle\boldsymbol{\epsilon}\rangle, \end{aligned} \qquad (5.61)$$

Since $\langle\boldsymbol{\epsilon}\rangle = 0$, we arrive at

$$\langle\hat{\mathbf{a}}\rangle = \mathbf{a}. \qquad (5.62)$$

Thus, the estimated parameters are unbiased.

Covariance Matrix: Although $\hat{a}_i$ is the best estimate for parameter a_i, it is nevertheless an estimate, and its value differs from the true value. The size of the difference can be characterized by the variance of the difference[2]

$$\sigma_{\hat{i}}^2 = \langle (\hat{a}_i - a_i)^2 \rangle. \tag{5.63}$$

In addition, the difference between $\hat{a}_i$ and a_i can be affected by differences between the estimated and true values of other parameters because of correlations. These correlations can be characterized by the covariances

$$\sigma_{\widehat{ij}} = \langle (\hat{a}_i - a_i)(\hat{a}_j - a_j) \rangle. \tag{5.64}$$

The variances and covariances can be collected together to form the covariance matrix for the estimated parameters:

$$\mathbf{C} = \begin{pmatrix} \sigma_{\hat{0}}^2 & \sigma_{\widehat{01}} & \cdots & \sigma_{\widehat{0m}} \\ \sigma_{\widehat{10}} & \sigma_{\hat{1}}^2 & \cdots & \sigma_{\widehat{1m}} \\ \vdots & \vdots & & \vdots \\ \sigma_{\widehat{m0}} & \sigma_{\widehat{m1}} & \cdots & \sigma_{\hat{m}}^2 \end{pmatrix}$$

$$= \begin{pmatrix} \langle (\hat{a}_0 - a_0)^2 \rangle & \langle (\hat{a}_0 - a_0)(\hat{a}_1 - a_1) \rangle & \cdots & \langle (\hat{a}_0 - a_0)(\hat{a}_m - a_m) \rangle \\ \langle (\hat{a}_1 - a_1)(\hat{a}_0 - a_0) \rangle & \langle (\hat{a}_1 - a_1)(\hat{a}_1 - a_1) \rangle & \cdots & \langle (\hat{a}_1 - a_1)(\hat{a}_m - a_m) \rangle \\ \vdots & \vdots & & \vdots \\ \langle (\hat{a}_m - a_m)(\hat{a}_0 - a_0) \rangle & \langle (\hat{a}_m - a_m)(\hat{a}_1 - a_1) \rangle & \cdots & \langle (\hat{a}_m - a_m)^2 \rangle \end{pmatrix}$$

$$= \left\langle \begin{pmatrix} \hat{a}_0 - a_0 \\ \hat{a}_1 - a_1 \\ \vdots \\ \hat{a}_m - a_m \end{pmatrix} \begin{pmatrix} \hat{a}_0 - a_0, & \hat{a}_1 - a_1, & \ldots, & \hat{a}_m - a_m \end{pmatrix} \right\rangle. \tag{5.65}$$

This can be written more succinctly as

$$\mathbf{C} = \langle (\hat{\mathbf{a}} - \mathbf{a})(\hat{\mathbf{a}} - \mathbf{a})^{\mathrm{T}} \rangle. \tag{5.66}$$

The covariance matrix is crucially important, because it quantifies how much the estimates of the parameters can be trusted. From equation 5.60 the covariance matrix is

$$\mathbf{C} = \left\langle \left\{ (\mathbf{X}^{\mathrm{T}}\mathbf{W}\mathbf{X})^{-1} \mathbf{X}^{\mathrm{T}}\mathbf{W}\epsilon \right\} \left\{ (\mathbf{X}^{\mathrm{T}}\mathbf{W}\mathbf{X})^{-1} \mathbf{X}^{\mathrm{T}}\mathbf{W}\epsilon \right\}^{\mathrm{T}} \right\rangle$$

$$= \left\langle (\mathbf{X}^{\mathrm{T}}\mathbf{W}\mathbf{X})^{-1} \mathbf{X}^{\mathrm{T}}\mathbf{W}\epsilon\epsilon^{\mathrm{T}}\mathbf{W}^{\mathrm{T}}\mathbf{X} \left\{ (\mathbf{X}^{\mathrm{T}}\mathbf{W}\mathbf{X})^{-1} \right\}^{\mathrm{T}} \right\rangle$$

$$= (\mathbf{X}^{\mathrm{T}}\mathbf{W}\mathbf{X})^{-1} \mathbf{X}^{\mathrm{T}}\mathbf{W} \langle \epsilon\epsilon^{\mathrm{T}} \rangle \mathbf{W}^{\mathrm{T}}\mathbf{X} \left\{ (\mathbf{X}^{\mathrm{T}}\mathbf{W}\mathbf{X})^{-1} \right\}^{\mathrm{T}}. \tag{5.67}$$

---

[2] The Greek letter sigma with a hat on the subscripts (e.g., $\sigma_{\widehat{12}}$) stands for the variances and covariances of fitted parameters, but without a hat on the subscripts (e.g., σ_{12}) it stands for the variances and covariances of the original data points, a subtle notational difference for two very different quantities.

We evaluate $\langle \epsilon \epsilon^\mathrm{T} \rangle$ as follows (note that the subscripts run from 1 to n, because the ϵ_i are the errors in the data points):

$$\langle \epsilon \epsilon^\mathrm{T} \rangle = \begin{pmatrix} \langle \epsilon_1 \epsilon_1 \rangle & \langle \epsilon_1 \epsilon_2 \rangle & \cdots & \langle \epsilon_1 \epsilon_n \rangle \\ \langle \epsilon_2 \epsilon_1 \rangle & \langle \epsilon_2 \epsilon_2 \rangle & \cdots & \langle \epsilon_2 \epsilon_n \rangle \\ \vdots & \vdots & & \vdots \\ \langle \epsilon_n \epsilon_1 \rangle & \langle \epsilon_n \epsilon_2 \rangle & \cdots & \langle \epsilon_n \epsilon_n \rangle \end{pmatrix} = \begin{pmatrix} \sigma_{11} & \sigma_{12} & \cdots & \sigma_{1n} \\ \sigma_{21} & \sigma_{22} & \cdots & \sigma_{2n} \\ \vdots & \vdots & & \vdots \\ \sigma_{n1} & \sigma_{n2} & \cdots & \sigma_{nn} \end{pmatrix}, \quad (5.68)$$

but since $\langle \epsilon_i \epsilon_j \rangle = \sigma_i^2 \delta_{ij}$, we have

$$\langle \epsilon \epsilon^\mathrm{T} \rangle = \begin{pmatrix} \sigma_1^2 & 0 & \cdots & 0 \\ 0 & \sigma_2^2 & \cdots & 0 \\ \vdots & \vdots & & \vdots \\ 0 & 0 & \cdots & \sigma_n^2 \end{pmatrix} = \mathbf{W}^{-1}. \quad (5.69)$$

Equation 5.67 now becomes

$$\begin{aligned} \mathbf{C} &= \left(\mathbf{X}^\mathrm{T} \mathbf{W} \mathbf{X} \right)^{-1} \mathbf{X}^\mathrm{T} \mathbf{W} \mathbf{W}^{-1} \mathbf{W}^\mathrm{T} \mathbf{X} \left\{ \left(\mathbf{X}^\mathrm{T} \mathbf{W} \mathbf{X} \right)^{-1} \right\}^\mathrm{T} \\ &= \left(\mathbf{X}^\mathrm{T} \mathbf{W} \mathbf{X} \right)^{-1} \mathbf{X}^\mathrm{T} \mathbf{W}^\mathrm{T} \mathbf{X} \left\{ \left(\mathbf{X}^\mathrm{T} \mathbf{W} \mathbf{X} \right)^{-1} \right\}^\mathrm{T}. \end{aligned} \quad (5.70)$$

Because $\mathbf{W}$ and $\mathbf{X}^\mathrm{T} \mathbf{W} \mathbf{X}$ are symmetric, they equal their transposes, so the covariance matrix becomes

$$\begin{aligned} \mathbf{C} &= \left(\mathbf{X}^\mathrm{T} \mathbf{W} \mathbf{X} \right)^{-1} \mathbf{X}^\mathrm{T} \mathbf{W} \mathbf{X} \left(\mathbf{X}^\mathrm{T} \mathbf{W} \mathbf{X} \right)^{-1} = \left(\mathbf{X}^\mathrm{T} \mathbf{W} \mathbf{X} \right)^{-1} \\ &= \mathbf{N}^{-1}. \end{aligned} \quad (5.71)$$

This is a remarkable result. The covariance matrix is equal to the inverse of the normal matrix! Since we must calculate the normal matrix and its inverse anyway, we get the covariance matrix for free.

Estimated Covariance Matrix: Despite its beauty, equation 5.71 is inadequate, because it assumes that the weights are correctly given by $w_i = 1/\langle \epsilon_i^2 \rangle = 1/\sigma_i^2$. In practice, the values of the σ_i supplied with the data points are often inaccurate. If so, then $\langle \epsilon_i^2 \rangle \neq \sigma_i^2$, and equation 5.69 is no longer correct. A common way to deal with this problem is to assume that the true weights are proportional to $1/\sigma_i^2$. Equation 5.59 still gives the correct values of $\hat{\mathbf{a}}$, because the proportionality constant cancels out in the product of the $\mathbf{W}$ and $\mathbf{W}^{-1}$ terms; but the covariance matrix given by equation 5.71 is no longer correct, because the proportionality constant does not cancel. Adopting this assumption, we replace equation 5.51 with

$$\langle \epsilon_i \epsilon_j \rangle = \sigma^2 \sigma_i^2 \delta_{ij}, \quad (5.72)$$

where σ^2 is the proportionality constant. The true weights are then given by

$$\text{true weights} = \frac{1}{\sigma^2 \sigma_i^2} = \frac{w_i}{\sigma^2}. \quad (5.73)$$

The proportionality constant is the same variance of unit weight we first met in equations 4.43 and 4.44. Equation 5.71 becomes

$$\mathbf{C} = \sigma^2 \left(\mathbf{X^T W X}\right)^{-1} = \sigma^2 \mathbf{N}^{-1}. \tag{5.74}$$

We now show how to estimate σ^2. All but the most dedicated readers may want to skip this unedifying derivation and jump to the summary near the end of this section. To find the estimates of the parameters we minimized, set

$$S = \boldsymbol{\epsilon}^T \mathbf{W} \boldsymbol{\epsilon} = (\mathbf{y} - \mathbf{Xa})^T \mathbf{W} (\mathbf{y} - \mathbf{Xa}). \tag{5.75}$$

The value of S at its minimum is

$$\hat{S}_{min} = (\mathbf{y} - \mathbf{X\hat{a}})^T \mathbf{W} (\mathbf{y} - \mathbf{X\hat{a}}). \tag{5.76}$$

Add and subtract $\mathbf{Xa}$ to $\mathbf{y} - \mathbf{X\hat{a}}$, and then rearrange the terms:

$$\begin{aligned}
\hat{S}_{min} &= (\mathbf{y} - \mathbf{X\hat{a}} - \mathbf{Xa} + \mathbf{Xa})^T \mathbf{W} (\mathbf{y} - \mathbf{X\hat{a}} - \mathbf{Xa} + \mathbf{Xa}) \\
&= [(\mathbf{y} - \mathbf{Xa}) - \mathbf{X}(\hat{a} - a)]^T \mathbf{W} [(\mathbf{y} - \mathbf{Xa}) - \mathbf{X}(\hat{a} - a)].
\end{aligned} \tag{5.77}$$

Remembering that $\mathbf{y} - \mathbf{Xa} = \boldsymbol{\epsilon}$ and using equation 5.60 to eliminate $(\hat{a} - a)$ we find

$$\begin{aligned}
\hat{S}_{min} &= \left[\boldsymbol{\epsilon} - \mathbf{X}(\mathbf{X^T W X})^{-1} \mathbf{X^T W} \boldsymbol{\epsilon} \right]^T \mathbf{W} \left[\boldsymbol{\epsilon} - \mathbf{X}(\mathbf{X^T W X})^{-1} \mathbf{X^T W} \boldsymbol{\epsilon} \right] \\
&= \boldsymbol{\epsilon}^T \mathbf{W} \boldsymbol{\epsilon} \\
&\quad - \boldsymbol{\epsilon}^T \mathbf{W^T X}[(\mathbf{X^T W X})^{-1}]^T \mathbf{X^T W} \boldsymbol{\epsilon} \\
&\quad - \boldsymbol{\epsilon}^T \mathbf{W X}(\mathbf{X^T W X})^{-1} \mathbf{X^T W} \boldsymbol{\epsilon} \\
&\quad + \boldsymbol{\epsilon}^T \mathbf{W^T X}[(\mathbf{X^T W X})^{-1}]^T \mathbf{X^T W X}(\mathbf{X^T W X})^{-1} \mathbf{X^T W} \boldsymbol{\epsilon}.
\end{aligned} \tag{5.78}$$

Because $\mathbf{W}$ and $\mathbf{X^T W X}$ are symmetric, this imposing equation collapses to

$$\hat{S}_{min} = \boldsymbol{\epsilon}^T \mathbf{W} \boldsymbol{\epsilon} - \boldsymbol{\epsilon}^T \mathbf{W X} \left(\mathbf{X^T W X}\right)^{-1} \mathbf{X^T W} \boldsymbol{\epsilon}. \tag{5.79}$$

Now take the mean value of $\hat{S}_{min}$:

$$\left\langle \hat{S}_{min} \right\rangle = \left\langle \boldsymbol{\epsilon}^T \mathbf{W} \boldsymbol{\epsilon} \right\rangle - \left\langle \boldsymbol{\epsilon}^T \mathbf{W X} \left(\mathbf{X^T W X}\right)^{-1} \mathbf{X^T W} \boldsymbol{\epsilon} \right\rangle. \tag{5.80}$$

The first term on the right-hand side of equation 5.80 becomes

$$\begin{aligned}
\left\langle \boldsymbol{\epsilon}^T \mathbf{W} \boldsymbol{\epsilon} \right\rangle &= \left\langle (\epsilon_1, \quad \epsilon_2, \quad \ldots, \quad \epsilon_n) \begin{pmatrix} w_1 \epsilon_1 \\ w_2 \epsilon_2 \\ \vdots \\ w_n \epsilon_n \end{pmatrix} \right\rangle \\
&= \left\langle \sum_{i=1}^n w_i \epsilon_i^2 \right\rangle = \sum_{i=1}^n w_i \left\langle \epsilon_i^2 \right\rangle = \sum_{i=1}^n \frac{1}{\sigma_i^2} \sigma^2 \sigma_i^2 = \sum_{i=1}^n \sigma^2 \\
&= n\sigma^2.
\end{aligned} \tag{5.81}$$

Note that the third equality in the penultimate line in equation 5.81 makes use of equation 5.72. Since $\mathbf{X}^{\mathrm{T}}\mathbf{W}\mathbf{X} = \mathbf{N}$, the second term on the right-hand side of equation 5.80 can be written

$$\left\langle \boldsymbol{\epsilon}^{\mathrm{T}}\mathbf{W}\mathbf{X}\left(\mathbf{X}^{\mathrm{T}}\mathbf{W}\mathbf{X}\right)^{-1}\mathbf{X}^{\mathrm{T}}\mathbf{W}\boldsymbol{\epsilon} \right\rangle = \left\langle \mathbf{b}^{\mathrm{T}}\mathbf{N}^{-1}\mathbf{b} \right\rangle, \tag{5.82}$$

where $\mathbf{b}$ is defined to be

$$\mathbf{b} = \mathbf{X}^{\mathrm{T}}\mathbf{W}\boldsymbol{\epsilon} = \begin{pmatrix} \sum_{i=1}^{n} w_i \epsilon_i \\ \sum_{i=1}^{n} w_i x_i \epsilon_i \\ \vdots \\ \sum_{i=1}^{n} w_i x_i^m \epsilon_i \end{pmatrix} = \begin{pmatrix} b_0 \\ b_1 \\ \vdots \\ b_m \end{pmatrix}. \tag{5.83}$$

Expanding equation 5.82, we have

$$\left\langle \mathbf{b}^{\mathrm{T}}\mathbf{N}^{-1}\mathbf{b} \right\rangle = \left\langle \sum_{j=0}^{m}\sum_{k=0}^{m} b_j (\mathbf{N}^{-1})_{jk} b_k \right\rangle = \sum_{j=0}^{m}\sum_{k=0}^{m} (\mathbf{N}^{-1})_{jk} \left\langle b_j b_k \right\rangle. \tag{5.84}$$

We must now evaluate $\langle b_j b_k \rangle$:

$$\langle b_j b_k \rangle = \left\langle \left(\sum_{i=1}^{n} w_i x^j \epsilon_i\right)\left(\sum_{\ell=1}^{n} w_\ell x^k \epsilon_\ell\right) \right\rangle = \sum_{i=1}^{n}\sum_{\ell=1}^{n} w_i w_\ell x^j x^k \langle \epsilon_i \epsilon_\ell \rangle$$

$$= \sum_{i=1}^{n}\sum_{\ell=1}^{n} w_i w_\ell x^j x^k \sigma^2 \sigma_i^2 \delta_{i\ell} = \sum_{i=1}^{n} w_i w_i x^j x^k \sigma^2 \sigma_i^2$$

$$= \sigma^2 \sum_{i=1}^{n} w_i x^j x^k. \tag{5.85}$$

Thus the $\langle b_j b_k \rangle$ are equal to σ^2 times the corresponding components of $\mathbf{N}$. Since $\mathbf{N}$ is symmetric,

$$\langle b_j b_k \rangle = \sigma^2 (\mathbf{N})_{jk} = \sigma^2 (\mathbf{N})_{kj}. \tag{5.86}$$

Returning to equation 5.84, we now have

$$\left\langle \mathbf{b}^{\mathrm{T}}\mathbf{N}^{-1}\mathbf{b} \right\rangle = \sigma^2 \sum_{j=0}^{m}\sum_{k=0}^{m} (\mathbf{N}^{-1})_{jk}(\mathbf{N})_{kj} = \sigma^2 \sum_{j=0}^{m} (\mathbf{I})_{jj}. \tag{5.87}$$

Thus we see that $\langle \mathbf{b}^{\mathrm{T}}\mathbf{N}^{-1}\mathbf{b} \rangle$ is just σ^2 times the trace of the $(m+1) \times (m+1)$ identity matrix:

$$\left\langle \mathbf{b}^{\mathrm{T}}\mathbf{N}^{-1}\mathbf{b} \right\rangle = \sigma^2 \mathrm{Trace}\,(\mathbf{I}) = (m+1)\sigma^2. \tag{5.88}$$

Returning finally to equation 5.80, we arrive at

$$\left\langle \hat{S}_{min} \right\rangle = n\sigma^2 - (m+1)\sigma^2 = (n-m-1)\sigma^2. \tag{5.89}$$

With this result, we are justified in taking the estimated value of σ^2 to be

$$\hat{\sigma}^2 = \frac{\hat{S}_{min}}{n-m-1}. \tag{5.90}$$

For the estimated covariance matrix, we can take[3]

$$\hat{\mathbf{C}} \equiv \begin{pmatrix} \hat{\sigma}_{\hat{0}}^2 & \hat{\sigma}_{\widehat{01}} & \cdots & \hat{\sigma}_{\widehat{0m}} \\ \hat{\sigma}_{\widehat{10}} & \hat{\sigma}_{\hat{1}}^2 & \cdots & \hat{\sigma}_{\widehat{1m}} \\ \vdots & \vdots & & \vdots \\ \hat{\sigma}_{\widehat{m0}} & \hat{\sigma}_{\widehat{m1}} & \cdots & \hat{\sigma}_{\hat{m}}^2 \end{pmatrix}$$

$$= \hat{\sigma}^2 \left(\mathbf{X}^{\mathbf{T}} \mathbf{W} \mathbf{X} \right)^{-1} = \frac{\hat{S}_{min}}{n - m - 1} \left(\mathbf{X}^{\mathbf{T}} \mathbf{W} \mathbf{X} \right)^{-1}, \tag{5.91}$$

where the $\hat{\sigma}_j^2$ and $\hat{\sigma}_{\widehat{ij}}$ are the estimated values of the variances and covariances, respectively. Equivalently,

$$\hat{\mathbf{C}} = \hat{\sigma}^2 \mathbf{N}^{-1} = \frac{\hat{S}_{min}}{n - m - 1} \mathbf{N}^{-1}. \tag{5.92}$$

Summary: We are given n independent data points (x_i, y_i, σ_i) and wish to fit them with the polynomial

$$y = a_0 + a_1 x + a_2 x^2 + \cdots + a_m x^m \tag{5.93}$$

by least squares. The normal equations are

$$\begin{pmatrix} \sum w_i & \sum w_i x_i & \cdots & \sum w_i x_i^m \\ \sum w_i x_i & \sum w_i x_i^2 & \cdots & \sum w_i x_i^{m+1} \\ \vdots & \vdots & & \vdots \\ \sum w_i x_i^m & \sum w_i x_i^{m+1} & \cdots & \sum w_i x_i^{2m} \end{pmatrix} \begin{pmatrix} \hat{a}_0 \\ \hat{a}_1 \\ \vdots \\ \hat{a}_m \end{pmatrix} = \begin{pmatrix} \sum w_i y_i \\ \sum w_i x_i y_i \\ \vdots \\ \sum w_i x_i^m y_i \end{pmatrix}, \tag{5.94}$$

where $w_i = 1/\sigma_i^2$, and the estimates of the a_j are given by

$$\begin{pmatrix} \hat{a}_0 \\ \hat{a}_1 \\ \vdots \\ \hat{a}_m \end{pmatrix} = \begin{pmatrix} \sum w_i & \sum w_i x_i & \cdots & \sum w_i x_i^m \\ \sum w_i x_i & \sum w_i x_i^2 & \cdots & \sum w_i x_i^{m+1} \\ \vdots & \vdots & & \vdots \\ \sum w_i x_i^m & \sum w_i x_i^{m+1} & \cdots & \sum w_i x_i^{2m} \end{pmatrix}^{-1} \begin{pmatrix} \sum w_i y_i \\ \sum w_i x_i y_i \\ \vdots \\ \sum w_i x_i^m y_i \end{pmatrix}. \tag{5.95}$$

In matrix form the normal equations are

$$\mathbf{N}\hat{\mathbf{a}} = \mathbf{Y}, \tag{5.96}$$

and the estimated parameters of the fit are

$$\hat{\mathbf{a}} = \mathbf{N}^{-1}\mathbf{Y}. \tag{5.97}$$

---

[3] One usually sees $n - m$ in the denominator, not $n - m - 1$. The intent is to subtract the number of degrees of freedom from n. For us, the number of degrees of freedom is $m + 1$, because the parameters are numbered from a_0 to a_m.

If the values of the σ_i are correct and the polynomial is a good description of the data, the covariance matrix for the estimated parameters is

$$
\mathbf{C} = \begin{pmatrix} \sigma_{\hat{0}}^2 & \sigma_{\widehat{01}} & \cdots & \sigma_{\widehat{0m}} \\ \sigma_{\widehat{10}} & \sigma_{\hat{1}}^2 & \cdots & \sigma_{\widehat{1m}} \\ \vdots & \vdots & & \vdots \\ \sigma_{\widehat{m0}} & \sigma_{\widehat{m1}} & \cdots & \sigma_{\hat{m}}^2 \end{pmatrix} = \mathbf{N}^{-1}, \tag{5.98}
$$

where $\sigma_{\hat{j}}^2$ is the variance of $\hat{a}_j$ and $\sigma_{\widehat{ij}}$ is the covariance between $\hat{a}_i$ and $\hat{a}_j$. Generally, however, one must use the estimated covariance matrix

$$
\hat{\mathbf{C}} = \begin{pmatrix} \hat{\sigma}_{\hat{0}}^2 & \hat{\sigma}_{\widehat{01}} & \cdots & \hat{\sigma}_{\widehat{0m}} \\ \hat{\sigma}_{\widehat{10}} & \hat{\sigma}_{\hat{1}}^2 & \cdots & \hat{\sigma}_{\widehat{1m}} \\ \vdots & \vdots & & \vdots \\ \hat{\sigma}_{\widehat{m0}} & \hat{\sigma}_{\widehat{m1}} & \cdots & \hat{\sigma}_{\hat{m}}^2 \end{pmatrix} = \frac{\hat{S}_{min}}{n-m-1}\mathbf{C} = \frac{\hat{S}_{min}}{n-m-1}\mathbf{N}^{-1}, \tag{5.99}
$$

where $(m+1)$ is the number of fitted parameters (see the footnote to equation 5.91), $\hat{\sigma}_{\hat{j}}^2$ and $\hat{\sigma}_{\widehat{ij}}$ are the estimated variances and covariances, respectively, and $\hat{S}_{min}$ is the minimum value of S:

$$
\hat{S}_{min} = \sum_{i=1}^{n} w_i (y_i - \hat{a}_0 - \hat{a}_1 x_i - \cdots - \hat{a}_m x_i^m)^2. \tag{5.100}
$$

Example: We continue the example of the straight-line least squares fit discussed in Section 5.3.1. We now wish to calculate variances and covariances for the parameters of fit. If the standard deviations of the data points are correct and if a straight line is a good description of the data, then the covariance matrix is equal to the inverse of the normal matrix:

$$
\mathbf{C} = \begin{pmatrix} \sigma_{\hat{0}}^2 & \sigma_{\widehat{01}} \\ \sigma_{\widehat{10}} & \sigma_{\hat{1}}^2 \end{pmatrix} = \mathbf{N}^{-1} = \begin{pmatrix} 3.7612 \times 10^{-2} & -5.7575 \times 10^{-3} \\ -5.7575 \times 10^{-3} & 9.0951 \times 10^{-4} \end{pmatrix}.
$$

This would suggest that the parameters and their standard deviations of the parameters are

$$
\hat{a}_0 \pm \sigma_{\hat{0}} = -3.45 \pm 0.19
$$
$$
\hat{a}_1 \pm \sigma_{\hat{1}} = 0.897 \pm 0.030.
$$

These variances and standard deviations are, however, surely too small. The data points and their error bars are replotted in Figure 5.4 along with the best fit straight line. If errors of the data points have a Gaussian distribution, we would expect roughly 1/3 of the data points to be more than one standard deviation from the fitted line. Instead, 7 of the 12 data points are more than a standard deviation from the line. It appears that the standard deviations of the data points have been underestimated. (This discrepancy is further quantified in Section 5.5.3, which discusses χ^2 as a test for goodness of fit.)

Continued on page 136

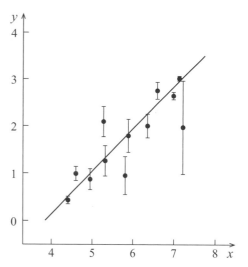

Figure 5.4: The weighted fit of a straight line to the data points. The figure is a repeat of the right panel in Figure 5.3, but the plot of the unweighted fit has been omitted for clarity. More than half the data points are more than one standard deviation from the fitted line, indicating that the errors on the data points are underestimated.

The covariances of the fitted parameters need to be adjusted to account for the inaccurate standard deviations of the data points. After additional calculations, we find

$$\hat{\sigma}^2 = \frac{\hat{S}_{min}}{n - m - 1} = \frac{\sum_{i=1}^{12} w_i(y_i + 3.45 - 0.897x_i)^2}{10} = 2.74.$$

The estimated covariance matrix is therefore

$$\hat{C} = \begin{pmatrix} \hat{\sigma}_{\hat{0}}^2 & \hat{\sigma}_{\hat{0}\hat{1}} \\ \hat{\sigma}_{\hat{1}\hat{0}} & \hat{\sigma}_{\hat{1}}^2 \end{pmatrix} = \hat{\sigma}^2 N^{-1} = \begin{pmatrix} 0.1029 & -0.01575 \\ -0.01575 & 0.002488 \end{pmatrix}.$$

The parameters and their revised standard deviations are

$$\hat{a}_0 \pm \hat{\sigma}_{\hat{0}} = -3.45 \pm 0.32$$
$$\hat{a}_1 \pm \hat{\sigma}_{\hat{1}} = 0.897 \pm 0.050$$

As expected, the standard deviations are larger than their unadjusted values, although the magnitude of the difference, $\sim 50\%$, may be surprising.

As a final point, the parameters derived from the unweighted fit to these same data points were $\hat{a}_0 = -2.43$ and $\hat{a}_1 = 0.707$. We now see that the differences between the weighted and unweighted parameters of fit are substantial and significant: 3.2 standard deviations for $\hat{a}_0$ and 3.8 standard deviations for $\hat{a}_1$.

5.3.4 Monte Carlo Error Analysis

The discussion of error analysis in the previous section assumed that the errors in the data either have Gaussian distributions or are well characterized by just their standard deviation. If, however, these assumptions are incorrect, the discussion the does not apply, and equations 5.98–5.100 cannot be used. Worse, the error analysis may not yield to analytic techniques.

A common way to handle this situation is to determine the probability distributions for the fitted parameters numerically by a Monte Carlo simulation. Let us assume the probability distributions that describe the data points and their errors are known (and not Gaussian). Create many artificial sets of data by generating random numbers from these probability distributions using the techniques discussed in Chapter 3. The point is to create artificial data whose properties, including their error distributions, are the same as the properties of the real data. Replace the real data with the artificial data, and for each set of artificial data, calculate the least squares (or maximum-likelihood) estimates of the fitting parameters. The resulting distribution of these artificial fitted parameters is proportional to the probability distribution for the real fitted parameters. Means and standard deviations of the fitted parameters can be extracted from these distributions using techniques similar to those outlined in Section 3.5.1.

Monte Carlo error analysis should, in principle, always work. In practice, the analysis may may require creating and analyzing many sets of artificial data. As a result, Monte Carlo data analysis is often computationally intensive.

5.4 Need for Covariances and Propagation of Errors

5.4.1 Need for Covariances

The need for the variances of the fitted parameters is clear. Here we show that the covariances among the fitted parameters are also needed. In Section 5.3.1 we fit a straight line $y = a_0 + a_1 x$ to a set of data points, finding

$$\hat{y} = \hat{a}_0 + \hat{a}_1 x. \qquad (5.101)$$

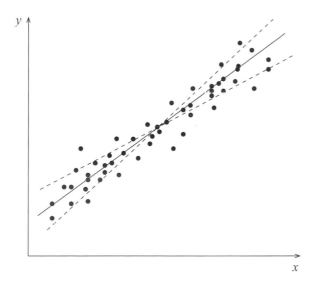

Figure 5.5: The solid line is $\hat{y} = \hat{a}_0 + \hat{a}_1 x$, the least squares fit to the data. Because of the scatter in the data, $\hat{a}_0$ and $\hat{a}_1$ are uncertain. There is a family of straight lines that fit the data almost as well as the least squares fit. Two members of the family are shown as dashed lines in the figure. They have a small but nonzero probability of being the correct fits to the data.

Note that there is a hat on $\hat{y}$, because equation 5.101 gives an estimate of y, not the true value of y. The fit is represented in Figure 5.5, where equation 5.101 is shown as a solid line. Since there is some uncertainty in $\hat{a}_0$ and $\hat{a}_1$, there is a family of straight lines that fit the data almost as well as equation 5.101, each member of the family having slightly different values for a_0 and a_1. Two members of the family are shown as the dashed lines in the figure. Thus the value of $\hat{y}$ has some uncertainty. The uncertainty can be characterized by the variance of $\hat{y}$,

$$\sigma_{\hat{y}}^2 = \left\langle (\hat{y} - y)^2 \right\rangle. \tag{5.102}$$

Substituting for $\hat{y}$ and y, we have

$$\sigma_{\hat{y}}^2 = \left\langle \left[(\hat{a}_0 + \hat{a}_1 x) - (a_0 + a_1 x) \right]^2 \right\rangle = \left\langle \left[(\hat{a}_0 - a_0) + (\hat{a}_1 - a_1)x \right]^2 \right\rangle$$
$$= \left\langle (\hat{a}_0 - a_0)^2 \right\rangle + 2 \left\langle (\hat{a}_0 - a_0)(\hat{a}_1 - a_1) \right\rangle x + \left\langle (\hat{a}_1 - a_1)^2 \right\rangle x^2$$
$$= \sigma_0^2 + 2\sigma_{\widehat{01}} x + \sigma_1^2 x^2. \tag{5.103}$$

Three points should be noted about this equation. First, calculation of $\sigma_{\hat{y}}^2$ requires the covariance $\sigma_{\widehat{01}}$ between $\hat{a}_0$ and $\hat{a}_1$, not just their variances. Second, the value of $\sigma_{\hat{y}}^2$ is not constant but depends on x. A plot of $\hat{y} \pm \sigma_{\hat{y}}$ is shown in Figure 5.6. One generally finds the smallest values of $\sigma_{\hat{y}}$ at values of x somewhere near the center of the cluster of data points. The standard deviation of $\hat{y}$ grows with distance from the center, ultimately growing linearly with x, demonstrating the danger of extrapolating a fitted function. Third, we generally know the measured values $\hat{\sigma}_0^2$, $\hat{\sigma}_{\widehat{01}}$, and $\hat{\sigma}_1^2$, not the true (unhatted) values. We must then be content with the estimated value of the variance of $\hat{y}$:

$$\hat{\sigma}_{\hat{y}}^2 = \hat{\sigma}_0^2 + 2\hat{\sigma}_{\widehat{01}} x + \hat{\sigma}_1^2 x^2. \tag{5.104}$$

Example: We return again to the first example in Section 5.3.1. In that example, we fit a straight line to data by weighted least squares. The data points from the example are replotted in Figure 5.6 along with their error bars. The best fit straight line is

$$\hat{y} = \hat{a}_0 + \hat{a}_1 x = -3.45 + 0.897x,$$

which is plotted as the solid line in the figure. The estimated covariance matrix for the parameters of fit (calculated near the end of Section 5.3.3) is

$$\hat{\mathbf{C}} = \begin{pmatrix} \hat{\sigma}_0^2 & \hat{\sigma}_{\widehat{01}} \\ \hat{\sigma}_{\widehat{10}} & \hat{\sigma}_1^2 \end{pmatrix} = \begin{pmatrix} 0.1029 & -0.01575 \\ -0.01575 & 0.002488 \end{pmatrix}.$$

We now wish to estimate the reliability of $\hat{y}$. From equation 5.104, the estimated variance of $\hat{y}$ is

$$\hat{\sigma}_{\hat{y}}^2 = 0.1029 - 2 \times 0.01575x + 0.002488x^2.$$

The two dotted lines in the figure show $\hat{y} \pm \hat{\sigma}_{\hat{y}}$. For values of x near the center of the cluster of points, $|\hat{\sigma}_{\hat{y}}| \approx 0.06$, so $\hat{y}$ is constrained to a relatively narrow range. This is largely a result of the small standard deviations of a few data points near the ends of the cluster. The standard deviation of $\hat{y}$ grows rapidly for values of x that are outside the

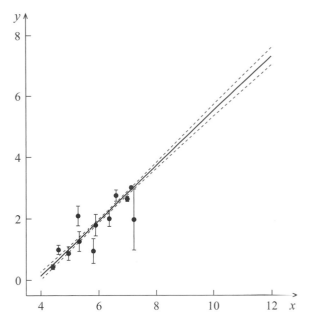

Figure 5.6: The solid line is the weighted least squares fit of a straight line $\hat{y} = \hat{a}_0 + \hat{a}_1 x$ to the data points shown in the figure. The dotted lines show $\hat{y} \pm \hat{\sigma}_{\hat{y}}$. The standard deviation of $\hat{y}$ is small near the cluster of data points but grows rapidly for values of x that are beyond the ends of the cluster. (See the first example in Section 5.3.1.)

cluster. When the best fit line is extrapolated to $x = 10$, a distance beyond the cluster that is roughly equal to the width of the cluster, the standard deviation has tripled to $|\hat{\sigma}_{\hat{y}}| \approx 0.18$.

5.4.2 Propagation of Errors

Any quantity that is calculated using fitted values of the parameters will be uncertain, because the fitted values themselves are uncertain. Suppose, for example, we have measured two quantities a_0 and a_1, finding $\hat{a}_0$ and $\hat{a}_1$ with variances and covariances $\sigma_{\hat{0}}$, $\sigma_{\hat{1}}$, and $\sigma_{\hat{1}\hat{2}}$. And suppose we wish to estimate a third quantity g that is the sum of a_0 and a_1. An unbiased estimate of g is given by

$$\hat{g} = \hat{a}_0 + \hat{a}_1. \tag{5.105}$$

The variance of $\hat{g}$ is given by

$$
\begin{aligned}
\sigma_{\hat{g}}^2 &= \langle (\hat{g} - g)^2 \rangle \\
&= \langle [(\hat{a}_0 + \hat{a}_1) - (a_0 + a_1)]^2 \rangle \\
&= \langle (\hat{a}_0 - a_0)^2 \rangle + 2 \langle (\hat{a}_0 - a_0)(\hat{a}_1 - a_1) \rangle + \langle (\hat{a}_1 - a_1)^2 \rangle \\
&= \sigma_{\hat{0}}^2 + 2\sigma_{\hat{0}\hat{1}} + \sigma_{\hat{1}}^2.
\end{aligned}
\tag{5.106}
$$

Even in this simplest of cases, calculation of the variance of $\hat{g}$ requires the variances and covariance of $\hat{a}_0$ and $\hat{a}_1$. Calculating the variance of a dependent quantity, such as $\hat{g}$, from the variances and covariances of independent quantities, such as $\hat{a}_0$ and $\hat{a}_1$, is called *propagation of errors*.

For the general case, let a quantity $g = g(a_0, a_1, \ldots, a_m)$ be a function of $m+1$ parameters a_j. Let the measured values of the parameters be $\hat{a}_j$, and expand $\hat{g}$ in a Taylor series about the true (unhatted) values:

$$\hat{g} = g(a_0, a_1, \ldots, a_m) + \sum_{j=0}^{m} \frac{\partial g}{\partial a_j}(\hat{a}_j - a_j) + \text{ higher order terms,} \qquad (5.107)$$

where the derivatives are evaluated at the a_j. The variance of $\hat{g}$ is

$$\sigma_{\hat{g}}^2 = \left\langle \left[\hat{g} - g(a_0, a_1, \ldots, a_m) \right]^2 \right\rangle$$

$$= \left\langle \left(\sum_{j=0}^{m} \frac{\partial g}{\partial a_j}(\hat{a}_j - a_j) \right)^2 \right\rangle$$

$$= \left\langle \sum_j \sum_k \frac{\partial g}{\partial a_j} \frac{\partial g}{\partial a_k}(\hat{a}_j - a_j)(\hat{a}_k - a_k) \right\rangle$$

$$= \sum_j \sum_k \frac{\partial g}{\partial a_j} \frac{\partial g}{\partial a_k} \left\langle (\hat{a}_j - a_j)(\hat{a}_k - a_k) \right\rangle$$

$$= \sum_j \sum_k \frac{\partial g}{\partial a_j} \frac{\partial g}{\partial a_k} \sigma_{\hat{j}\hat{k}}, \qquad (5.108)$$

where $\sigma_{\hat{j}\hat{k}}$ is the true covariance between $\hat{a}_j$ and $\hat{a}_k$, and $\sigma_{\hat{j}\hat{j}} = \sigma_j^2$ is the true variance of $\hat{a}_j$. We generally know neither the true values of the a_j nor the true variances and covariances, and must be content with their estimated values:

$$\hat{\sigma}_{\hat{g}}^2 = \sum_j \sum_k \frac{\partial g}{\partial a_j} \frac{\partial g}{\partial a_k} \hat{\sigma}_{\hat{j}\hat{k}}, \qquad (5.109)$$

where the derivatives are evaluated at the estimated values $\hat{a}_j$, and the variances and covariances are taken from the estimated covariance matrix $\hat{\mathbf{C}}$.

The following example shows how to propagate errors when taking the ratio of two measured quantities $\hat{a}_0$ and $\hat{a}_1$.

Example: Suppose we have measured two parameters, finding $\hat{a}_0$ and $\hat{a}_1$ with estimated variances $\hat{\sigma}_{\hat{0}}^2$ and $\hat{\sigma}_{\hat{1}}^2$ and covariance $\hat{\sigma}_{\hat{0}\hat{1}}$. A third quantity g is the ratio of the measured parameters

$$\hat{g} = \frac{\hat{a}_1}{\hat{a}_0}.$$

Because $\hat{a}_0$ and $\hat{a}_1$ are uncertain, $\hat{g}$ is also uncertain. Its variance is given by

$$\sigma_{\hat{g}}^2 = \left\langle (\hat{g} - g)^2 \right\rangle = \left\langle \Delta g^2 \right\rangle.$$

Expanding g in a Taylor series and retaining only the first derivatives, we have

$$\Delta g = \frac{\partial g}{\partial a_0} \Delta a_0 + \frac{\partial g}{\partial a_1} \Delta a_1 = -\frac{a_1}{a_0^2} \Delta a_0 + \frac{1}{a_0} \Delta a_1,$$

where $\Delta a_0 = \hat{a}_0 - a_0$, and $\Delta a_1 = \hat{a}_1 - a_1$. The variance of $\hat{g}$ is then

$$\sigma_{\hat{g}}^2 = \left\langle \frac{a_1^2}{a_0^4} \Delta a_0^2 - 2\frac{a_1}{a_0^3} \Delta a_0 \Delta a_1 + \frac{1}{a_0^2} \Delta a_1^2 \right\rangle$$

$$= \frac{a_1^2}{a_0^4} \left\langle \Delta a_0^2 \right\rangle - 2\frac{a_1}{a_0^3} \left\langle \Delta a_0 \Delta a_1 \right\rangle + \frac{1}{a_0^2} \left\langle \Delta a_1^2 \right\rangle$$

$$= \frac{a_1^2}{a_0^4} \sigma_{\hat{0}}^2 - 2\frac{a_1}{a_0^3} \sigma_{\widehat{01}} + \frac{1}{a_0^2} \sigma_{\hat{1}}^2.$$

We know the measured values $\hat{a}_0$, $\hat{a}_1$, $\hat{\sigma}_{\hat{0}}^2$, $\hat{\sigma}_{\widehat{01}}$, and $\hat{\sigma}_{\hat{1}}^2$, not the true values, so we must take

$$\hat{\sigma}_{\hat{g}}^2 = \frac{\hat{a}_1^2}{\hat{a}_0^4} \hat{\sigma}_{\hat{0}}^2 - 2\frac{\hat{a}_1}{\hat{a}_0^3} \hat{\sigma}_{\widehat{01}} + \frac{1}{\hat{a}_0^2} \hat{\sigma}_{\hat{1}}^2.$$

The expression for $\hat{\sigma}_{\hat{g}}^2$ can be put in a somewhat more enlightening form by dividing both sides by $\hat{g}^2$:

$$\frac{\hat{\sigma}_{\hat{g}}^2}{\hat{g}^2} = \frac{\hat{\sigma}_{\hat{0}}^2}{\hat{a}_0^2} - 2\frac{\hat{\sigma}_{\widehat{01}}}{\hat{a}_0 \hat{a}_1} + \frac{\hat{\sigma}_{\hat{1}}^2}{\hat{a}_1^2}.$$

Even in this simple case, the variance of $\hat{g}$ is a fairly complicated function of the parameters and their variances and covariances.

Example: Let us apply the result from the previous example to the example we have been following since Section 5.3.1. Figure 5.6 shows the least squares fit of a straight line to the data points. The best fit line is

$$\hat{y} = \hat{a}_0 + \hat{a}_1 x = -3.45 + 0.897x,$$

and the estimated covariance matrix for the parameters of fit is

$$\begin{pmatrix} \hat{\sigma}_{\hat{0}}^2 & \hat{\sigma}_{\widehat{01}} \\ \hat{\sigma}_{\widehat{10}} & \hat{\sigma}_{\hat{1}}^2 \end{pmatrix} = \begin{pmatrix} 0.1029 & -0.01575 \\ -0.01575 & 0.002488 \end{pmatrix}.$$

We wish to determine the intercept of the fitted line with the x-axis and the variance of the intercept. To find the intercept, set $\hat{y} = 0$ and solve for x:

$$x_{int} = -\frac{\hat{a}_0}{\hat{a}_1} = -\frac{(-3.45)}{0.897} = 3.85$$

Continued on page 142

Noting that the subscripts here are the reverse of the subscripts in the previous example, we write

$$\hat{\sigma}^2_{x_{int}} = \frac{\hat{a}_0^2}{\hat{a}_1^4}\hat{\sigma}_1^2 - 2\frac{\hat{a}_0}{\hat{a}_1^3}\hat{\sigma}_{01} + \frac{1}{\hat{a}_1^2}\hat{\sigma}_0^2$$

$$= \frac{(-3.45)^2}{0.897^4} \times 0.002488 - 2 \times \frac{(-3.45)}{0.897^3} \times (-0.01575) + \frac{1}{0.897^2} \times 0.1029$$

$$= 0.0457 - 0.1506 + 0.1279$$

$$= 0.02306$$

Thus, we find

$$x_{int} \pm \hat{\sigma}_{x_{int}} = 3.85 \pm 0.15.$$

In this case, including the covariance between $\hat{a}_0$ and $\hat{a}_1$ substantially reduced the estimated uncertainty of x_{int}.

5.4.3 Monte Carlo Error Propagation

Under many conditions, the high-order terms in equation 5.107 cannot be ignored. If they cannot, it is better to give up the analytic approach to error analysis and propagate the errors numerically by a Monte Carlo simulation. Let the probability distribution function for the a_j be the multivariate Gaussian distribution

$$f(\mathbf{a}) = \frac{1}{(2\pi)^{(m+1)/2}|\hat{\mathbf{C}}|^{1/2}} \exp\left[-\frac{1}{2}(\mathbf{a} - \hat{\mathbf{a}})^{\mathrm{T}}\hat{\mathbf{C}}^{-1}(\mathbf{a} - \hat{\mathbf{a}})\right], \tag{5.110}$$

where $(\mathbf{a} - \hat{\mathbf{a}})^{\mathrm{T}}$ is the vector $(a_0 - \hat{a}_0, a_1 - \hat{a}_1, \ldots, a_m - \hat{a}_m)$, and $\hat{\mathbf{C}}$ is the estimated covariance matrix for the $\hat{a}_j$. Generate many random samples of the parameter set $(a_0, a_1, \ldots, a_m)$ from $f(\mathbf{a})$. For each random sample, calculate g, producing many values g_k. One may then characterize the distribution of the g_k in the usual ways, perhaps by calculating its mean and standard deviation.

Generating random deviates from $f(\mathbf{a})$ is not a trivial chore because of the covariances among the parameters. A default way to do this is to generate MCMC samples of $f(\mathbf{a})$ using the techniques discussed in Section 3.5. While MCMC sampling should always work, it can be computationally intensive, especially if there are strong correlations among the parameters. The discussion of principle component analysis in Section 4.6 suggests an alternative way to generate vectors of artificial data. The reader may wish to review that section before continuing. The basic idea is to rotate the covariance matrix into a coordinate system in which the covariances disappear. Since the covariances are all equal to 0, the artificial data can be generated from independent one-dimensional Gaussian distributions. To reestablish the covariances, the artificial data is rotated back to the original coordinate system.

To diagonalize the covariance matrix, we must first find its eigenvalues and eigenvectors. Let $\mathbf{v}$ be a vector satisfying the matrix equation

$$\hat{\mathbf{C}}\mathbf{v} = \lambda\mathbf{v}, \tag{5.111}$$

where λ is a scalar constant. The $m + 1$ vectors $\mathbf{v}_j$ and corresponding scalars λ_j are the eigenvectors and eigenvalues, respectively, of the matrix. The normalized eigenvector corresponding to eigenvalue λ_j is $\mathbf{e}_j = \mathbf{v}_j/|\mathbf{v}_j|$. The $\mathbf{e}_j$ together make up a complete set of orthonormal basis vectors. Create the square matrix $\mathbf{U}$ by setting its rows equal to the transposed basis vectors:

$$\mathbf{U} = \begin{bmatrix} \mathbf{e}_0^{\mathrm{T}} \\ \mathbf{e}_1^{\mathrm{T}} \\ \vdots \\ \mathbf{e}_m^{\mathrm{T}} \end{bmatrix}. \tag{5.112}$$

By construction, $\mathbf{U}$ is an orthogonal matrix, so its inverse is equal to its transpose: $\mathbf{U}^{-1} = \mathbf{U}^{\mathrm{T}}$. The diagonal matrix corresponding to $\hat{\mathbf{C}}$ is calculated from the similarity transformation

$$\hat{\mathbf{C}}' = \mathbf{U}\mathbf{C}\mathbf{U}^{-1}, \tag{5.113}$$

and the diagonal elements of $\hat{\mathbf{C}}'$ are the eigenvalues

$$(\hat{\mathbf{C}}')_{jj} = \lambda_j. \tag{5.114}$$

Because $\hat{\mathbf{C}}'$ is diagonal, its inverse is also diagonal. As a result, the multivariate Gaussian distribution of equation 5.110 reduces to the product of $m + 1$ independent Gaussian distributions. The diagonal components of the inverse are just $1/\lambda_j$, so we can now identify the λ_j as the variances along the corresponding axis:

$$\lambda_j = (\sigma')^2. \tag{5.115}$$

We are now equipped to generate the artificial data sets. Let $f_j(x', \lambda_j)$ be the one-dimensional Gaussian probability distributions in which the variances are the λ_j:

$$f_j(x', \lambda_j) \propto \exp\left[-\frac{(x')^2}{2\lambda_j}\right]. \tag{5.116}$$

Generate one random deviate ζ'_j from each of these Gaussians, and create a vector $\mathbf{z}'$ from the deviates:

$$\mathbf{z}' = \begin{pmatrix} \zeta'_0 \\ \zeta'_1 \\ \vdots \\ \zeta'_m \end{pmatrix}. \tag{5.117}$$

Transform $\mathbf{z}$ back to the original coordinate system by taking

$$\mathbf{z} = \begin{pmatrix} \zeta_0 \\ \zeta_1 \\ \vdots \\ \zeta_m \end{pmatrix} = \mathbf{U}^{-1}\mathbf{z}'. \tag{5.118}$$

One set of random deviates generated from the original multivariate Gaussian distribution is then given by $\mathbf{z} + \hat{\mathbf{a}}$. Since λ_j and matrix $\mathbf{U}$ need to be calculated only once, this is an efficient algorithm for producing Gaussian deviates with correlations.

5.5 General Linear Least Squares

5.5.1 Linear Least Squares with Nonpolynomial Functions

Although most discussions of linear least squares begin with fits of polynomials to data, there is nothing special about polynomials. As long as the function being fitted to the data is linear in the parameters of fit, one is dealing with linear least squares optimization. Suppose that we wish to fit n data points (x_i, y_i, σ_i) with a linear combination of two functions $f_0(x)$ and $f_1(x)$,

$$y = a_0 f_0(x) + a_1 f_1(x), \tag{5.119}$$

where the parameters of fit are a_0 and a_1. For example, one might wish to fit

$$y = a_0 \exp[-x] + a_1 \sin(x). \tag{5.120}$$

Even though $f_0(x)$ and $f_1(x)$ are nonlinear functions of x, this is a linear least squares fit, because y is a linear function of a_0 and a_1. To fit the data, form the weighted sum of the squared residuals

$$S = \sum_{i=1}^{n} w_i [y_i - a_0 f_0(x_i) - a_1 f_1(x_i)]^2, \tag{5.121}$$

where, as usual, $w_i = 1/\sigma_i^2$. To find the values of the parameters that minimize S, set the derivatives of S with respect to a_0 and a_1 equal to zero:

$$\frac{\partial S}{\partial a_0} = -2 \sum_{i=1}^{n} w_i f_0(x_i)[y_i - \hat{a}_0 f_0(x_i) - \hat{a}_1 f_1(x_i)] = 0 \tag{5.122}$$

$$\frac{\partial S}{\partial a_1} = -2 \sum_{i=1}^{n} w_i f_1(x_i)[y_i - \hat{a}_0 f_0(x_i) - \hat{a}_1 f_1(x_i)] = 0. \tag{5.123}$$

Rearranging these equations, we obtain the normal equations

$$\hat{a}_0 \sum_i w_i f_0(x_i) f_0(x_i) + \hat{a}_1 \sum_i w_i f_0(x_i) f_1(x_i) = \sum_i w_i f_0(x_i) y_i \tag{5.124}$$

$$\hat{a}_0 \sum_i w_i f_1(x_i) f_0(x_i) + \hat{a}_1 \sum_i w_i f_1(x_i) f_1(x_i) = \sum_i w_i f_1(x_i) y_i, \tag{5.125}$$

which are equivalent to the matrix equation

$$\begin{pmatrix} \sum w_i f_0(x_i) f_0(x_i) & \sum w_i f_0(x_i) f_1(x_i) \\ \sum w_i f_1(x_i) f_0(x_i) & \sum w_i f_1(x_i) f_1(x_i) \end{pmatrix} \begin{pmatrix} \hat{a}_0 \\ \hat{a}_1 \end{pmatrix} = \begin{pmatrix} \sum w_i f_0(x_i) y_i \\ \sum w_i f_1(x_i) y_i \end{pmatrix}. \tag{5.126}$$

The solution is

$$\begin{pmatrix} \hat{a}_0 \\ \hat{a}_1 \end{pmatrix} = \begin{pmatrix} \sum w_i f_0(x_i) f_0(x_i) & \sum w_i f_0(x_i) f_1(x_i) \\ \sum w_i f_1(x_i) f_0(x_i) & \sum w_i f_1(x_i) f_1(x_i) \end{pmatrix}^{-1} \begin{pmatrix} \sum w_i f_0(x_i) y_i \\ \sum w_i f_1(x_i) y_i \end{pmatrix}. \tag{5.127}$$

Equations 5.126 and 5.127 can be put in more compact form by setting

$$\mathbf{N} = \begin{pmatrix} \sum w_i f_0(x_i) f_0(x_i) & \sum w_i f_0(x_i) f_1(x_i) \\ \sum w_i f_1(x_i) f_0(x_i) & \sum w_i f_1(x_i) f_1(x_i) \end{pmatrix}, \tag{5.128}$$

$$\hat{\mathbf{a}} = \begin{pmatrix} \hat{a}_0 \\ \hat{a}_1 \end{pmatrix}, \quad \text{and} \quad \mathbf{Y} = \begin{pmatrix} \sum w_i f_0(x_i) y_i \\ \sum w_i f_1(x_i) y_i \end{pmatrix}. \tag{5.129}$$

The normal equations become

$$\mathbf{N}\hat{\mathbf{a}} = \mathbf{Y}, \tag{5.130}$$

and the solution for the best fitting parameters is

$$\hat{\mathbf{a}} = \mathbf{N}^{-1}\mathbf{Y}. \tag{5.131}$$

The generalization to an arbitrary number of parameters can now be written down by inspection. Suppose we wish to fit a set of n data points (x_i, y_i, σ_i) with a linear combination of $m+1$ functions $f_j(x)$ of x:

$$y = a_0 f_0(x) + a_1 f_1(x) + \cdots + a_m f_m(x). \tag{5.132}$$

The normal matrix is

$$\mathbf{N} = \begin{pmatrix} \sum w_i f_0(x_i) f_0(x_i) & \sum w_i f_0(x_i) f_1(x_i) & \cdots & \sum w_i f_0(x_i) f_m(x_i) \\ \sum w_i f_1(x_i) f_0(x_i) & \sum w_i f_1(x_i) f_1(x_i) & \cdots & \sum w_i f_1(x_i) f_m(x_i) \\ \vdots & \vdots & & \vdots \\ \sum w_i f_m(x_i) f_0(x_i) & \sum w_i f_m(x_i) f_1(x_i) & \cdots & \sum w_i f_m(x_i) f_m(x_i) \end{pmatrix} \tag{5.133}$$

and the vector $\mathbf{Y}$ is

$$\mathbf{Y} = \begin{pmatrix} \sum w_i f_0(x_i) y_i \\ \sum w_i f_1(x_i) y_i \\ \vdots \\ \sum w_i f_m(x_i) y_i \end{pmatrix}, \tag{5.134}$$

where the sums are taken over the n data points. The normal equations are

$$\mathbf{N}\hat{\mathbf{a}} = \mathbf{Y}, \tag{5.135}$$

and the estimated parameters of the fit are

$$\hat{\mathbf{a}} = \mathbf{N}^{-1}\mathbf{Y}. \tag{5.136}$$

All the derivations of Section 5.3.3 proceed in the same way, so the estimated covariance matrix is

$$\hat{\mathbf{C}} = \frac{\hat{S}_{min}}{n - m - 1}\mathbf{N}^{-1}, \tag{5.137}$$

where $(m+1)$ is the number of fitted parameters (see the footnote to equation 5.91), and

$$\hat{S}_{min} = \sum_{i=1}^{n} w_i \left[y_i - \sum_{j=0}^{m} \hat{a}_j f_j(x_i) \right]^2. \tag{5.138}$$

In matrix notation, the normal equations and their solution have precisely the same form as for the polynomial fit. Only the components of the normal matrix change. Equations 5.132–5.138 are the version of linear least squares that should be used by most people most of the time.

Even when the dependence on the parameters appears to be nonlinear, it may be a linear dependence in disguise. The following example shows how to fit a sine curve of known period to data by linear least squares.

Example: Suppose we wish to fit a sine curve to the time series data $(t_i, y_i, \sigma_i), i = 1, \ldots, n$, shown in Figure 5.7. The variation of the signal with time is known to be a sine curve with angular frequency ω, but the amplitude and phase of the sine curve are unknown. The equation to be fit to the data is then

$$y = A\sin(\omega t + \theta),$$

where A is the amplitude, and θ is the phase of the sine curve. At first sight this is a nonlinear least squares problem, because the dependence of y on θ is nonlinear. However, expanding the sine function, we find

$$y = A\cos\theta \sin\omega t + A\sin\theta \cos\omega t$$
$$= a_0 \sin\omega t + a_1 \cos\omega t,$$

Which is linear in a_0 and a_1. We proceed by minimizing the weighted sum of the squared residuals:

$$S = \sum_i w_i(y_i - a_0 \sin\omega t_i - a_1 \cos\omega t_i)^2.$$

The derivatives of S by a_0 and a_1 yield the normal equations:

$$0 = \frac{\partial S}{\partial a_0} = -2\sum_i w_i \sin\omega t_i(y_i - \hat{a}_0 \sin\omega t_i - \hat{a}_1 \cos\omega t_i)$$

$$0 = \frac{\partial S}{\partial a_1} = -2\sum_i w_i \cos\omega t_i(y_i - \hat{a}_0 \sin\omega t_i - \hat{a}_1 \cos\omega t_i).$$

Rearranging these equations and converting to matrix notation, we have

$$\begin{pmatrix} \sum w_i \sin^2\omega t_i & \sum w_i \sin\omega t_i \cos\omega t_i \\ \sum w_i \sin\omega t_i \cos\omega t_i & \sum w_i \cos^2\omega t_i \end{pmatrix} \begin{pmatrix} \hat{a}_0 \\ \hat{a}_1 \end{pmatrix} = \begin{pmatrix} \sum w_i y_i \sin\omega t_i \\ \sum w_i y_i \cos\omega t_i \end{pmatrix},$$

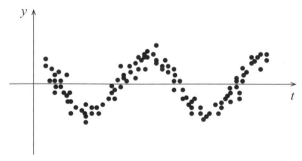

Figure 5.7: The time series data $(t_i, y_i, \sigma_i), i = 1, \ldots, n$. The variation of the signal with time is known to be a sine curve with angular frequency ω.

which we recognize as the matrix equation

$$\mathbf{N}\hat{\mathbf{a}} = \mathbf{Y}.$$

This is solved for $\hat{a}_0$ and $\hat{a}_1$ and their covariances in the usual way.

Now calculate $\hat{A}$ and $\hat{\theta}$ from

$$\hat{A} = \left(\hat{a}_0^2 + \hat{a}_1^2\right)^{1/2}$$

$$\hat{\theta} = \tan^{-1}\left(\frac{\hat{a}_1}{\hat{a}_0}\right),$$

taking care to use the correct branch of the inverse tangent. Finally, the variances and covariances of $\hat{a}_0$ and $\hat{a}_1$ can be propagated through to the variances of $\hat{A}$ and $\hat{\theta}$ using the techniques developed in Section 5.4.

5.5.2 Fits with Correlations among the Measurement Errors

One final generalization may be needed for some applications of linear least squares. Suppose the data points are $\{x_i, y_i\}$, $i = 1,\ldots,n$, but now the errors in the y_i are correlated, so that $\sigma_{ij} \neq 0$. We assume the correlations are symmetric so that $\sigma_{ji} = \sigma_{ij}$. As before, we wish to fit the data with a function of the form

$$y = a_0 f_0(x) + a_1 f_1(x) + \cdots + a_m f_m(x), \tag{5.139}$$

where the $f_j(x)$ are functions of x but not of the a_j. Set

$$\mathbf{F} = \begin{pmatrix} f_0(x_1) & f_1(x_1) & \cdots & f_m(x_1) \\ f_0(x_2) & f_1(x_2) & \cdots & f_m(x_2) \\ \vdots & \vdots & & \vdots \\ f_0(x_n) & f_1(x_n) & \cdots & f_m(x_n) \end{pmatrix}, \quad \mathbf{a} = \begin{pmatrix} a_1 \\ a_2 \\ \vdots \\ a_m \end{pmatrix}, \quad \text{and} \quad \mathbf{y} = \begin{pmatrix} y_1 \\ y_2 \\ \vdots \\ y_n \end{pmatrix}. \tag{5.140}$$

It is not correct to set $\mathbf{y} = \mathbf{Fa}$, because the function does not go through all the data points. Instead we have

$$\mathbf{y} = \mathbf{Fa} + \boldsymbol{\epsilon}, \tag{5.141}$$

where $\boldsymbol{\epsilon}$ is the vector of residuals. Equation 5.141 is the generalization of equation 5.56.

We wish to find the best fitting values of $\mathbf{a}$. Let us assume that we know the probability distribution function for $\boldsymbol{\epsilon}$ and that it is the multivariate Gaussian distribution (equation 2.80):

$$g(\boldsymbol{\epsilon}) \propto \exp\left[-\frac{1}{2}\boldsymbol{\epsilon}^{\mathrm{T}}\mathbf{W}\boldsymbol{\epsilon}\right], \tag{5.142}$$

where the weight matrix $\mathbf{W}$ is

$$\mathbf{W} = \begin{pmatrix} w_{11} & w_{12} & \cdots & w_{1n} \\ w_{21} & w_{22} & \cdots & w_{2n} \\ \vdots & \vdots & & \vdots \\ w_{n1} & w_{n2} & & w_{nn} \end{pmatrix} = \begin{pmatrix} \sigma_{11} & \sigma_{12} & \cdots & \sigma_{1n} \\ \sigma_{21} & \sigma_{22} & \cdots & \sigma_{2n} \\ \vdots & \vdots & & \vdots \\ \sigma_{n1} & \sigma_{n2} & \cdots & \sigma_{nn} \end{pmatrix}^{-1}. \tag{5.143}$$

This is equivalent to assuming that

$$\langle \epsilon_i \rangle = 0 \tag{5.144}$$

$$\langle \epsilon_i \epsilon_j \rangle = \sigma_{ij}. \tag{5.145}$$

Since we know the probability distribution, we can apply the maximum likelihood principle. Using $\epsilon = \mathbf{y} - \mathbf{Fa}$, we convert $g(\epsilon)$ to

$$g(\vec{x}, \vec{a}) \propto \exp\left[-\frac{1}{2}(\mathbf{y} - \mathbf{Fa})^{\mathrm{T}} \mathbf{W}(\mathbf{y} - \mathbf{Fa}) \right]. \tag{5.146}$$

Since the data points are given and the parameters are unknown, this is the likelihood function for the parameters of fit, $L(\vec{x}, \vec{a}) \propto g(\vec{x}, \vec{a})$. According to the maximum likelihood principle, the best estimate of the components of $\mathbf{a}$ is obtained by maximizing the log likelihood function or, equivalently, by minimizing

$$S = (\mathbf{y} - \mathbf{Fa})^{\mathrm{T}} \mathbf{W}(\mathbf{y} - \mathbf{Fa}) \tag{5.147}$$

$$= \sum_i \sum_j w_{ij} \left[y_i - \sum_{k=0}^{m} a_k f_k(x_i) \right] \left[y_j - \sum_{k=0}^{m} a_k f_k(x_j) \right]. \tag{5.148}$$

Had we adopted a least squares approach, we could have written down equation 5.147 without assuming that ϵ has a multivariate Gaussian distribution. It might not have been so easy, however, to decide that the weights are given by equation 5.143. We also now recognize that S is a χ^2 variable (see equation 2.126), so minimizing S is the same as χ^2 minimization.

We now need to find the values of the components of $\mathbf{a}$ that minimize S. To do this, we set the derivatives of S with respect to the a_j equal to 0:

$$\frac{\partial S}{\partial a_j} = 0. \tag{5.149}$$

Let δ_j be a column vector in which the jth element is equal to 1 and all the other elements are equal to 0. Then

$$\frac{\partial \mathbf{a}}{\partial a_j} = \delta_j. \tag{5.150}$$

The partial derivatives of S by a_j lead to the $m+1$ equations:

$$0 = -(\mathbf{F}\delta_j)^{\mathrm{T}} \mathbf{W}(\mathbf{y} - \mathbf{F}\hat{\mathbf{a}}) - (\mathbf{y} - \mathbf{F}\hat{\mathbf{a}})^{\mathrm{T}} \mathbf{WF}\delta_j. \tag{5.151}$$

Note that $\mathbf{a}$ has become $\hat{\mathbf{a}}$. Since S is a scalar, its derivatives are scalars, and both terms in equation 5.151 must be scalars. The transpose of a scalar is equal to itself, so equation 5.151 becomes

$$0 = -2(\mathbf{F}\delta_j)^{\mathrm{T}} \mathbf{W}(\mathbf{y} - \mathbf{F}\hat{\mathbf{a}}). \tag{5.152}$$

Again, equation 5.152 is actually $m+1$ equations, one for each derivative of S by a_j. Note that

$$\mathbf{F}\delta_j = \begin{pmatrix} f_j(x_1) \\ f_j(x_2) \\ \vdots \\ f_j(x_n) \end{pmatrix}, \tag{5.153}$$

which is the jth column of $\mathbf{F}$. We can combine the $m + 1$ equations into a single equation, obtaining

$$0 = \mathbf{F}^{\mathrm{T}}\mathbf{W}(\mathbf{y} - \mathbf{F}\hat{\mathbf{a}}). \tag{5.154}$$

Expanding and rearranging, we arrive at the normal equations:

$$(\mathbf{F}^{\mathrm{T}}\mathbf{W}\mathbf{F})\hat{\mathbf{a}} = \mathbf{F}^{\mathrm{T}}\mathbf{W}\mathbf{y}. \tag{5.155}$$

The reader may verify that the components of the normal matrix $\mathbf{N} = \mathbf{F}^{\mathrm{T}}\mathbf{W}\mathbf{F}$ are

$$(\mathbf{F}^{\mathrm{T}}\mathbf{W}\mathbf{F})_{pq} = \sum_{i=1}^{n}\sum_{j=1}^{n} w_{ij} f_p(x_i) f_q(x_j), \tag{5.156}$$

and the components of $\mathbf{F}^{\mathrm{T}}\mathbf{W}\mathbf{y}$ are

$$(\mathbf{F}^{\mathrm{T}}\mathbf{W}\mathbf{y})_p = \sum_{i=1}^{n}\sum_{j=1}^{n} w_{ij} y_i f_p(x_i). \tag{5.157}$$

At this point we have connected with the formalism of Section 5.3 and can write down by inspection the solution for the estimated parameters

$$\hat{\mathbf{a}} = (\mathbf{F}^{\mathrm{T}}\mathbf{W}\mathbf{F})^{-1}\mathbf{F}^{\mathrm{T}}\mathbf{W}\mathbf{y}, \tag{5.158}$$

and the estimated covariance matrix

$$\hat{\mathbf{C}} = \frac{\hat{S}_{min}}{n - m - 1}\left(\mathbf{F}^{\mathrm{T}}\mathbf{W}\mathbf{F}\right)^{-1}, \tag{5.159}$$

where $(m + 1)$ is the number of fitted parameters (see Section 5.3.3 and the footnote to equation 5.91), and

$$\hat{S}_{min} = (\mathbf{y} - \mathbf{F}\hat{\mathbf{a}})^{\mathrm{T}}\mathbf{W}(\mathbf{y} - \mathbf{F}\hat{\mathbf{a}}) \tag{5.160}$$

$$= \sum_{i=1}^{n}\sum_{j=1}^{n} w_i w_j \left[y_i - \sum_{k=0}^{m} \hat{a}_k f_k(x_i)\right]\left[y_j - \sum_{k=0}^{m} \hat{a}_k f_k(x_j)\right]. \tag{5.161}$$

5.5.3 χ^2 Test for Goodness of Fit

The least squares method will cheerfully fit nearly any function to nearly any set of data points. Consider the data and the least squares fits shown in Figure 5.8. The fit in the middle panel looks good, but the fits in the two end panels do not—the left panel because the function is too simple to fit the data, the right panel because the fitted function is too complicated. The χ^2 distribution provides a way to quantify these impressions.

Suppose the function $f = f(x, a_0, a_1, \ldots, a_m)$ has been fitted to n data points (x_i, y_i, σ_i) by minimizing the weighted sum of the squared residuals

$$S = \sum_{i=1}^{n} w_i [y_i - f(x_i, a_0, a_1, \ldots, a_m)]^2, \tag{5.162}$$

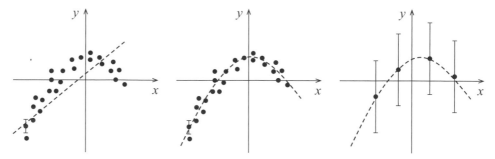

Figure 5.8: The left panel shows the fit of a straight line to a set of data points. All the points have the same small error, plotted only for the leftmost point. The straight line is a poor fit, because the differences between the fitted line and the data are often much larger than the errors in the data. The middle panel shows the fit of a quadratic polynomial to the same data. This fit is much better, because the differences between the line and the data are comparable to the errors. The right panel shows the fit of a quadratic polynomial to data points with large errors. The fit is excellent, but the measurement errors are large, suggesting that the data could be fit adequately by a simpler function. The value of χ^2 provides a way to quantify these impressions.

where $w_i = 1/\sigma_i^2$. As before, the fitted values for the a_j are $\hat{a}_j$, the fitted function is $f = f(x, \hat{a}_0, \hat{a}_1, \dots, \hat{a}_m)$, and the value of S at its minimum is

$$S_{min} = \sum_i w_i [y_i - f(x_i, \hat{a}_0, \hat{a}_1, \dots, \hat{a}_m)]^2. \tag{5.163}$$

Let $\epsilon_i = y_i - f(x_i, \hat{a}_0, \hat{a}_1, \dots, \hat{a}_m)$ be the residual for point i, and rewrite S_{min} as

$$S_{min} = \sum_i \frac{\epsilon_i^2}{\sigma_i^2}. \tag{5.164}$$

If the ϵ_i have a Gaussian distribution, then S_{min} is a χ^2 variable and has a χ_k^2 distribution with $k = n - m - 1$ degrees of freedom.

From our discussion of the χ^2 probability distribution in Section 2.6, we expect the value of χ_k^2 to lie in the range

$$\langle \chi_k^2 \rangle \pm \sigma_{\chi_k^2} = k \pm \sqrt{2k}. \tag{5.165}$$

There are 25 data points in the left and middle panels of Figure 5.8, and the fitted functions have either 2 or 3 free parameters (linear or quadratic fits), so $\chi^2 = S_{min}$ should lie between about 15 and 35. In the left panel, many of the residuals are 2 or 3 times greater than the errors, and each of these contributes $\epsilon_i^2/\sigma_i^2 \approx 4-9$ to the sum in equation 5.164. As a result, χ^2 is much greater than 30, and we conclude that the residuals are too large to be due to random errors. The fit is poor. In the middle panel, the residuals to the fit are comparable to the errors. Each point contributes an amount $\epsilon_i^2/\sigma_i^2 \approx 1$ to the sum, so $\chi^2 \approx 25$. Since this is the expected value of χ^2, the quadratic polynomial is a good representation of the data.

There are only four data points in the right panel of Figure 5.8, and they have been fitted with a quadratic polynomial, so we expect χ^2 to be near $k = 4 - 3 = 1$. In contrast to the left and middle panels, the residuals to the fitted curve are much less than the errors. As a result, each ϵ_i/σ_i is nearly 0, so $\chi^2 = S_{min} \approx 0$. The fitted line yields residuals that are too small to be consistent with the errors in the data. A straight line or even a simple average would yield a value of χ^2 closer to its expected value and, therefore, would have a much higher probability of being the correct representation of the data than the quadratic polynomialis.

More precisely, the probability that a value of χ_k^2 will be greater than or equal to some value b is

$$P(\chi_k^2 \geq b) = \int_b^\infty f_k(\chi^2) d(\chi^2), \tag{5.166}$$

where $f_k(\chi^2)$ is the χ^2 distribution with k degrees of freedom. If we set b equal to S_{min}, then $P(\chi_k^2 \geq S_{min})$ is the probability of getting $\chi_k^2 \geq S_{min}$ purely by chance. If the probability is small, the residuals are too large to be consistent with the errors, and the fit is poor. Likewise, the probability that a value of χ_k^2 will be less than or equal to some value b is

$$P(\chi_k^2 \leq b) = \int_0^b f_k(\chi^2) d(\chi^2). \tag{5.167}$$

If we set b equal to S_{min}, then $P(\chi_k^2 \leq S_{min})$ is the probability of getting $\chi_k^2 \leq S_{min}$ purely by chance. If this probability is small, the residuals are too small to be consistent with the errors, and a fit with a function having fewer parameters is warranted.

For the χ^2 test to give a meaningful result, the values of the σ_i^2 in equation 5.164 must be correct. In the following example, we reverse this requirement to to tighten our conclusion that the error bars on the data points shown in Figure 5.4 are too small.

Example: We return one last time to the numerical example we have been following since Section 5.3.1. (Although we will use the same data for an example in Chapter 7 on Bayesian statistics.) In the example, we have been fitting a straight line to data by weighted least squares. The data points and their error bars are plotted in Figure 5.4 along with the the straight line fitted by weighted least squares.

In the example accompanying the figure, we noted that 7 of the 12 data points were more than 1 of their standard deviations away from the line. If the errors in the data points had a Gaussian distribution, only 1/3 of the points should have departed from the line by this much. We cited that as the reason the variances of the fitted parameters needed to be adjusted.

We can now quantify that conclusion. In the process of correcting the variance, we calculated

$$\hat{\sigma}^2 = \frac{\hat{S}_{min}}{n - m - 1} = \frac{\sum_{i=1}^{12} w_i (y_i + 3.45 - 0.897 x_i)^2}{10} = 2.74,$$

and $n - m - 1 = 10$ is the number of degrees of freedom. In the present context we turn this around and find

$$S_{min} = 10\hat{\sigma}^2 = 27.4 = \chi_{10}^2.$$

Using equation 5.166, the probability that this would occur by chance is (from tables)

$$P(\chi_{10}^2 \geq 27.4) = \int_{27.4}^\infty f_{10}(\chi^2) d(\chi^2) = 0.0023.$$

Continued on page 152

We interpret this to mean that there is only about 1 chance in 400 that the distribution of data points in Figure 5.4 would arise by chance if the data points were selected from Gaussian distributions with the tabulated standard deviations. We conclude now as we did earlier that the tabulated errors are too small and that the standard deviations of the fitted parameters must be adjusted.

This example also gives another way to understand the nature of the adjustment being made in equation 5.99. The effect of the adjustment is to force S_{min} to equal its expected mean value, $S_{min} \rightarrow \langle \chi_k^2 \rangle = k$.

5.6 Fits with More Than One Dependent Variable

One often needs to fit data with more than one dependent variable. Consider n data points (t_i, x_i, y_i, z_i), where t is the independent variable and is known without error, and the dependent variables x, y, and z have uncorrelated measurement errors $(\sigma_{x_i}, \sigma_{x_i}, \sigma_{z_i})$. For example, these might be measurements of the positions of an object at different times as it moves through space. We wish to fit a curve that is a function of t to the data points (see Figure 5.9); that is, we wish to fit a curve for which t is an independent variable and x, y, and z are dependent variables.

Let us assume that the curve to be fit is described by the parametric equations

$$x = f_x(t, a_0, a_1, \ldots, a_{m_x}) = f_x(t, \vec{a}) \tag{5.168}$$

$$y = f_y(t, b_0, b_1, \ldots, b_{m_y}) = f_y(t, \vec{b}) \tag{5.169}$$

$$z = f_z(t, c_0, c_1, \ldots, c_{m_z}) = f_z(t, \vec{c}), \tag{5.170}$$

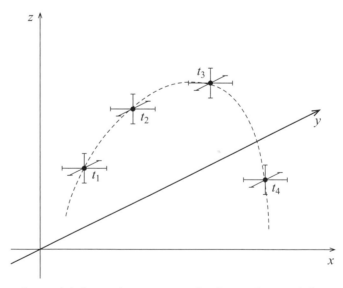

Figure 5.9: The points and their error bars represent the data set $\{t_i, x_i, y_i, z_i\}$, $i = 1, \ldots, n$, where the measurement errors of t are negligibly small, and x, y, and z have uncorrelated measurement errors $(\sigma_{x_i}, \sigma_{x_i}, \sigma_{z_i})$. The data might, for example, correspond to the measured positions of an object as it moves through space. The dashed line is the fitted curve, which is a function of t.

where $\vec{a}$, $\vec{b}$, and $\vec{c}$ are the parameters of fit. We use the maximum likelihood principle to determine the best parameters. If the measurement errors have a Gaussian distribution, the likelihood function is

$$L = \frac{1}{\sqrt{2\pi}\,\sigma_{x_1}} \exp\left\{ -\frac{\left[x_1 - f_x(t_1,\vec{a})\right]^2}{2\sigma_{x_1}^2} \right\} \cdots \frac{1}{\sqrt{2\pi}\,\sigma_{x_n}} \exp\left\{ -\frac{\left[x_n - f_x(t_n,\vec{a})\right]^2}{2\sigma_{x_n}^2} \right\}$$

$$\times \frac{1}{\sqrt{2\pi}\,\sigma_{y_1}} \exp\left\{ -\frac{\left[y_1 - f_y(t_1,\vec{b})\right]^2}{2\sigma_{y_1}^2} \right\} \cdots \frac{1}{\sqrt{2\pi}\,\sigma_{y_n}} \exp\left\{ -\frac{\left[y_n - f_y(t_n,\vec{b})\right]^2}{2\sigma_{y_n}^2} \right\}$$

$$\times \frac{1}{\sqrt{2\pi}\,\sigma_{z_1}} \exp\left\{ -\frac{\left[z_1 - f_z(t_1,\vec{c})\right]^2}{2\sigma_{z_1}^2} \right\} \cdots \frac{1}{\sqrt{2\pi}\,\sigma_{z_n}} \exp\left\{ -\frac{\left[z_n - f_z(t_n,\vec{c})\right]^2}{2\sigma_{z_n}^2} \right\} \quad (5.171)$$

The log-likelihood function is then

$$\ell(t,\vec{a},\vec{b},\vec{c}) = -\frac{1}{2}\sum_{i=1}^{n}\left\{ \frac{\left[x_i - f_x(t_i,\vec{a})\right]^2}{\sigma_{x_i}^2} + \frac{\left[y_i - f_y(t_i,\vec{b})\right]^2}{\sigma_{y_i}^2} + \frac{\left[z_i - f_z(t_i,\vec{c})\right]^2}{\sigma_{z_i}^2} \right\}, \quad (5.172)$$

where we have ignored additive constants that are independent of the parameters of fit. The parameters of fit are determined by maximizing the log-likelihood.

We enter the realm of least squares by setting $S = -2\ell(t,\vec{a},\vec{b},\vec{c})$ and adjusting the parameters to minimize S. For convenience set

$$w_{x_i} = \frac{1}{\sigma_{x_i}^2}, \quad w_{y_i} = \frac{1}{\sigma_{y_i}^2}, \quad w_{z_i} = \frac{1}{\sigma_{z_i}^2}, \quad (5.173)$$

so that

$$S = \sum_{i=1}^{n}\left\{ w_{x_i}\left[x_i - f_x(t_i,\vec{a})\right]^2 + w_{y_i}\left[y_i - f_y(t_i,\vec{b})\right]^2 + w_{z_i}\left[z_i - f_z(t_i,\vec{c})\right]^2 \right\}. \quad (5.174)$$

Equation 5.174 can be rewritten as

$$S = S_x(\vec{a}) + S_y(\vec{b}) + S_z(\vec{c}), \quad (5.175)$$

where

$$S_x(\vec{a}) = \sum_{i=1}^{n} w_{x_i}\left[x_i - f_x(t_i,\vec{a})\right]^2 \quad (5.176)$$

$$S_y(\vec{b}) = \sum_{i=1}^{n} w_{y_i}\left[y_i - f_y(t_i,\vec{b})\right]^2 \quad (5.177)$$

$$S_z(\vec{c}) = \sum_{i=1}^{n} w_{z_i}\left[z_i - f_z(t_i,\vec{c})\right]^2. \quad (5.178)$$

Thus S is the sum of the weighted residuals squared for each individual dependent parameter. The techniques developed in the preceding sections can be used to minimize S.

One special case merits attention. If the parameters of fit are independent of one another (i.e., $\vec{a}$, $\vec{b}$, and $\vec{c}$ are not functions of one another), $S_x(\vec{a})$, $S_y(\vec{b})$, and $S_z(\vec{c})$ can be minimized separately. The minimum of $S_x(\vec{a})$ yields the best values of $\vec{a}$, the minimum of $S_y(\vec{b})$ yields the best values of $\vec{b}$, and the minimum of $S_z(\vec{c})$ yields the best values of $\vec{c}$. In other words, the fit becomes three independent fits, one for each coordinate.

Nonlinear Least Squares Estimation

6.1 Introduction

Chapter 5 discusses how to fit a function to data by least squares (or maximum likelihood) when the function is linear in the parameters of fit. If the function is nonlinear in the parameters, the likelihood equations (e.g., equation 5.8) are usually also nonlinear in the parameters, and the elementary methods for solving them described in Chapter 5 will fail. The following example shows that the normal equations defy analytic solution when fitting a function even so simple as the sum of two exponentials.

Example: Suppose we wish to fit the function

$$y = \exp[-a_0 x] + \exp[-a_1 x]$$

to n equally weighted data points (x_i, y_i) by least squares. The sum of the squared residuals is

$$S = \sum_i \left(y_i - \exp[-a_0 x_i] - \exp[-a_1 x_i]\right)^2.$$

To find the values of a_0 and a_1 that minimize S, set the derivative of S with respect to a_0 and a_1 equal to zero:

$$0 = \frac{\partial S}{\partial a_0} = 2\sum_i x_i \exp[-\hat{a}_0 x_i]\left(y_i - \exp[-\hat{a}_0 x_i] - \exp[-\hat{a}_1 x_i]\right)$$

$$0 = \frac{\partial S}{\partial a_1} = 2\sum_i x_i \exp[-\hat{a}_1 x_i]\left(y_i - \exp[-\hat{a}_0 x_i] - \exp[-\hat{a}_1 x_i]\right),$$

where, following the conventions of Chapter 5, the hats on a_0 and a_1 denote fitted values. These equations are nonlinear in $\hat{a}_0$ and $\hat{a}_1$ and cannot be solved analytically.

The difficulties are not merely computational. In linear least squares there is a unique solution for the parameters of fit. If, however, the function is nonlinear in the parameters,

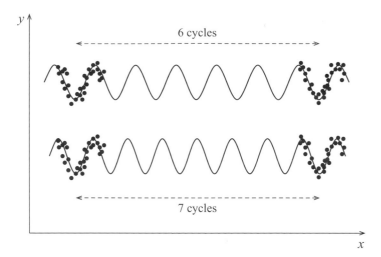

Figure 6.1: A set of data points (x_i, y_i) separated into two groups by a large gap in x have been fitted by the sine curve $y = a_0 \sin(a_1 x + a_2)$. The sine curve is nonlinear in the parameter a_1. Two different values of a_1 give good fits to the data, one value producing 6 cycles of the sine function over the gap, the other 7 cycles. The data have been duplicated with a vertical shift to show the two fits more clearly.

more than one set of values for the parameters can yield good fits. Figure 6.1 shows 57 data points (x_i, y_i) that are distributed in two clumps separated by a large gap in x. They have been fitted by the sine curve $y = a_0 \sin(a_1 x + a_2)$. The sine curve is linear in a_0, and the nonlinearity in a_2 can be removed by expanding the sine (see the example at the end of Section 5.5.1), but the nonlinearity in a_1 cannot be avoided. Despite the large number of data points, two disparate values of a_1 yield good fits to the data, one value producing 6 cycles of the sine function over the gap, the other 7 cycles. The two fits correspond to two different minima of S, called *local minima*. Finding all the local minima and deciding which is the correct minimum can challenging.

Finally, in linear least squares, the covariance matrix is equal or proportional to the inverse of the matrix for the normal equations (equation 5.117). The matrix formalism yields the variances and covariances of the fitted parameters with little extra work. Most methods for finding the parameters of fit in nonlinear least squares problems do not yield the covariance matrix. It must be calculated separately, sometimes greatly increasing the computational effort.

The subject of this chapter is nonlinear least squares estimation. The goal is to fit the function $f(x, a_1, a_2, \ldots, a_m) = f(x, \vec{a})$ to data by least squares, where the $\vec{a}$ are the parameters of fit, and the function is nonlinear in the parameters. If $f(x, \vec{a})$ is well behaved near the minimum of S, it is often possible to convert a nonlinear least squares fit to a linear fit by expanding the function as a Taylor series in the $\vec{a}$ and retaining only the first-order terms. Section 6.2 discusses linearization of nonlinear fits. When the fit cannot be linearized or when this method gives poor results, other methods for finding the parameters that minimize S must be used. Section 6.3 discusses the method of steepest descent, Newton's method, and Marquardt's method, which are related techniques that expand S instead of $f(x, \vec{a})$ in a Taylor series. This section also covers exhaustive grid mapping, simplex optimization, and simulated annealing, which avoid taking derivatives altogether. Section 6.4 discusses ways to calculate the covariance matrix for the parameters and Section 6.5 discusses confidence limits. Finally, up to now we have assumed that the errors in the x-coordinate are small enough to be ignored. Section 6.6 discusses the simplest case of fits when there are errors in both coordinates.

6.2 Linearization of Nonlinear Fits

One way to circumvent the complications of a nonlinear least squares fit is to convert the fit to linear least squares, a process called linearization. Linearization assumes that preliminary guesses for the values of the parameters of fit are available. One expands the function to be fitted in a Taylor series about the guessed values. The expansion is truncated after the first-order terms, thus replacing the original function with a new function that is linear in the parameters of fit but valid only near the guessed values of the parameters. Because it is linear in the parameters, this approximate function can be fit to the data using the techniques of linear least squares developed in Chapter 5. Linearization has the considerable advantage that it yields the covariance matrix as well as estimates of the parameters.

The essential features of linearized least squares optimization can be seen in a simple unweighted, one-parameter fit. Suppose we wish to fit the function $f(x,a)$ to n data points (x_i, y_i). More specifically, we wish to find $\hat{a}$, the value of a that minimizes the sum of the squared residuals

$$S = \sum_{i=1}^{n} \left[y_i - f(x_i, a) \right]^2. \tag{6.1}$$

The procedure is:

1. Make an initial guess for $\hat{a}$, and call the guess a_g. Expand the function in a Taylor series about a_g:

$$f(x,a) = f(x,a_g) + \left. \frac{df(x,a)}{da} \right|_{a_g} \Delta a + \cdots, \tag{6.2}$$

where

$$\Delta a = a - a_g. \tag{6.3}$$

Note that $f(x,a)$ has been expanded in a, not in x.

2. If a_g is a good guess, the higher-order terms in equation 6.2 can be ignored, and $f(x,a)$ becomes linear in a. The sum of the squared residuals becomes

$$S = \sum_i \left[y_i - f(x_i, a_g) - \left. \frac{df(x_i,a)}{da} \right|_{a_g} \Delta a \right]^2. \tag{6.4}$$

3. Find the minimum of S by setting its derivative with respect to Δa equal to 0:

$$0 = \frac{\partial S}{\partial \Delta a} = -2 \sum_i \left. \frac{df(x_i,a)}{da} \right|_{a_g} \left[y_i - f(x_i, a_g) - \left. \frac{df(x_i,a)}{da} \right|_{a_g} \Delta \hat{a} \right], \tag{6.5}$$

where $\Delta \hat{a} = \hat{a} - a_g$.

4. Equation 6.5 is linear in $\Delta \hat{a}$ and is easily solved:

$$\Delta \hat{a} = \left\{ \sum_i \left. \frac{df(x_i,a)}{da} \right|_{a_g} \left[y_i - f(x_i, a_g) \right] \right\} \bigg/ \sum_i \left[\left. \frac{df(x_i,a)}{da} \right|_{a_g} \right]^2. \tag{6.6}$$

5. The revised guess for $\hat{a}$ is

$$\hat{a} = a_g + \Delta\hat{a}. \tag{6.7}$$

6. The revised value of $\hat{a}$ is not exactly correct because of the higher-order terms that were dropped from the Taylor series expansion. It should, however, be closer to the correct value than a_g was. To find a value yet closer to the correct value, set $a_g = \hat{a}$ and return to step 3, iterating this loop until the changes in $\hat{a}$ are small.

The following example shows how to linearize the fit of an exponential function.

Example: The Taylor series expansion of the exponential function is

$$f(x, a) = \exp[-ax]$$
$$= \exp[-a_g x] - x\exp[-a_g x]\Delta a + \cdots.$$

Again, note that the function is expanded in a about a_g, not in x. After the higher-order terms in the Taylor series have been discarded, the sum of the squared residuals becomes

$$S = \sum_i \left(y_i - \exp[-a_g x_i] + x_i \exp[-a_g x_i]\Delta a\right)^2.$$

Find the minimum of S by setting its derivative with respect to Δa equal to 0, giving

$$0 = \frac{\partial S}{\partial \Delta a} = 2\sum_i x_i \exp[-a_g x_i]\left(y_i - \exp[-a_g x_i] + x_i \exp[-a_g x_i]\Delta\hat{a}\right).$$

The solution for $\Delta\hat{a}$ is

$$\Delta\hat{a} = -\frac{\sum_i x_i \exp[-a_g x_i]\left(y_i - \exp[-a_g x_i]\right)}{\sum_i x_i^2 \exp[-2a_g x_i]},$$

and the revised estimate of $\hat{a}$ is

$$\hat{a} = a_g + \Delta\hat{a}.$$

If desired, set $a_g = \hat{a}$ and repeat the fit.

6.2.1 Data with Uncorrelated Measurement Errors

We now generalize to data with measurement errors and functions with $m + 1$ parameters of fit. The data points are (x_i, y_i, σ_i), $i = 1, \ldots, n$, and the function to be fitted to the data is $f(x, a_0, a_1, \ldots, a_m) = f(x, \vec{a})$. We wish to adjust the parameters of fit to minimize the sum of the weighted residuals

$$S = \sum_{i=1}^{n} w_i \left[y_i - f(x_i, \vec{a})\right]^2, \tag{6.8}$$

where $w_i = 1/\sigma_i^2$ are the weights. The guessed values of the parameters are $\vec{a}_g$. The Taylor series expansion of $f(x, \vec{a})$ about the $\vec{a}_g$ is

$$f(x, \vec{a}) = f(x, \vec{a}_g) + \sum_{j=0}^{m} \left. \frac{\partial f(x, \vec{a})}{\partial a_j} \right|_{\vec{a}_g} \Delta a_j + \cdots, \tag{6.9}$$

where $\Delta a_j = a_j - a_{j_g}$, and a_{j_g} is the preliminary guess for the value of $\hat{a}_j$. With only the first-order terms in the expansion retained, the weighted sum of the squared residuals becomes

$$S = \sum_{i=1}^{n} w_i \left[y_i - f(x_i, \vec{a}_g) - \sum_{j=0}^{m} \left. \frac{\partial f(x_i, \vec{a})}{\partial a_j} \right|_{\vec{a}_g} \Delta a_j \right]^2. \tag{6.10}$$

To minimize S, set its $m + 1$ derivatives with respect to the Δa_k equal to 0:

$$0 = \frac{\partial S}{\partial \Delta a_k} = -2 \sum_{i=1}^{n} w_i \left. \frac{\partial f(x_i, \vec{a})}{\partial a_k} \right|_{\vec{a}_g} \left[y_i - f(x_i, \vec{a}_g) - \sum_{j=0}^{m} \left. \frac{\partial f(x_i, \vec{a})}{\partial a_j} \right|_{\vec{a}_g} \Delta \hat{a}_j \right]. \tag{6.11}$$

Expanding this equation, we have

$$0 = \sum_{i=1}^{n} w_i \left. \frac{\partial f(x_i, \vec{a})}{\partial a_k} \right|_{\vec{a}_g} \left[y_i - f(x_i, \vec{a}_g) \right] - \sum_{i=1}^{n} w_i \left. \frac{\partial f(x_i, \vec{a})}{\partial a_k} \right|_{\vec{a}_g} \left(\sum_{j=0}^{m} \left. \frac{\partial f(x_i, \vec{a})}{\partial a_j} \right|_{\vec{a}_g} \Delta \hat{a}_j \right),$$
$$\tag{6.12}$$

and then reversing the order of the summation in the last term on the right, we arrive at the desired form of the normal equations:

$$\sum_{j=0}^{m} \left(\sum_{i=1}^{n} w_i \left. \frac{\partial f(x_i, \vec{a})}{\partial a_k} \right|_{\vec{a}_g} \left. \frac{\partial f(x_i, \vec{a})}{\partial a_j} \right|_{\vec{a}_g} \right) \Delta \hat{a}_j = \sum_{i=1}^{n} w_i \left[y_i - f(x_i, \vec{a}_g) \right] \left. \frac{\partial f(x_i, \vec{a})}{\partial a_k} \right|_{\vec{a}_g}. \tag{6.13}$$

This can be rewritten as the matrix equation

$$\begin{pmatrix} \sum_i w_i \dfrac{\partial f(x_i)}{\partial a_0} \dfrac{\partial f(x_i)}{\partial a_0} & \sum_i w_i \dfrac{\partial f(x_i)}{\partial a_0} \dfrac{\partial f(x_i)}{\partial a_1} & \cdots & \sum_i w_i \dfrac{\partial f(x_i)}{\partial a_0} \dfrac{\partial f(x_i)}{\partial a_m} \\ \sum_i w_i \dfrac{\partial f(x_i)}{\partial a_1} \dfrac{\partial f(x_i)}{\partial a_0} & \sum_i w_i \dfrac{\partial f(x_i)}{\partial a_1} \dfrac{\partial f(x_i)}{\partial a_1} & \cdots & \sum_i w_i \dfrac{\partial f(x_i)}{\partial a_1} \dfrac{\partial f(x_i)}{\partial a_m} \\ \vdots & \vdots & & \vdots \\ \sum_i w_i \dfrac{\partial f(x_i)}{\partial a_m} \dfrac{\partial f(x_i)}{\partial a_0} & \sum_i w_i \dfrac{\partial f(x_i)}{\partial a_m} \dfrac{\partial f(x_i)}{\partial a_1} & \cdots & \sum_i w_i \dfrac{\partial f(x_i)}{\partial a_m} \dfrac{\partial f(x_i)}{\partial a_m} \end{pmatrix} \begin{pmatrix} \Delta \hat{a}_0 \\ \Delta \hat{a}_1 \\ \vdots \\ \Delta \hat{a}_m \end{pmatrix}$$

$$= \begin{pmatrix} \sum_i w_i \left[y_i - f(x_i, \vec{a}_g) \right] \dfrac{\partial f(x_i)}{\partial a_0} \\ \sum_i w_i \left[y_i - f(x_i, \vec{a}_g) \right] \dfrac{\partial f(x_i)}{\partial a_1} \\ \vdots \\ \sum_i w_i \left[y_i - f(x_i, \vec{a}_g) \right] \dfrac{\partial f(x_i)}{\partial a_m} \end{pmatrix}, \tag{6.14}$$

where the derivatives are understood to be evaluated at $\vec{a}_g$, and the sums over i run from 1 to n. In matrix notation, the normal equations compact to

$$\mathbf{N}\Delta\hat{\mathbf{a}} = \mathbf{Y}, \tag{6.15}$$

where the components of $\mathbf{N}$, $\mathbf{Y}$, and $\Delta\hat{\mathbf{a}} = \hat{\mathbf{a}} - \mathbf{a_g}$ are

$$(\mathbf{N})_{jk} = \sum_{i=1}^{n} w_i \left.\frac{\partial f(x_i,\vec{a})}{\partial a_j}\right|_{\vec{a}_g} \left.\frac{\partial f(x_i,\vec{a})}{\partial a_k}\right|_{\vec{a}_g} \tag{6.16}$$

$$(\mathbf{Y})_k = \sum_{i=1}^{n} w_i \left[y_i - f(x_i,\vec{a}_g)\right] \left.\frac{\partial f(x_i,\vec{a})}{\partial a_k}\right|_{\vec{a}_g}, \tag{6.17}$$

and

$$\Delta\hat{\mathbf{a}} = \begin{pmatrix} \Delta\hat{a}_0 \\ \Delta\hat{a}_1 \\ \vdots \\ \Delta\hat{a}_m \end{pmatrix}, \quad \hat{\mathbf{a}} = \begin{pmatrix} \hat{a}_0 \\ \hat{a}_1 \\ \vdots \\ \hat{a}_m \end{pmatrix}, \quad \mathbf{a_g} = \begin{pmatrix} a_{0_g} \\ a_{1_g} \\ \vdots \\ a_{m_g} \end{pmatrix}. \tag{6.18}$$

The solution for $\Delta\hat{\mathbf{a}}$ is

$$\Delta\hat{\mathbf{a}} = (\mathbf{N})^{-1}\mathbf{Y}, \tag{6.19}$$

and the revised parameters of fit are

$$\hat{\mathbf{a}} = \mathbf{a_g} + \Delta\hat{\mathbf{a}}. \tag{6.20}$$

If further iterations are necessary, replace $\mathbf{a_g}$ with $\hat{\mathbf{a}}$ and repeat the solution.

As the solution converges and $\mathbf{a_g}$ approaches $\hat{\mathbf{a}}$, the first-order Taylor series expansion becomes a progressively better approximation to $f(x,\hat{a}_0,\hat{a}_1,\ldots,\hat{a}_m)$. Under these conditions, the formalism for calculating the covariance matrix in linear least squares also applies to linearized least squares. The covariance matrix is then

$$\mathbf{C} = (\mathbf{N})^{-1}. \tag{6.21}$$

Strictly speaking, this is the covariance matrix for the $\Delta\hat{a}_j$, but it is also the covariance matrix for $\hat{a}_j$, since a small error in $\Delta\hat{a}_j$ produces an equal error in $\hat{a}_j$. If the estimates of the measurement errors are thought to be incorrect, one should take

$$\widehat{\mathbf{C}} = \frac{S_{min}}{n-m-1}(\mathbf{N})^{-1} \tag{6.22}$$

(see Section 5.2.3), where $(m+1)$ is the number of fitted parameters,[1] and

$$S_{min} = \sum_i w_i \left[y_i - f(x_i,\hat{a}_0,\hat{a}_1,\ldots,\hat{a}_m)\right]^2. \tag{6.23}$$

---

[1] One usually sees $n-m$ in the denominator, not $n-m-1$. The intent is to subtract the number of degrees of freedom from n. For us, the number of degrees of freedom is $m+1$, because the parameters are numbered from a_0 to a_m.

6.2.2 Data with Correlated Measurement Errors

Suppose now that the measurement errors are correlated, so that $\sigma_{ij} \neq 0$. Following the discussion in Section 5.5.2, the weight matrix is

$$\mathbf{W} = \begin{pmatrix} w_{11} & w_{12} & \cdots & w_{1n} \\ w_{21} & w_{22} & \cdots & w_{2n} \\ \vdots & \vdots & & \vdots \\ w_{n1} & w_{n2} & & w_{nn} \end{pmatrix} = \begin{pmatrix} \sigma_{11} & \sigma_{12} & \cdots & \sigma_{1n} \\ \sigma_{21} & \sigma_{22} & \cdots & \sigma_{2n} \\ \vdots & \vdots & & \vdots \\ \sigma_{n1} & \sigma_{n2} & \cdots & \sigma_{nn} \end{pmatrix}^{-1}. \tag{6.24}$$

Let the vectors $\mathbf{f}$ and $\mathbf{y}$ be

$$\mathbf{y} = \begin{pmatrix} y_1 \\ y_2 \\ \vdots \\ y_n \end{pmatrix} \quad \text{and} \quad \mathbf{f} = \begin{pmatrix} f(x_1, \vec{a}) \\ f(x_2, \vec{a}) \\ \vdots \\ f(x_n, \vec{a}) \end{pmatrix}. \tag{6.25}$$

We wish to minimize

$$S = (\mathbf{y} - \mathbf{f})^{\mathrm{T}} \mathbf{W} (\mathbf{y} - \mathbf{f}) \tag{6.26}$$

Linearize by expanding $\mathbf{f}$ to first order about $\vec{a}_g$,

$$f(x, \vec{a}) = f(x, \vec{a}_g) + \sum_{j=0}^{m} \frac{\partial f(x, \vec{a})}{\partial a_j} \Delta a_j, \tag{6.27}$$

so that $\mathbf{f}$ becomes

$$\mathbf{f} = \begin{pmatrix} f(x_1, \vec{a}_g) + \sum_j \dfrac{\partial f(x_1, \vec{a})}{\partial a_j} \Delta a_j \\ f(x_2, \vec{a}_g) + \sum_j \dfrac{\partial f(x_2, \vec{a})}{\partial a_j} \Delta a_j \\ \vdots \\ (f(x_n, \vec{a}_g) + \sum_j \dfrac{\partial f(x_n, \vec{a})}{\partial a_j} \Delta a_j \end{pmatrix}. \tag{6.28}$$

Set the derivatives of S by each Δa_k equal to 0. After some algebra, the normal equations, become

$$\left(\mathbf{F}^{\mathrm{T}} \mathbf{W} \mathbf{F} \right) \Delta \hat{\mathbf{a}} = \mathbf{F}^{\mathrm{T}} \mathbf{W} (\mathbf{y} - \mathbf{f}), \tag{6.29}$$

where the $(m+1) \times n$ matrix $\mathbf{F}$ is

$$\mathbf{F} = \begin{pmatrix} \dfrac{\partial f(x_1, \vec{a})}{\partial a_0}\bigg|_{\vec{a}_g} & \dfrac{\partial f(x_1, \vec{a})}{\partial a_1}\bigg|_{\vec{a}_g} & \cdots & \dfrac{\partial f(x_1, \vec{a})}{\partial a_m}\bigg|_{\vec{a}_g} \\ \dfrac{\partial f(x_2, \vec{a})}{\partial a_0}\bigg|_{\vec{a}_g} & \dfrac{\partial f(x_2, \vec{a})}{\partial a_1}\bigg|_{\vec{a}_g} & \cdots & \dfrac{\partial f(x_2, \vec{a})}{\partial a_m}\bigg|_{\vec{a}_g} \\ \vdots & \vdots & & \vdots \\ \dfrac{\partial f(x_n, \vec{a})}{\partial a_0}\bigg|_{\vec{a}_g} & \dfrac{\partial f(x_n, \vec{a})}{\partial a_1}\bigg|_{\vec{a}_g} & \cdots & \dfrac{\partial f(x_n, \vec{a})}{\partial a_m}\bigg|_{\vec{a}_g} \end{pmatrix}, \tag{6.30}$$

and the normal matrix is

$$\mathbf{N} = \mathbf{F}^{\mathrm{T}} \mathbf{W} \mathbf{F}. \tag{6.31}$$

Note that this definition of $\mathbf{F}$ does reduce to the definition given in equation 5.140 when the dependence of the fitted function on the parameters is linear. The solution for $\Delta \hat{\mathbf{a}}$ is

$$\Delta \hat{\mathbf{a}} = \left(\mathbf{F}^{\mathrm{T}} \mathbf{W} \mathbf{F} \right)^{-1} \mathbf{F}^{\mathrm{T}} \mathbf{W} (\mathbf{y} - \mathbf{f}), \tag{6.32}$$

and the revised parameters of fit are

$$\hat{\mathbf{a}} = \mathbf{a_g} + \Delta \hat{\mathbf{a}}. \tag{6.33}$$

If further iterations are necessary, replace $\mathbf{a_g}$ with $\hat{\mathbf{a}}$ and repeat the solution.

Drawing again on our discussion of linear least squares with correlated measurement errors in Section 5.5.2, we can write down the covariance matrix by inspection:

$$\mathbf{C} = \left(\mathbf{F}^{\mathrm{T}} \mathbf{W} \mathbf{F} \right)^{-1}. \tag{6.34}$$

As usual, if the estimates of the measurement errors are thought to be incorrect, one may want to take (see the discussion of the estimated covariance matrix in Section 5.2.3 and the footnote following equation 6.22)

$$\widehat{\mathbf{C}} = \frac{S_{min}}{n - m - 1} \left(\mathbf{F}^{\mathrm{T}} \mathbf{W} \mathbf{F} \right)^{-1}, \tag{6.35}$$

where

$$S_{min} = (\mathbf{y} - \hat{\mathbf{f}})^{\mathrm{T}} \mathbf{W} (\mathbf{y} - \hat{\mathbf{f}}), \tag{6.36}$$

and where the symbol $\hat{\mathbf{f}}$ means that $\mathbf{f}$ is evaluated at $\hat{\mathbf{a}}$.

6.2.3 Practical Considerations

If S has more than one minimum, then it will have saddle points, places where one or more of the derivatives of S are equal to 0 but S is at a local *maximum* with respect to those coordinates. The normal matrices $\mathbf{N}$ or $\mathbf{F}^{\mathrm{T}} \mathbf{W} \mathbf{F}$ are singular at saddle points. In the neighborhood of saddle points, their inverses then have large (or infinite) terms, and the vector of corrections $\Delta \hat{\mathbf{a}}$ can become large, throwing the solution wildly off and into the wrong minimum of S. More generally, if the initial guesses for the parameters of fit are not close to their values at the deepest minimum of S, the solution may converge to an incorrect local minimum. The initial guesses $\mathbf{a_g}$ must therefore be chosen with care. A more serious annoyance is that, if S has multiple minima of nearly equal depth, noise in the data can introduce uncertainty about which minimum is the true minimum. The properties of S must be understood well enough to deal with these situations.

Nonlinear least squares solutions can converge slowly and can yield values of $\hat{\mathbf{a}}$ that oscillate or even diverge. One reason this can happen is that $\Delta \hat{\mathbf{a}}$ often points in the right direction but is too long, overshooting the minimum. The value of S should be checked after every iteration to make sure it has become smaller. One way to handle this bad behavior is to make corrections to $\hat{\mathbf{a}}$ that are only a fraction of $\Delta \hat{\mathbf{a}}$:

$$\hat{\mathbf{a}} = \mathbf{a_g} + \alpha \Delta \hat{\mathbf{a}}, \tag{6.37}$$

where α is a constant less than 1. It is often adequate to take α in the range 0.1–0.5, but smaller values of α may occasionally be needed.

There is no single best convergence criterion. Perhaps the most sensible criterion is to continue iterating until $\Delta\hat{a}_k$ is much less than $\sigma_{\hat{a}_k}$ for all k; that is, until the changes in $\hat{a}_k$ are much smaller than their uncertainties. One could also consider iterating until the changes in S are much smaller than S. Neither of these criteria is foolproof, though, because even small values for Δa_k or S could add up to large total changes with many additional iterations.

6.3 Other Methods for Minimizing S

The linearization technique discussed in Section 6.2 is based on a Taylor series expansion of $f(x,\vec{a})$. If analytic expressions for the first derivatives of $f(x,\vec{a})$ with respect to the parameters of fit are known or if the derivatives can be easily calculated numerically, linearization is often the method of choice, because it automatically yields the covariance matrix. Linearization is not likely to be a good choice if the derivatives are difficult to evaluate. It may also be a poor choice if S has a complicated topology with numerous local minima, or a pathological topology that requires many iterations to find the minimum (e.g., suppose the minimum of S lies somewhere in a tightly wound spiral-shaped valley).

This section describes several useful alternatives to linearizing $f(x,\vec{a})$. Recall that the goal is to find the values of $\vec{a}$ that minimize S. This is a specific case of the more general problem of minimizing a function, a common problem for which there is a huge literature.[2] The minimization techniques described here are those that are well adapted to least squares problems. Exhaustive grid mapping is computationally inefficient but can illuminate the geography of S and may be a required first step to find the deepest minimum of S. Steepest descent, Newton's method, and Marquardt's method are related techniques that use derivatives of S instead of $f(x,\vec{a})$ to find the minimum of S. They are easy to understand and together provide an effective set of tools. Their disadvantage is that they require derivatives of S, which generally must be calculated numerically. Simplex optimization avoids the use of derivatives altogether, making it more robust against numerical instabilities. This section also includes a short discussion of simulated annealing. While simulated annealing is not guaranteed to find the best possible fit, it can find good fits when finding the best fit is difficult or impossible. It can also be an effective tool for finding local minima worth exploring further.

These alternative methods do not automatically yield the covariance matrix, so the variances and covariances of the fitted parameters must be calculated separately. Section 6.4 describes several different ways to do this.

6.3.1 Grid Mapping

Figure 6.2 shows a surface S whose true minimum is surrounded by local minima in the shape of rings. Techniques for finding the minimum of S that look only at local values of S, such as Marquardt's method or the simplex algorithm (see sections 6.3.2 and 6.3.3), can become trapped in one of the local minima. They may even fail to converge, because the local minima are closed loops with same value of S around the entire loop. Now imagine the surface turned upside down. The deepest minimum is itself a ring, and the minimization techniques discussed up to now are doomed. There is no substitute for understanding the geography of S!

[2] See, for example, Press et al. (2007).

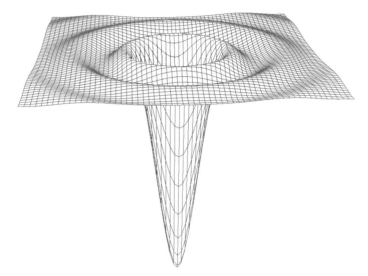

Figure 6.2: A pathological example of S, where $S = 1 - \sin(r)/r$, and $r = (x^2 + y^2)^{1/2}$. The true minimum of S is surrounded by local minima in the shape of rings. Techniques for finding the minimum of S by looking at local values of S (e.g., the simplex algorithm) or by calculating local gradients (e.g., the method of steepest decent) can become trapped in one of the local minima.

If a sufficiently powerful computer is available, it may be possible and even desirable to map the interesting part of parameter space by covering it with a grid of points and calculating S for each point. There is much to recommend this method, especially if little is known about the character of S, because a grid search can map the idiosyncrasies of S, at least if the grid is fine enough. If the parameter space is large or there are many parameters, calculating a full map of S can bring even powerful computers to their knees, though.

6.3.2 Method of Steepest Descent, Newton's Method, and Marquardt's Method

Method of Steepest Descent: We wish to find the values of $\vec{a}$ that minimize the weighted sum of the squares of the residuals

$$S = \sum_{i=1}^{n} w_i [y_i - f(x_i, a_0, a_1, \ldots, a_m)]^2. \tag{6.38}$$

In the method of steepest descent, Newton's method, and Marquardt's method, S is visualized as a surface in $(m+2)$-dimensional space lying above the $(m+1)$-dimensional space whose coordinates are the parameters of fit, $(a_0, a_1, \ldots, a_m)$. The best fit values for the parameters are their values at the deepest minimum in the surface.

In the method of steepest descent—also known as the gradient-search method—one begins at a convenient position in parameter space, specified by guessed values $\vec{a}_g$ of the parameters of fit, and then moves along the surface in the direction it decreases most rapidly. Expanding S as a Taylor series in $\vec{a}$ about $\vec{a}_g$ and retaining only first-order terms, we have

$$S = S(\vec{a}_g) + \sum_{j=0}^{m} \left. \frac{\partial S}{\partial a_j} \right|_{\vec{a}_g} \Delta a_j, \tag{6.39}$$

where $\Delta a_j = a_j - a_{j_g}$. Equation 6.39 can be written in vector notation as

$$S = S(\vec{a}_g) + \Delta \mathbf{a}^{\mathrm{T}} \nabla S, \qquad (6.40)$$

where ∇S is the gradient of S. The direction in which S decreases most rapidly (steepest descent) is $-\nabla S$, so the steepest-descent corrections to $\vec{a}_g$ are given by $\Delta \mathbf{a} \propto -\nabla S$ or

$$\Delta a_j = -\alpha \left. \frac{\partial S}{\partial a_j} \right|_{\vec{a}_g}, \qquad (6.41)$$

where α specifies the length of the vector of corrections. The revised values for the parameters are then $a_j = a_{j_g} + \Delta a_j$. Since equation 6.40 is only an approximation to the dependence of S on the parameters of fit, the revised parameters are still only approximations to their values at the true minimum of S. To find a better approximation, replace $\vec{a}_g$ the newly revised values for the parameters, and calculate new corrections, repeating this cycle until the fit is satisfactory.

There is no single best value for α. A common way to choose its value is to use the second derivatives of S, as these determine the amount of curvature and therefore set length scales for the parameters. To do this, calculate the second derivatives $\partial^2 S / \partial a_j^2$ and set α equal to the inverse of the largest of the second derivatives:

$$\alpha = \min \left[\left(\left. \frac{\partial^2 S}{\partial a_j^2} \right|_{\vec{a}_g} \right)^{-1} \right]. \qquad (6.42)$$

The value of α can be adjusted up or down as desired to improve the speed of convergence.

In a modified version of the method of steepest descent, a different value of α is assigned to each of the $m + 1$ individual equations in equation 6.41, so that

$$\Delta a_j = -\alpha_j \left. \frac{\partial S}{\partial a_j} \right|_{\vec{a}_g}, \qquad (6.43)$$

where

$$\alpha_j = \alpha \left(\left. \frac{\partial^2 S}{\partial a_j^2} \right|_{\vec{a}_g} \right)^{-1} \qquad (6.44)$$

and α is a constant. The point of equations 6.43 and 6.44 is to make the distance moved along each direction in parameter space inversely proportional to the curvature in that direction. One generally starts with $\alpha \approx 1$ and then allows α to increase or decrease as needed to improve convergence.

Newton's Method: The method of steepest descent is effective when the guessed values for the parameters are far from their values at the minimum of S. When close to the minimum, however, the method of steepest descent often becomes inefficient, taking small zig-zag steps and requiring many iterations. Newton's method usually gives excellent results when the parameters are close to their values at the minimum of S, as this method converges quadratically toward the minimum. Like the method of steepest descent, Newton's method starts with guessed values $\vec{a}_g$ for the parameters but then approximates S in the neighborhood of $\vec{a}_g$ with a paraboloid. The revised values for the parameters are their values at the minimum of the parabola.

To use Newton's method, expand S as a Taylor series in $\vec{a}$ about $\vec{a}_g$, and drop terms higher than quadratic:

$$S = S(\vec{a}_g) + \sum_{i=0}^{m} \left.\frac{\partial S}{\partial a_i}\right|_{\vec{a}_g} \Delta a_i + \frac{1}{2} \sum_{i=0}^{m} \sum_{j=0}^{m} \left.\frac{\partial^2 S}{\partial a_i \partial a_j}\right|_{\vec{a}_g} \Delta a_i \Delta a_j. \tag{6.45}$$

This is the paraboloid that approximates S near $\vec{a}_g$. To find the minimum of the paraboloid, set the derivatives of S with respect to the Δa_k equal to 0:

$$\frac{\partial S}{\partial \Delta a_k} = 0 = \left.\frac{\partial S}{\partial a_k}\right|_{\vec{a}_g} + \sum_{j=0}^{m} \left.\frac{\partial^2 S}{\partial a_k \partial a_j}\right|_{\vec{a}_g} \Delta \hat{a}_j. \tag{6.46}$$

Equation 6.46 is equivalent to the matrix equation

$$\begin{pmatrix} \dfrac{\partial^2 S}{\partial a_0 \partial a_0} & \cdots & \dfrac{\partial^2 S}{\partial a_0 \partial a_m} \\ \vdots & & \vdots \\ \dfrac{\partial^2 S}{\partial a_m \partial a_0} & \cdots & \dfrac{\partial^2 S}{\partial a_m \partial a_m} \end{pmatrix} \begin{pmatrix} \Delta \hat{a}_0 \\ \vdots \\ \Delta \hat{a}_m \end{pmatrix} = - \begin{pmatrix} \dfrac{\partial S}{\partial a_0} \\ \vdots \\ \dfrac{\partial S}{\partial a_m} \end{pmatrix}, \tag{6.47}$$

where the derivatives are all evaluated at $\vec{a}_g$. The matrix of second derivatives of a function is called the *Hessian matrix*. The Hessian matrix is symmetric and has a nonnegative determinant near minima in S. The vector on the right-hand side of equation 6.46 is just the gradient of S. If we set

$$\mathbf{H} = \begin{pmatrix} \dfrac{\partial^2 S}{\partial a_0 \partial a_0} & \cdots & \dfrac{\partial^2 S}{\partial a_0 \partial a_m} \\ \vdots & & \vdots \\ \dfrac{\partial^2 S}{\partial a_m \partial a_0} & \cdots & \dfrac{\partial^2 S}{\partial a_m \partial a_m} \end{pmatrix} \quad \text{and} \quad \nabla S = \begin{pmatrix} \dfrac{\partial S}{\partial a_0} \\ \vdots \\ \dfrac{\partial S}{\partial a_m} \end{pmatrix}, \tag{6.48}$$

equations 6.45 and 6.47 can be written more compactly as

$$S = S(\vec{a}_g) + \Delta \mathbf{a}^{\mathrm{T}} \nabla S + \frac{1}{2} \Delta \mathbf{a}^{\mathrm{T}} \mathbf{H} \Delta \mathbf{a} \tag{6.49}$$

and

$$\mathbf{H} \Delta \hat{\mathbf{a}} = -\nabla S. \tag{6.50}$$

The corrections to the guessed values of the parameters of fit are

$$\Delta \hat{\mathbf{a}} = -\mathbf{H}^{-1} \nabla S, \tag{6.51}$$

and the revised parameters of fit are $\hat{\mathbf{a}} = \mathbf{a_g} + \Delta \hat{\mathbf{a}}$.

While the paraboloid is a better approximation to the shape of S than the linear approximation used in the method of steepest descent, the location of the minimum in the paraboloid is unlikely to lie at the true minimum of S, so the revised values for the parameters are approximations to their best fit values. To find better approximations, replace $\vec{a}_g$ with the revised values for the parameters and repeat the fit, repeating the cycle as necessary. The rate of convergence can sometimes be improved by correcting the parameters by a fraction of $\Delta\hat{\mathbf{a}}$,

$$\hat{\mathbf{a}} = \mathbf{a_g} + \alpha\Delta\hat{\mathbf{a}}, \tag{6.52}$$

where $\alpha < 1$.

Marquardt's Method: Newton's method often gives good results when the $\vec{a}_g$ are near their best values, while the method of steepest descent often gives good results when far from the minimum of S. Marquardt's method is a convenient way to combine the best of both. In Newton's method the corrections to $\vec{a}_g$ are given by the solution to $\mathbf{H}\Delta\hat{\mathbf{a}} = -\nabla S$ (equation 6.50). Marquardt's method replaces $\mathbf{H}$ by a new matrix $\mathbf{H}'$, whose components are

$$\left(\mathbf{H}'\right)_{jk} = \begin{cases} (\mathbf{H})_{jk}\,(1+\alpha), & j=k \\ (\mathbf{H})_{jk}, & j \neq k \end{cases} \tag{6.53}$$

and calculates the corrections to $\vec{a}_g$ from

$$\Delta\hat{\mathbf{a}} = -(\mathbf{H}')^{-1}\nabla S. \tag{6.54}$$

When the constant α is large, equation 6.54 reduces to m separate equations:

$$\Delta a_j \approx \frac{1}{1+\alpha}\left(\frac{\partial^2 S}{\partial a_j^2}\right)^{-1}\frac{\partial S}{\partial a_j}. \tag{6.55}$$

Equation 6.55 has the same form as equations 6.43 and 6.44 taken together, which are the modified version of steepest descent. When α is small, $\mathbf{H}' \approx \mathbf{H}$, and equation 6.54 reduces to $\Delta\hat{\mathbf{a}} = -(\mathbf{H})^{-1}\nabla S$, becoming Newton's method (equation 6.50).

Marquardt[3] and others recommend beginning with a small value for α, say, $\alpha \approx 0.01$. Then increase or decrease α by successive factors of 10 to improve convergence. Larger values of α are generally needed far from convergence, smaller values close to convergence.

The method of steepest descent, Newton's method, and Marquardt's method all require derivatives of S. The derivatives must generally be calculated numerically from, for example,

$$\frac{\partial S}{\partial a_j} \approx \frac{S(a_j+\delta_j) - S(a_j-\delta_j)}{2\delta_j} \tag{6.56}$$

$$\frac{\partial^2 S}{\partial a_j^2} \approx \frac{S(a_j+\delta_j) - 2S(a_j) + S(a_j-\delta_j)}{\delta_j^2} \tag{6.57}$$

$$\frac{\partial^2 S}{\partial a_j\partial a_k} \approx \frac{S(a_j+\delta_j,a_k+\delta_k) - S(a_j-\delta_j,a_k+\delta_k) - S(a_j+\delta_j,a_k-\delta_k) + S(a_j-\delta_j,a_k-\delta_k)}{4\delta_j\delta_k},$$
$$\tag{6.58}$$

where δ_j and δ_k are small offsets in a_j and a_k. If there is any noise in the calculated values of S, the numerical derivatives will be prone to instability and must be handled with care.

---

[3] D. W. Marquardt. 1963. "An Algorithm for Least-Squares Estimation of Nonlinear Parameters". *Journal of the Society of Industrial and Applied Mathematics* vol. 11, p. 431.

6.3.3 Simplex Optimization

Again think of S as a surface in $(m + 2)$-dimensional space lying above the $(m+1)$-dimensional parameter space with coordinates $(a_0, a_1, \ldots, a_m)$. Simplex optimization is somewhat like placing an amoeba on the surface and letting it crawl downhill to a minimum. Instead of an amoeba, the algorithm uses a simplex, the geometrical figure that generalizes the concept of a triangle in two-dimensional space or a tetrahedron in three-dimensional space to higher dimensions. Formally, a simplex is the minimum convex surface that encloses $m + 2$ vertices in $(m + 1)$-dimensional space. For our purposes, though, we can ignore the enclosing surface and concentrate on the points at the vertices.

The $m + 2$ points defining the initial simplex must be placed close enough to the deepest minimum of S to prevent the simplex from migrating toward and falling into a shallower local minimum. The points must also span the $(m + 1)$-dimensional parameter space. One way to satisfy these constraints is to choose the first point close to the expected minimum. Calling this location $(a_{0_0}, a_{1_0}, \ldots, a_{m_0})$, distribute the other $m + 1$ points by adding small offsets ϵ_j to each of the parameters in turn, forming $(a_{0_0} + \epsilon_0, a_{1_0}, \ldots, a_{m_0})$, $(a_{0_0}, a_{1_0} + \epsilon_1, \ldots, a_{m_0})$, and so on.

The simplex migrates to the minimum of S by a series of operations called *reflection, expansion, contraction,* and *scaling.* Evaluate S at each of the points in the simplex, and note the points giving the highest (worst) and lowest (best) values of S (see Figure 6.3). Find the arithmetic mean position—the centroid—of all the points other than the worst. One now looks for a new point to replace the worst point. The first new point to try is the reflection of the worst point through the centroid. If the value of S at the reflected point is lower than the worst value of S, the reflected point replaces the worst point, forming a new simplex. If the value of S at the reflected point is lower (better) than the very lowest value of S, a second point is tried in the same direction as the first trial but to twice the distance from the centroid. If this second trial point yields a yet lower value of S, it replaces the first trial point, forming an expanded simplex.

If neither of the first two trial points yields an improved value of S, try a third point that is half the distance from the worst point to the centroid. If the value of S at the third point is better than at the worst point, replace the worst point with the third point, forming a contracted simplex. If none of the three trial points yields an improved value of S, scale the simplex. The vector from the centroid to each vertex of the simplex is either doubled or halved, either expanding or shrinking the simplex by a factor of two. The choice of expansion or contraction is up to the user. Expanding the simplex enlarges the region of parameter space being searched. Shrinking it allows regions where S is complicated to be explored in more detail.

Whatever the result of these operations, a new simplex is formed, and the search for a point giving a yet better value of S repeats with the new simplex. One iterates until the convergence criterion is satisfied. The possible convergence criteria are the same as for Newton's method, but common sense must be used. For example, if the first three trial points did not improve S and the simplex has to be expanded or contracted, a criterion based on S may declare the search has converged when it has not.

6.3.4 Simulated Annealing

Some least squares problems can defeat even the most concerted effort to find the single best fit. This can happen when, for example, there are many parameters of fit and many narrow local minima in S. If one is willing to accept a good fit instead of the single best fit, several fitting techniques are available. Simulated annealing is one of the most attractive.

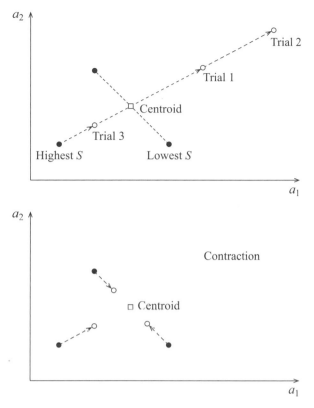

Figure 6.3: Operations for simplex optimization of two parameters. The three filled circles in the top panel are the initial points in the parameter space (a_1, a_2); they define the initial simplex. The points that give the highest and lowest values for S are labeled. The open square is the average position (the "centroid") of all the points except the one with the highest value of S. The open circles mark the locations of the first three trials for a point to replace the highest point. If none of the three trial points give a value for S lower than the highest point, the simplex is either contracted toward the centroid (shown in the bottom panel) or expanded away from the centroid.

Simulated annealing takes its inspiration from the physical process of annealing. The probability that a physical system will find itself in a state with energy E is proportional to the Boltzmann factor

$$f(E) \propto \exp[-E/kT], \tag{6.59}$$

where T is the temperature, and k is Boltzmann's constant. The higher the temperature, the greater the probability that a system will be in a high energy state. Material is annealed by raising its temperature until its molecules are likely to have enough energy to overcome the binding energy holding them in place. The molecules can begin to migrate, both individually and in bulk, and with time can explore many different configurations. When the body is cooled, the molecules have less energy and become bound in place again. If given enough time, the molecules can settle into lower-energy configurations than the ones in which they started.

Simulated annealing refers to a class of iterative techniques in which the new parameters of fit at each iteration are determined in part by a deterministic algorithm and in part by random noise. In early iterations, the random noise is large; but as iterations proceed, the random noise is reduced, eventually leaving just a deterministic approach to the minimum. The point of the random noise is to bump fits out of shallow local minima in S, providing a chance for the fit to find a deeper minimum. As the amplitude of the noise is decreased, fits are more likely to be confined to regions around a smaller number of minima and are more likely to be near deeper rather than shallower minima. When the noise goes to 0, the solution stays within the region around a single minimum, finding the deepest part of that minimum.

An analogy is provided by a vibrating table whose top has many dents. Place a small steel ball on the table. When the amplitude of the vibrations is large, the steel ball bounces over the entire surface of the table. As the amplitude of the vibrations is decreased, the ball bounces less, tending to spend more time within dents. When the amplitude of the vibration becomes small, there is a good chance the steel ball will wind up in one of the deeper dents; and when the vibration is turned off altogether, the ball will roll to the deepest part of the dent in which it currently finds itself. If this experiment is repeated many times, the steel ball will eventually wind up in one of the deepest dents, although not necessarily the deepest dent.

Many implementations of simulated annealing have been proposed.[4] Press et al. (2007) recommend an implementation based on simplex optimization. In the standard simplex method, the value of S at each vertex of the simplex is recorded, and the value of S at each trial point is compared to the recorded values of S. In the simulated annealing version of simplex optimization (different) random numbers are added to each of the stored values of S and subtracted from the values of S at every trial point. The movement of the simplex is determined by the noisy values of S, not the true values of S. Subtracting random numbers from S at the trial points means that points with high S will occasionally look like they have a low S, and the simplex will move there. In effect, the simplex will sometimes move uphill. The amount by which it can move uphill depends on the amplitude of the noise. One starts with high noise, allowing the simplex to wander widely about the surface of S, then slowly turns down the noise, progressively limiting the motion of the simplex, finally turning off the noise altogether and forcing the simplex to migrate downhill to the nearest minimum. The algorithm must be rerun many times starting the simplex at different places to ensure that the minima of S have been well sampled.

To maintain the analogy with physical annealing, the random numbers ϵ_s added to S at the stored points and subtracted at the trial points are generated from the exponential distribution

$$f(\epsilon_s) = \frac{1}{kT} \exp[-\epsilon_s / kT]. \tag{6.60}$$

The amplitude of the noise depends of the temperature. When T is large, the amplitude is large; when T is small, the amplitude is small. One begins a fitting run with a large value of T and then slowly turns T down. The efficiency of simulated annealing depends sensitively on the cooling schedule and the number of reruns. There is little guidance for choosing among the suggestions that have been made for each. The user will have to experiment with both for the specific problem at hand.

---

[4] For example, P.J.M. van Laarhoven and E.H.L. Aarts. 1987. *Simulated Annealing: Theory and Applications.* New York: Springer; P. Salamon, P. Sibani, and R. Frost. 2002. *Facts, Conjectures, and Improvements for Simulated Annealing.* Philadelphia: SIAM Press.

6.4 Error Estimation

Because minimization techniques of the kind described in Section 6.3 do not automatically yield a covariance matrix, the variances and covariances of the fitted parameters must be calculated separately. Many investigators resort to brute-force Monte Carlo error analysis. This is usually not necessary. One obvious way to calculate the covariance matrix is to calculate the normal matrix using equation 6.16 or 6.31, even though the normal matrix is not used to optimize the parameters. The covariance matrix is the inverse of the normal matrix.

In this section we describe two additional ways to find the covariance matrix, neither of which requires explicit calculation of the normal matrix. The first calculates the covariance matrix from the inverse of the Hessian matrix. The second is a direct, albeit cumbersome, calculation of the covariance matrix. We finish with a discussion of confidence limits.

6.4.1 Inversion of the Hessian Matrix

We begin again with the weighted sum of the squares of the residuals

$$S = \sum_{i=1}^{n} w_i [y_i - f(x_i, \vec{a})]^2. \tag{6.61}$$

Assume that the values $\hat{a}_j$ that minimize S have already been found. At those values, we have

$$S = S_{min} = \sum_i w_i \left[y_i - f(x_i, \vec{\hat{a}}) \right]^2. \tag{6.62}$$

Expand $f(x, \vec{a})$ to first order in a Taylor series about the $\hat{a}_j$ to get

$$f(x, \vec{a}) = f(x, \vec{\hat{a}}) + \sum_{j=0}^{m} \left. \frac{\partial f(x, \vec{a})}{\partial a_j} \right|_{\vec{\hat{a}}} (a_j - \hat{a}_j)$$

$$= f(x, \vec{\hat{a}}) + \sum_{j=0}^{m} \frac{\partial f(x, \vec{a})}{\partial a_j} \Delta a_j, \tag{6.63}$$

where $\Delta a_j = a_j - \hat{a}_j$. Here and from now on, the partial derivatives of $f(x, \vec{a})$ are understood to be evaluated at the $\hat{a}_j$. Inserting this expression into equation 6.61 and expanding, we have

$$S = \sum_i w_i \left[y_i - f(x_i, \vec{\hat{a}}) - \sum_{j=0}^{m} \frac{\partial f(x_i, \vec{a})}{\partial a_j} \Delta a_j \right]^2$$

$$= \sum_i w_i \left[y_i - f(x_i, \vec{\hat{a}}) \right]^2 - 2 \sum_i \left\{ w_i \left[y_i - f(x_i, \vec{\hat{a}}) \right] \sum_{j=0}^{m} \frac{\partial f(x_i, \vec{a})}{\partial a_j} \Delta a_j \right\}$$

$$+ \sum_i w_i \left(\sum_{j=0}^{m} \frac{\partial f(x_i, \vec{a})}{\partial a_j} \Delta a_j \right)^2. \tag{6.64}$$

The first term on the right-hand side of equation 6.64 is S_{min}. Since S is at a minimum when $\Delta a_j = 0$, its first derivatives there are equal to 0:

$$0 = \left. \frac{\partial S}{\partial \Delta a_j} \right|_{\Delta a_j = 0} = \sum_i w_i \left[y_i - f(x_i, \vec{a}) \right] \frac{\partial f(x_i, \vec{a})}{\partial a_j}. \tag{6.65}$$

Thus the second term on the right-hand side of equation 6.65 is 0 for all j, and equation 6.64 simplifies to[5]

$$S = S_{min} + \sum_i w_i \left[\sum_j \frac{\partial f(x_i, \vec{a})}{\partial a_j} \Delta a_j \right]^2. \tag{6.66}$$

Expanding equation 6.66 and rearranging a bit, we find

$$S = S_{min} + \sum_j \sum_k \Delta a_j \Delta a_k \left[\sum_i w_i \frac{\partial f(x_i, \vec{a})}{\partial a_j} \frac{\partial f(x_i, \vec{a})}{\partial a_k} \right]$$
$$= S_{min} + \Delta \mathbf{a}^{\mathrm{T}} \mathbf{N} \Delta \mathbf{a}, \tag{6.67}$$

where we have recognized that the quantities inside the large square bracket are the components of the normal matrix $\mathbf{N}$ with the derivatives of $f(x, \vec{a})$ evaluated at the converged values $\hat{a}_j$ (see equation 6.16).

Now assume that the minimum of S has been found by one of the iterative techniques discussed in Section 6.3. As the iterative solution approaches S_{min}, the linear expansion in equation 6.63 becomes a successively better approximation. The nonlinear problem becomes a linear problem, merging with the formalism for linear least squares. In linear least squares, the covariance matrix is given by the inverse of the normal matrix, $\mathbf{C} = \mathbf{N}^{-1}$, and when the solution has converged to S_{min}, this is also true for the nonlinear least squares. This was the logic used to obtain equations 6.21 and 6.34. Here, however, we are assuming that we cannot or do not want to calculate $\mathbf{N}$ from the derivatives of $f(x, \vec{a})$. We do, though, know how to calculate S as a function of $\vec{a}$ and can use this to calculate $\mathbf{N}$. Comparing equation 6.49 (evaluated at $\hat{a}_j$) with equation 6.67 and remembering that $\nabla S = 0$ at S_{min}, we can write

$$\frac{1}{2} \Delta \mathbf{a}^{\mathrm{T}} \mathbf{H} \Delta \mathbf{a} = \Delta \mathbf{a}^{\mathrm{T}} \mathbf{N} \Delta \mathbf{a}. \tag{6.68}$$

Since $\Delta \mathbf{a}$ is a free variable, this can only be true if the normal matrix is equal to half the Hessian matrix:

$$\mathbf{N} = \frac{1}{2} \mathbf{H}. \tag{6.69}$$

We can now specify how to calculate the covariance matrix. If the minimum of S has been found using Newton's or Marquardt's method, the Hessian matrix is already available, and the covariance matrix can be calculated from

$$\mathbf{C} = \mathbf{N}^{-1} = 2 \mathbf{H}^{-1}. \tag{6.70}$$

---

[5] Some authors expand the Taylor series in equation 6.63 to second order. The second-order expansion yields additional second-order terms in equation 6.66. The additional terms are usually small and almost universally ignored.

If the Hessian matrix is not already available, calculate all the second derivatives of S numerically by, for example, using equations 6.57 and 6.58. The components of $\mathbf{H}$ are equal to the second derivatives

$$(\mathbf{H})_{jk} = \frac{\partial^2 S}{\partial \Delta a_j \partial \Delta a_k}, \tag{6.71}$$

and again the covariance matrix is given by $\mathbf{C} = 2\mathbf{H}^{-1}$.

6.4.2 Direct Calculation of the Covariance Matrix

It may be preferable to calculate the variances and covariances directly. The way to do this is to offset a_k by a small amount ϵ_k from its best fit value $\hat{a}_k$, reoptimize S by letting all the other a_i vary freely, and then determine the variance from the amount by which S changes. The need to reoptimize S is illustrated in Figure 6.4, which shows a contour plot of S for two parameters of fit, a_1 and a_2. The best fit value S_{min} lies in a long narrow valley. In the left panel of the figure, S increases rapidly when a_1 is displaced by ϵ_1, because the path of the displacement crawls up the side wall of the valley. The rapid increase suggests that a_1 is restricted to a small range of values and has a small error. In fact, it is possible to displace a_1 by ϵ_1 without a large increase in S by also increasing a_2 and moving along the floor of the valley. Because S increases only slowly along the valley floor, the variance of a_1 is large. When calculating the covariance matrix from the inverse of the normal matrix or the Hessian matrices, the off-diagonal elements of the matrices automatically include this effect. When calculating the covariance matrix directly, the effect is included by reoptimizing S.

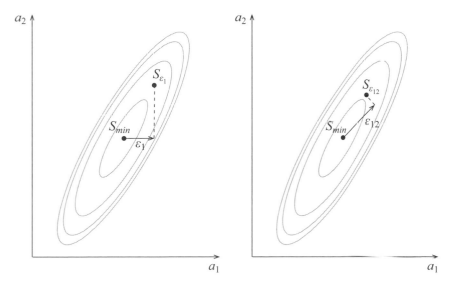

Figure 6.4: A contour plot of S for two parameters of fit, a_1 and a_2. The best fit value S_{min} lies in a long narrow valley. If a_1 is displaced by ϵ_1 without allowing a_2 to vary, as shown in the left panel, the path crawls up the side of the valley, and S increases rapidly, suggesting that the variance of a_2 is small. It is, however, possible to displace a_1 without a large increase in S by moving along the floor of the valley. Because S increases only slowly along the valley floor, the variance of a_1 is large. The right panel shows a similar effect when the sum of a_1 and a_2 is displaced by ϵ_{12}.

A convenient way to pick out one parameter for error analysis is to use the vector $\boldsymbol{\delta}_k$, where $\boldsymbol{\delta}_k$ is all zeros except for the kth element, which is set equal to 1:

$$\boldsymbol{\delta}_k = \begin{pmatrix} 0 \\ 0 \\ \vdots \\ 1 \\ \vdots \\ 0 \end{pmatrix}. \tag{6.72}$$

The following equation offsets a_k from $\hat{a}_k$ by an amount ϵ_k without constraining the values of the other parameters:

$$\boldsymbol{\delta}_k{}^{\mathrm{T}}\Delta\mathbf{a} = \epsilon_k. \tag{6.73}$$

We now seek to reminimize

$$S = S_{min} + \Delta\mathbf{a}^{\mathrm{T}}\mathbf{N}\Delta\mathbf{a} \tag{6.74}$$

(equation 6.67), subject to the constraint

$$g(\Delta\mathbf{a}) = \boldsymbol{\delta}_k{}^{\mathrm{T}}\Delta\mathbf{a} - \epsilon_k = 0. \tag{6.75}$$

This problem can be solved with Lagrange multipliers. Setting the gradients of S and g proportional to each other, we have

$$\Delta\mathbf{a}^{\mathrm{T}}\mathbf{N} = \lambda\boldsymbol{\delta}_k{}^{\mathrm{T}} \tag{6.76}$$

$$\Delta\mathbf{a}^{\mathrm{T}} = \lambda\boldsymbol{\delta}_k{}^{\mathrm{T}}\mathbf{N}^{-1} \tag{6.77}$$

$$\Delta\mathbf{a} = \lambda\left(\mathbf{N}^{-1}\right)^{\mathrm{T}}\boldsymbol{\delta}_k = \lambda\mathbf{N}^{-1}\boldsymbol{\delta}_k, \tag{6.78}$$

where λ is the undetermined multiplier, and the last equality holds because $\mathbf{N}^{-1}$ is symmetric. Plugging this result back into the constraint equation, we have

$$\lambda\boldsymbol{\delta}_k{}^{\mathrm{T}}\mathbf{N}^{-1}\boldsymbol{\delta}_k = \epsilon_k, \tag{6.79}$$

which yields

$$\lambda = \frac{\epsilon_k}{\boldsymbol{\delta}_k{}^{\mathrm{T}}\mathbf{N}^{-1}\boldsymbol{\delta}_k}. \tag{6.80}$$

Equations 6.77 and 6.78 for $\Delta\mathbf{a}^{\mathrm{T}}$ and $\Delta\mathbf{a}$ become

$$\Delta\mathbf{a} = \mathbf{N}^{-1}\boldsymbol{\delta}_k\frac{\epsilon_k}{\boldsymbol{\delta}_k{}^{\mathrm{T}}\mathbf{N}^{-1}\boldsymbol{\delta}_k} \tag{6.81}$$

$$\Delta\mathbf{a}^{\mathrm{T}} = \boldsymbol{\delta}_k{}^{\mathrm{T}}\mathbf{N}^{-1}\frac{\epsilon_k}{\boldsymbol{\delta}_k{}^{\mathrm{T}}\mathbf{N}^{-1}\boldsymbol{\delta}_k}, \tag{6.82}$$

Denote the reoptimized value of S by S_{ϵ_k}. Plugging equations 6.81 and 6.82 into equation 6.74, we find

$$S_{\epsilon_k} = S_{min} + \boldsymbol{\delta}_k{}^{\mathrm{T}}\mathbf{N}^{-1}\mathbf{N}\mathbf{N}^{-1}\boldsymbol{\delta}_k\left(\frac{\epsilon_k}{\boldsymbol{\delta}_k{}^{\mathrm{T}}\mathbf{N}^{-1}\boldsymbol{\delta}_k}\right)^2$$

$$= S_{min} + \frac{\epsilon_k^2}{\boldsymbol{\delta}_k{}^{\mathrm{T}}\mathbf{N}^{-1}\boldsymbol{\delta}_k},$$

or

$$\delta_k{}^T N^{-1} \delta_k = \frac{\epsilon_k^2}{S_{\epsilon_k} - S_{min}}. \tag{6.83}$$

Remembering that the covariance matrix is $C = N^{-1}$, we have

$$\delta_k{}^T C \delta_k = \frac{\epsilon_k^2}{S_{\epsilon_k} - S_{min}}. \tag{6.84}$$

Now note that $\delta_k{}^T C \delta_k$ picks out the kth diagonal member of C:

$$\delta_k{}^T C \delta_k = \sigma_{\hat{k}}^2, \tag{6.85}$$

which is the desired variance of $\hat{a}_k$. The final result is

$$\sigma_{\hat{k}}^2 = \frac{\epsilon_k^2}{S_{\epsilon_k} - S_{min}}. \tag{6.86}$$

This relation is less easy to apply than it looks, because a substantial amount of work may be needed to reoptimize S, and S must be reoptimized anew for each variance and covariance.

To find the covariances $\sigma_{\hat{kj}}^2$, transform to a new set of variables $\Delta a_i'$, where $\Delta a_i' = \Delta a_i$ for all i except

$$\Delta a_k' = (\Delta a_k + \Delta a_j)/\sqrt{2} \tag{6.87}$$

$$\Delta a_j' = (\Delta a_k - \Delta a_j)/\sqrt{2}. \tag{6.88}$$

As before, $S = S_{min}$ when all the $\Delta a_i' = 0$. Now set $\Delta a_k' = \epsilon_k'$ (or, equivalently, set $a_k + a_j = \epsilon_k'$, as shown in Figure 6.4). Using the same logic as before, we reoptimize S by letting all the other parameters (including $\Delta a_j'$) vary freely and arrive at

$$(\sigma_{\hat{k}}')^2 = \frac{(\epsilon_k')^2}{S_{\epsilon_k'} - S_{min}}, \tag{6.89}$$

where $(\sigma_{\hat{k}}')^2$ is the variance of $\Delta a_k'$, and $S_{\epsilon_k'}$ is the reoptimized value of S. From standard results for the propagation of errors, we have

$$(\sigma_{\hat{k}}')^2 = \frac{1}{2}\left(\sigma_{\hat{k}}^2 + 2\sigma_{\hat{kj}} + \sigma_{\hat{j}}^2\right), \tag{6.90}$$

so the desired expression for the covariance between Δa_k and Δa_j is

$$\sigma_{\hat{kj}} = (\sigma_{\hat{k}}')^2 - \frac{1}{2}(\sigma_{\hat{k}}^2 + \sigma_{\hat{j}}^2)$$

$$= \frac{(\epsilon_k')^2}{S_{\epsilon_k'} - S_{min}} - \frac{1}{2}(\sigma_{\hat{k}}^2 + \sigma_{\hat{j}}^2). \tag{6.91}$$

6.4.3 Summary and the Estimated Covariance Matrix

We have provided four ways to calculate the covariance matrix in nonlinear least squares problems:

1. If the time and a sufficiently powerful computer are available, use the Monte Carlo error analysis.
2. If one is prepared to calculate the normal matrix from the first derivatives of $f(x, \vec{a})$, use equations 6.16 or 6.31 to calculate the components of $\mathbf{N}$. The covariance matrix is the inverse of $\mathbf{N}$. This method can be used even if the best fit parameters were found by, for example, simplex optimization.
3. Alternatively it may be easier to calculate the second derivatives of S. If so, calculate the components of the Hessian matrix using equation 6.48. The covariance matrix is equal to twice the inverse of the Hessian matrix (equation 6.70).
4. To avoid numerical second derivatives of S, calculate the components of the covariance matrix directly from equations 6.86 and 6.91. The penalty paid in this fourth approach is that S must be reoptimized for each variance and covariance.

All four techniques for calculating the covariance matrix assume that the variances of the original data points are correct. If they are incorrect, the covariance matrix is incorrect. To compensate for possibly incorrect variances, one should generally use the estimated covariance matrix $\hat{\mathbf{C}}$ (see the discussion of the estimated covariance matrix in Section 5.2.3 and the footnote following equation 6.22),

$$\hat{\mathbf{C}} = \frac{S_{min}}{n - m - 1} \mathbf{C}, \qquad (6.92)$$

where n is the number of data points, and $(m + 1)$ is the number of fitted parameters.

6.5 Confidence Limits

Suppose that an experiment yields an estimate $\hat{a}_{obs}$ for a parameter whose true value is a. If $\hat{a}_{obs}$ is a continuous variable, the probability that it exactly equals a is zero! A nonzero probability can be assigned to the result only by specifying a range of values in which a might lie. A typical way to do this is to report the standard deviation $\hat{\sigma}_{\hat{a}}$. But how should we interpret $\hat{\sigma}_{\hat{a}}$ or, indeed, any other range of values? In Bayesian statistics, the posterior probability distribution can be used to infer such statements as "There is a 90% probability that the true value of parameter a lies between a_1 and a_2." The interval from a_1 and a_2 is called a *credible interval* (see Section 7.3.5). Statements like this are not possible in frequentist statistics. To a frequentist, only one true value exists for a parameter even if the value is unknown, so it makes no sense to talk about its probability distribution. Instead, one must talk about the probability that a range of values, called the *confidence limits*, encompasses the true value. Taking liberties with a standard metaphor, we can compare confidence limits to a game of garden quoits played in dim light. In garden quoits a stake is driven into the ground, and players toss a rope hoop, attempting to encircle the stake. The stake represents the single true value of a parameter, and the rope loop represents the confidence limits derived from measurements. One asks "What is the probability that a quoit will encircle the stake?" Several problems arise when attempting to derive confidence limits. First, the true value of the parameter is not known; second, the probability distribution for

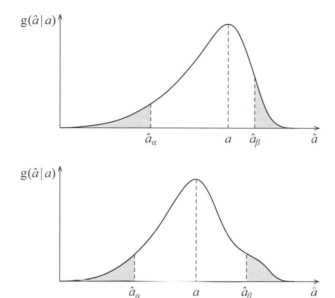

Figure 6.5: Probability distribution $g(\hat{a}|a)$ for measurements $\hat{a}$ of a parameter whose true value is a. In general the distribution will depend on a. We assume that a is unknown, but the dependence of the distribution on a is known. The top and bottom panels show how the probability distribution might look for two different values of a. The shaded areas in the figures are integrals of the distribution function. They represent the probabilities α that $\hat{a}$ is less than $\hat{a}_\alpha$, and β that $\hat{a}$ is greater than $\hat{a}_\beta$. Note that $\hat{a}_\alpha$ and $\hat{a}_\beta$ depend on a. The probability that $\hat{a}$ lies in the unshaded region is $P(\hat{a}_\alpha < \hat{a} < \hat{a}_\beta) = 1 - \alpha - \beta$.

$\hat{a}$ may not be precisely specified; and third, $\hat{\sigma}_{\hat{a}}$ is only an estimated value of the standard deviation. We must deal with all these issues.

Let $g(\hat{a}|a)$ be the probability distribution for measurements $\hat{a}$ of a parameter whose true value is a. Since the purpose of our experiment is to estimate a, its value is unknown, but, at least initially, we will assume that the functional form of $g(\hat{a}|a)$ is known. Figure 6.5 shows how $g(\hat{a}|a)$ might appear for two different values of a. The shaded areas in the figures are integrals of the distribution function. They represent the probabilities α that $\hat{a}$ is less than $\hat{a}_\alpha$, and β that $\hat{a}$ is greater than $\hat{a}_\beta$:

$$\alpha = P(\hat{a} \leq \hat{a}_\alpha) = \int_{-\infty}^{\hat{a}_\alpha} g(\hat{a}|a)d\hat{a} \tag{6.93}$$

$$\beta = P(\hat{a} \geq \hat{a}_\beta) = \int_{\hat{a}_\beta}^{\infty} g(\hat{a}|a)d\hat{a}. \tag{6.94}$$

The probability that $\hat{a}$ lies in the unshaded area is

$$P(\hat{a}_\alpha \leq \hat{a} \leq \hat{a}_\beta) = 1 - \alpha - \beta. \tag{6.95}$$

In the present context one chooses the probabilities α and β, and then equations 6.93 and 6.94 become implicit equations for $\hat{a}_\alpha$ and $\hat{a}_\beta$. The two limits depend on a. Figure 6.6 shows plots of $\hat{a}_\alpha$ and $\hat{a}_\beta$ as functions of a. The band between them is called the *confidence belt*.

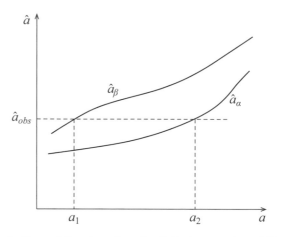

Figure 6.6: As shown in Figure 6.5 and as given implicitly by equations 6.93 and 6.94, $\hat{a}_\alpha$ and $\hat{a}_\beta$ depend on the true value a of the parameter. The solid curves are plots of $\hat{a}_\alpha$ and $\hat{a}_\beta$ as functions of a. The band between them is the confidence belt. The horizontal dashed line marks $\hat{a}_{obs}$, a single determination of the parameter. It intersects $\hat{a}_\beta$ at a_1 and $\hat{a}_\alpha$ at a_2. The confidence interval for a is a_1 to a_2.

Suppose one performs an experiment that yields the value $\hat{a}_{obs}$ for the parameter. This is plotted as the horizontal dashed line in Figure 6.6. The line intersects $\hat{a}_\beta$ at $a = a_1$ and $\hat{a}_\alpha$ at $a = a_2$. If a is less than a_1, then $\hat{a}_{obs}$ must be greater than $\hat{a}_\beta$, which happens with probability β, so

$$P(a \leq a_1) = \beta. \tag{6.96}$$

If a is greater than a_2, then $\hat{a}_{obs}$ must be less than $\hat{a}_\alpha$, which happens with probability α, so

$$P(a \geq a_2) = \alpha. \tag{6.97}$$

Together these equations yield

$$P(a_1 \leq a \leq a_2) = 1 - \alpha - \beta. \tag{6.98}$$

The interval $[a_1, a_2]$ is called the confidence interval, and a_1 and a_2 are the confidence limits. Note that the random variable used when deriving equation 6.98 is $\hat{a}_{obs}$, not a. Therefore, the proper interpretation of confidence limits is as follows. If the experiment were to be repeated, it would yield different values for $\hat{a}_{obs}$ and, therefore, of a_1 and a_2. If the experiment were repeated many times, a would lie between a_1 and a_2 with frequency $1 - \alpha - \beta$.

The analytic equivalent of the graphical construction in Figure 6.6 is obtained by setting the intersection points to $\hat{a}_\alpha = \hat{a}_{obs}$ at $a = a_2$ and $\hat{a}_\beta = \hat{a}_{obs}$ at $a = a_1$, so that equations 6.93 and 6.94 become

$$\alpha = \int_{-\infty}^{\hat{a}_{obs}} g(\hat{a}|a_2) d\hat{a} \tag{6.99}$$

$$\beta = \int_{\hat{a}_{obs}}^{\infty} g(\hat{a}|a_1) d\hat{a}. \tag{6.100}$$

These are now understood to be implicit equations for a_1 and a_2. The practitioner is free to choose values for α and β. A typical choice is $\alpha = \beta$ so that, for example, $\alpha = \beta = 0.025$

would yield a 95% confidence interval. Equations 6.99 and 6.100 do not depend on the unknown true value of the parameter, so we have sidestepped the first problem.

The second problem is that the functional form of $g(\hat{a}|a)$ is often not known. If not, and if one is willing to expend considerable computational effort, a Monte Carlo approach may be a feasible way to solve this problem. Create many sets of artificial data whose noise properties are the same as the real data set, allowing a to vary from set to set. Determine $\hat{a}_{obs}$ for each set. The distribution of the $\hat{a}_{obs}$ traces out $g(\hat{a}|a)$.

In practice, however, motivated by the central limit theorem and by the simplicity of the resulting calculations, one often assumes that the distribution is Gaussian. Let $g(\hat{a}|a)$ be the Gaussian distribution

$$g(\hat{a}|a) = \frac{1}{\sqrt{2\pi}\,\sigma_{\hat{a}}} \exp\left[-\frac{1}{2}\frac{(\hat{a}-a)^2}{\sigma_{\hat{a}}^2}\right]. \tag{6.101}$$

For the moment, we use the real variance $\sigma_{\hat{a}}^2$ and not the estimated variance $\hat{\sigma}_{\hat{a}}^2$. Equations 6.99 and 6.100 become

$$\alpha = \frac{1}{\sqrt{2\pi}\,\sigma_{\hat{a}}} \int_{-\infty}^{\hat{a}_{obs}} \exp\left[-\frac{1}{2}\frac{(\hat{a}-a_2)^2}{\sigma_{\hat{a}}^2}\right] d\hat{a} = \frac{1}{\sqrt{2\pi}\,\sigma_{\hat{a}}} \int_{-\infty}^{\hat{a}_{obs}-a_2} \exp\left[-\frac{1}{2}\frac{x^2}{\sigma_{\hat{a}}^2}\right] dx \tag{6.102}$$

$$\beta = \frac{1}{\sqrt{2\pi}\,\sigma_{\hat{a}}} \int_{\hat{a}_{obs}}^{\infty} \exp\left[-\frac{1}{2}\frac{(\hat{a}-a_1)^2}{\sigma_{\hat{a}}^2}\right] d\hat{a} = \frac{1}{\sqrt{2\pi}\,\sigma_{\hat{a}}} \int_{\hat{a}_{obs}-\hat{a}_1}^{\infty} \exp\left[-\frac{1}{2}\frac{x^2}{\sigma_{\hat{a}}^2}\right] dx. \tag{6.103}$$

These can be combined into the single equation:

$$1 - \alpha - \beta = \frac{1}{\sqrt{2\pi}\,\sigma_{\hat{a}}} \int_{\hat{a}_{obs}-a_2}^{\hat{a}_{obs}-a_1} \exp\left[-\frac{1}{2}\frac{x^2}{\sigma_{\hat{a}}^2}\right] dx. \tag{6.104}$$

Because the Gaussian distribution is symmetric about its center point, it is appropriate to set $\alpha = \beta$ and to set the limits of this integral to be symmetric about $x = 0$. Let $Q = 1 - \alpha - \beta = 1 - 2\alpha$ be the confidence level. Calculation of the confidence interval reduces to calculating the value of Δx for which

$$Q = \frac{1}{\sqrt{2\pi}\,\sigma_{\hat{a}}} \int_{-\Delta x = \hat{a}_{obs}-a_2}^{+\Delta x = \hat{a}_{obs}-a_1} \exp\left[-\frac{1}{2}\frac{x^2}{\sigma_{\hat{a}}^2}\right] dx. \tag{6.105}$$

The confidence limits are then

$$[a_1, a_2] = [\hat{a}_{obs} - \Delta x, \hat{a}_{obs} + \Delta x]. \tag{6.106}$$

Equations 6.105 and 6.106 are the desired result for this specific case. The are easy to apply. Choose a confidence limit Q. Then from tables for integrals of the Gaussian distribution, find the value of Δx for which the integral from $-\Delta x$ to $+\Delta x$ is equal to Q. The 68.3% confidence interval ($Q = 0.683$), for example, is $[a_1, a_2] = [\hat{a}_{obs} - \sigma_{\hat{a}}, \hat{a}_{obs} + \sigma_{\hat{a}}]$. The 95.4% confidence interval is $[\hat{a}_{obs} - 2\sigma_{\hat{a}}, \hat{a}_{obs} + 2\sigma_{\hat{a}}]$.

The third problem is that we do not generally know the true variance $\sigma_{\hat{a}}^2$. The default way to handle this is to assume that the estimated variance is a good approximation for the true variance and set $\sigma_{\hat{a}}^2 = \hat{\sigma}_{\hat{a}}^2$. The 95% confidence interval, for example, becomes

$$[a_1, a_2] = [\hat{a}_{obs} - 2\hat{\sigma}_{\hat{a}}, \hat{a}_{obs} + 2\hat{\sigma}_{\hat{a}}]. \tag{6.107}$$

For small sample sizes, this may not be a good approximation (see Section 4.5). If not, one will likely need to resort to a Monte Carlo calculation.

Equations 6.105 and 6.106 are simple and intuitive. The logic that leads to them is anything but. A quote from Kendall would seem appropriate at this point:

> The reader to whom this approach is new will probably ask: but is this not a roundabout way of using the standard error to set limits to an estimate of the mean? In a way it is. In effect, what we have done ... is to show how the use of the standard error of the mean in normal samples may be justified on logical grounds without appeal to new principles of inference other than those incorporated in the theory of probability itself. In particular we make no use of Bayes' postulate.[6]

The extension from one-parameter to multiparameter confidence limits is relatively easy if the distribution of the measured parameters is a multiparameter Gaussian. Let the multiparameter generalization of $g(\hat{a}|a)$ be

$$g(\hat{\mathbf{a}}|\mathbf{a}) = \frac{1}{(2\pi)^{n/2}|\mathbf{C}|^{1/2}} \exp\left[-\frac{1}{2}(\hat{\mathbf{a}}-\mathbf{a})^{\mathrm{T}}\mathbf{C}^{-1}(\hat{\mathbf{a}}-\mathbf{a})\right], \tag{6.108}$$

where $\hat{\mathbf{a}}$ is the vector of measurements of the parameters, $\mathbf{a}$ is the vector of their true values, and $\mathbf{C}$ is the covariance matrix (see Section 2.5). If G is a constant, the expression

$$G = (\hat{\mathbf{a}}-\mathbf{a})^{\mathrm{T}}\mathbf{C}^{-1}(\hat{\mathbf{a}}-\mathbf{a}) \tag{6.109}$$

defines the surface of a multidimensional ellipsoid centered on $\mathbf{a}$. If $\hat{\mathbf{a}}_{obs}$ is a vector of realized measurements of the parameters, then

$$G = (\hat{\mathbf{a}}-\hat{\mathbf{a}}_{obs})^{\mathrm{T}}\mathbf{C}^{-1}(\hat{\mathbf{a}}-\hat{\mathbf{a}}_{obs}) \tag{6.110}$$

defines the surface of an identical multidimensional ellipsoid centered on $\hat{\mathbf{a}}_{obs}$. The integral of $g(\hat{\mathbf{a}}|\hat{\mathbf{a}}_{obs})$ over the volume V_G bounded by this surface is

$$\int_{V_G} g(\hat{\mathbf{a}}|\hat{\mathbf{a}}_{obs})d\hat{\mathbf{a}}.$$

To extend confidence limits for the one-parameter case to the multiparameter case, let Q be the confidence level. Our goal is to find the value G_Q such that the multidimensional integral of $g(\hat{\mathbf{a}}|\hat{\mathbf{a}}_{obs})$ over the volume V_{G_Q} is equal to Q:

$$Q = \int_{V_{G_Q}} g(\hat{\mathbf{a}}|\hat{\mathbf{a}}_{obs})d\hat{\mathbf{a}}. \tag{6.111}$$

To simplify this equation, recognize that G is a χ^2 variable with $m+1$ degrees of freedom, where $m+1$ is the number of parameters (see Section 2.6). Equation 6.111 can therefore be rewritten as the one-dimensional integral

$$Q = \int_0^{G_Q} f_{m+1}(\chi^2)d(\chi^2), \tag{6.112}$$

---

[6] M. G. Kendall. (1969), p. 64.

where $f_{m+1}(\chi^2)$ is the χ^2 distribution for $m+1$ degrees of freedom. As before, this is an implicit equation for G_Q, but unlike equation 6.111, it can be solved numerically using readily available tables of integrals for the χ^2 distribution. The confidence region is the volume bounded by the surface

$$G_Q = (\hat{\mathbf{a}} - \hat{\mathbf{a}}_{obs})^\mathrm{T} \mathbf{C}^{-1} (\hat{\mathbf{a}} - \hat{\mathbf{a}}_{obs}). \tag{6.113}$$

As for the one-dimensional case, the true covariance matrix is unknown, so one generally substitutes the estimated covariance matrix $\hat{\mathbf{C}}$ for $\mathbf{C}$. To accommodate the limited ability of humans to visualize higher-dimensional spaces, one usually plots confidence regions for just two or three parameters at a time. The two- and three-dimensional confidence regions are projections of the multidimensional confidence region, not crosscuts through it. If the confidence region is an ellipsoid as it is here, the projection is easy to calculate.[7]

We are now in a position to generalize these results. Suppose one has fit a model with $m+1$ parameters $\mathbf{a}$ to data by minimizing $\chi^2(\mathbf{a})$. If the minimum value of χ^2 is χ^2_{min} and the best fit values of the parameters are $\hat{\mathbf{a}}$, then $\chi^2_{min} = \chi^2(\hat{\mathbf{a}})$. Even if we no longer assume that $\chi^2(\mathbf{a})$ is a multidimensional Gaussian, equation 6.112 is still valid. Once again, let Q be the confidence level. We choose a value for Q and then write

$$Q = \int_0^{\chi^2_Q} f_{m+1}(\chi^2) d(\chi^2), \tag{6.114}$$

which is to be understood as an implicit equation for χ^2_Q. Equation 6.113 is now replaced by the constraint

$$\chi^2_Q = \chi^2(\mathbf{a}). \tag{6.115}$$

This equation defines the surface of constant χ^2_Q in $(m+1)$-dimensional parameter space. The region enclosed by the surface is the confidence region for the $\mathbf{a}$ that corresponds to confidence level Q. Since the confidence region is no longer an ellipsoid, calculating the projection can be a challenge.

The final generalization is to maximum likelihood fits for which the log-likelihood function is not simply related to χ^2. These cases must usually be handled numerically. One good way is to treat the likelihood function as an unnormalized probability distribution and extract its properties using MCMC sampling. The two properties of interest here are the surfaces of constant likelihood and the fraction of the probability (the value of Q) that is included within the contours. The practitioner chooses a value of Q, which then constrains the contour.

6.6 Fits with Errors in Both the Dependent and Independent Variables

Least squares with errors in both coordinates is surprisingly complex, and the general problem was not solved satisfactorily until 1980 by Jefferys.[8] Here we describe only the fit of a straight line. The account is a simplified version of that given by Jefferys.

---

[7] See, for example, Press et al. (2007), p. 815.

[8] W. H. Jefferys. 1980. "On the Method of Least Squares." *Astronomical Journal* vol. 85, p.177; W. H. Jefferys. 1981. "On the Method of Least Squares. II." *Astronomical Journal* vol. 86, p.149.

6.6.1 Data with Uncorrelated Errors

Consider n data points (x_i, y_i) with uncorrelated errors in both coordinates $(\sigma_{x_i}, \sigma_{y_i})$, as shown in Figure 6.7. We wish to fit the data with a straight line of the form

$$y = a_0 + a_1 x. \tag{6.116}$$

The complication here is that the measured values of both x_i and y_i differ from their true values x_{t_i} and y_{t_i} because of measurement errors ϵ_{x_i} and ϵ_{y_i}:

$$x_{t_i} = x_i + \epsilon_{x_i} \tag{6.117}$$

$$y_{t_i} = y_i + \epsilon_{y_i}. \tag{6.118}$$

Furthermore, we only have access to the measured values of the coordinates, not their true values, so we are forced to fit

$$y_i + \epsilon_{y_i} = a_0 + a_1(x_i + \epsilon_{x_i}). \tag{6.119}$$

The amounts by which the line misses the data points are unknown in advance, so there are actually $2n + 2$ unknowns to the problem: $2n$ values for the ϵ_{x_i} and ϵ_{y_i}, plus one each for a_0 and a_1.

One can adopt a maximum likelihood approach to this problem by assuming that the ϵ_{x_i} and ϵ_{y_i} have a multivariate Gaussian distribution with variances $\sigma_{x_i}^2$ and $\sigma_{y_i}^2$, respectively, and covariances $\langle \epsilon_{x_i} \epsilon_{y_i} \rangle = \sigma_{xy_i}$ that are equal to 0. With these assumptions, the maximum likelihood values of a_0 and a_1 are those that maximize the log-likelihood function

$$\ell = -\frac{1}{2} \sum_{i=1}^{n} \left(\frac{\epsilon_{x_i}^2}{\sigma_{x_i}^2} + \frac{\epsilon_{y_i}^2}{\sigma_{y_i}^2} \right). \tag{6.120}$$

Alternatively we can simply adopt the least squares formalism and minimize the quantity

$$S = \sum_{i=1}^{n} \left(\frac{\epsilon_{x_i}^2}{\sigma_{x_i}^2} + \frac{\epsilon_{y_i}^2}{\sigma_{y_i}^2} \right). \tag{6.121}$$

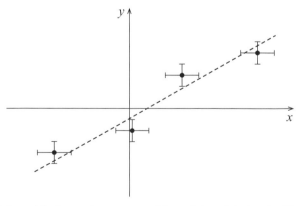

Figure 6.7: The points and their error bars represent the n data points (x_i, y_i) with uncorrelated errors $(\sigma_{x_i}, \sigma_{y_i})$. The dashed line is the fit of $y = a_0 + a_1 x$ to the data.

Then we wish to minimize S subject to the n constraint equations

$$g_j(\epsilon_{x_j}, \epsilon_{y_j}) = y_j + \epsilon_{y_j} - a_0 - a_1(x_j + \epsilon_{x_j}) = 0, \tag{6.122}$$

which force all the ϵ_{x_j} and ϵ_{y_j} to be consistent with a single line with parameters a_0 and a_1.

We solve this problem by the method of Lagrange multipliers (see Appendix B). For this application, equation B.7 become the $2n$ equations

$$\frac{\partial S}{\partial \epsilon_{x_i}} + \sum_{j=1}^{n} \lambda_j \frac{\partial g_j(\epsilon_{x_j}, \epsilon_{y_j})}{\partial \epsilon_{x_i}} = 0 \tag{6.123}$$

$$\frac{\partial S}{\partial \epsilon_{y_i}} + \sum_{j=1}^{n} \lambda_j \frac{\partial g_j(\epsilon_{x_j}, \epsilon_{y_j})}{\partial \epsilon_{y_i}} = 0, \tag{6.124}$$

where the λ_j are the Lagrange multipliers. The solutions to equations 6.123 and 6.124 yield estimates for ϵ_{x_j} and ϵ_{y_j} in terms of the λ_j. From equations 6.122, we have

$$\frac{\partial g_j(\epsilon_{x_j}, \epsilon_{y_j})}{\partial \epsilon_{x_i}} = -a_1 \delta_j^i \tag{6.125}$$

$$\frac{\partial g_j(\epsilon_{x_j}, \epsilon_{y_j})}{\partial \epsilon_{y_i}} = \delta_j^i, \tag{6.126}$$

where δ_j^i is the Kronecker delta, so equations 6.123 and 6.124 become

$$2 \frac{1}{\sigma_{x_i}^2} \hat{\epsilon}_{x_i} - a_1 \lambda_i = 0 \tag{6.127}$$

$$2 \frac{1}{\sigma_{y_i}^2} \hat{\epsilon}_{y_i} + \lambda_i = 0. \tag{6.128}$$

Since these are now estimates, we have equipped the epsilons with hats. Solving for the $\hat{\epsilon}_{x_i}$ and $\hat{\epsilon}_{y_i}$, we find

$$\hat{\epsilon}_{x_i} = \frac{1}{2} a_1 \sigma_{x_i}^2 \lambda_i \tag{6.129}$$

$$\hat{\epsilon}_{y_i} = -\frac{1}{2} \sigma_{y_i}^2 \lambda_i. \tag{6.130}$$

Substituting these estimates back into the constraint equations (equation 6.122), we have

$$y_i - \frac{1}{2} \sigma_{y_i}^2 \lambda_i = a_0 + a_1 \left(x_i + \frac{1}{2} \sigma_{x_i}^2 a_1 \lambda_i \right), \tag{6.131}$$

and then solving for λ_i, we find

$$\lambda_i = \frac{2}{\sigma_{y_i}^2 + a_1^2 \sigma_{x_i}^2} (y_i - a_0 - a_1 x_i). \tag{6.132}$$

The estimated errors become

$$\hat{\epsilon}_{x_i} = \frac{a_1 \sigma_{x_i}^2}{\sigma_{y_i}^2 + a_1^2 \sigma_{x_i}^2}(y_i - a_0 - a_1 x_i) \tag{6.133}$$

$$\hat{\epsilon}_{y_i} = -\frac{\sigma_{y_i}^2}{\sigma_{y_i}^2 + a_1^2 \sigma_{x_i}^2}(y_i - a_0 - a_1 x_i). \tag{6.134}$$

We use these estimates to eliminate ϵ_{x_i} and ϵ_{y_i} in the expression for S (equation 6.121), finding

$$S = \sum_{i=1}^{n} \left\{ \frac{1}{\sigma_{x_i}^2} \left[\frac{a_1 \sigma_{x_i}^2}{\sigma_{y_i}^2 + a_1^2 \sigma_{x_i}^2}(y_i - a_0 - a_1 x_i) \right]^2 \right.$$
$$\left. + \frac{1}{\sigma_{y_i}^2} \left[\frac{\sigma_{y_i}^2}{\sigma_{y_i}^2 + a_1^2 \sigma_{x_i}^2}(y_i - a_0 - a_1 x_i) \right]^2 \right\} \tag{6.135}$$

$$S = \sum_{i=1}^{n} \frac{1}{\sigma_{y_i}^2 + a_1^2 \sigma_{x_i}^2}(y_i - a_0 - a_1 x_i)^2. \tag{6.136}$$

Equation 6.136 now gives S as a function of a_0 and a_1 and known quantities. The dependence on the unknown values of ϵ_{x_i} and ϵ_{y_i} has been eliminated. Note that if there is no error in the x_i (i.e., if $\sigma_{x_i}^2 = 0$), equation 6.136 reduces to

$$S = \sum_{i=1}^{n} \frac{1}{\sigma_{y_i}^2}(y_i - a_0 - a_1 x_i)^2; \tag{6.137}$$

and if there are errors in the x_i but not in the y_i (if $\sigma_{y_i}^2 = 0$), it reduces to

$$S = \sum_{i=1}^{n} \frac{1}{\sigma_{x_i}^2}(x_i - b_0 - b_1 y_i)^2, \tag{6.138}$$

where $b_0 = -a_0/a_1$, and $b_1 = 1/a_1$, both of which are the correct expressions for S.

The least squares estimates of a_0 and a_1, which here are also their maximum likelihood estimates, are those that minimize S as given by equation 6.136. Note that the investigator still has much work to do! One must still find the values of a_0 and a_1 that minimize S, and even in this simplest of cases, the fit is nonlinear, because a_1 appears in the denominator inside the summation. We spent considerable effort developing methods for solving nonlinear least squares problems earlier in this chapter. Any and all of them should be adequate for minimizing S and determining a_0 and a_1. Likewise the methods for calculating the covariance matrix developed in Section 6.4 should work well here.

6.6.2 Data with Correlated Errors

Equation 6.121 for S assumes that the errors in x_i and y_i are uncorrelated. In effect the axes of the error ellipses for the data points are aligned parallel to the coordinate axes. It can happen that the errors are correlated and the error ellipses are tilted. In this case $\sigma_{xy_i} \neq 0$, and the errors in the data are described by the covariance matrices

$$\boldsymbol{\sigma}_i = \begin{pmatrix} \sigma_{x_i}^2 & \sigma_{xy_i} \\ \sigma_{xy_i} & \sigma_{y_i}^2 \end{pmatrix}. \tag{6.139}$$

The weighted sum of the squares or, if one prefers, the log-likelihood function, become (see the discussion of the multivariate Gaussian distribution in Chapter 2, especially equations 2.80–2.83)

$$S = \sum_{i=1}^{n} \left(w_{x_i} \epsilon_{x_i}^2 + 2 w_{xy_i} \epsilon_{x_i} \epsilon_{y_i} + w_{y_i} \epsilon_{y_i}^2 \right) = -2\ell, \tag{6.140}$$

where the weights for point i are the components of the weight matrix

$$\mathbf{w}_i = \begin{pmatrix} w_{x_i} & w_{xy_i} \\ w_{xy_i} & w_{y_i} \end{pmatrix} = (\boldsymbol{\sigma}_i)^{-1}. \tag{6.141}$$

We proceed in much the same way as for uncorrelated errors (but with more algebra!). Equations 6.127 and 6.128 become

$$2 w_{x_i} \hat{\epsilon}_{x_1} + 2 w_{xy_i} \hat{\epsilon}_{y_i} - a_1 \lambda_i = 0 \tag{6.142}$$

$$2 w_{x_i} \hat{\epsilon}_{y_i} + 2 w_{xy_i} \hat{\epsilon}_{x_i} + \lambda_i - 0 \tag{6.143}$$

and equations 6.129 and 6.130 become

$$\hat{\epsilon}_{x_i} = \frac{1}{2} \frac{w_{xy_i} + w_{y_i} a_1}{w_{x_i} w_{y_i} - w_{xy_i}^2} \lambda_i \tag{6.144}$$

$$\hat{\epsilon}_{y_i} = -\frac{1}{2} \frac{w_{x_i} + w_{xy_i} a_1}{w_{x_i} w_{y_i} - w_{xy_i}^2} \lambda_i. \tag{6.145}$$

Inserting these expressions for $\hat{\epsilon}_{x_i}$ and $\hat{\epsilon}_{y_i}$ into the constraint equation and solving for λ_i, we find

$$\lambda_i = \frac{2(w_{x_i} w_{y_i} - w_{xy_i}^2)}{w_{x_i} + 2 w_{xy_i} a_1 + w_{y_i} a_1^2} (y_i - a_0 - a_1 x_i), \tag{6.146}$$

and then equations 6.133 and 6.134 become

$$\hat{\epsilon}_{x_i} = \frac{w_{xy_i} + w_{y_i} a_i}{w_{x_i} + 2 w_{xy_i} a_1 + w_{y_i} a_1^2} (y_i - a_0 - a_1 x_i) \tag{6.147}$$

$$\hat{\epsilon}_{y_i} = -\frac{w_{x_i} + w_{xy_i} a_1}{w_{x_i} + 2 w_{xy_i} a_1 + w_{y_i} a_1^2} (y_i - a_0 - a_1 x_i). \tag{6.148}$$

Using these expressions to eliminate ϵ_{x_i} and ϵ_{y_i} in equation 6.140, we find

$$S = \sum_{i=1}^{n} \frac{w_{x_i} w_{y_i} - w_{xy_i}^2}{w_{x_i} + 2 w_{xy_i} a_1 + w_{y_i} a_1^2} (y_i - a_0 - a_1 x_i)^2. \tag{6.149}$$

Finally, since the inverses of the weight matrices are

$$\mathbf{w}_i = \frac{1}{w_{x_i} w_{y_i} - w_{xy_i}^2} \begin{pmatrix} w_{y_i} & -w_{xy_i} \\ -w_{xy_i} & w_{x_i} \end{pmatrix} = \boldsymbol{\sigma}_i, \tag{6.150}$$

equation 6.149 can be simplified to

$$S = \sum_{i=1}^{n} \frac{1}{\sigma_{y_i}^2 - 2\sigma_{xy_i}a_1 + \sigma_{x_i}^2 a_1^2} (y_i - a_0 - a_1 x_i)^2. \tag{6.151}$$

Note that equation 6.151 reduces correctly to equation 6.136 when the errors in the data are uncorrelated and $\sigma_{xy_i} = 0$.

Equation 6.151 is the desired result. As before, though, the investigator still has much work ahead. The equation is an expression for S as a function of a_0 and a_1. The investigator must find the values of a_0 and a_1 that minimize S. Equation 6.151, like equation 6.136, is nonlinear in the parameters of fit, but also like equation 6.136, should yield to the techniques discussed earlier in this chapter.

Bayesian Statistics

7.1 Introduction to Bayesian Statistics

Bayes's theorem plays only a minor role in the frequentist approach to statistics described in the previous chapters. This chapter describes Bayesian statistics, an approach to statistics in which Bayes's theorem dominates, and equation 4.2, which defines probability as a frequency, loses its fundamental role. The change is not merely mathematical. The use of Bayes's theorem implies a shift in one's perception of the meaning of probability and statistics. The spreading use of Bayesian statistics and the implied change in our understanding of probability is arguably the most important development in statistics in the past half century.

We begin with Bayes's theorem. The Venn diagram in Figure 7.1 shows the overlap $A \cap B$ between two sets A and B contained in set S. The probability that an element of set S lies in the overlap region can be written in two ways:

$$P(A \cap B) = P(B|A)\,P(A) \tag{7.1}$$

$$P(A \cap B) = P(A|B)\,P(B). \tag{7.2}$$

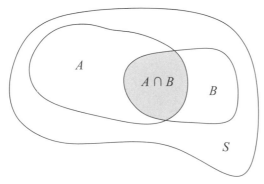

Figure 7.1: Sets A and B are subsets of the set S. The shaded region is $P(A \cap B)$, the intersection of A and B.

Eliminating $P(A \cap B)$ and rearranging, we find

$$P(B|A) = \frac{P(A|B)\,P(B)}{P(A)}. \tag{7.3}$$

Now let $A = D$ be the outcome of an experiment (D for data!), and let the set B have a finite number of discrete elements B_j. Equation 7.3 becomes

$$P(B_j|D) = \frac{P(D|B_j)\,P(B_j)}{P(D)}. \tag{7.4}$$

If D can occur only if one of the B_j occurs, the total probability that D occurs is

$$P(D) = \sum_k P(D|B_k)\,P(B_k). \tag{7.5}$$

Using this relation to eliminate $P(D)$ from equation 7.4, we arrive at one of the standard forms of Bayes's theorem:

$$P(B_j|D) = \frac{P(D|B_j)\,P(B_j)}{\sum_k P(D|B_k)\,P(B_k)}. \tag{7.6}$$

The fundamental objects of Bayesian statistics are probability distributions. Equation 7.6 describes how probability distributions evolve as new data become available. The distribution $P(B_j)$ is the probability for B_j based on existing knowledge. It is called the *prior probability distribution* or just the *prior*. The new data appears in the conditional probability $P(D|B_j)$, the probability that an experiment will yield result D if B_j occurs. It is identical to the likelihood function used in maximum likelihood calculations (see Section 5.2), and is called the *likelihood*. The conditional probability $P(B_j|D)$ is the revised probability that event B_j will occur based on the new data D. It is called the *posterior probability distribution* or just the *posterior*. The probability $P(D)$ is the total probability that D occurs; it acts as a normalization factor.

If B is continuous, Bayes's theorem takes the form

$$f_1(B|D) = \frac{L(D|B)\,f_0(B)}{\displaystyle\int L(D|B)f_0(B)dB}, \tag{7.7}$$

where $L(D|B)$ is the likelihood, $f_0(B)$ is the prior, $f_1(B|D)$ is the posterior, and subscripts have been added to emphasize that f_0 and f_1 are different distributions. We need not worry that $L(D|B)$ is not generally normalized, because the normalization factor cancels out. As in any probability distribution, the B_j in equation 7.6 and B in equation 7.7 can have many different meanings depending on the application. They can be numerical parameters; they can be states; or even, as in hypothesis testing, logical statements.

The following elementary example highlights the differences between Bayesian and frequentist statistics.

Example: Suppose a disease occurs in 0.01% of young adults. The disease is initially symptomless, so it is impossible to know whether someone has the disease without testing for it. The prior probability that a young adult has the disease is then

$$P(\text{diseased}) = 0.0001$$

$$P(\text{healthy}) = 0.9999.$$

Also suppose there is a diagnostic test for the disease, and the test is highly but not perfectly reliable. It gives the right answer 99% of the time, the wrong answer 1% of the time, and the wrong answers occur randomly. The conditional probabilities for the diagnostic test are

$$P(\text{positive}|\text{diseased}) = 0.99$$

$$P(\text{negative}|\text{diseased}) = 0.01$$

$$P(\text{positive}|\text{healthy}) = 0.01$$

$$P(\text{negative}|\text{healthy}) = 0.99,$$

where "positive" and "negative" mean the test reports that one does or does not have the disease, respectively.

Suppose a young adult (let's call this person Dana) is tested for the disease, and the test is positive. What is the probability that Dana really has the disease? Many people would be tempted to say there is a 99% probability that Dana has the disease. Bayesian statistics gives a different answer. According to equation 7.6, the posterior probability that Dana has the disease is

$$P(\text{diseased}|\text{positive})$$

$$= \frac{P(\text{positive}|\text{diseased})P(\text{diseased})}{P(\text{positive}|\text{diseased})P(\text{diseased}) + P(\text{positive}|\text{healthy})P(\text{healthy})}$$

$$= \frac{0.99 \times 0.0001}{0.99 \times 0.0001 + 0.01 \times .9999}$$

$$= 0.01.$$

Although Dana tested positive on a highly reliable test, there is only 1 chance in 100 that Dana has the disease!

For many people, this may be a counterintuitive result. To better understand what it means, let us modify the example to make it is amenable to frequentist thinking. The University of Texas at Austin has roughly 50,000 students, mostly young adults. Since the probability of occurrence is 10^{-4}, we expect about 5 of the students to have the disease. Suppose the university administration were to require all 50,000 students to be tested for the disease. Because the test has a 1% false positive rate, about 500 students would test positive—but 495 of these students do not really have the disease! So, there is only a 1% chance that the students who test positive for the disease really have it. This is what Bayesian statistics was telling us. Note, however, that Bayesian statistics allowed us to calculate a valid probability from a single test of a single student.

Continued on page 190

After seeing the results of the first test, Dana decides to take the test a second time. It is positive again. The prior probabilities have changed and now are

$$P(\text{diseased}) = 0.01$$

$$P(\text{healthy}) = 0.99,$$

so the posterior probability is

$$P(\text{diseased}|\text{positive}) = \frac{0.99 \times 0.01}{0.99 \times 0.01 + 0.01 \times .99}$$

$$= 0.50.$$

There is now 1 chance in 2 that Dana has the disease, a much higher probability than before. But even after two positive results on the test, the probability is still far from certainty.

The previous example displays several important features of Bayesian statistics. First, Bayes's theorem specifies how to merge new information with existing knowledge to yield new probability distributions. It can be viewed as giving meaning to inductive logic, making inductive logic a well-defined, quantitative procedure.

Second, Bayesian induction forces one to assess, quantify, and use preexisting knowledge. The preexisting knowledge is incorporated in two ways. There must be a prior probability distribution function $P(B_j)$, and it must be possible to calculate the likelihood function $P(D|B_j)$. Bayesian statistics cannot proceed without both.

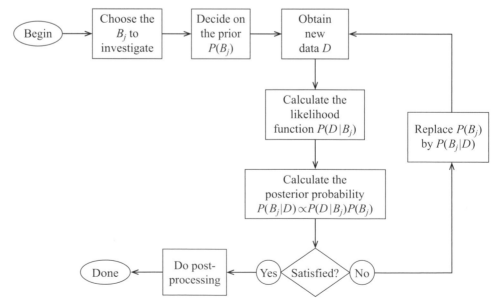

Figure 7.2: A flow chart for the Bayesian statistics.

Third, any amount of new data, even a single measurement, can be used to revise the probability distribution function. There is no need to measure frequencies of occurrences, although knowledge often improves with repeated measurements.

Finally, there is no need to prespecify the end point of an experiment or group of experiments. Data are incorporated as they becomes available and one can terminate the experiment whenever one sees fit. Figure 7.2 describes how to continue incorporating new data into existing data until one is satisfied with the results. In essence, one replaces the prior probability distribution $P(B_j)$ with the posterior probability distribution $P(B_j|D)$ and then obtains new data, cycling through the loop as often as desired.

7.2 Single-Parameter Estimation: Means, Modes, and Variances

7.2.1 Introduction

Parameter estimation follows a different logical path in Bayesian statistics than it does in frequentist statistics. To estimate a single continuous parameter, we rewrite equation 7.7 in the form

$$f_1(a|D) = \frac{L(D|a)f_0(a)}{\int L(D|a)f_0(a)da},$$
(7.8)

where a is the parameter to be estimated. The prior probability distribution $f_0(a)$ should include all knowledge about a before any new data are obtained. The likelihood $L(D|a)$ looks like it should be understood as the probability of obtaining data D given a value of a. In fact, though, its meaning is the reverse: the data are given, a is not, so it is the likelihood of a given the measured data D. The result of the Bayesian calculation is the posterior probability distribution function $f_1(a|D)$. This is the revised probability distribution for a, given the new data. If the data consist of n measurements x_i of a with standard deviations σ_i, the posterior probability distribution can be rewritten as

$$f_1(a|\vec{x},\vec{\sigma}) = \frac{L(\vec{x},\vec{\sigma}|a)f_0(a)}{\int L(\vec{x},\vec{\sigma}|a)f_0(a)da},$$
(7.9)

where $\vec{x}$ stands for $x_1, x_2, \ldots, x_n$, and $\vec{\sigma}$ stands for $\sigma_1, \sigma_2, \ldots, \sigma_n$.

Bayes's theorem gives the posterior probability distribution for a, not an explicit estimate of a. The posterior distribution contains all available information about the parameter and, indeed, any attempt to distill the information into, for example, a single characteristic number decreases the amount of information. Nevertheless, one typically does want to characterize the distribution by just a few numbers, perhaps the mean value of the parameter and its standard deviation, or a credible interval. These must be extracted from the posterior probability distribution by additional calculations. This is done in the usual way. The mean value of a is given by

$$\hat{a} = \langle a \rangle = \int af_1(a|D)da,$$
(7.10)

and the variance is given by

$$\sigma_{\hat{a}}^2 = \langle a^2 \rangle - \langle a \rangle^2,$$
(7.11)

where

$$\langle a^2 \rangle = \int a^2 f_1(a|D)da.$$
(7.12)

As we shall see in the following sections, if the prior and the likelihood are both Gaussian distributions, Bayes's theorem gives results that seem intuitively reasonable. The results for other distributions are less so, sometimes much less so.

7.2.2 Gaussian Priors and Likelihood Functions

Single-parameter estimation in Bayesian statistics is straightforward if the likelihood and prior are both Gaussians. The following is a distinctively Bayesian example: revise the estimated value of a parameter given a prior value and a single new measurement of the parameter.

Example: The catalog value for the distance to a galaxy is $\mu_0 = 10$ million parsecs (mpc) with a standard deviation $\sigma_0 = \pm 2$ Mpc. Suppose we make a new measurement of its distance, finding d_1 with a standard deviation σ_{d_1}. Assuming the probability distributions are all Gaussian, the prior probability distribution for the distance is

$$f_0(d) = \frac{1}{\sqrt{2\pi}\,\sigma_0} \exp\left[-\frac{1}{2}\frac{(d-\mu_0)^2}{\sigma_0^2}\right],$$

and the likelihood function is

$$L(d_1, \sigma_{d_1}|d) = \frac{1}{\sqrt{2\pi}\,\sigma_{d_1}} \exp\left[-\frac{1}{2}\frac{(d_1-d)^2}{\sigma_{d_1}^2}\right].$$

The unnormalized posterior probability distribution is

$$f_1(d|d_1, \sigma_{d_1}) \propto L(d_1, \sigma_{d_1}|d)\, f_0(d)$$

$$\propto \frac{1}{\sqrt{2\pi}\,\sigma_{d_1}} \exp\left[-\frac{1}{2}\frac{(d_1-d)^2}{\sigma_{d_1}^2}\right] \frac{1}{\sqrt{2\pi}\,\sigma_0} \exp\left[-\frac{1}{2}\frac{(d-\mu_0)^2}{\sigma_0^2}\right].$$

The product of two Gaussians is a Gaussian (see Section C.2 in Appendix C), so we can rewrite the posterior probability distribution as

$$f_1(d|d_1, \sigma_{d_1}) = \frac{1}{\sqrt{2\pi}\,\sigma_1} \exp\left[-\frac{1}{2}\frac{(d-\mu_1)^2}{\sigma_1^2}\right],$$

where

$$\mu_1 = \frac{w_0\mu_0 + w_{d_1}d_1}{w_0 + w_{d_1}} = \frac{\sigma_{d_1}^2\mu_0 + \sigma_0^2 d_1}{\sigma_0^2 + \sigma_{d_1}^2}$$

$$\sigma_1^2 = \frac{1}{w_0 + w_{d_1}} = \frac{\sigma_0^2\sigma_{d_1}^2}{\sigma_0^2 + \sigma_{d_1}^2},$$

and the weights are $w_0 = 1/\sigma_0^2$ and $w_{d_1} = 1/\sigma_{d_1}^2$. Because the posterior distribution is a Gaussian, the mean value and standard deviation of d can be extracted from the posterior by inspection: $\langle d\rangle \pm \sigma_d = \mu_1 \pm \sigma_1$. Thus the mean is the weighted average of the prior value and the new measurement.

> To be specific, let the new measurement be $d_1 = 16$ Mpc. If the new measurement has the same error as the catalog value, $\sigma_{d_1} = \pm 2$ Mpc, the revised value of the distance is 13 ± 1.4 Mpc. If the new measurement is much more reliable than the catalog value, $\sigma_{d_1} = \pm 0.5$ Mpc, the revised value of the distance is 15.6 ± 0.5 Mpc.

We now generalize to many measurements of the parameter. Suppose that the data consist of n independent measurements of a, the individual measurements are x_i, the measurement errors are σ_i, and the errors have a Gaussian distribution. The likelihood function is

$$L(\vec{x}, \vec{\sigma} \,|\, a) = \prod_{i=1}^{n} \frac{1}{\sqrt{2\pi}\,\sigma_i} \exp\left[-\frac{1}{2} \frac{(x_i - a)^2}{\sigma_i^2} \right]. \tag{7.13}$$

Suppose also that the prior knowledge of a can be encoded as a Gaussian distribution with mean a_0 and standard deviation σ_{a_0}. The prior probability distribution function is

$$f_0(a) = \frac{1}{\sqrt{2\pi}\,\sigma_{a_0}} \exp\left[-\frac{1}{2} \frac{(a - a_0)^2}{\sigma_{a_0}^2} \right]. \tag{7.14}$$

The unnormalized posterior probability distribution is

$$f_1(a|\vec{x}, \vec{\sigma}) \propto \left\{ \prod_{i=1}^{n} \frac{1}{\sqrt{2\pi}\,\sigma_i} \exp\left[-\frac{1}{2} \frac{(x_i - a)^2}{\sigma_i^2} \right] \right\} \frac{1}{\sqrt{2\pi}\,\sigma_{a_0}} \exp\left[-\frac{1}{2} \frac{(a - a_0)^2}{\sigma_{a_0}^2} \right]. \tag{7.15}$$

The product of many Gaussian probability distributions is another Gaussian (see Section C.2 in Appendix C), so we can rewrite the posterior probability distribution as

$$f_1(a|\vec{x}, \vec{\sigma}) = \frac{1}{\sqrt{2\pi}\,\sigma_{a_1}} \exp\left[-\frac{1}{2} \frac{(a - a_1)^2}{\sigma_{a_1}^2} \right], \tag{7.16}$$

where

$$a_1 = \left(w_{a_0} a_0 + \sum_{i=1}^{n} w_i x_i \right) \Big/ \left(w_{a_0} + \sum_{i=1}^{n} w_i \right) \tag{7.17}$$

$$\sigma_{a_1}^2 = 1 \Big/ \left(w_{a_0} + \sum_{i=1}^{n} w_i \right), \tag{7.18}$$

and the weights are $w_{a_0} = 1/\sigma_{a_0}^2$ and $w_i = 1/\sigma_i^2$. By inspection, the revised mean value and standard deviation of a are $\langle a \rangle \pm \sigma_a = a_1 \pm \sigma_{a_1}$.

Equations 7.17 and 7.18 are almost identical to, say, equations 4.55 and 4.57 for weighted means and variances in frequentist statistics. The difference is that the summations in equations 7.17 and 7.18 include a_0 weighted by w_{a_0}. Thus the prior value is included exactly as if it were another measurement. If a_0 is poorly known—if σ_{a_0} is large—then w_{a_0} is small, and a_0 has little influence on the posterior mean value of a. In the limit as w_{a_0} goes to 0, equations 7.17 and 7.18 become identical to the expressions for mean values in frequentist statistics.

7.2.3 Binomial and Beta Distributions

The binomial and beta distributions are appropriate when there are only two possible outcomes of an experiment, such as heads or tails for the flip of a coin, or a white or black ball drawn blindly from a container containing a mix of the two. The following example applies these two distributions to the flip of a bagel: bagelsian statistics.

Example: Unlike coins, bagels are asymmetric. When flipped, bagels do not land on their two sides with equal probability. Suppose from prior experience one believes that the probability a bagel will land round side up is $u_0 = 1/4$ and one's confidence in this number is $\sigma_{u_0} = 1/8$ or $\sigma_{u_0}^2 = 1/64$. Suppose a bagel is flipped $n = 20$ times, and it lands round side up $k_1 = 12$ times. What is the posterior probability that a bagel flipped once will land with its round side up?

Bagel flips can be described by the binomial probability distribution. The probability of getting k flips with round side up out of n trials is

$$P(k) \propto u^k (1-u)^{n-k},$$

where u is the probability that a bagel flipped just once will land round side up. The likelihood function is therefore

$$L(k_1|u) \propto u^{k_1}(1-u)^{n-k_1},$$

that is, $L(k_1|u)$ is the likelihood of a particular value of u, given that the bagel was observed to land round side up k_1 times in n flips.

The beta function (see Section 2.7) is a convenient way to describe the prior probability distribution for u. The beta function is

$$\beta(u) \propto u^{a-1}(1-u)^{b-1},$$

where u lies in the range $0 \leq u \leq 1$, and a and b are positive integers. We set $f_0(u) = \beta(u)$ and choose values of a and b so that $\langle u \rangle = u_0$ and $\sigma_u^2 = \sigma_{u_0}^2$. From equations 2.133 and 2.134, we have

$$a = u_0 \left[\frac{u_0(1-u_0)}{\sigma_{u_0}^2} - 1 \right]$$

$$b = a \frac{1-u_0}{u_0}.$$

The posterior probability distribution function for u becomes

$$f_1(u|k_1) \propto L(k_1|u) f_0(u) \propto u^{k_1}(1-u)^{n-k_1} u^{a-1}(1-u)^{b-1}$$

$$\propto u^{k_1+a-1}(1-u)^{n-k_1+b-1}.$$

We recognize that $f_1(u|k_1)$ is also a beta function, so the posterior mean and variance of u can be calculated from it using equations 2.131 and 2.132

$$u_1 = \langle u \rangle = \frac{a+k_1}{a+b+n}$$

$$\sigma_{u_1}^2 = \frac{(a+k_1)(n+b-k_1)}{(a+b+n)^2(a+b+n+1)}.$$

For $u_0 = 1/4$ and $\sigma_{u_0}^2 = 1/64$, the values of the constants in the prior are $a \approx 3$ and $b \approx 9$. With $n = 20$ and $k_1 = 12$, the revised estimate for the probability of landing round side up is

$$u_1 = \frac{3+12}{3+9+20} = 0.47,$$

and

$$\sigma_{u_1}^2 = \frac{(3+12)(20+9-12)}{(3+9+20)^2(3+9+20+1)} = 7.55 \times 10^{-3}$$

$$\sigma_{u_1} = 0.087.$$

Who would have thought that calculating probabilities for a flipped bagel could be so complicated!

A frequentist might have calculated $u_1 = k_1/n \pm \sqrt{k_1}/n = 12/20 \pm \sqrt{12}/20 = 0.60 \pm 0.17$ in the previous example, quite different from the result of the Bayesian calculation. The Bayesian estimate does approach the frequentist estimate as k_1 and n become large, since

$$\lim_{n,k_1 \gg a,b} \langle u \rangle = \lim_{n,k_1 \gg a,b} \frac{a+k_1}{a+b+n} - \frac{k_1}{n}. \tag{7.19}$$

However, if k_1 and n are not much larger than a and b, the difference between the Bayesian estimate and the frequentist estimate can be substantial, as it was in the example.

Suppose one wants to use the results of the Bayesian calculation to predict the results of a new experiment, say, the probability of the same flipped bagel landing round side up m times in r flips. It is tempting to use the binomial probability distribution with the posterior mean value $\langle u \rangle = u_1$ for the probability distribution for m:

$$P(m) = \binom{r}{m} u_1^m (1-u_1)^{r-m} \;\; ?? \tag{7.20}$$

This is not correct, because the Bayesian calculation gave a probability distribution for u. The mean is merely a one-number characterization of the full distribution. It is better to take

$$P_1(m) = \int_{u=0}^{1} P(m) f_1(u|k_1)\,du$$

$$= A \int_{u=0}^{1} u^m (1-u)^{r-m} u^{k_1+a-1} (1-u)^{n-k_1+b-1}\,du, \tag{7.21}$$

where A is the normalization constant for the posterior probability distribution. We leave this integral as an exercise for the reader, but is relatively easy to evaluate, since the integrand is just the beta function yet again.

7.2.4 Poisson Distribution and Uniform Priors

The Poisson distribution is often the appropriate likelihood function when an experiment can yield only integers. Uniform distributions are useful when one needs a prior corresponding to no prior information. The following example shows how to apply Bayesian statistics to a single measurement of an integer drawn from a Poisson distribution when there is no prior information.

Example: Suppose an experiment yields the datum k_1, a single integer drawn from a Poisson distribution. What is the posterior probability distribution for the mean value of k? The probability of getting k from a Poisson distribution with mean μ is

$$P(k) = \frac{\mu^k}{k!} \exp[-\mu],$$

where μ is the mean value of k. The likelihood function for the specific sample k_1 is then

$$L(k_1|\mu) = \frac{\mu^{k_1}}{k_1!} \exp[-\mu].$$

Let us suppose we know nothing about μ except that it is greater than or equal to 0. We can represent this information by a flat prior probability distribution function for μ:

$$f_0(\mu) = 1/A, \quad 0 \le \mu \le A.$$

Since k_1 and μ are unrestricted, we must allow A to be very large. The denominator in Bayes's theorem is

$$\int L(k_1|\mu)f_0(\mu)d\mu = \int_0^A \frac{1}{A}\frac{\mu^{k_1}}{k_1!} \exp[-\mu]d\mu = \frac{1}{Ak_1!} \int_0^A \mu^{k_1} \exp[-\mu]d\mu.$$

For $A \gg k_1$, the integral approaches the Γ function (see Appendix A), so the denominator approaches

$$\int L(k_1|\mu)f_0(\mu)d\mu = \frac{1}{Ak_1!}\Gamma(k_1+1) = \frac{1}{A},$$

and the normalized posterior probability distribution becomes

$$f_1(\mu|k_1) = A\frac{\mu^{k_1}}{k_1!} \exp[-\mu]\frac{1}{A} = \frac{\mu^{k_1}}{k_1!} \exp[-\mu].$$

Although $f_1(\mu|k_1)$ has the same form as a Poisson distribution, it has a different meaning: it is the probability of μ given k_1, not of k_1 given μ. Note that as long as A is large, its precise value is unimportant, because it cancels out of the final result.

Let us generalize the previous example to many measurements. Suppose we have n independent measurements k_i, all sampled from the same Poisson distribution. The likelihood function is

$$L(\vec{k}|\mu) = \prod_{i=1}^n \frac{\mu^{k_i}}{k_i!} \exp[-\mu] = \frac{\mu^{\sum k_i}}{\prod k_i!} \exp[-n\mu]. \tag{7.22}$$

As before, we have no prior knowledge about μ and adopt the flat prior probability distribution

$$f_0(\mu) = 1/A, \quad 0 < \mu \le A.$$

The denominator in Bayes's theorem is

$$\int L(\vec{k}|\mu)f_0(\mu)d\mu = \int_0^A \frac{1}{A} \frac{\mu^{\sum k_i}}{\prod k_i!} \exp[-n\mu]d\mu. \quad (7.23)$$

For convenience, set $K = \sum k_i$. Then we have

$$\int L(\vec{k}|\mu)f_0(\mu)d\mu = \frac{1}{A\prod k_i!} \int_0^A \mu^K \exp[-n\mu]d\mu$$

$$= \frac{1}{n^{(K+1)}A\prod k_i!} \int_0^A (n\mu)^K \exp[-n\mu]d(n\mu). \quad (7.24)$$

If we set $t = n\mu$ and let A become large, the integral approaches the Γ function, and

$$\int L(\vec{k}|\mu)f_0(\mu)d\mu = \frac{1}{n^{(K+1)}A\prod k_i!}\Gamma(K+1) = \frac{K!}{n^{(K+1)}A\prod k_i!}. \quad (7.25)$$

The normalized posterior probability distribution becomes

$$f_1(\mu|\vec{k}) = \frac{L(\vec{k}|\mu)f_0(\mu)}{\int L(\vec{k}|\mu)f_0(\mu)d\mu} = \frac{n^{(K+1)}A\prod k_i!}{K!} \frac{1}{A} \frac{\mu^K}{\prod k_i!}\exp[-n\mu]$$

$$= n\frac{(n\mu)^K}{K!}\exp[-n\mu]. \quad (7.26)$$

Although the posterior probability looks like a Poisson distribution, it is the probability distribution function for μ, not K. The mean value of μ is

$$\langle\mu\rangle = \int \mu f_1(\mu|\vec{k})d\mu = \int_0^\infty \mu n\frac{(n\mu)^K}{K!}\exp[-n\mu]d\mu$$

$$= \frac{1}{nK!}\int (n\mu)^{K+1}\exp[-n\mu]d(n\mu) = \frac{1}{nK!}\Gamma(K+2) = \frac{1}{nK!}(K+1)!$$

$$= \frac{K+1}{n} = \frac{1+\sum k_i}{n}. \quad (7.27)$$

This is an odd result. If we have just one measurement k_1, the estimated mean is $\langle\mu\rangle = k_1 + 1$. The frequentist approach gave $\langle\mu\rangle = \sum k_i/n$ (see the last example in Section 5.2.2). Bayes adds 1 to the sum. To calculate the variance, we need $\langle\mu^2\rangle$. Leaving some steps to the reader, we find

$$\langle\mu^2\rangle = \int \mu^2 f_1(\mu|\vec{k})d\mu = \frac{1}{n^2K!}\Gamma(K+3) = \frac{(K+1)(K+2)}{n^2}.$$

The variance is

$$\sigma_\mu^2 = \langle\mu^2\rangle - \langle\mu\rangle^2 = \frac{(K+1)(K+2)}{n^2} - \left(\frac{K+1}{n}\right)^2 = \frac{K+1}{n^2}$$

$$= \frac{1+\sum_i k_i}{n^2}. \quad (7.28)$$

Once again, this is a rather odd result, because the Bayesian analysis adds 1 to the sum.

The mean of a distribution is, however, just one of several ways to characterize a distribution by a single number. If we decide to use the mode as the single-number characterization of equation 7.26, we have

$$\left. \frac{df_1(\mu|\vec{k})}{d\mu} \right|_{\mu_{mode}} = 0. \tag{7.29}$$

The derivative is easy to calculate and yields

$$\mu_{mode} = \frac{\sum k_i}{n}, \tag{7.30}$$

a more intuitive result than the mean value.

7.2.5 More about the Prior Probability Distribution

Inevitably there is a subjective aspect to the choice of the prior probability distribution. Different people can legitimately have different assessments of existing knowledge and legitimately choose different priors. Even if everyone chooses a prior of the same form, perhaps a Gaussian prior, they may choose different means and variances for the Gaussian. These lead to different posterior distributions and, as we saw from equations 7.17 and 7.18, different posterior means and variances. While sometimes seen as a defect in Bayesian statistics, the subjectivity of the prior is actually a strength. Bayesian statistics forces one to recognize and deal with the uncertainties in prior knowledge. Perhaps the best way to deal with the uncertainties is for investigators to discuss and converge on a common prior. If this cannot be done, one should consider taking enough new data so that the differences among the various priors has little effect on the posterior distribution.

The uniform prior invoked in the discussion of the Poisson likelihood function in Section 7.2.4 is peculiar. The prior was supposed to represent a total lack of information about the parameter and should have been a flat distribution from 0 to infinity. Such a distribution cannot, however, be normalized; it is an "improper distribution." Instead, the prior was chosen to be a flat distribution with a width much larger than the width of the likelihood function. The width divided out after normalization, so the finite width of the prior did not affect the posterior distribution. Another way to handle a flat prior is to start with a normalized flat prior of finite width, calculate the posterior probability using the finite-width prior, and then let the width go to infinity. The limiting posterior probability distribution will usually be well behaved.

One should, however, recognize that flatness is not an invariant property of a probability distribution. Nonlinear coordinate transformations destroy flatness. It can be unclear in which coordinate system a prior should be flat. Suppose, for example, you would like to acknowledge that the σ of a parameter in the prior is often poorly determined and want to give σ its own probability distribution in the prior. If you want its prior distribution to be flat, should it be flat in σ or in σ^2? Context can often provide an answer, but not always.

One can choose any function for the prior as long as it correctly describes the prior knowledge. But it is worth choosing a prior that simplifies the posterior distribution. For the Gaussian likelihood functions discussed in Section 7.2.2, a Gaussian prior produces a Gaussian posterior that was easily interpreted and yielded intuitively reasonable results. For the binomial likelihood function discussed in Section 7.2.3, the beta distribution was a much better choice for a prior than a Gaussian distribution was, because the beta distribution easily incorporates the restriction that u must lie between 0 and 1. It also led to tractable calculations of means and variances. In both these cases, the prior had the same functional form as the likelihood function. Probability distribution functions with the same form as the likelihood function are called *conjugate distributions*. Conjugate distributions are commonly employed in Bayesian statistics. Although we chose a flat prior distribution when discussing the Poisson likelihood in Section 7.2.4, the Poisson distribution does have a conjugate distribution, called the gamma distribution (do not confuse this with the gamma function). The gamma distribution can be written as

$$f(x) = \frac{1}{\Gamma(k)\theta^k} x^{k-1} \exp\left[-\frac{x}{\theta}\right], \tag{7.31}$$

where $f(x)$ is the probability distribution for x, and k and θ are free parameters. The mean and variance of x are $\langle x \rangle = k\theta$ and $\sigma_x^2 = k\theta^2$. If one is given prior values for the mean and variance and wants to use the gamma distribution for the prior, one can invert these equations to find appropriate values for k and θ.

7.3 Multiparameter Estimation

7.3.1 Formal Description of the Problem

We now generalize to the estimation of $m + 1$ parameters a_j from n data points (x_i, y_i, σ_i), where the σ_i are the measurement errors on the y_i. In practice, this usually means fitting a function $y = g(x, a_0, a_1, \ldots, a_m)$ to the data. Bayes's theorem can be rewritten as

$$f_1(\vec{a}|D) = f_1(\vec{a}|\vec{x}, \vec{y}, \vec{\sigma}) = \frac{L(\vec{x}, \vec{y}, \vec{\sigma}\,|\vec{a})f_0(\vec{a})}{\int_{a_0} \cdots \int_{a_m} L(\vec{x}, \vec{y}, \vec{\sigma}\,|\vec{a})f_0(\vec{a})da_0 \cdots da_m}, \tag{7.32}$$

where the arrow accent has been used to write the equation more compactly (e.g., $\vec{x} = x_1, x_2, \ldots, x_n$ and $\vec{a} = a_0, a_1, \ldots, a_m$). The multiple integral in the denominator normalizes the posterior distribution. The following example expands equation 7.32 explicitly for the case of a two-parameter fit with Gaussian likelihood and priors.

Example: Suppose we wish to fit the two-parameter function $y = g(x, a_0, a_1)$ to the n data points (x_i, y_i, σ_i). Assume that the residuals to the fit have a Gaussian distribution. Also assume that the prior probability distribution for the two parameters are Gaussians with means and variances $(\bar{a}_0, \bar{\sigma}_{a_0}^2)$ and $(\bar{a}_1, \bar{\sigma}_{a_1}^2)$.

The prior probability distribution is

$$f_0(\vec{a}) \propto \exp\left[-\frac{1}{2}\frac{(a_0 - \bar{a}_0)^2}{\bar{\sigma}_{a_0}^2}\right] \exp\left[-\frac{1}{2}\frac{(a_1 - \bar{a}_1)^2}{\bar{\sigma}_{a_1}^2}\right],$$

and the likelihood function is

$$L(\vec{x}, \vec{y}, \vec{\sigma}^2|\vec{a}) \propto \prod_{i=1}^{n} \exp\left[-\frac{1}{2}\frac{[y_i - g(x_i, a_0, a_1)]^2}{\sigma_i^2}\right],$$

Continued on page 200

so the posterior probability distribution is

$$f_1(\vec{a}|\vec{x},\vec{y},\vec{\sigma}) \propto \left\{ \prod_{i=1}^{n} \exp\left[-\frac{1}{2} \frac{[y_i - g(x_i, a_0, a_1)]^2}{\sigma_i^2} \right] \right\} \exp\left[-\frac{1}{2} \frac{(a_0 - \bar{a}_0)^2}{\bar{\sigma}_{a_0}^2} \right]$$

$$\times \exp\left[-\frac{1}{2} \frac{(a_1 - \bar{a}_1)^2}{\bar{\sigma}_{a_1}^2} \right].$$

The posterior distribution contains all the information about the parameters, but it often defies comprehension. One can make the information about the parameters more intelligible by extracting their means and standard deviations from the posterior distribution or perhaps by calculating credible intervals. And before trusting any of these quantities, one should examine the marginal probability distributions for the individual parameters, where the marginal distributions are

$$f(a_j) = \int_{a_0} \cdots \int_{a_{j-1}} \int_{a_{j+1}} \cdots \int_{a_m} f_1(\vec{a}|\vec{x},\vec{y},\vec{\sigma}) da_0 \cdots da_{j-1} da_{j+1} \cdots da_m. \tag{7.33}$$

It can be useful to calculate bivariate marginal distributions to determine whether there are strong covariances between any of the parameters. For the two parameters a_j and a_r, the bivariate marginal distribution is

$$f(a_j, a_r) = \int_{a_0} \cdots \int_{a_{j-1}} \int_{a_{j+1}} \cdots \int_{a_{r-1}} \int_{a_{r+1}} \cdots \int_{a_m} f_1(\vec{a}|\vec{x},\vec{y},\vec{\sigma})$$

$$\times da_0 \cdots da_{j-1} da_{j+1} \cdots da_{r-1} da_{r+1} \cdots da_m. \tag{7.34}$$

Unfortunately, calculating any of these integrals is often difficult or time consuming. Indeed, calculating the normalization integral in the denominator of equation 7.32 can be extremely difficult. As a result, most of multiparameter Bayesian statistics revolves around ways to extract meaningful information from the posterior distribution, even when it is not properly normalized.

The rest of this section discusses two ways to extract information from the posterior probability distribution. The first is the Laplace approximation (Section 7.3.2), which fits a multivariate Gaussian distribution to the highest peak of the posterior distribution. The Laplace approximation also leads to a deeper insight into the relation between frequentist and Bayesian multiparameter estimation (Section 7.3.3). The second is Monte Carlo sampling of the posterior distribution (Section 7.3.4). Both of these avoid the need to normalize the posterior distribution. The section ends with a short discussion of credible intervals (Section 7.3.5), which are the Bayesian equivalent of frequentist confidence intervals.

7.3.2 Laplace Approximation

One-Parameter Laplace Approximation: Suppose the posterior probability distribution resulting from a one-parameter Bayesian analysis is

$$f_1(a|\vec{x}) = AL(\vec{x}|a)f_0(a), \tag{7.35}$$

where A is an unknown normalization constant. Suppose further that the posterior distribution is dominated by a single peak much higher than any other peaks. One may then be able to approximate the posterior distribution as a Gaussian distribution centered on the peak. The peak of the posterior distribution occurs at the same place as the peak of the log of the distribution,

$$\ell = \ln f_1(a|\vec{x}) = \ln\left[L(\vec{x}|a)f_0(a)\right] + \ln A, \tag{7.36}$$

so the location of the peak can be found by setting the derivative of ℓ equal to 0,

$$\left.\frac{\partial\ell}{\partial a}\right|_{\hat{a}} = 0, \tag{7.37}$$

and then solving for $\hat{a}$. The location of the peak is independent of the normalization constant. To find the variance, expand ℓ in a Taylor series around $\hat{a}$:

$$\ell = \ell(\hat{a}|\vec{x}) + \left.\frac{\partial\ell}{\partial a}\right|_{\hat{a}}(a - \hat{a}) + \frac{1}{2}\left.\frac{\partial^2\ell}{\partial a^2}\right|_{\hat{a}}(a - \hat{a})^2 + \cdots. \tag{7.38}$$

From equation 7.37, the first derivative of ℓ is 0 at $a = \hat{a}$, so the second term on the right-hand side of equation 7.38 is 0. In the neighborhood of $\hat{a}$ the posterior probability distribution is therefore

$$f_1(a|\vec{x}) = \exp[\ell] = \exp[\ell(\hat{a}|\vec{x})]\exp\left[\frac{1}{2}\left.\frac{\partial^2\ell}{\partial a^2}\right|_{\hat{a}}(a - \hat{a})^2\right] \times \text{(higher-order terms)}. \tag{7.39}$$

Dropping the higher-order terms and noting that $\exp[\ell(\hat{a}|\vec{x})]$ is independent of a, the posterior can be approximated by

$$f_1(a|\vec{x}) \propto \exp\left[\frac{1}{2}\left.\frac{\partial^2\ell}{\partial a^2}\right|_{\hat{a}}(a - \hat{a})^2\right], \tag{7.40}$$

which is a Gaussian distribution with variance

$$\sigma_{\hat{a}}^2 = -\left(\left.\frac{\partial^2\ell}{\partial a^2}\right|_{\hat{a}}\right)^{-1}. \tag{7.41}$$

Equation 7.40 with $\hat{a}$ given by equation 7.37 and σ^2 given by equation 7.41 is the Laplace approximation to the posterior distribution. Since the approximation is a Gaussian distribution, the mean and standard deviation of a are, by inspection, $\hat{a} \pm \sigma_{\hat{a}}$

Example: Let us return to the case of the Poisson distribution with a uniform prior discussed in Section 7.2.4. The posterior probability distribution for μ given n measurements k_i is

$$f_1(\mu|\vec{k}) = n\frac{(n\mu)^K}{K!}\exp[-n\mu]$$

(see equation 7.26), where $K = \sum k_i$. Extracting the mean value and variance of μ from this distribution was cumbersome, and the results were peculiar: $\langle\mu\rangle = (1 + \sum_i k_i)/n$ and $\sigma^2 = (1 + \sum_i k_i)/n^2$ (equations 7.27 and 7.28).

Continued on page 202

To apply the Laplace approximation, we first take the log of the posterior distribution:

$$\ell = \ln f_1(\mu|\vec{k}) = \ln\left(\frac{n}{K!}\right) + K\ln(n\mu) - n\mu.$$

The maximum value of the posterior occurs at

$$0 = \left.\frac{\partial\ell}{\partial\mu}\right|_{\hat{\mu}} = \frac{K}{\hat{\mu}} - n$$

or

$$\hat{\mu} = \frac{K}{n} = \frac{\sum_{i=1}^{n} k_i}{n}.$$

To calculate the variance we first need

$$\left.\frac{\partial^2\ell}{\partial\mu^2}\right|_{\hat{\mu}} = \left.\frac{\partial}{\partial\mu}\left(\frac{K}{\mu} - n\right)\right|_{\hat{\mu}} = -\frac{K}{\hat{\mu}^2},$$

and with this the variance becomes

$$\sigma_{\hat{\mu}}^2 = -\left(\left.\frac{\partial^2\ell}{\partial\mu^2}\right|_{\hat{\mu}}\right)^{-1} = \frac{\hat{\mu}^2}{K} = \frac{K}{n^2} = \frac{1}{n^2}\sum_{i-1}^{n} k_i.$$

Note that $\hat{\mu}$ and $\sigma_{\hat{\mu}}^2$ have lost the strange 1 added to the sum.

Multiparameter Laplace Approximation: We now generalize the Laplace approximation to the multiparameter case. The logarithm of the multiparameter posterior distribution is

$$\ell(\vec{a}|\vec{x},\vec{y},\vec{\sigma}) = \ln f_1(\vec{a}|\vec{x},\vec{y},\vec{\sigma}) = \ln L(\vec{x},\vec{y},\vec{\sigma}|\vec{a}) + \ln f_0(\vec{a}) + \text{constant}. \qquad (7.42)$$

Since the maximum of $f_1(\vec{a}|\vec{x},\vec{y},\vec{\sigma})$ is at the same location as the maximum of $\ell(\vec{a}|\vec{x},\vec{y},\vec{\sigma})$, the location of the maximum is found by setting the $m+1$ derivatives of $\ell(\vec{a}|\vec{x},\vec{y},\vec{\sigma})$ equal to 0:

$$\left.\frac{\partial\ell(\vec{a}|\vec{x},\vec{y},\vec{\sigma})}{\partial a_j}\right|_{\hat{a}} = 0, \qquad (7.43)$$

where the values of a_j that satisfy these equations are denoted by $\hat{a}_j$. It is worth noting that the derivatives of $\ell(\vec{a}|\vec{x},\vec{y},\vec{\sigma})$ with respect to the a_j also appeared in maximum likelihood estimation of frequentist statistics (see equation 5.8), but now $\ell(\vec{a}|\vec{x},\vec{y},\vec{\sigma})$ includes the prior probability distribution $f_0(\vec{a})$, so the derivatives do not produce the same equations as they do in maximum likelihood techniques, although we will see that they can be similar.

Let us assume that equation 7.43 has been solved and we have values for the $\hat{a}_j$. To find the variances and covariances of the $\hat{a}_j$, expand $\ell(\vec{a}|\vec{x},\vec{y},\vec{\sigma})$ in a Taylor series about the $\hat{a}_j$, discarding terms higher than quadratic:

$$\ell(\vec{a}) = \ell(\hat{\vec{a}}) + \sum_{j=0}^{m}\left.\frac{\partial\ell(\vec{a})}{\partial a_j}\right|_{\hat{a}}(a_j - \hat{a}_j) + \frac{1}{2}\sum_{j=0}^{m}\sum_{k=0}^{m}\left.\frac{\partial^2\ell(\vec{a})}{\partial a_j\partial a_k}\right|_{\hat{a}}(a_j - \hat{a}_j)(a_k - \hat{a}_k), \qquad (7.44)$$

where for compactness we are abbreviating $\ell(\vec{a}|\vec{x},\vec{y},\vec{\sigma})$ by $\ell(\vec{a})$ and $\ell(\hat{\vec{a}}|\vec{x},\vec{y},\vec{\sigma})$ by $\ell(\hat{\vec{a}})$. Recognizing that the first derivatives of $\ell(\vec{a})$ are all equal to 0 at $\vec{a} = \hat{\vec{a}}$, we have

$$\ell(\vec{a}) = \ell(\hat{\vec{a}}) + \frac{1}{2}\sum_{j=0}^{m}\sum_{k=0}^{m} \left.\frac{\partial^2 \ell(\vec{a})}{\partial a_j \partial a_k}\right|_{\hat{a}} (a_j - \hat{a}_j)(a_k - \hat{a}_k). \tag{7.45}$$

This equation can be put in a more concise form using matrix notation. Let the components of the matrix $\mathbf{Q}$ be

$$(\mathbf{Q})_{jk} = -\left.\frac{\partial^2 \ell(\vec{a})}{\partial a_j \partial a_k}\right|_{\hat{a}}, \tag{7.46}$$

and let the column vector $\Delta\hat{\boldsymbol{a}}$ be

$$\Delta\hat{\boldsymbol{a}} = \begin{pmatrix} a_0 - \hat{a}_0 \\ a_1 - \hat{a}_1 \\ \vdots \\ a_m - \hat{a}_m \end{pmatrix}. \tag{7.47}$$

Equation 7.45 can be written as

$$\ell(\vec{a}) = \ell(\hat{\vec{a}}) - \frac{1}{2}\Delta\hat{\mathbf{a}}^{\mathsf{T}}\mathbf{Q}\Delta\hat{\mathbf{a}}. \tag{7.48}$$

Returning to $f_1(\vec{a}|\vec{x},\vec{y},\vec{\sigma})$, we have

$$f_1(\vec{a}|\vec{x},\vec{y},\vec{\sigma}) = \exp\left[\ell(\vec{a})\right] \propto \exp\left(-\frac{1}{2}\Delta\hat{\mathbf{a}}^{\mathsf{T}}\mathbf{Q}\Delta\hat{\mathbf{a}}\right). \tag{7.49}$$

We recognize this as a multivariate Gaussian distribution. From our earlier work on multivariate Gaussian distributions, we know that if $\mathbf{C}$ is the covariance matrix for the parameters, the multivariate Gaussian can be written as

$$f_1(\vec{a}|\vec{x},\vec{y},\vec{\sigma}) \propto \exp\left(-\frac{1}{2}\Delta\hat{\mathbf{a}}^{\mathsf{T}}\mathbf{C}^{-1}\Delta\hat{\mathbf{a}}\right), \tag{7.50}$$

so we identify $\mathbf{C}^{-1} = \mathbf{Q}$, or

$$\mathbf{C} = \mathbf{Q}^{-1}. \tag{7.51}$$

Equation 7.50 along with equations 7.43, 7.46, 7.47, and 7.51 constitute the multiparameter Laplace approximation.[1]

7.3.3 Gaussian Likelihoods and Priors: Connection to Least Squares

Because multiparameter fits in which both the likelihood and the prior are Gaussian distributions are so common in Bayesian statistics, and because they show the connection to ordinary least-squares estimation so clearly, let us discuss this case in more detail.

---

[1] The alert reader may note that the calculation of the covariance matrix in Chapter 6 using equations 6.70 and 6.71 is essentially a Laplace approximation for S, but equation 6.70 differs from equation 7.51 by a factor of 2. The difference arises because our definition of S differs from the definition of ℓ in, for example, equation 7.55 by a factor of $-1/2$.

Suppose one has n data points (x_i, y_i, σ_i), where the σ_i are uncorrelated measurement errors on the y_i, and one wants to fit a function $g(x, \vec{a})$ to the data. The residuals between the function and the data are $\epsilon_i = y_i - g(x_i, \vec{a})$. If the residuals have Gaussian distributions with variances σ_i^2, the likelihood function is

$$L(\vec{x}, \vec{y}, \vec{\sigma} | \vec{a}) \propto \prod_{i=1}^{n} \exp\left[-\frac{1}{2}\frac{\epsilon_i^2}{\sigma_i^2}\right] = \prod_{i=1}^{n} \exp\left[-\frac{1}{2}\frac{(y_i - g(x_i, \vec{a}))^2}{\sigma_i^2}\right]. \tag{7.52}$$

Let us also assume that our prior information about the parameters can be encoded as Gaussian distributions with mean values $\bar{a}_i$ and standard deviations $\bar{\sigma}_{a_i}^2$. The prior becomes

$$f_0(\vec{a}) \propto \prod_{j=0}^{m} \exp\left[-\frac{1}{2}\frac{(a_i - \bar{a}_i)}{\bar{\sigma}_{a_i}^2}\right]. \tag{7.53}$$

The posterior distribution is

$$f_1(\vec{a} | \vec{x}, \vec{y}, \vec{\sigma}) \propto \prod_{i=1}^{n} \exp\left[-\frac{1}{2}\frac{(y_i - g(x_1, \vec{a}))^2}{\sigma_i^2}\right] \prod_{j=0}^{m} \exp\left[-\frac{1}{2}\frac{(a_i - \bar{a}_i)^2}{\bar{\sigma}_{a_i}^2}\right], \tag{7.54}$$

and the log of the posterior distribution is

$$\ell(\vec{a} | \vec{x}, \vec{y}, \vec{\sigma}) = -\frac{1}{2}\sum_{i=1}^{n} w_i [y_i - g(x_i, \vec{a})]^2 - \frac{1}{2}\sum_{j=0}^{m} \bar{w}_{a_i}(a_i - \bar{a}_i)^2 + \text{constant}, \tag{7.55}$$

where the weights are $w_i = 1/\sigma_i^2$, and $\bar{w}_{a_i} = 1/\bar{\sigma}_{a_i}^2$. The values of the a_j at the maximum of the posterior probability distribution are given by setting the derivatives of $\ell(\vec{a} | \vec{x}, \vec{y}, \vec{\sigma})$ equal to 0:

$$\frac{\partial \ell(\vec{a})}{\partial a_j}\bigg|_{\hat{a}} = 0 = \sum_{i=1}^{n} w_i \frac{\partial g(x_i, \vec{a})}{\partial a_j}\bigg|_{\hat{a}}\left[y_i - g(\hat{\vec{a}}, x_i)\right] - \bar{w}_{a_i}(a_i - \bar{a}_i). \tag{7.56}$$

Equation 7.55 is the same as equation 6.8 in least squares estimation except for the sum over terms involving the $\bar{a}_j$; and equation 7.56 is identical to the normal equations in least squares except that the term $\bar{w}_{a_i}(a_i - \bar{a}_i)$ has been added to each of the equations.

If $g(x_i, \vec{a})$ is nonlinear in the parameters, the assumption that the posterior distribution can be approximated by a Gaussian may not be warranted. The properties of the posterior should be investigated more fully, perhaps by MCMC sampling, before proceeding. However, if $g(x_i, \vec{a})$ is linear in the parameters, the Laplace approximation is fully justified. The solution to equation 7.56 is nearly identical to the solution of the normal equations in linear least squares. The fit of a straight line to data illustrates these principles.

Example: Fit the straight line $y = a_0 + a_1 x$ to n data points (x_i, y_i, σ_i) using Bayesian statistics.

From equation 7.52, the likelihood function is

$$L(\vec{x}, \vec{y}, \vec{\sigma} | a_0, a_1) \propto \prod_{i=1}^{n} \exp\left[-\frac{1}{2}\frac{(y_i - a_0 - a_1 x_i)^2}{\sigma_i^2}\right], \tag{7.57}$$

and from equation 7.53, the prior probability distribution is

$$f_0(a_0, a_1) \propto \exp\left[-\frac{1}{2}\frac{(a_0 - \bar{a}_0)^2}{\bar{\sigma}_{a_0}^2}\right]\exp\left[-\frac{1}{2}\frac{(a_1 - \bar{a}_1)^2}{\bar{\sigma}_{a_1}^2}\right]. \tag{7.58}$$

The posterior probability distribution for a_0 and a_1 is

$$f_1(a_0, a_1 | \vec{x}, \vec{y}, \vec{\sigma}) \propto \prod_{i=1}^{n}\exp\left[-\frac{1}{2}\frac{(y_i - a_0 - a_1 x_i)^2}{\sigma_i^2}\right]$$

$$\times \exp\left[-\frac{1}{2}\frac{(a_0 - \bar{a}_0)^2}{\bar{\sigma}_{a_0}^2}\right]\exp\left[-\frac{1}{2}\frac{(a_1 - \bar{a}_1)^2}{\bar{\sigma}_{a_1}^2}\right], \tag{7.59}$$

and the log of the posterior distribution (with additive constants ignored) is

$$\ell(a_0, a_1 | \vec{x}, \vec{y}, \vec{\sigma}) = -\frac{1}{2}\sum_{i=1}^{n}w_i(y_i - a_0 - a_1 x_i)^2 - \frac{1}{2}\bar{w}_{a_0}(a_0 - \bar{a}_0)^2 - \frac{1}{2}\bar{w}_{a_1}(a_1 - \bar{a}_1)^2. \tag{7.60}$$

The posterior is centered at $(\hat{a}_0, \hat{a}_1)$, which are given by the solution to the equations

$$0 = \frac{\partial\ell}{\partial a_0} = \sum_i w_i(y_i - \hat{a}_0 - \hat{a}_1 x_i) - \bar{w}_{a_0}(\hat{a}_0 - \bar{a}_0) \tag{7.61}$$

$$0 = \frac{\partial\ell}{\partial a_1} = \sum_i w_i x_i(y_i - \hat{a}_0 - \hat{a}_1 x_i) - \bar{w}_{a_1}(\hat{a}_1 - \bar{a}_1). \tag{7.62}$$

After rearrangement, these equations become

$$\hat{a}_0\left(\bar{w}_{a_0} + \sum_i w_i\right) + \hat{a}_1\sum_i w_i x_i = \bar{w}_{a_0}\bar{a}_0 + \sum_i w_i y_i \tag{7.63}$$

$$\hat{a}_0\sum_i w_i x_i + \hat{a}_1\left(\bar{w}_{a_1} + \sum_i w_i x_i^2\right) = \bar{w}_{a_1}\bar{a}_1 + \sum_i w_i x_i y_i \tag{7.64}$$

or, in matrix form,

$$\begin{pmatrix} \left(\bar{w}_{a_0} + \sum_i w_i\right) & \sum_i w_i x_i \\ \sum_i w_i x_i & \left(\bar{w}_{a_1} + \sum_i w_i x_i^2\right) \end{pmatrix}\begin{pmatrix} \hat{a}_0 \\ \hat{a}_1 \end{pmatrix} = \begin{pmatrix} \bar{w}_{a_0}\bar{a}_0 + \sum_i w_i y_i \\ \bar{w}_{a_1}\bar{a}_1 + \sum_i w_i x_i y_i \end{pmatrix}. \tag{7.65}$$

With an obvious notation, this equation can be put in the form

$$\mathbf{N\hat{a}} = \mathbf{Y}, \tag{7.66}$$

which has the solution

$$\mathbf{\hat{a}} = \mathbf{N}^{-1}\mathbf{Y}. \tag{7.67}$$

Continued on page 206

To find the covariance matrix, we begin with the second derivatives of $\ell(a_0, a_1 | \vec{x}, \vec{y}, \vec{\sigma})$

$$\frac{\partial^2 \ell}{\partial a_0 \partial a_0} = -\bar{w}_{a_0} - \sum_i w_i \tag{7.68}$$

$$\frac{\partial^2 \ell}{\partial a_0 \partial a_1} = -\sum_i w_i x_i \tag{7.69}$$

$$\frac{\partial^2 \ell}{\partial a_1 \partial a_1} = -\bar{w}_{a_1} - \sum_i w_i x_i^2. \tag{7.70}$$

The matrix $\mathbf{Q}$ is

$$\mathbf{Q} = \begin{pmatrix} \left(\bar{w}_{a_0} + \sum_i w_i \right) & \sum_i w_i x_i \\ \sum_i w_i x_i & \left(\bar{w}_{a_1} + \sum_i w_i x_i^2 \right) \end{pmatrix} = \mathbf{N}, \tag{7.71}$$

so the covariance matrix is

$$\mathbf{C} = \begin{pmatrix} \sigma_{\hat{0}}^2 & \sigma_{\hat{0}\hat{1}} \\ \sigma_{\hat{0}\hat{1}} & \sigma_{\hat{1}}^2 \end{pmatrix} = \mathbf{Q}^{-1} = \mathbf{N}^{-1}. \tag{7.72}$$

The only difference between equations 7.65, 7.71, and 7.72 and the equivalent equations for the least squares fit of a straight line to data in frequentist statistics is a few additional terms involving $(\bar{a}_0, \bar{w}_{a_0})$ and $(\bar{a}_1, \bar{w}_{a_1})$. In this case the jump from the frequentist calculation to a Bayesian calculation is trivially easy.

Assume our prior knowledge of the parameters in the previous example is poor. We would assign large values to $\bar{\sigma}_{a_0}$ and $\bar{\sigma}_{a_1}$ or, equivalently, low values to $\bar{w}_{a_0}$ and $\bar{w}_{a_1}$. As the weights go to zero, equation 7.65 approaches

$$\begin{pmatrix} \sum_i w_i & \sum_i w_i x_i \\ \sum_i w_i x_i & \sum_i w_i x_i^2 \end{pmatrix} \begin{pmatrix} \hat{a}_0 \\ \hat{a}_1 \end{pmatrix} = \begin{pmatrix} \sum_i w_i y_i \\ \sum_i w_i x_i y_i \end{pmatrix}, \tag{7.73}$$

which is now identical to the equation for a least squares fit in frequentist statistics. Likewise, equations 7.71 and 7.72 for the covariance matrix are similar to equations 5.45 and 5.98 for the covariance matrix in frequentist statistics, but the Bayesian version contains additional terms involving $(\bar{a}_0, \bar{w}_{a_0})$ and $(\bar{a}_1, \bar{w}_{a_1})$. As before, if the weights on the prior estimates of the parameters go to 0, the covariance matrix becomes

$$\mathbf{C} = \begin{pmatrix} \sum_i w_i & \sum_i w_i x_i \\ \sum_i w_i x_i & \sum_i w_i x_i^2 \end{pmatrix}^{-1}, \tag{7.74}$$

which is identical to the covariance matrix in frequentist statistics.

This is more generally true. If the prior is so broad as to be essentially constant, $\ell(\vec{a})$ becomes in effect

$$\ell(\vec{a} | \vec{x}, \vec{y}, \vec{\sigma}) = \ln L(\vec{x}, \vec{y}, \vec{\sigma} | \vec{a}) + \text{constant}, \tag{7.75}$$

which is identical to the definition for $\ell(\vec{a})$ in the maximum likelihood method. Bayesian statistics then yields the same values for the parameters as the maximum likelihood or least squares estimation. There is, nevertheless, a fundamental difference in viewpoint between Bayesian statistics and these frequentist techniques, even when the mathematics is identical. In Bayesian statistics the true reality is the posterior probability distribution, $f_1(\vec{a}|\vec{x},\vec{y},\vec{\sigma})$. The maximum likelihood values of the parameters are merely convenient single-number ways to characterize the distribution.

The covariance matrix highlights another difference between Bayesian and frequentist statistics. Frequentist statistics can often deal comfortably with data in which the the σ_i are incorrect. It does so by assuming the σ_i are proportional to the true errors and then calculating the estimated covariance matrix by, for example, equations 5.99 and 5.100 (see Section 5.2.3). This is a relatively benign abuse of the data, since only the covariance matrix is changed; the estimated values of the parameters remain unchanged. The abuse is *not* benign in Bayesian statistics, because changing the values of the σ_i changes the relative weights of the prior and the likelihood function. The estimated values of the parameters *do* change. While it is possible to include uncertainties about the values of the σ_i in the prior, a more typical Bayesian approach is to demand that the errors in the data be correct.

The following extended example illustrates these issues. The example shows the fit of a straight line to data using Bayesian statistics and the Laplace approximation, which, for this example, is entirely adequate. The data in this example are identical to the data in the example discussed extensively in Section 5.2, which introduced linear least squares. The reader is encouraged to compare the (frequentist) least squares fit in that example to the Bayesian fit here.

Example: Fit the straight line

$$y = a_0 + a_1 x$$

to the data points listed in Table 7.1 using Bayesian statistics and the Laplace approximation. Unlike ordinary least squares fitting, Bayesian methods require the probability distributions of the errors to be known. We specify that the distributions are Gaussians with variances σ_i^2. The likelihood function is therefore given by equation 7.57.

Let us assume that the prior probability distributions for the two parameters are independent Gaussian distributions, so the joint prior distribution is the product of the individual distributions (equation 7.58). Specify that the prior mean values of the parameters are

$$\bar{a}_0 = 1.0$$

$$\bar{a}_1 = 0.125.$$

Table 7.1: Data points for fit to straight line

x_i	y_i	σ_i	x_i	y_i	σ_i	x_i	y_i	σ_i
4.41	0.43	0.08	5.32	1.26	0.32	6.60	2.75	0.18
4.60	0.99	0.15	5.81	0.95	0.40	6.99	2.64	0.08
4.95	0.87	0.22	5.89	1.79	0.35	7.13	3.01	0.05
5.28	2.09	0.32	6.36	2.00	0.25	7.22	1.97	0.99

Continued on page 208

A plot of the data points and their error bars, and of the line

$$y = \bar{a}_0 + \bar{a}_1 x = 1.0 + 0.125x$$

is shown in the left panel of Figure 7.3.

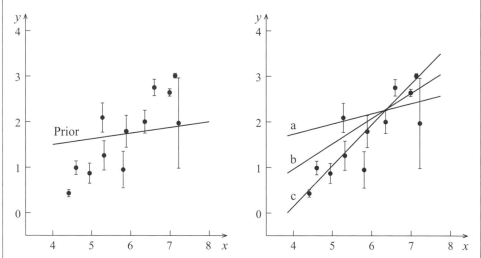

Figure 7.3: Fits of a straight line to data using Bayesian statistics and the Laplace approximation. The data points and the error bars are identical in the two panels. The line labeled "Prior" in the left panel is $y = \bar{a}_0 + \bar{a}_1 x$, where $\bar{a}_0$ and $\bar{a}_1$ are the mean values of a_0 and a_1 in the prior probability distribution. The right panel shows three versions of the line $y = \hat{a}_0 + \hat{a}_1 x$, where $\hat{a}_0$ and $\hat{a}_1$ are the mean values determined from the posterior probability distribution. The line labeled "a" gives high weight to the prior values of the parameters, and the line labeled "b" gives intermediate weight to the prior values. The line labeled "c" gives low weight to the prior values and is essentially identical the least squares fit discussed in Section 5.2.

Inspection of the figure suggests that the new data are not well described by the prior and that the parameters need to be updated. We will calculate three cases, one corresponding to a high confidence in the prior values of the parameters, one to an intermediate confidence, and one to a low confidence.

Case I: High confidence in the prior values of the parameters. This case corresponds to small standard deviations for the prior values:

$$\bar{a}_0 \pm \bar{\sigma}_{a0} = 1.00 \pm 0.1000$$

$$\bar{a}_1 \pm \bar{\sigma}_{a_1} = 0.125 \pm 0.0125$$

The normal matrix is (equation 7.65)

$$\mathbf{N} = \begin{pmatrix} \left(\bar{w}_{a_0} + \sum_i w_i\right) & \sum_i w_i x_i \\ \sum_i w_i x_i & \left(\bar{w}_{a_1} + \sum_i w_i x_i^2\right) \end{pmatrix} = \begin{pmatrix} 100.00 + 859.43 & 5440.707 \\ 5440.707 & 6400.00 + 35,542.25 \end{pmatrix}$$

$$= \begin{pmatrix} 959.43 & 5440.707 \\ 5440.707 & 41,942.25 \end{pmatrix}.$$

The **Y** vector is (equation 7.65)

$$\mathbf{Y} = \begin{pmatrix} \bar{w}_{a_0}\bar{a}_0 + \sum_i w_i y_i \\ \bar{w}_{a_1}\bar{a}_1 + \sum_i w_i x_i y_i \end{pmatrix} = \begin{pmatrix} 100.00 + 1917.81 \\ 800.00 + 13{,}127.58 \end{pmatrix} = \begin{pmatrix} 2017.81 \\ 13927.58 \end{pmatrix}.$$

The posterior probability distribution for the parameters is a multivariate Gaussian distributions with mean values

$$\begin{pmatrix} \hat{a}_0 \\ \hat{a}_1 \end{pmatrix} = \hat{\mathbf{a}} = \mathbf{N}^{-1}\mathbf{Y} = \begin{pmatrix} 0.832 \\ 0.2241 \end{pmatrix}.$$

The covariance matrix for the posterior distribution is (equation 7.72)

$$\mathbf{C} = \mathbf{N}^{-1} = \begin{pmatrix} 3.942 \times 10^{-3} & -5.114 \times 10^{-4} \\ -5.114 \times 10^{-4} & 9.018 \times 10^{-5} \end{pmatrix}.$$

As always, it is the full posterior probability distribution that is the true output of a Bayesian analysis, but one might want to report the results of this analysis as means and standard deviations of the parameters:

$$\hat{a}_0 \pm \hat{\sigma}_{\hat{0}} = 0.832 \pm 0.063$$
$$\hat{a}_1 \pm \hat{\sigma}_{\hat{1}} = 0.224 \pm 0.009.$$

The line

$$y = \hat{a}_0 + \hat{a}_1 = 0.832 + 0.224x$$

is plotted in the right panel of Figure 7.3 with the label "a."

Case II: Moderate confidence in the prior values of the parameters. This case corresponds to moderate standard deviations for the prior values:

$$\bar{a}_0 \pm \bar{\sigma}_{a0} = 1.00 \pm 0.333$$
$$\bar{a}_1 \pm \bar{\sigma}_{a_1} = 0.125 \pm 0.04167$$

We find

$$\mathbf{N} = \begin{pmatrix} 868.43 & 5440.707 \\ 5440.707 & 36118.25 \end{pmatrix}, \qquad \mathbf{Y} = \begin{pmatrix} 1926.81 \\ 13{,}199.58 \end{pmatrix},$$

and

$$\mathbf{C} = \mathbf{N}^{-1} = \begin{pmatrix} 2.046 \times 10^{-2} & -3.082 \times 10^{-3} \\ -3.082 \times 10^{-3} & 4.920 \times 10^{-4} \end{pmatrix},$$

which yields the parameters:

$$\hat{a}_0 \pm \hat{\sigma}_{\hat{0}} = -1.26 \pm 0.14$$
$$\hat{a}_1 \pm \hat{\sigma}_{\hat{1}} = 0.555 \pm 0.022.$$

Continued on page 210

The line

$$y = \hat{a}_0 + \hat{a}_1 = -1.26 + 0.555x$$

is plotted in the right panel of Figure 7.3 with the label "b."

Case III: Low confidence in the prior values of the parameters. This case corresponds to large standard deviations for the prior values:

$$\bar{a}_0 \pm \bar{\sigma}_{a0} = 1.00 \pm 10.0$$

$$\bar{a}_1 \pm \bar{\sigma}_{a_1} = 0.125 \pm 1.25.$$

The resulting fitted the parameters and their variances are

$$\hat{a}_0 \pm \hat{\sigma}_{\hat{0}} = -3.45 \pm 0.19$$

$$\hat{a}_1 \pm \hat{\sigma}_{\hat{1}} = 0.897 \pm 0.030.$$

The line

$$y = \hat{a}_0 + \hat{a}_1 = -3.45 + 0.897x$$

is plotted in the right panel of Figure 7.3 with the label "c."

Several things are notable about this example. The first is that, despite the rather messy mathematical background, the Laplace approximation is easy to implement. It is a minor chore to modify existing computer codes for calculating (frequentist) linear least squares fits to Bayesian fits with the Laplace approximation. A massive MCMC analysis of the posterior probability distribution is not needed.

The second, is the sensitivity of the posterior distribution to the variances of the prior values of the parameters. The prior values of the parameters themselves are the same for all three cases, but the posterior values differ substantially, because the variances of the prior values are different. In Case I, the posterior distribution was dominated almost entirely by the prior distribution. In Case III, the posterior was dominated almost entirely by the data points and, indeed, the mean values of the parameters were essentially identical to the values determined from the ordinary least squares fit discussed in Section 5.3.1. This sensitivity to the reliability of the prior is an innate feature of Bayesian statistics, not of the Laplace approximation. The sensitivity is not encouraging, because in real applications variances are perforce *estimated* variances, not true variances. As the discussion in Section 4.5 makes clear, estimated variances are often highly uncertain.

Finally, when fitting these same data points by ordinary least squares, we discovered that the standard deviations of the data points are probably too small, and therefore the standard deviations of the fitted parameters are probably too small. We increased the variances of the fitted parameters to offset the too-small data errors (see the discussion in Section 5.3.3 and the example near the end of that section). Recognizing that ex post facto adjustment of the data errors is problematic in Bayesian analysis, we have not adjusted the posterior distribution in the current example. As a result, the posterior standard deviations of the fitted parameters in Case III are the same as the unadjusted standard deviations in ordinary least squares.

The fit of a straight line is easily extended to the Bayesian equivalent of generalized linear least squares. Suppose we wish to fit n data points (x_i, y_i, σ_i) with a linear combination of

$m + 1$ functions $g_j(x)$ of x:

$$y = g(x, \vec{a}) = a_0 g_0(x) + a_1 g_1(x) + \cdots + a_m g_m(x): \tag{7.76}$$

If, for example, one wants to fit a polynomial to the data, the functions are $g_j(x) = x^j$. The likelihood function is

$$L(\vec{x}, \vec{y}, \vec{\sigma} \,|\, \vec{a}) \propto \prod_{i=1}^{n} \exp\left\{ -\frac{1}{2} \frac{[y_i - a_0 g_0(x_i) - a_1 g_1(x_i) - \cdots - a_m g_m(x_i)]^2}{\sigma_i^2} \right\}, \tag{7.77}$$

the prior probability distribution function is

$$f_0(\vec{a}) \propto \prod_{j=0}^{m} \exp\left[-\frac{1}{2} \frac{(a_j - \bar{a}_j)^2}{\bar{\sigma}_{a_j}^2} \right], \tag{7.78}$$

and the log of the posterior probability distribution is

$$\ell(\vec{a} \,|\, \vec{x}, \vec{y}, \vec{\sigma}) = -\frac{1}{2} \sum_{i=1}^{n} w_i [y_i - a_0 g_0(x_i) - a_1 g_1(x_i) - \cdots - a_m g_m(x_i)]^2 - \frac{1}{2} \sum_{j=0}^{m} \bar{w}_{a_j} (a_j - \bar{a}_j)^2, \tag{7.79}$$

where the weights are $w_i = 1/\sigma_i^2$, and $\bar{w}_{a_j} = 1/\bar{\sigma}_{a_j}^2$. The normal matrix is

$$\mathbf{N} = \begin{pmatrix} (\bar{w}_{a_0} + \sum w_i g_0(x_i) g_0(x_i)) & \sum w_i g_0(x_i) g_1(x_i) & \cdots & \sum w_i g_0(x_i) g_m(x_i) \\ \sum w_i g_1(x_i) g_0(x_i) & (\bar{w}_{a_1} + \sum w_i g_1(x_i) g_1(x_i)) & \cdots & \sum w_i g_1(x_i) g_m(x_i) \\ \vdots & \vdots & & \vdots \\ \sum w_i g_m(x_i) g_0(x_i) & \sum w_i g_m(x_i) g_1(x_i) & \cdots & (\bar{w}_{a_m} + \sum w_i g_m(x_i) g_m(x_i)) \end{pmatrix}, \tag{7.80}$$

and the $\hat{\mathbf{a}}$ and $\mathbf{Y}$ vectors are

$$\hat{\mathbf{a}} = \begin{pmatrix} \hat{a}_0 \\ \hat{a}_1 \\ \vdots \\ \hat{a}_m \end{pmatrix}, \qquad \mathbf{Y} = \begin{pmatrix} \bar{w}_{a_0} \bar{a}_0 + \sum w_i g_0(x_i) y_i \\ \bar{w}_{a_1} \bar{a}_1 + \sum w_i g_1(x_i) y_i \\ \vdots \\ w_{a_m} \bar{a}_m + \sum w_i g_m(x_i) y_i \end{pmatrix}, \tag{7.81}$$

where all sums are taken over the n data points. The solution for the $\hat{a}_j$ is

$$\hat{\mathbf{a}} = \mathbf{N}^{-1} \mathbf{Y}, \tag{7.82}$$

and the covariance matrix is

$$\mathbf{C} = \mathbf{N}^{-1}. \tag{7.83}$$

These equations are strikingly similar to the normal equations in generalized linear least squares (see Section 5.5). The only difference is the additive factors involving the prior mean values and weights of the parameters. Even for this more general case, the leap from a frequentist calculation to a Bayesian calculation remains easy. One large change, however, is that there is no equivalent of equation 5.92 for the estimated covariance matrix.

7.3.4 Difficult Posterior Distributions: Markov Chain Monte Carlo Sampling

Suppose one has fitted a function $g(x, \vec{a})$ to data D using Bayesian techniques, producing a posterior probability distribution function $f_1(\vec{a} | D)$ for the parameters $\vec{a}$. The Laplace approximation is a good way to extract information from the posterior when $g(x, \vec{a})$ is linear in the parameters and the prior and likelihood are both Gaussians. It is may not be a good approximation to the posterior if these conditions do not hold. In extreme cases the posterior can become so complicated that neither analytic nor numerical integrations are feasible. MCMC techniques come to the rescue. MCMC sampling can generate random deviates from the posterior distribution efficiently, and, of crucial importance, MCMC sampling can generate the deviates even if the posterior is not normalized. The deviates can then be used for post-processing the distribution.

Suppose, for example, that a Markov chain has been generated from $f_1(\vec{a}|D)$ and that the chain consists of N states $s^{(j)} = (u_0^{(j)}, u_1^{(j)}, \ldots, u_m^{(j)})$, where $u_k^{(j)}$ is the jth deviate for parameter a_k (the reader might find it useful to review Sections 3.4 and 3.5 at this point). It is not unusual to generate 10^6 or more states in a chain. If a quantity $h(a_0, a_1, \ldots, a_m)$ is a function of the parameters, its mean value can be calculated from

$$\langle h(\vec{a}) \rangle \approx \frac{1}{N} \sum_{j=1}^{N} h(u_0^{(j)}, u_1^{(j)}, \ldots, u_m^{(j)}), \qquad (7.84)$$

where the accuracy of the approximation improves as N increases. Thus, to calculate the mean value of parameter a_k, set $h(\vec{a}) = a_k$. Then

$$\langle a_k \rangle \approx \frac{1}{N} \sum_{j=1}^{N} u_k^{(j)}. \qquad (7.85)$$

The standard deviation of $\langle a_k \rangle$ can be calculated from $\sigma_{a_k}^2 = \langle a_k^2 \rangle - \langle a_k \rangle^2$, where

$$\langle a_k^2 \rangle \approx \frac{1}{N} \sum_{j=1}^{N} \left[u_k^{(j)} \right]^2. \qquad (7.86)$$

Marginal distributions are easy to extract from the Markov chain. The marginal distribution for parameter a_k is just the histogram of the $u_k^{(j)}$ with the number of deviates in each bin divided by N, so that the sum over all bins is equal to 1. The cumulative distribution function for a_k can be calculated from the normalized marginal distribution by adding the bins together. The bivariate marginal distributions are also easy to extract from the Markov chain. The bivariate marginal distribution for parameters a_k and a_r is the two-dimensional histogram for the deviate pairs $(u_k^{(j)}, u_r^{(j)})$, again normalized by dividing by N.

7.3.5 Credible Intervals

In Bayesian statistics one is allowed to make such statements as "there is a 90% probability that parameter a lies between a_1 and a_2." The interval $a_1 < a < a_2$ is called a *credible interval*, in this case a 90% credible interval. Credible intervals replace the confidence intervals used in frequentist statistics (although, at risk of confusion, some practitioners of Bayesian statistics use the term "confidence interval" for what is really a credible interval).

To formalize credible intervals, suppose that parameter a has a posterior probability distribution $f_1(a)$. We choose a probability Q and then wish to determine two values a_1 and a_2 such that the probability is Q that $a_1 < a < a_2$. If the probability that a lies between

a_1 and a_2 is $P(a_1 < a < a_2)$, the limits of the credible interval are given implicitly by

$$Q = P(a_1 < a < a_2) = \int_{a_1}^{a_2} f_1(a)\, da. \tag{7.87}$$

Equation 7.87 is not enough to uniquely define a credible interval, because it can be satisfied by many different values of a_1 and a_2. The most typical ways to fully define the credible interval are the following.

1. Place the center point of the interval a_c at the mean, median, or mode of a; set the half-width of the interval equal to $\Delta a = (a_2 - a_1)/2$. The limits of the credible interval are given implicitly by

$$Q = \int_{a_c - \Delta a}^{a_c + \Delta a} f_1(a)\, da. \tag{7.88}$$

2. Set the limits of the interval so that the probability that a lies below the interval equals the probability that it lies above the interval:

$$\frac{1}{2}(1 - Q) = P(a < a_1) = \int_{-\infty}^{a_1} f_1(a)\, da \tag{7.89}$$

and

$$\frac{1}{2}(1 - Q) = P(a > a_2) = \int_{a_2}^{\infty} f_1(a)\, da. \tag{7.90}$$

3. Choose values for a_1 and a_2 that minimize the width of the interval. If the posterior probability distribution is dominated by just one peak, the expression for the location of the minimum width limits is simple and intuitive. Suppose the limits of the integral in equation 7.87 are changed by small amounts da_1 and da_2. The change to the integrated probability is

$$\delta P = f_1(a_2)da_2 - f_1(a_1)da_1, \tag{7.91}$$

where, because the changes are infinitesimal, we can ignore the second-order effects introduced by the slope of the probability distribution. Because the total probability integrated over the credible interval is fixed, changes to a_1 and a_2 must leave the probability unchanged, so we impose the constraint $\delta P = 0$, and equation 7.91 becomes

$$f_1(a_2)da_2 = f_1(a_1)da_1. \tag{7.92}$$

Let $w = a_2 - a_1$ be the width of the interval. The width of the interval is at an extremum when small changes da_1 and da_2 that satisfy this constraint do not change the width of the interval, which means $dw = da_2 - da_1 = 0$, or $da_2 = da_1$. Therefore, the minimum width credible interval occurs at

$$f_1(a_2) = f_1(a_1). \tag{7.93}$$

Figure 7.4 shows two possible 90% credible intervals on the same probability distribution. The left panel in the figure shows a 90% credible interval defined by setting $f_1(a_1) = f_1(a_2)$ in accord with definition 3. The right panel shows a 90% credible interval defined by $P(x < a_1) = P(x > a_2) = 0.05$ in accord with definition 2.

None of the definitions of the credible interval is obviously better than the others. Definition 1 is easy to apply and can be generalized to multiparameter distributions: The

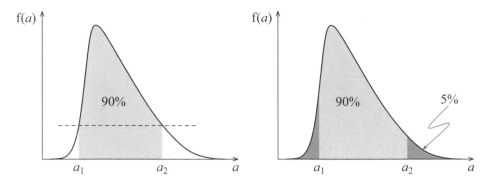

Figure 7.4: Credible intervals are not uniquely defined. The left panel shows a 90% credible interval $a_1 < a < a_2$ defined by setting $f(a_1) = f(a_2)$ to give the minimum-width credible interval. The right panel shows a 90% credible interval defined by $P(a < a_1) = P(a > a_2) = 0.05$. The left panel corresponds to the more commonly used definition.

credible interval becomes the credible region inside a multidimensional sphere centered on a_c. The definition is, however, misleading for complicated or asymmetric distributions. Definition 2 is easy to apply and works for any distribution, but it cannot be generalized directly to multiparameter distributions. The multiparameter distribution must first be marginalized to a single parameter. Definition 3 is also easy to generalize to multiparameter distributions. If the posterior probability distribution for the multiparameter fit is $f_1(\vec{a})$, the credible region is the volume interior to the surface for which $f_1(\vec{a}) = $ constant. Thus, credible regions are the volumes inside level contours. If, in addition, the posterior distribution can be approximated by a multiparameter Gaussian distribution, calculation of credible intervals simplifies greatly, becoming identical to the calculation of confidence intervals for multiparameter Gaussian distributions in frequentist statistics (see the discussion accompanying equations 6.111–6.113).

While definition 3 is appealing, it too has defects. Level contours can be confusing or inappropriate if the probability distribution has many nearly equal peaks. If the posterior distribution must be sampled by Monte Carlo techniques, determining the precise location of level contours may require a large number of samples.

7.4 Hypothesis Testing

To make Bayes's theorem useful for testing hypotheses, we restate it in the form

$$P_1(H_i|D) = \frac{P(D|H_i)\,P_0(H_i)}{\sum_k P(D|H_k)\,P_0(H_k)}, \tag{7.94}$$

where H_i is hypothesis i and D is new data. In this form Bayes's theorem reads "the probability that hypothesis H_i is correct in light of new data D is proportional to the probability of getting data D if H_i is correct times the prior probability that H_i is correct." Hypotheses should not be tested in isolation. Two or more hypotheses should be compared to decide which is more likely to be true. One compares hypothesis H_i to H_j by comparing the posterior probabilities that they are correct: If $P_1(H_i|D)$ is greater than $P_1(H_j|D)$, then hypothesis H_i is more likely to be correct than H_j, and vice versa.

To avoid evaluating the sum in the denominator of equation 7.94, one typically uses the ratio of the probabilities, called the *odds ratio*:

$$\frac{P_1(H_i|D)}{P_1(H_j|D)} = \frac{P(D|H_i)\,P_0(H_i)}{P(D|H_j)\,P_0(H_j)}. \tag{7.95}$$

The odds ratio can be interpreted qualitatively to mean

$$\frac{P_1(H_i|D)}{P_1(H_j|D)} > 1 \quad \Rightarrow \quad \text{hypothesis } H_i \text{ is favored}$$

$$\frac{P_1(H_i|D)}{P_1(H_j|D)} < 1 \quad \Rightarrow \quad \text{hypothesis } H_j \text{ is favored.} \tag{7.96}$$

The odds ratio also lends itself to quantitative interpretation. For example, if the ratio of the posterior probabilities is 2, we say the odds are 2:1 in favor of hypothesis H_i. It is generally a mistake to turn odds into absolute probabilities. If the odds are 2:1 in favor of hypothesis H_i, it is incorrect to say there is a 67% probability that hypothesis H_i is correct and 33% probability that hypothesis H_j is correct unless H_i and H_j are the only two hypotheses possible.

Example: Suppose that tomatoes come in only two types. One type is organically grown, natural tomatoes, which are rather small and variable in size. Their mean diameter is 2.00 inches, and the standard deviation of their size is 1.00 inches. We can describe the distribution of their diameters d by a Gaussian probability distribution:

$$f(d|\text{organic}) = \frac{1}{\sqrt{2\pi}\,1.0} \exp\left[-\frac{1}{2}\left(\frac{d-2.0}{1.0}\right)^2\right]. \tag{7.97}$$

The other type is hydroponically grown, genetically modified (GM) tomatoes, which are larger and more uniform in size. Their mean diameter is 3.00 inches, and the standard deviation of their size is 0.1 inches. We can describe the distribution of their diameters d by a second Gaussian probability distribution:

$$f(d|\text{GM}) = \frac{1}{\sqrt{2\pi}\,0.1} \exp\left[-\frac{1}{2}\left(\frac{d-3.0}{0.1}\right)^2\right]. \tag{7.98}$$

The two Gaussians are plotted in Figure 7.5.

It so happens that you walk by a grocery story, and a store employee standing out front hands you a tomato, saying "Here, try our new line of tomatoes. You will love them!" Being a curious type, you wonder whether the tomatoes are organic or GM; and being a dedicated empiricist, you pull out your calipers (which you carry with you everywhere) and measure the size of the tomato. Its diameter is 2.97 inches. From equations 7.97 and 7.98 you calculate the likelihood function to be

$$L(2.97|\text{organic}) = f(2.97|\text{organic}) = 0.249$$
$$L(2.97|\text{GM}) = f(2.97|\text{GM}) = 3.814.$$

Continued on page 216

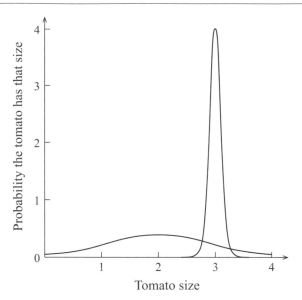

Figure 7.5: The distribution in sizes of the two types of tomatoes. The broad distribution corresponds to organically grown tomatoes, the narrow distribution to the chemically grown, genetically modified tomatoes.

At this point a frequentist might calculate the likelihood ratio

$$LR = \frac{L(2.97|\text{GM})}{L(2.97|\text{organic})} = \frac{3.814}{0.249} = 15.3,$$

and conclude that this is a GM tomato. Is this enough for a Bayesian to conclude it is a GM tomato? No! A dedicated Bayesian must assess the prior probabilities that the tomato is one type or the other, and must then use Bayes's theorem to find the posterior probabilities. Take two possible cases for comparison.

Case I: The store is a mainstream grocery store that sells both organic and genetically modified tomatoes. Lacking any additional information, you assign equal prior probabilities to the two hypotheses:

$$P_0(\text{organic}) = 0.5$$
$$P_0(\text{GM}) = 0.5.$$

Calculate the odds that it is a genetically modified tomato from

$$\frac{P_1(\text{GM}|2.97)}{P_1(\text{organic}|2.97)} = \frac{L(2.97|\text{GM})\,P_0(\text{GM})}{L(2.97|\text{organic})\,P_0(\text{organic})} = \frac{3.814 \times 0.5}{0.249 \times 0.5}$$

$$= 15.3.$$

So the odds are 15.3:1 that the tomato is genetically modified. This is, of course, the same as the frequentist result.

Case II: The store specializes in organic foods and claims to carry only organic products. In this case the prior probability that the tomato is genetically modified is low. Assign a probability of 0.01 that the tomato is genetically modified, not a smaller number and certainly not 0, because there is always a chance that someone at the store made a mistake or the store was misinformed by its supplier. The prior probabilities are

$$P_0(\text{organic}) = 0.99$$

$$P_0(\text{GM}) = 0.01.$$

The odds that it is a genetically modified tomato become

$$\frac{P_1(\text{GM}|2.97)}{P_1(\text{organic}|2.97)} = \frac{L(2.97|\text{GM})\,P_0(\text{GM})}{L(2.97|\text{organic})\,P_0(\text{organic})} = \frac{3.814 \times 0.01}{0.249 \times 0.99}$$

$$= 0.155$$

The odds are now 6:1 that the tomato is an organic tomato.

You might want to measure a *second* tomato, because 6:1 odds might not be enough to convince you that the tomato is organic. If the second tomato has a diameter of 3.03 inches, the posterior odds now become

$$\frac{P_1(\text{GM}|3.03)}{P_1(\text{organic}|3.03)} = \frac{3.814}{0.234} \times 0.155 = 2.5$$

and thus the odds are 2.5:1 that they are GM tomatoes. Just two tomatoes were enough to overcome your strong prejudice that the organic food store has only organic tomatoes.

The previous example demonstrates again that Bayesian statistics can derive meaningful results from small samples, and it shows how Bayesian statistics allows one to add yet more new data until one is satisfied with the result. The example also illustrates a deep danger in hypothesis testing—a danger not limited to Bayesian statistics. We should also have considered a third hypothesis about the tomato diameters: perhaps the store employee, anxious to give shoppers a good experience, selected uniformly large and good looking tomatoes from a large pile of organic tomatoes! The apparent similarity to genetically modified tomatoes would then be the result of selection by the employee, not a result of the tomatoes' DNA. Inclusion of this hypothesis in the odds calculation would surely have affected our conclusions. Depending on the prior, the results would likely be less conclusive and could even shift the odds back to organic in Case II. One must be careful to include all viable hypotheses when testing which is correct.

7.5 Discussion

7.5.1 Prior Probability Distribution

Bayesian statistics demands a prior probability distribution that encodes all existing knowledge about the problem being analyzed. The prior can dominate the posterior distribution if the existing knowledge is firmly established. Consider the example in Section 7.1. Even though a highly reliable test indicated that Dana had the disease, the posterior probability that Dana had the disease remained low, because the prior probability was low. If the prior probability of some result is 0, the posterior probability remains 0 even if that result occurs—good reason to be suspicious of priors that can go to 0 anywhere in the possible domain of the parameters.

Additional data dilutes the effect of the prior on the posterior distribution. With much new data, the effect of the prior tends to 0, and the difference between Bayesian statistics and frequentist statistics fades away. This point is pertinent, because an experimental scientist can usually opt to obtain more data. It is not always possible to obtain more data, though. The Kamiokande experiment detected only ~11 neutrinos from supernova 1987a in the Large Magellanic Cloud.[2] While it would be good to observe neutrinos from more supernovae, one may need to wait for decades, perhaps generations, for another nearby supernova.

7.5.2 Likelihood Function

Bayesian statistics demands a likelihood function. Discussions about prior information in Bayesian statistics usually focus on the prior probability distribution, but the choice of a likelihood function also depends on prior information, and an incorrect choice can yield results that are simply wrong. Consider again the tomato example of Section 7.4. Suppose that the employee had given you a tomato that was 1.0 inches in diameter. A Bayesian analysis gives huge odds ($\sim 10^{86}$) that the tomato is organic, no matter what kind of store it came from, and one would not bother with additional samples to determine the nature of the tomatoes. In fact, all the tomatoes handed out by the store employee were close to 1.0 inches in diameter. The tomato was a cherry tomato, and it was genetically modified and chemically grown. The Bayesian analysis gave the wrong answer.

What went wrong? Relying on an incorrect understanding of genetically modified tomatoes, the likelihood function $L(d|GM)$ included only one kind of genetically modified tomato, those with diameters near 3 inches. It failed to include other kinds of genetically modified tomatoes. A probability distribution that included the possibility of genetically modified cherry tomatoes might be

$$P(d|\text{GM}) = \frac{0.5}{\sqrt{2\pi}\,0.1} \exp\left[-\frac{1}{2}\left(\frac{d-3.0}{0.1}\right)^2\right] + \frac{0.5}{\sqrt{2\pi}\,0.1} \exp\left[-\frac{1}{2}\left(\frac{d-1.0}{0.1}\right)^2\right]. \quad (7.99)$$

The use of this distribution to calculate likelihood function would yield more reasonable odds. One cannot avoid noticing that a frequentist shopper would want to sample several tomatoes, not just one, and would have a better chance of discovering that the 1.0-inch tomatoes were genetically modified.

7.5.3 Posterior Distribution Function

The output of a Bayesian analysis is a posterior probability distribution function. It encodes all available information from both prior knowledge and new data. Any characterization of the posterior by just a few numbers generally has less information than the distribution itself. One does, nevertheless, often want to characterize the posterior more compactly, perhaps by its marginal distributions or even more compactly by means and standard deviations or credible intervals for its parameters. This is perfectly acceptable, but it

---

[2] W. D. Arnett, J. N. Bahcall, R. P. Kirshner, and S. E. Woosley. 1989. "Supernova 1987a." *Annual Review of Astronomy and Astrophysics* vol. 27, p. 629.

should always be remembered that in Bayesian statistics, the fundamental objects are the distributions, not the parameters. Values for parameters can change as a result of an experiment, but only indirectly and only because the posterior distribution changes. This is evident when calculating functions of parameters. Suppose the posterior distribution for a single parameter a given data D is $f_1(a|D)$ and that the mean value of a derived from the posterior is $\hat{a}$. Let a quantity g be a function of the parameter, $g = g(a)$. The mean value of g is emphatically not $g(\hat{a})$. It is $\int g(a)f_1(a|D)da$.

7.5.4 Meaning of Probability

Chapter 4 opened with a frequentist definition of probability: If an event A is observed to occur k times in n trials, the posterior probability of A is (see equation 4.2)

$$P(A) = \lim_{n \to \infty} \frac{k}{n}. \tag{7.100}$$

The ratio k/n is a frequency. In contrast, Bayesian statistics allows probability distributions to be revised on the basis of single measurements of single events. It happily answers such questions as: What is the probability that a specific young adult really is ill after a single positive test? What is the probability that a specific candidate for president of the United States will be elected? An interpretation of probability in terms of frequencies is meaningless when dealing with single events. Frequentist statistics can sometimes deal with such questions by inventing ensembles, large sets of imaginary similar systems. The concept of ensemble has been highly successful in, for example, statistical mechanics and demonstrably produces reliable predictions, but these ensembles do not really exist. Worse, the concept of an ensemble loses meaning when dealing with truly unique events, such as the election of a specific candidate to the presidency.

In Bayesian statistics probability is interpreted as plausibility, not frequency. Plausibility can be (and often is) thought of as "degree of belief", which emphasizes the subjectivity of probability and has led some to call this "subjective probability." Plausibility can also be thought of as a "state of knowledge," which has a more objective feel and suggests kinship to such concepts as entropy. The objectivity is only apparent, though. Two different people applying Bayesian inference to precisely the same data can find different posterior probabilities, because they start with different priors.

7.5.5 Thoughts

It appears to your author that the concerns of frequentists about the meaning of probability in Bayesian statistics are legitimate. Nevertheless, the inclusion of prior knowledge, formalization of inductive logic, and expression of results as a posterior probability distribution make Bayesian statistics overwhelmingly attractive. Unfortunately, the requirements for a reliable Bayesian analysis are strict and often not fulfilled. The true functional forms of the likelihood function and the prior probability distribution are rarely known. Most practitioners, like Procrustes, simply force the data and preexisting knowledge to fit assumed likelihoods and priors. A poorly chosen likelihood or prior can seriously compromise a Bayesian analysis. And, of course, a prior chosen without good judgement invites mere prejudice to crash the party.

If a full Bayesian analysis is impossible, likelihood statistics provides a first fallback position. In practice, likelihood statistics is essentially Bayesian statistics without a prior. If even the likelihood function is unknown, frequentist statistics provides a final haven. Most of the techniques discussed in Chapters 4 and 5 do not require knowledge of the underlying probability distributions and do not explicitly require the use of prior knowledge. It is enough for means, variances, and covariances to exist. The penalty for using frequentist techniques is the introduction of uncertainty about the logical meaning of statistical results. In compensation, even nonexperts have a decent intuitive understanding of means and standard deviations.

Finally, to an experimental scientist the posterior probability distribution should be considered a challenge, not an answer. Many things could have gone wrong—the prior probability distribution could have been chosen incorrectly, or the likelihood function could have been chosen incorrectly, or the new data could be defective. The challenge is to decide whether the posterior distribution is really correct. The posterior distribution attaches probabilities to many different possible events. A good way to test the posterior distribution is to make many measurements, enough to test whether events occur in the proper ratio. This is frequentism. It may be that Bayesian statistics cannot be entirely divorced from frequentist statistics.

Introduction to Fourier Analysis

8.1 Introduction

Fourier analysis is the decomposition of a function into sines and cosines. It plays a fundamental role in the analysis of sequences of data, particularly if the sequence is periodic or quasiperiodic. Fourier analysis is also intimately related to convolution, autocorrelation, and cross correlation. All these play a fundamental role in the investigation of nonperiodic and stochastic sequences of data.

This section introduces Fourier analysis, laying the foundations for Chapters 9 and 10, which cover the analysis of sequences of data. It begins with an introduction to complete sets of orthonormal functions, of which the sines and cosine functions are an example. It then covers the Fourier series, the Fourier transform, and the discrete Fourier transform, introducing the power spectrum along the way. Finally, it covers convolution and the convolution theorem, which relates convolution to the Fourier transform.

8.2 Complete Sets of Orthonormal Functions

We begin with a brief review of orthonormal functions and complete sets. The inner product of two continuous real functions $f(x)$ and $g(x)$ is defined to be

$$\int_{x_1}^{x_2} f(x)g(x)dx, \tag{8.1}$$

where the limits of the integral depend on the functions. Two functions are orthogonal to each other if their inner product is 0,

$$\int_{x_1}^{x_2} f(x)g(x)dx = 0, \tag{8.2}$$

and a function is normalized if the inner product of the function with itself is equal to 1,

$$\int_{x_1}^{x_2} f(x)f(x)dx = 1. \tag{8.3}$$

Two functions are orthonormal to each other if they are orthogonal and both are normalized. More generally, the functions $h_j(x)$ are orthonormal if

$$\int_{x_1}^{x_2} h_j(x)h_k(x)dx = \delta_{jk}, \tag{8.4}$$

where δ_{jk} is the Kronecker delta function. The inner product of two complex functions is

$$\int_{x_1}^{x_2} f^*(x)g(x)dx, \tag{8.5}$$

where $f^*(x)$ is the complex conjugate of $f(x)$. The orthonormality relations for a set of complex functions is

$$\int_{x_1}^{x_2} h_j^*(x)h_k(x)dx = \delta_{jk}. \tag{8.6}$$

Example: Let the two functions be

$$f(x) = A\sin(x)$$
$$g(x) = B\sin(2x),$$

and let the interval over which their inner product is defined be $0 \leq x \leq 2\pi$. The inner product of the two functions is

$$\int_0^{2\pi} A\sin(x)B\sin(2x)dx = AB\int_0^{2\pi} \sin(x) \times 2\sin(x)\cos(x)dx$$

$$= 2AB\int_0^{2\pi} \sin^2(x)\,d[\sin(x)] = \frac{2}{3}AB\sin^3 x\Big|_0^{2\pi}$$

$$= 0.$$

The two functions are therefore orthogonal. Figure 8.1 illustrates the symmetries that cause $f(x)$ and $g(x)$ to be orthogonal.

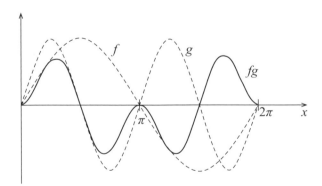

Figure 8.1: Functions $f = \sin(x)$, $g = \sin(2x)$, and their product fg. The product is antisymmetric around $\pi/2$ over the interval $0 \leq x \leq \pi$, and it is also antisymmetric around $3\pi/2$ over the interval $\pi \leq x \leq 2\pi$. Consequently the integral of fg over the entire interval is 0.

The value of A that normalizes $f(x)$ is given by

$$1 = \int_0^{2\pi} A^2 \sin^2(x)\, dx = A^2 \int_0^{2\pi} \frac{1}{2}[1 - \cos(2x)]\, dx$$

$$= \frac{A^2}{4}[2x - \sin(2x)]\Big|_0^{2\pi} = \pi A^2.$$

Thus $A = 1/\sqrt{\pi}$. By similar logic the normalization constant for $g(x)$ is $B = 1/\sqrt{\pi}$, so the functions

$$f(x) = \frac{1}{\sqrt{\pi}} \sin(x)$$

$$g(x) = \frac{1}{\sqrt{\pi}} \sin(2x)$$

are an orthonormal pair over the interval $0 \le x \le 2\pi$. The reader can easily verify that the two functions are orthonormal over any interval whose width is 2π.

Example: Now let $f(x)$ and $g(x)$ be the two complex functions

$$f = \frac{1}{\sqrt{2\pi}} \exp[ix]$$

$$g = \frac{1}{\sqrt{2\pi}} \exp[i2x],$$

where henceforth, $i = \sqrt{-1}$. Both functions are normalized over the interval $0 \le x \le 2\pi$ since, for example,

$$\int_0^{2\pi} g^*(x) g(x)\, dx = \int_0^{2\pi} \frac{1}{\sqrt{2\pi}} \exp[-i2x] \frac{1}{\sqrt{2\pi}} \exp[i2x]\, dx$$

$$= \frac{1}{2\pi} \int_0^{2\pi} dx = 1.$$

The inner product of the two functions is

$$\int_0^{2\pi} f^*(x) g(x)\, dx = \int_0^{2\pi} \frac{1}{\sqrt{2\pi}} \exp[-ix] \frac{1}{\sqrt{2\pi}} \exp[i2x]\, dx$$

$$= \frac{1}{2\pi} \int_0^{2\pi} \exp[ix]\, dx = \frac{1}{2\pi i} \exp[ix]\Big|_0^{2\pi}.$$

Since $\exp[0] = \exp[i2\pi] = 1$, the inner product is

$$\int_{x_1}^{x_2} f(x) g^*(x)\, dx = 0$$

and the functions are orthogonal. Again, the two functions are orthonormal over any interval whose width is 2π.

Let the functions $h_j(x)$ be defined over the interval $x_1 \leq x \leq x_2$, and let $f(x)$ be another function defined over the same interval. Consider the integral

$$R = \int_{x_1}^{x_2} \left(f - \sum_j a_j h_j \right)^2 dx, \tag{8.7}$$

where the a_j are constants. The quantity inside the parentheses is the difference between $f(x)$ and a linear combination of the $h_j(x)$. When this quantity is squared, it is always 0 or positive, so if $\sum_j a_j h_j$ fails to match $f(x)$ anywhere in the interval, R will be positive. Conversely a perfect fit between $f(x)$ and the linear combination of the $h_j(x)$ will yield

$$R = \int_{x_1}^{x_2} \left(f - \sum_j a_j h_j \right)^2 dx = \int_{x_1}^{x_2} 0\, dx = 0. \tag{8.8}$$

When equation 8.8 is satisfied

$$f(x) = \sum_j a_j h_j(x), \tag{8.9}$$

and we say that $f(x)$ has been decomposed or expanded into functions $h_j(x)$. The functions $h_j(x)$ are called a *complete set* if, for any well behaved function $f(x)$, there is a set of numbers a_j such that equation 8.8 holds. There must generally be an infinite number of functions $h_j(x)$ for a set to be complete, but it may not be necessary to use all the $h_j(x)$ to decompose a particular $f(x)$, so the sum may or may not have an infinite number of terms.

If the $h_j(x)$ are orthonormal, the constants a_j are unique and relatively easy to calculate. To show this, let the constants c_j be the inner products of the $h_j(x)$ with $f(x)$:

$$c_j = \int_{x_1}^{x_2} f(x) h_j(x) dx. \tag{8.10}$$

Now consider the expansion of equation 8.7:

$$R = \int_{x_1}^{x_2} f^2 dx - 2 \int_{x_1}^{x_2} \left(f \sum_j a_j h_j \right) dx + \int_{x_1}^{x_2} \left(\sum_j \sum_k a_j a_k h_j h_k \right) dx. \tag{8.11}$$

The second term on the right-hand side is

$$-2 \int_{x_1}^{x_2} \left(f \sum_j a_j h_j \right) dx = -2 \sum_j a_j \int_{x_1}^{x_2} f h_j dx = -2 \sum_j a_j c_j, \tag{8.12}$$

and the third term on the right-hand side is

$$\int_{x_1}^{x_2} \left(\sum_j \sum_k a_j a_k h_j h_k \right) dx = \sum_j \sum_k a_j a_k \int_{x_1}^{x_2} h_j h_k dx = \sum_j \sum_k a_j a_k \delta_{jk} = \sum_j a_j^2. \tag{8.13}$$

Equation 8.11 becomes

$$R = \int_{x_1}^{x_2} f^2 dx - 2 \sum_j a_j c_j + \sum_j a_j^2. \tag{8.14}$$

Add and subtract the quantity $\sum_j c_j^2$ to this equation, and rearrange terms to find

$$R = \int_{x_1}^{x_2} f^2 dx + \sum_j \left(a_j - c_j\right)^2 - \sum_j c_j^2. \tag{8.15}$$

Since $\left(a_j - c_j\right)^2$ is always positive, the minimum value of R occurs when $a_j = c_j$ for all j. In other words, the best values of the a_j are

$$a_j = c_j = \int_{x_1}^{x_2} f(x) h_j(x) dx. \tag{8.16}$$

Note that this result depends explicitly on the orthonormality of the $h_j(x)$.

When the a_j have been set equal to the c_j but not enough terms have been included in the sum to give a perfect fit, the second term on the right-hand side in equation 8.15 is zero, but R is not zero, so

$$R = \int_{x_1}^{x_2} f^2 dx - \sum_{j=0}^{n} c_j^2 \geq 0 \tag{8.17}$$

or

$$\int_{x_1}^{x_2} f^2 dx \geq \sum_{j=0}^{n} c_j^2. \tag{8.18}$$

Equation 8.18 is called *Bessel's inequality*. It guarantees that one never "overshoots" $f(x)$ in the least squares sense. When decomposing a function into orthonormal functions, adding more terms always improves a fit and never degrades it. The values of the c_j are independent of n, so one always uses the same values for the c_j no matter how many terms are included in the sum. The c_j are unique. This is what makes orthonormal functions so important.

Example: By definition, any function $f(x)$ that is analytic near x_1 can be expanded in a Taylor series around x_1:

$$f(x) = f(x_1) + \left.\frac{\partial f(x)}{\partial x}\right|_{x_1} (x - x_1) + \frac{1}{2} \left.\frac{\partial^2 f(x)}{\partial x^2}\right|_{x_1} (x - x_1)^2 + \cdots$$

$$= a_0 + a_1 x + a_2 x^2 + \cdots = \sum_{j=0}^{\infty} a_j x^j.$$

Therefore, the polynomials p_j,

$$p_0 = 1, \quad p_1 = x, \quad p_2 = x^2, \quad \ldots$$

make up a complete set for analytic functions. This particular set of polynomials is neither normalized nor orthogonal, but one can construct linear combinations of the p_j that are orthonormal. One example is a modified form of the Legendre polynomials, which are orthonormal over the interval $-1 \leq x \leq 1$. The first three are

$$p_0 = \sqrt{1/2}, \quad p_1 = \sqrt{3/2}\, x, \quad p_2 = \frac{\sqrt{5/2}}{2}(3x^2 - 1).$$

8.3 Fourier Series

We commence our discussion of Fourier analysis with a derivation of the orthogonality relations for sines and cosines. Consider the sines and cosines with an integral number of cycles over the interval $-\pi \leq x \leq \pi$,

$$\cos(nx), \quad n = 0, \ldots, \infty \tag{8.19}$$

$$\sin(mx), \quad m = 1, \ldots, \infty, \tag{8.20}$$

where we follow the convention of starting n at zero but m at 1. There are three orthogonality relations. First, the inner product of any sine with any cosine is

$$\int_{-\pi}^{\pi} \cos(nx) \sin(mx) dx = \frac{1}{2} \int_{-\pi}^{\pi} \{\sin([m+n]x) + \sin([m-n]x)\} dx. \tag{8.21}$$

Since both $\sin([m+n]x)$ and $\sin([m-n]x)$ either equal 0 or have an integral number of cycles in 2π, their integrals over 2π are 0, so

$$\int_{-\pi}^{\pi} \cos(nx) \sin(mx) dx = 0. \tag{8.22}$$

Second, the inner product of a sine with another sine is

$$\int_{-\pi}^{\pi} \sin(nx) \sin(mx) dx = \frac{1}{2} \int_{-\pi}^{\pi} \{\cos([m-n]x) - \cos([m+n]x)\} dx. \tag{8.23}$$

If $m \neq n$, $\cos([m-n]x)$ and $\cos([m+n]x)$ both have an integral number of cycles in 2π, and their integrals are 0. If $m = n$, the integral becomes

$$\int_{-\pi}^{\pi} \sin(nx) \sin(mx) dx = \frac{1}{2} \int_{-\pi}^{\pi} \{\cos(0x) - \cos(2mx)\} dx = \pi. \tag{8.24}$$

Third, the inner product of a cosine with another cosine is

$$\int_{-\pi}^{\pi} \cos(nx) \cos(mx) dx = \frac{1}{2} \int_{-\pi}^{\pi} \{\cos([m-n]x) + \cos([m+n]x)\} dx. \tag{8.25}$$

As before, if $m \neq n$, $\cos([m-n]x)$ and $\cos([m+n]x)$ both have an integral number of cycles in 2π and their integrals are 0. For $m = n$, the inner product becomes

$$\int_{-\pi}^{\pi} \cos(nx) \cos(mx) dx = \frac{1}{2} \int_{-\pi}^{\pi} \{\cos(0x) + \cos(2mx)\} dx. \tag{8.26}$$

For $m = n = 0$, the integral equals 2π; for $m = n \neq 0$, it equals π.

The orthonormality relations are generally summarized and written in the form

$$\frac{1}{\pi} \int_{-\pi}^{\pi} \cos(nx) \sin(mx) dx = 0 \tag{8.27}$$

$$\frac{1}{\pi} \int_{-\pi}^{\pi} \sin(nx) \sin(mx) dx = \begin{cases} 0, & m \neq n \\ 1, & m = n \end{cases} \tag{8.28}$$

$$\frac{1}{\pi} \int_{-\pi}^{\pi} \cos(nx) \cos(mx) dx = \begin{cases} 0, & m \neq n \\ 1, & m = n \neq 0 \\ 2, & m = n = 0 \end{cases}. \tag{8.29}$$

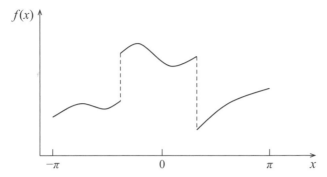

Figure 8.2: The function $f(x)$ is finite but discontinuous over the interval $-\pi \leq x \leq \pi$. As long as it has a finite number of discontinuities, it can be decomposed into a Fourier series.

The normalized sines and cosines are therefore

$$\frac{1}{\sqrt{2\pi}}, \ \frac{1}{\sqrt{\pi}}\cos(nx), \ \frac{1}{\sqrt{\pi}}\sin(mx), \quad n, m = 1, \ldots, \infty. \tag{8.30}$$

The sines and cosines also make up a complete set. This is easy to demonstrate for any function that can be expanded in polynomials (see Courant and Hilbert (1989), vol. I, Chapter II). In fact, though, a broader range of functions can be decomposed into sines and cosines. Any function $f(x)$ that (1) can be integrated over the interval $x_1 \leq x \leq x_2$, (2) has a finite number of maxima and minima in the interval, and (3) has a finite number of discontinuous jumps in the interval can be decomposed into sines and cosines (see Figure 8.2). The function need not even be finite everywhere. Delta functions and linear combinations of delta functions can also be decomposed into sines and cosines. Essentially all functions of physical interest satisfy these mild constraints.[1] The reader should consult monographs on Fourier analysis for the proof of this more general statement.

Let a function $f(x)$ be defined over the interval $-\pi \leq x \leq \pi$. Since the sines and cosines are a complete orthogonal set, equations 8.9 and 8.16 become

$$f(x) = \frac{A_0}{2\sqrt{\pi}} + \frac{1}{\sqrt{\pi}}\sum_{n=1}^{\infty}[A_n\cos(nx) + B_n\sin(nx)], \tag{8.31}$$

where

$$A_n = \frac{1}{\sqrt{\pi}}\int_{-\pi}^{\pi} f(x)\cos(nx)dx \tag{8.32}$$

$$B_n = \frac{1}{\sqrt{\pi}}\int_{-\pi}^{\pi} f(x)\sin(nx)dx. \tag{8.33}$$

The expansion of $f(x)$ in equation 8.31 is called a *Fourier series* after Jean-Baptiste Joseph Fourier (1768–1830), and the quantities A_n and B_n are the *Fourier coefficients*. The factors of $1/\sqrt{\pi}$ are sometimes combined into a single factor of $1/\pi$ in equation 8.31 or in equations 8.32 and 8.33, although this hides the fact that the $f(x)$ is being expanded in a series of orthonormal functions.

---

[1] $f(x) = 1/\sin(1/x)$ is an example of a function that fails to satisfy the constraints.

To understand why both sines and cosines are needed in a Fourier series, expand a sine function with arbitrary amplitude α and phase β:

$$\alpha \sin(nx + \beta) = \alpha \sin(\beta)\cos(nx) + \alpha \cos(\beta)\sin(nx)$$

$$= A\cos(nx) + B\sin(nx), \tag{8.34}$$

where $A = \alpha \sin(\beta)$ and $B = \alpha \cos(\beta)$. Thus two constants A and B are needed to account for both the phase and the amplitude of the sine. The amplitude and phase of component n in a Fourier series can be recovered from

$$\alpha = \frac{1}{\sqrt{\pi}}\left(A_n^2 + B_n^2\right)^{1/2} \tag{8.35}$$

$$\beta = \tan^{-1}\left(\frac{A_n}{B_n}\right), \tag{8.36}$$

taking care that β is in the correct quadrant.

Since the integrals in equations 8.32 and 8.33 for A_n and B_n are calculated over the interval $-\pi \leq x \leq \pi$, the reconstruction of $f(x)$ given by equation 8.31 is strictly defined only for that interval. The sines and cosines in the equation can nevertheless be evaluated outside the interval. By inspection we see that $f(x + 2m\pi) = f(x)$ for any integer m, so outside the original interval the series merely replicates $f(x)$ over and over with period 2π. This property can be used to decompose periodic functions of infinite length: the decomposition is the Fourier series for one cycle of the periodic function. The term with period 2π is often called the *fundamental term*, and the remaining terms are called *overtones* or *harmonics*. The following example expands a square wave into a Fourier series.

Example: One cycle of a periodic square wave is shown in Figure 8.3. The square wave can be expanded in a Fourier series, even though it has discontinuities and infinite length. From equations 8.32 and 8.33, the Fourier coefficients are

$$A_n = \frac{1}{\sqrt{\pi}}\int_{-\pi}^{0}(-1)\cos(nx)dx + \frac{1}{\sqrt{\pi}}\int_{0}^{\pi}(+1)\cos(nx)dx$$

$$= -\frac{1}{\sqrt{\pi}}\int_{0}^{\pi}\cos(nx)dx + \frac{1}{\sqrt{\pi}}\int_{0}^{\pi}\cos(nx)dx$$

$$= 0$$

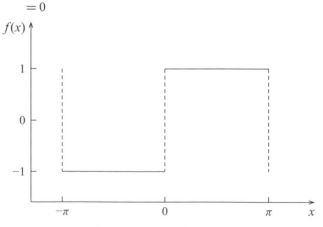

Figure 8.3: A square wave.

and

$$B_n = \frac{1}{\sqrt{\pi}} \int_{-\pi}^{0} (-1)\sin(nx)dx + \frac{1}{\sqrt{\pi}} \int_{0}^{\pi} (+1)\sin(nx)dx$$

$$= \frac{1}{\sqrt{\pi}} \int_{0}^{\pi} \sin(nx)dx + \frac{1}{\sqrt{\pi}} \int_{0}^{\pi} \sin(nx)dx \;=\; \frac{2}{\sqrt{\pi}} \int_{0}^{\pi} \sin(nx)dx$$

$$= -\frac{2}{\sqrt{\pi}} \left[\frac{1}{n} \cos(nx) \right]_{0}^{\pi}$$

$$= \frac{2}{n\sqrt{\pi}} [1 - \cos(n\pi)].$$

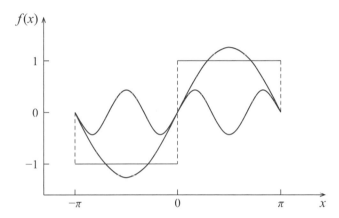

Figure 8.4: First two components in the Fourier series expansion for a square wave.

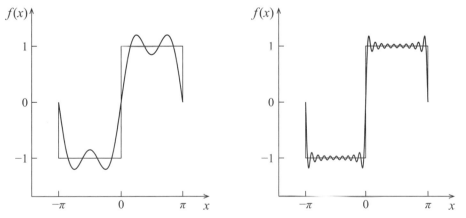

Figure 8.5: The sum of the first two components in the Fourier series for a square wave (left panel). The sum of the first 10 components (right panel). The overshoot and ringing are generic phenomena, not specific to a square wave. The ringing is caused by truncation of the Fourier series. The overshoot, called the Gibbs phenomenon, is caused by discontinuities.

Continued on page 230

If n is zero or even, $\cos(n\pi) = 1$ and $B_n = 0$. If n is odd, $\cos(n\pi) = -1$ and $B_n = 4/n\sqrt{\pi}$. From equation 8.31, the Fourier series expansion for the square wave is then

$$f(x) = \frac{1}{\sqrt{\pi}} \left[\frac{4}{\sqrt{\pi}} \sin(x) + \frac{4}{3\sqrt{\pi}} \sin(3x) + \frac{4}{5\sqrt{\pi}} \sin(5x) + \cdots \right]$$

$$= \frac{4}{\pi} \left[\sin(x) + \frac{1}{3} \sin(3x) + \frac{1}{5} \sin(5x) + \cdots \right].$$

Figure 8.4 shows the first two members of the Fourier series overplotted on the square wave. Figure 8.5 shows the sum of the first two and the first 10 members overplotted on the square wave.

Note the overshoot and ringing in Figure 8.5. The ringing is an artifact of using just the first few members of an infinite series; the ringing becomes smaller and has a higher frequency as more members of the series are included. The overshoot, called the *Gibbs phenomenon*, becomes narrower but does not disappear as more members of the Fourier series are included, instead approaching an overshoot of about 18% of the square wave's peak-to-peak amplitude. The Gibbs phenomenon arises at discontinuities and is always present there.

It is often helpful to display the Fourier coefficients as a plot of $P_n = A_n^2 + B_n^2$ against n. The plot is called a *power spectrum* by analogy to the power in an electrical circuit with an alternating current, in which the power is proportional to the square of the amplitude of the sine curve. One can also display the coefficients as an amplitude spectrum by plotting $\sqrt{P_n/\pi}$ against n. The amplitude spectrum for a square wave with an amplitude of $\sqrt{\pi}/4$ is shown in Figure 8.6.

With the help of equations 8.27–8.29 one readily derives Bessel's inequality for the Fourier series:

$$\int_{-\pi}^{\pi} f^2(x)\,dx \geq \frac{A_0^2}{2} + \sum_{n=1}^{m} \left(A_n^2 + B_n^2 \right) = \frac{1}{2}P_0 + \sum_{n=1}^{m} P_n. \tag{8.37}$$

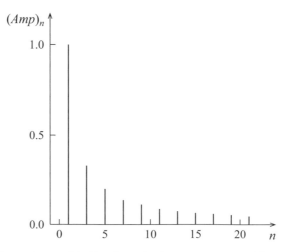

Figure 8.6: Amplitude spectrum of the Fourier series for a square wave with an amplitude of $\sqrt{\pi}/4$.

As m goes to infinity, this must become

$$\int_{-\pi}^{\pi} f^2(x)\,dx = \frac{A_0^2}{2} + \sum_{n=1}^{\infty}\left(A_n^2 + B_n^2\right) = \frac{1}{2}P_0 + \sum_{n=1}^{\infty} P_n,\tag{8.38}$$

because the sines and cosines are a complete set. Equation 8.38 is *Parseval's theorem*. It says that the power in a function is equal to the sum of the power in the individual components of the Fourier series for the function.

It is often useful to express the Fourier series in complex form. Let the complex Fourier coefficient be

$$C_n = A_n - iB_n.\tag{8.39}$$

Then from equations 8.32 and 8.33,

$$C_n = \frac{1}{\sqrt{\pi}} \int_{-\pi}^{\pi} f(x)\,[\cos(nx) - i\sin(nx)]\,dx\tag{8.40}$$

$$= \frac{1}{\sqrt{\pi}} \int_{-\pi}^{\pi} f(x)\exp[-inx]\,dx.\tag{8.41}$$

If $f(x)$ is real, then

$$C_{-n} = \frac{1}{\sqrt{\pi}} \int_{-\pi}^{\pi} f(x)\exp[inx]\,dx = C_n^*\tag{8.42}$$

$$C_0 = \frac{1}{\sqrt{\pi}} \int_{-\pi}^{\pi} f(x)\,dx = C_0^*.\tag{8.43}$$

The individual terms of the Fourier series become

$$A_n\cos(nx) + B_n\sin(nx) = \frac{1}{2}\left(C_n + C_n^*\right)\frac{1}{2}\Big(\exp[inx] + \exp[-inx]\Big)$$

$$+ \frac{(-1)}{2i}\left(C_n - C_n^*\right)\frac{1}{2i}\Big(\exp[inx] - \exp[-inx]\Big)$$

$$= \frac{1}{2}C_n\exp[inx] + \frac{1}{2}C_n^*\exp[-inx]$$

$$= \frac{1}{2}C_n\exp[inx] + \frac{1}{2}C_{-n}\exp[-inx],\tag{8.44}$$

so the Fourier expansion of $f(x)$ is

$$f(x) = \frac{C_0}{2\sqrt{\pi}} + \frac{1}{2\sqrt{\pi}}\sum_{n=1}^{\infty}\left(C_n\exp[inx] + C_{-n}\exp[-inx]\right)$$

$$= \frac{1}{2\sqrt{\pi}}\sum_{n=-\infty}^{\infty} C_n\exp[inx].\tag{8.45}$$

To summarize, in complex form the equations for the Fourier series of a real function are

$$f(x) = \frac{1}{2\sqrt{\pi}} \sum_{n=-\infty}^{\infty} C_n \exp[inx] \tag{8.46}$$

$$C_n = \frac{1}{\sqrt{\pi}} \int_{-\pi}^{\pi} f(x) \exp[-inx] dx. \tag{8.47}$$

As before, the factor $1/2\pi$ can be redistributed between the two equations at will. A common form is to have the same factor $1/\sqrt{2\pi}$ on both equations.

Since $A_n^2 + B_n^2 = C_n^* C_n = |C_n|^2$, Parseval's theorem takes on a particularly pleasing form in complex notation:

$$\int_{-\pi}^{\pi} |f(x)|^2 dx = \frac{1}{2} \sum_{n=-\infty}^{\infty} |C_n|^2. \tag{8.48}$$

If $f(x)$ is a real function, Parseval's theorem can be written as

$$\int_{-\pi}^{\pi} f^2(x) dx = \frac{C_0^2}{2} + \sum_{n=1}^{\infty} |C_n|^2. \tag{8.49}$$

The power spectrum is normally defined to be

$$P_n = \frac{1}{2} |C_n|^2, \tag{8.50}$$

but if $f(x)$ a real function, the power can be defined as

$$P_n = |C_n|^2, \quad n \geq 1 \tag{8.51}$$

$$P_0 = \frac{1}{2} C_0^2. \tag{8.52}$$

This is sometimes called a *one-sided power spectrum* to distinguish it from from the usual power spectrum.

Finally, with an appropriate change of variables, all the previous results hold over any finite interval, not just $-\pi$ to π. It will later be convenient to make the interval run from $-T/2$ to $T/2$. With the transformation $x = 2\pi t/T$ and $dx = (2\pi/T)dt$ equations 8.31–8.33 become

$$f(t) = \frac{A_0}{2\sqrt{\pi}} + \frac{1}{\sqrt{\pi}} \sum_{n=1}^{\infty} \left[A_n \cos\left(\frac{2\pi nt}{T}\right) + B_n \sin\left(\frac{2\pi nt}{T}\right) \right] \tag{8.53}$$

$$A_n = \frac{2\sqrt{\pi}}{T} \int_{-T/2}^{T/2} f(t) \cos\left(\frac{2\pi nt}{T}\right) dt \tag{8.54}$$

$$B_n = \frac{2\sqrt{\pi}}{T} \int_{-T/2}^{T/2} f(t) \sin\left(\frac{2\pi nt}{T}\right) dt, \tag{8.55}$$

and equations 8.46 and 8.47 become

$$f(t) = \frac{1}{2\sqrt{\pi}} \sum_{n=-\infty}^{\infty} C_n \exp\left[i\frac{2\pi nt}{T}\right] \tag{8.56}$$

$$C_n = \frac{2\sqrt{\pi}}{T} \int_{-T/2}^{T/2} f(t) \exp\left[-i\frac{2\pi nt}{T}\right] dt. \tag{8.57}$$

8.4 Fourier Transform

Equations 8.56 and 8.57 give the Fourier series for a function defined over a finite interval from $-T/2$ to $T/2$. The *Fourier transform* is the limit of Fourier series as the interval goes to infinity. To calculate the limit, define

$$\nu = \frac{n}{T}, \quad \Delta\nu = \frac{1}{T}, \tag{8.58}$$

and define the discrete function $F(\nu)$ to be

$$F(\nu) = \frac{T}{2\sqrt{\pi}} C_n = \int_{-T/2}^{T/2} f(t) \exp[-i2\pi \nu t] dt. \tag{8.59}$$

Then multiply equation 8.56 by T/T, and the Fourier series becomes

$$f(t) = \sum_{n=-\infty}^{\infty} \left(\frac{T}{2\sqrt{\pi}} C_n\right) \exp[i2\pi \nu t] \frac{1}{T}$$

$$= \sum_{n=-\infty}^{\infty} F(\nu) \exp[i2\pi \nu t] \Delta\nu. \tag{8.60}$$

Now let T go to infinity. In the limit, $\Delta\nu \to d\nu$, the sum becomes an integral, and $F(\nu)$ becomes continuous, yielding

$$f(t) = \int_{-\infty}^{\infty} F(\nu) \exp[i2\pi \nu t] d\nu \tag{8.61}$$

$$F(\nu) = \int_{-\infty}^{\infty} f(t) \exp[-i2\pi \nu t] dt. \tag{8.62}$$

The function $F(\nu)$ is called the *Fourier transform* of $f(t)$, and $F(\nu)$ and $f(t)$ are called *Fourier transform pairs*. By convention equation 8.62 is usually called the forward transform and equation 8.61 the reverse transform. The two are distinguished by the sign of the exponent. If t is time, ν can be identified with frequency, but other identifications are possible, most notably distance and wavenumber.

The Fourier transform of $f(at + b)$ is

$$G(\nu) = \int_{-\infty}^{\infty} f(at + b) \exp[-i2\pi\nu t] dt$$

$$= \frac{1}{a} \exp\left[i2\pi\frac{\nu}{a}b\right] \int_{-\infty}^{\infty} f(at + b) \exp\left[-i2\pi\frac{\nu}{a}(at + b)\right] d(at + b)$$

$$= \frac{1}{a} \exp\left[i2\pi\frac{\nu}{a}b\right] F\left(\frac{\nu}{a}\right). \tag{8.63}$$

Thus, $f(at)$ transforms to $F(\nu/a)$. As $f(at)$ becomes broader, $F(\nu/a)$ becomes proportionally narrower. The Fourier transform is a linear operation, so if $f(t)$ is the linear combination of two functions

$$f(t) = a_1 f_1(t) + a_2 f_2(t), \tag{8.64}$$

the Fourier transform of $f(t)$ is the same linear combination of the individual Fourier transforms:

$$F(\nu) = a_1 F_1(\nu) + a_2 F_{(2}(\nu). \tag{8.65}$$

In complex notation Parseval's theorem becomes

$$\int_{-\infty}^{\infty} |f(t)|^2 dt = \int_{-\infty}^{\infty} F^*(\nu) F(\nu) d\nu \tag{8.66}$$

The power spectrum becomes a plot of $P(\nu) = |F(\nu)|^2$, although, since $P(\nu)$ is a continuous function, the plot is sometimes called a power density spectrum. If $f(t)$ is real, $F(-\nu) = F^*(\nu)$ and $F^*(-\nu) = F(\nu)$, so that

$$P(-\nu) = P(\nu). \tag{8.67}$$

As a result, one often works with the one-sided power spectrum:

$$P_1(\nu) = 2P(\nu), \quad \nu > 0. \tag{8.68}$$

8.4.1 Fourier Transform Pairs

The literature on Fourier transforms is littered with long tables of Fourier transform pairs. The following pairs will be particularly useful to us. They are summarized in our own long table(!), at the end of Section 8.4.

Symmetric Exponential Function: The Fourier transform of the symmetric exponential function

$$f(t) = \exp[-|t|] \tag{8.69}$$

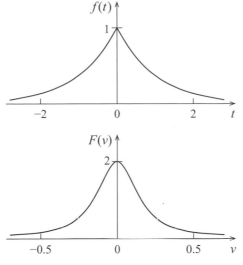

Figure 8.7: Fourier transform pairs $f(t) = \exp[-|t|]$ (top panel) and $F(v) = 2/[1 + (2\pi v)^2]$ (bottom panel).

is

$$F(v) = \int_{-\infty}^{\infty} f(t) \exp[-i2\pi vt] dt$$

$$= \int_{-\infty}^{\infty} \exp[-|t|] \exp[-i2\pi vt] dt$$

$$= \int_{-\infty}^{0} \exp[t] \exp[-i2\pi vt] dt + \int_{0}^{\infty} \exp[-t] \exp[-i2\pi vt] dt$$

$$= \frac{1}{1 - i2\pi v} + \frac{1}{1 + i2\pi v}$$

$$= \frac{2}{1 + (2\pi v)^2}, \tag{8.70}$$

which is the Lorentzian function (see Figure 8.7).

Gaussian Function: The Fourier transform of the Gaussian function

$$f(t) = \exp\left[-\pi t^2\right] \tag{8.71}$$

is

$$F(v) = \int_{-\infty}^{\infty} \exp\left[-\pi t^2\right] \exp[-i2\pi vt] dt = \int_{-\infty}^{\infty} \exp\left[-\pi t^2 - i2\pi vt\right] dt$$

$$= \exp[-\pi v^2] \int_{-\infty}^{\infty} \exp\left[-\pi(t^2 + i2vt - v^2)\right] dt$$

$$= \exp[-\pi v^2] \int_{-\infty}^{\infty} \exp\left[-\pi(t + iv)^2\right] dt. \tag{8.72}$$

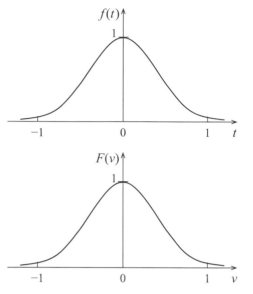

Figure 8.8: Fourier transform pairs $f(t) = \exp[-\pi t^2]$ (top panel) and $F(\nu) = \exp[-\pi \nu^2]$ (bottom panel).

With a change of variables to $x = t + i\nu$, equation 8.72 becomes

$$F(\nu) = \exp[-\pi \nu^2] \int_{-\infty}^{\infty} \exp\left[-\pi x^2\right] dx. \qquad (8.73)$$

The integral equals 1 (see Appendix C). The Fourier transform of the Gaussian is then

$$F(\nu) = \exp[-\pi \nu^2], \qquad (8.74)$$

so the Fourier transform of a Gaussian function is a Gaussian function (see Figure 8.8).

Rectangle Function: The Fourier transform of the rectangle function

$$f(t) = \begin{cases} 1, & |t| \leq b/2 \\ 0, & |t| > b/2 \end{cases} \qquad (8.75)$$

is

$$F(\nu) = \int_{-b/2}^{b/2} \exp[-i2\pi \nu t] dt = -\frac{1}{i2\pi \nu} \left[\exp[-i2\pi \nu t] \right]_{-b/2}^{b/2}$$

$$= -\frac{1}{i2\pi \nu} \left(\exp[-i\pi \nu b] - \exp[i\pi \nu b] \right) = \frac{1}{i2\pi \nu} 2i \sin(\pi \nu b)$$

$$= b \frac{\sin(\pi \nu b)}{\pi \nu b}. \qquad (8.76)$$

The Fourier transform of the rectangle function is called the *sinc function* (see Figure 8.9). The limit of $F(\nu)$ as ν goes to 0 does exist and is

$$\lim_{\nu \to 0} F(\nu) = b \qquad (8.77)$$

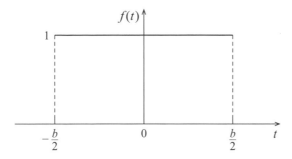

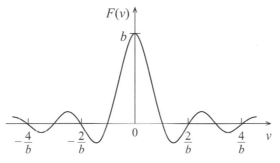

Figure 8.9: Rectangle function (top panel) and its Fourier transform $F(\nu) = b\sin(\pi\nu b)/\pi\nu b$ (bottom panel).

(to see this, either expand $\sin(\pi\nu b)$ in a Taylor series or apply l'Hôpital's rule). The sinc function is equal to 0 at $\pi\nu b = \pm n\pi$, so

$$\nu = \pm\frac{n}{b}, \quad n \neq 0. \tag{8.78}$$

Between the zeros, it oscillates from positive to negative, the extrema of the oscillations dropping off as $1/\nu$.

Triangle Function: The Fourier transform of the triangle function

$$f(t) = \begin{cases} 1 - |t|/b, & |t| \leq b \\ 0, & |t| > b \end{cases} \tag{8.79}$$

is

$$
\begin{aligned}
F(\nu) &= \int_{-b}^{b} \left(1 - \frac{|t|}{b}\right) \exp[-i2\pi\nu t]dt \\
&= \int_{-b}^{0} \left(1 + \frac{t}{b}\right) \exp[-i2\pi\nu t]dt + \int_{0}^{b} \left(1 - \frac{t}{b}\right) \exp[-i2\pi\nu t]dt \\
&= \int_{0}^{b} \left(1 - \frac{t}{b}\right) \left\{\exp[+i2\pi\nu t] + \exp[-i2\pi\nu t]\right\}dt = 2\int_{0}^{b} \left(1 - \frac{t}{b}\right)\cos(2\pi\nu t)dt \\
&= \frac{\sin(2\pi\nu b)}{\pi\nu} - \frac{2}{b}\int_{0}^{b} t\cos(2\pi\nu t)dt. \tag{8.80}
\end{aligned}
$$

Recognizing that $\theta \cos(\theta)d\theta = d[\theta \sin(\theta)] - \sin(\theta)d\theta$, we can integrate by parts to get

$$F(v) = \frac{\sin(2\pi vb)}{\pi v} - \frac{2}{b}\frac{1}{(2\pi v)^2}\left\{(2\pi vt)\sin(2\pi vt)\,\Big|_0^b + \cos(2\pi vt)\,\Big|_0^b\right\}$$

$$= \frac{\sin(2\pi vb)}{\pi v} - \frac{\sin(2\pi vb)}{\pi v} + \frac{2}{b}\frac{1}{(2\pi v)^2}\{1 - \cos(2\pi vb)\}$$

$$= \frac{2}{b}\frac{1}{(2\pi v)^2}2\sin^2(\pi vb) = b\frac{\sin^2(\pi vb)}{(\pi vb)^2}. \tag{8.81}$$

This is the square of the sinc function, a result that could have been derived in just one or two lines using the convolution theorem (see Section 8.6).

Constant Function: The Fourier transform of the constant function $f(t) = 1$ can be found by letting the width b of the rectangle function go to infinity. The Fourier transform of the rectangle function is a sinc function. The width of the central lobe of the sinc function decreases as $1/b$, and the amplitude of its side lobes decreases as $1/vb$. In the limit as $b \to \infty$, the width of the sinc function goes to 0 and its amplitude goes to infinity. One can show (see Appendix A) that

$$\int_{-\infty}^{\infty} \frac{\sin x}{x}dx = \pi, \tag{8.82}$$

so the integral over all v of the Fourier transform of the rectangle function is

$$\int_{-\infty}^{\infty} F(v)dv = \int_{-\infty}^{\infty} b\frac{\sin(\pi vb)}{\pi vb}dv$$

$$= \frac{1}{\pi}\int_{-\infty}^{\infty} \frac{\sin(\pi vb)}{\pi vb}d(\pi vb) = \frac{1}{\pi}\pi$$

$$= 1. \tag{8.83}$$

Thus, the integral of $F(v)$ over v is independent of the width the rectangle function.

We recognize this function as the normalized Dirac δ function:[2]

$$F(v) = \delta(v) = \begin{cases} 0, & v \neq 0 \\ \infty, & v = 0 \end{cases} \tag{8.84}$$

and

$$\int_{-\infty}^{\infty} F(v)dv = \int_{-\infty}^{\infty} \delta(v)dv = 1. \tag{8.85}$$

Sine and Cosine Functions: The Fourier transform of a truncated cosine function with frequency v_0,

$$f(t) = \begin{cases} \cos(2\pi v_0 t), & |t| \leq b/2 \\ 0, & |t| > b/2 \end{cases}, \tag{8.86}$$

---

[2] To be complete, one should also prove that the limit of the sinc function satisfies equation 1.28. The proof can be found in standard texts on Fourier analysis.

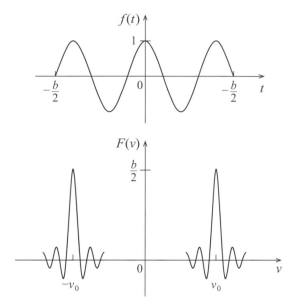

Figure 8.10: Fourier transform of the truncated cosine function $f(t) = \cos(2\pi \nu_0 t)$, $|t| \leq b/2$ (top panel). It is a pair of sinc functions, one centered on $\nu = \nu_0$ and the other on $\nu = -\nu_0$ (bottom panel).

is

$$F(\nu) = \int_{-b/2}^{b/2} \cos(2\pi \nu_0 t) \exp[-i2\pi \nu t] dt$$

$$= \int_{-b/2}^{b/2} \frac{1}{2} \left(\exp[i2\pi \nu_0 t] + \exp[-i2\pi \nu_0 t] \right) \exp[-i2\pi \nu t] dt$$

$$= \frac{1}{2} \int_{-b/2}^{b/2} \exp[-i2\pi (\nu - \nu_0) t] dt + \frac{1}{2} \int_{-b/2}^{b/2} \exp[-i2\pi (\nu + \nu_0) t] dt$$

$$= \frac{b}{2} \frac{\sin(\pi [\nu - \nu_0] b)}{\pi [\nu - \nu_0] b} + \frac{b}{2} \frac{\sin(\pi [\nu + \nu_0] b)}{\pi [\nu + \nu_0] b}. \tag{8.87}$$

This is the sum of two sinc functions, one centered on $\nu = \nu_0$ and the other on $\nu = -\nu_0$, as shown in Figure 8.10.

By inspection, the Fourier transform of a truncated sine function with frequency ν_0 is

$$F(\nu) = -\frac{ib}{2} \frac{\sin(\pi [\nu - \nu_0] b)}{\pi [\nu - \nu_0] b} + \frac{ib}{2} \frac{\sin(\pi [\nu + \nu_0] b)}{\pi [\nu + \nu_0] b}. \tag{8.88}$$

As before, as $b \to \infty$, the sinc functions become δ functions. The Fourier transform of the infinite cosine function $f(t) = \cos(2\pi \nu_0 t)$ is then

$$F(\nu) = \frac{1}{2}\delta(\nu - \nu_0) + \frac{1}{2}\delta(\nu + \nu_0), \tag{8.89}$$

and the Fourier transform of the infinite sine function $f(t) = \sin(2\pi \nu_0 t)$ is

$$F(\nu) = -\frac{i}{2}\delta(\nu - \nu_0) + \frac{i}{2}\delta(\nu + \nu_0). \tag{8.90}$$

Delta Function: While the result is obvious from the previous examples, the Fourier transform of the delta function $\delta(t - t_0)$ is worth stating explicitly:

$$F(\nu) = \int_{-\infty}^{\infty} \delta(t - t_0) \exp[-i2\pi \nu t] dt = \exp[-i2\pi \nu t_0]. \qquad (8.91)$$

The power spectrum of the delta function is

$$|F(\nu)|^2 = 1. \qquad (8.92)$$

Note that the Fourier transform of $\delta(t)$ (i.e., a delta function at $t_0 = 0$) is $\exp[0] = 1$.

Comb Function: Let $f(t)$ be an infinite string of delta functions separated by Δt,

$$f(t) = \sum_{n=-\infty}^{\infty} \delta(t - n\Delta t). \qquad (8.93)$$

This is sometimes called the *comb function*. The Fourier transform of the comb function is

$$F(\nu) = \int_{-\infty}^{\infty} \left[\sum_{n=-\infty}^{\infty} \delta(t - n\Delta t) \right] \exp[-i2\pi \nu t] dt$$

$$= \sum_{n=-\infty}^{\infty} \exp[-i2\pi \nu n\Delta t]. \qquad (8.94)$$

This expression is not amenable to direct evaluation. Instead, we evaluate the quantity

$$\mathsf{F}(\nu) = \sum_{n=-N}^{N} \exp[-i2\pi \nu n\Delta t] \qquad (8.95)$$

and then let $N \to \infty$ so that $\mathsf{F}(\nu) \to F(\nu)$. Remembering the expansion

$$\frac{1 - a^s}{1 - a} = 1 + a + a^2 + \cdots + a^{s-1} \qquad (8.96)$$

and indulging in some index manipulation, we find

$$\mathsf{F}(\nu) = \exp[i2\pi \nu N\Delta t] \sum_{k=0}^{2N} \exp[-i2\pi \nu k\Delta t]$$

$$= \exp[i2\pi \nu N\Delta t] \frac{1 - \exp[-i2\pi \nu(2N+1)\Delta t]}{1 - \exp[-i2\pi \nu \Delta t]}$$

$$= \frac{\exp[i2\pi \nu N\Delta t] - \exp[-i2\pi \nu(N+1)\Delta t]}{1 - \exp[-i2\pi \nu \Delta t]}. \qquad (8.97)$$

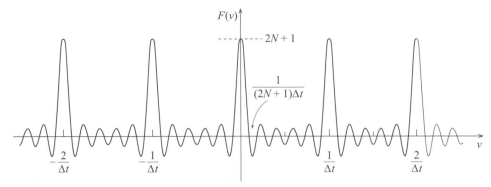

Figure 8.11: The Function F(ν) given by equation 8.98. The function has larger peaks regularly spaced at intervals $1/\Delta t$. The height of the peaks is $2N + 1$, and their half width to the first zero is $1/([2N+1]\Delta t)$.

Multiplying the numerator and denominator of equation 8.98 by $1 + \exp[i2\pi \nu \Delta t]$, we arrive at

$$
\begin{aligned}
\mathsf{F}(\nu) &= \frac{\sin(2\pi \nu N \Delta t) + \sin(2\pi \nu[N+1]\Delta t)}{\sin(2\pi \nu \Delta t)} \\
&= \frac{2\sin(2\pi \nu[N+1/2]\Delta t)\cos(\pi \nu \Delta t)}{2\sin(\pi \nu \Delta t)\cos(\pi \nu \Delta t)} \\
&= \frac{\sin(\pi \nu[2N+1]\Delta t)}{\sin(\pi \nu \Delta t)} \\
&= (2N+1)\left\{\frac{\sin(\pi \nu[2N+1]\Delta t)}{\pi \nu[2N+1]\Delta t}\right\}\left\{\frac{\pi \nu \Delta t}{\sin(\pi \nu \Delta t)}\right\}.
\end{aligned}
\tag{8.98}
$$

We recognize $\mathsf{F}(\nu)$ as a rapidly oscillating sinc function divided by a slowly oscillating sinc function. Despite appearances, it is finite where $\sin(\pi \nu \Delta t) = 0$. To see this, apply l'Hôpital's rule to equation 8.98 and find $\mathsf{F}(\nu) = 2N + 1$ when $\nu \Delta t$ is an integer. Figure 8.11 shows a plot of $\mathsf{F}(\nu)$. The function has larger peaks regularly spaced at intervals $1/\Delta t$ with heights equal to $2N + 1$ and half widths to the first zero of $1/([2N+1]\Delta t)$. The product of their height times their width is $1/\Delta t$. Between the large peaks $\mathsf{F}(\nu)$ oscillates rapidly.

Now let $N \to \infty$ while keeping Δt constant. The height of the large peaks goes to infinity, and their width goes to 0 but the product of their height times their width remains constant. The rapid oscillations die off infinitely rapidly, so $\mathsf{F}(\nu)$ goes to 0 between the peaks. We recognize this as an evenly spaced series of delta functions separated by $\Delta \nu = k/\Delta t$, so

$$
F(\nu) = \sum_{k=-\infty}^{\infty} \delta(\nu - k/\Delta t).
\tag{8.99}
$$

Note the symmetry: $f(t)$ and $F(\nu)$ are both infinite strings of evenly spaced delta functions.

8.4.2 Summary of Useful Fourier Transform Pairs

The Fourier transform pairs discussed in this section are summarized in Table 8.1.

Table 8.1: Some Fourier transform pairs

Function name	Function	Fourier transform
General function f	$f(t) = \int_{-\infty}^{\infty} F(\nu)\exp[i2\pi\nu t]d\nu$	$F(\nu) = \int_{-\infty}^{\infty} f(t)\exp[-i2\pi\nu t]dt$
Symmetric exponential	$\exp[-|t|]$	$\dfrac{1}{1+(2\pi\nu)^2}$
Gaussian	$\exp[-\pi t^2]$	$\exp[-\pi\nu^2]$
Rectangle	$\begin{cases} 1, & |t| \le b/2 \\ 0, & |t| > b/2 \end{cases}$	$b\dfrac{\sin(\pi\nu b)}{\pi\nu b}$
Triangle	$\begin{cases} 1-|t|/b, & |t| \le b \\ 0, & |t| > b \end{cases}$	$b\dfrac{\sin^2(\pi\nu b)}{(\pi\nu b)^2}$
Constant	1	$\delta(\nu)$
Truncated cosine	$\begin{cases} \cos(2\pi\nu_0 t), & |t| \le b/2 \\ 0, & |t| > b/2 \end{cases}$	$\dfrac{b}{2}\dfrac{\sin(n\pi[\nu-\nu_0]b)}{\pi[\nu-\nu_0]b} + \dfrac{b}{2}\dfrac{\sin(\pi[\nu+\nu_0]b)}{\pi[\nu+\nu_0]b}$
Cosine	$\cos(2\pi\nu_0 t)$	$\dfrac{1}{2}\delta(\nu-\nu_0) + \dfrac{1}{2}\delta(\nu+\nu_0)$
Truncated sine	$\begin{cases} \sin(2\pi\nu_0 t), & |t| \le b/2 \\ 0, & |t| > b/2 \end{cases}$	$-\dfrac{ib}{2}\dfrac{\sin(\pi[\nu-\nu_0]b)}{\pi[\nu-\nu_0]b} + \dfrac{ib}{2}\dfrac{\sin(\pi[\nu+\nu_0]b)}{\pi[\nu+\nu_0]b}$
Sine	$\sin(2\pi\nu_0 t)$	$-\dfrac{i}{2}\delta(\nu-\nu_0) + \dfrac{i}{2}\delta(\nu+\nu_0)$
Delta	$\delta(t-t_0)$	$\exp[-i2\pi\nu t_o]$
Comb	$\sum_{n=-\infty}^{\infty} \delta(t-n\Delta t)$	$\sum_{k=-\infty}^{\infty} \delta\left(\nu - \dfrac{k}{\Delta t}\right)$

8.5 Discrete Fourier Transform

Suppose a function $f(t)$ is sampled N times at equal intervals Δt, as shown in Figure 8.12. For mathematical convenience, we assume that there is an even number of samples $N = 2n$,

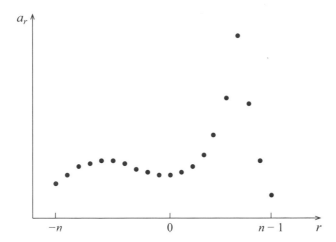

Figure 8.12: The function $f(t)$ is evaluated N times at equal intervals Δt. There are an even number of points $N = 2n$, and the points are numbered from $r = -n$ to $n - 1$. The value of the function at point r is $a_r = f(r\Delta t)$.

that the samples are numbered from $r = -n$ to $n - 1$, and that $t = r\Delta t$. At the sample point r the value of $f(t)$ is

$$f(r\Delta t) = a_r. \tag{8.100}$$

The samples can be decomposed into a finite-length series of N sines and cosines of the form $\sin(2\pi mr/N)$ and $\cos(2\pi mr/N)$, where the m are integers between 0 and $n - 1$. The decomposition is called the *discrete Fourier transform*. There are two common approaches to the discrete Fourier transform, one as a special case of the continuous Fourier transform and one that begins with the orthogonality relations for discretely sampled sine and cosine functions.

8.5.1 Derivation from the Continuous Fourier Transform

The function $f(t)$ can be represented by a sum of delta functions with amplitudes a_r:

$$f(t) = \sum_{r=-n}^{n-1} a_r \delta(t - r\Delta t). \tag{8.101}$$

The Fourier transform of $f(t)$ is

$$F(v) = \int_{-\infty}^{\infty} f(t) \exp[-i2\pi vt]dt = \int_{-\infty}^{\infty} \left\{ \sum_{r=-n}^{n-1} a_r \delta(t - r\Delta t) \right\} \exp[-i2\pi vt]dt$$

$$= \sum_{r=-n}^{n-1} a_r \left\{ \int_{-\infty}^{\infty} \delta(t - r\Delta t) \exp[-i2\pi vt]dt \right\}$$

$$= \sum_{r=-n}^{n-1} a_r \exp[-i2\pi vr\Delta t]. \tag{8.102}$$

Equation 8.102 underdetermines $F(v)$, because $F(v)$ is a continuous function but is calculated from a finite number of discrete points. It can be fully determined by adding extra constraints. One way to do this is to make $f(t)$ periodic with period N, so that $f(r\Delta t) = f([r+N]\Delta t)$. This boundary condition imposes the requirement that

$$\exp[-i2\pi\nu r\Delta t] = \exp[-i2\pi\nu(r+N)\Delta t]$$

$$1 = \exp[-i2\pi\nu N\Delta t]$$

$$2\pi\nu N\Delta t = 2m\pi$$

$$\nu\Delta t = \frac{m}{N}, \tag{8.103}$$

where m is an integer in the range $-n \leq m \leq (n-1)$. One can interpret equation 8.103 to mean that $F(\nu)$ exists only at those frequencies where $\nu_m = m/(N\Delta t)$. With this constraint, equation 8.102 becomes the discrete Fourier transform:

$$F_m \equiv \frac{F(\nu_m)}{N} = \frac{1}{N}\sum_{r=-n}^{n-1} a_r \exp\left[-i\frac{2\pi m}{N}r\right]. \tag{8.104}$$

Alternatively, one can represent $F(\nu)$ as a continuous function made of a sum of delta functions with amplitudes F_m located at $\nu_m = m/N\Delta t$:

$$F(\nu) = \sum_{m=-n}^{n-1} F_m \delta\left(\nu - \frac{m}{N\Delta t}\right). \tag{8.105}$$

In either case, the transform of a series of discrete points evenly spaced in time is a series of discrete points evenly spaced in frequency. The reverse transform is

$$f(t) = \int_{-\infty}^{\infty} F(\nu)\exp[i2\pi\nu t]d\nu = \int_{-\infty}^{\infty}\left\{\sum_{m=-n}^{n-1} F_m \delta\left(\nu - \frac{m}{N\Delta t}\right)\right\}\exp[i2\pi\nu t]d\nu$$

$$= \sum_{m=-n}^{n-1} F_m\left\{\int_{-\infty}^{\infty}\delta\left(\nu - \frac{m}{N\Delta t}\right)\exp[i2\pi\nu t]d\nu\right\}$$

$$= \sum_{m=-n}^{n-1} F_m \exp\left[i2\pi\frac{m}{N\Delta t}t\right]. \tag{8.106}$$

At $t = r\Delta t, f(t)$ is

$$f(r\Delta t) = a_r = \sum_{m=-n}^{n-1} F_m \exp\left[i\frac{2\pi m}{N}r\right]. \tag{8.107}$$

In summary, the equations for the discrete Fourier transform are (equations 8.104 and 8.107)

$$a_r = \sum_{m=-n}^{n-1} F_m \exp\left[i\frac{2\pi m}{N}r\right] \tag{8.108}$$

$$F_m = \frac{1}{N}\sum_{r=-n}^{n-1} a_r \exp\left[-i\frac{2\pi m}{N}r\right]. \tag{8.109}$$

It is often useful to represent both a_r and F_m as delta functions, in which case the discrete Fourier transform is written as

$$f(t) = \sum_{r=-n}^{n-1} a_r \delta(t - r\Delta t) \tag{8.110}$$

$$F(\nu) = \sum_{m=-n}^{n-1} F_m \delta\left(\nu - \frac{m}{N\Delta t}\right) \tag{8.111}$$

The values of F_m and a_r are still calculated from equations 8.108 and 8.109.

8.5.2 Derivation from the Orthogonality Relations for Discretely Sampled Sine and Cosine Functions

Consider the sum

$$S = \sum_{r=-n}^{n-1} \exp\left[i\frac{2\pi k}{N}r\right]^* \exp\left[i\frac{2\pi m}{N}r\right], \tag{8.112}$$

where $N = 2n$, and k and m are positive or negative integers or 0. To evaluate the sum, first perform a change of variables to $q = r + n$. Then we have

$$S = \sum_{q=0}^{2n-1} \exp\left[i\frac{2\pi k}{N}q\right]^* \exp\left[i\frac{2\pi m}{N}q\right] \exp[-i\pi k]^* \exp[-i\pi m]$$

$$= (-1)^{(m-k)} \sum_{q=0}^{2n-1} \exp\left[i\frac{2\pi(m-k)}{N}q\right], \tag{8.113}$$

where we have used $\exp[-i\pi(m-k)] = (-1)^{m-k}$. For $m = k$, the sum becomes

$$S = (-1)^0 \sum_{q=0}^{2n-1} \exp[0] = 2n = N. \tag{8.114}$$

To evaluate the case $m \neq k$, use the identity

$$\frac{1 - a^s}{1 - a} = 1 + a + a^2 + \cdots + a^{s-1}. \tag{8.115}$$

With this expansion, the sum becomes

$$S = (-1)^{(m-k)} \sum_{q=0}^{2n-1} \left(\exp\left[i\frac{2\pi(m-k)}{N}\right]\right)^q$$

$$= (-1)^{(m-k)} \frac{1 - \exp[i2\pi(m-k)]}{1 - \exp[i2\pi(m-k)/N]}. \tag{8.116}$$

For $m \neq k$, the numerator is zero. Thus we have derived the orthogonality relations

$$\sum_{r=-n}^{n-1} \exp\left[i\frac{2\pi k}{N}r\right]^* \exp\left[i\frac{2\pi m}{N}r\right] = \begin{cases} 0, & m \neq k \\ N, & m = k \end{cases} \tag{8.117}$$

$$= N\delta_{mk}.$$

If one is working with the complex representation of the Fourier series, the orthogonality relations expressed by Equation 8.117 are the only ones needed. If one is working with sines and cosines, the required orthogonally relations are

$$\sum_{r=-n}^{n-1} \sin\left(\frac{2\pi kr}{N}\right) \cos\left(\frac{2\pi mr}{N}\right) = 0 \tag{8.118}$$

$$\sum_{r=-n}^{n-1} \sin\left(\frac{2\pi kr}{N}\right) \sin\left(\frac{2\pi mr}{N}\right) = \begin{cases} 0, & k \neq m \\ 0, & k \text{ or } m = 0, n \\ N/2, & k = m \neq 0, n \end{cases} \tag{8.119}$$

$$\sum_{r=-n}^{n-1} \cos\left(\frac{2\pi kr}{N}\right) \cos\left(\frac{2\pi mr}{N}\right) = \begin{cases} 0, & k \neq m \\ N, & k = m = 0, n, \\ N/2, & k = m \neq 0, n \end{cases} \tag{8.120}$$

where, again, $N = 2n$. We derive only equation 8.119; the others can be derived in a similar way. There are three cases: k or m equals 0 or n; $k \neq m$; and $k = m$.

Case I (k or $m = 0, n$): One or the other of the sines in equation 8.119 is 0 for all r, so the sum is 0.

To evaluate the remaining two cases, express the sines in complex form:

$$\sum_{r=-n}^{n-1} \sin\left(\frac{2\pi kr}{N}\right) \sin\left(\frac{2\pi mr}{N}\right)$$

$$= \sum_{r=-n}^{n-1} \frac{1}{2i}\left\{\exp\left[i\frac{2\pi k}{N}r\right] - \exp\left[-i\frac{2\pi k}{N}r\right]\right\} \frac{1}{2i}\left\{\exp\left[i\frac{2\pi m}{N}r\right] - \exp\left[-i\frac{2\pi m}{N}r\right]\right\}$$

$$= -\frac{1}{4}\sum_{r=-n}^{n-1} \left\{\exp\left[i\frac{2\pi k}{N}r\right]\exp\left[i\frac{2\pi m}{N}r\right] - \exp\left[i\frac{2\pi k}{N}r\right]\exp\left[-i\frac{2\pi m}{N}r\right]\right.$$

$$\left. - \exp\left[-i\frac{2\pi k}{N}r\right]\exp\left[i\frac{2\pi m}{N}r\right] + \exp\left[-i\frac{2\pi k}{N}r\right]\exp\left[-i\frac{2\pi m}{N}r\right]\right\}. \tag{8.121}$$

Each of the four terms in this equation has a form that allows us to directly apply equation 8.117.

Case II ($m \neq k$): From equation 8.117 the sums are all identically 0.

Case III ($m = k \neq 0, n$): Because of the complex conjugate in equation 8.117, the only terms in equation 8.121 that contribute are the second and third, in which k and m have opposite signs. The sum over these two terms is $-N$, so the total sum is $-(1/4)(0 - N - N - 0) = N/2$.

We now take as given that the sines and cosines in equations 8.118–8.120 make up a complete set. The discrete samples a_r can therefore be represented by the sum

$$a_r = A_0 + 2\sum_{m=1}^{n} \left\{A_m \cos\left(\frac{2\pi mr}{N}\right) + B_m \sin\left(\frac{2\pi mr}{N}\right)\right\}. \tag{8.122}$$

Note that for $m = n$, $\sin(2\pi mr/N) = \sin(\pi r) = 0$, so B_n is irrelevant. Thus equation 8.122 uses N sines and cosines and their N coefficients to represent the N values of a_r.

To find a particular B_k, multiply equation 8.122 by $\sin(2\pi kr/N)$ and sum over r:

$$\sum_{r=-n}^{n-1} a_r \sin\left(\frac{2\pi kr}{N}\right) = \sum_{r=-n}^{n-1} A_0 \sin\left(\frac{2\pi kr}{N}\right)$$
$$+ \sum_{r=-n}^{n-1} \left[2 \sum_{m=1}^{n} \left\{ A_m \cos\left(\frac{2\pi mr}{N}\right) \right. \right.$$
$$\left. \left. + B_m \sin\left(\frac{2\pi mr}{N}\right) \right\} \right] \sin\left(\frac{2\pi kr}{N}\right). \quad (8.123)$$

The term with A_0 on the right-hand side of equation 8.123 is essentially equation 8.118 with $m = 0$, so that term equals 0. Reverse the order of the summation in the remaining terms, and apply equations 8.118 and 8.119 to find

$$\sum_{r=-n}^{n-1} a_r \sin\left(\frac{2\pi kr}{N}\right) = 2 \sum_{m=1}^{n} \left[\sum_{r=-n}^{n-1} A_m \cos\left(\frac{2\pi mr}{N}\right) \sin\left(\frac{2\pi kr}{N}\right) \right]$$
$$+ 2 \sum_{m=1}^{n} \left[\sum_{r=-n}^{n-1} B_m \sin\left(\frac{2\pi mr}{N}\right) \sin\left(\frac{2\pi kr}{N}\right) \right]$$
$$= 2 \sum_{m=1}^{n} B_m \left(\frac{N}{2}\right) \delta_{km} \quad (8.124)$$
$$= NB_k. \quad (8.125)$$

To find a_k, multiply equation 8.122 by $\cos(2\pi kr/N)$ and sum over r. In summary, the equations for the discrete Fourier transform for A_k and B_k are

$$a_r = A_0 + 2 \sum_{m=1}^{n} \left\{ A_m \cos\left(\frac{2\pi mr}{N}\right) + B_m \sin\left(\frac{2\pi mr}{N}\right) \right\} \quad (8.126)$$

$$A_k = \frac{1}{N} \sum_{r=-n}^{n-1} a_r \cos\left(\frac{2\pi kr}{N}\right) \quad (8.127)$$

$$B_k = \frac{1}{N} \sum_{r=-n}^{n-1} a_r \sin\left(\frac{2\pi kr}{N}\right), \quad (8.128)$$

where $N = 2n$. Note that A_0 is the mean value of a_r, and that B_0 and B_n are always 0. With a small amount of extra work, one can set $F_k = A_k - iB_k$ and convert equations 8.126–8.128 to the complex equations 8.108 and 8.109.

8.5.3 Parseval's Theorem and the Power Spectrum

It is easiest to derive Parseval's theorem for discrete Fourier transforms by starting with equation 8.109. The sum of $|F_m|^2$ over m is

$$\sum_{m=-n}^{n-1} |F_m|^2 = \sum_{m=-n}^{n-1} \left\{ \frac{1}{N^2} \left(\sum_{r=-n}^{n-1} a_r^* \exp\left[i\frac{2\pi m}{N}r \right] \right) \left(\sum_{q=-n}^{n-1} a_q \exp\left[-i\frac{2\pi m}{N}q \right] \right) \right\}$$

$$= \frac{1}{N^2} \sum_{r=-n}^{n-1} \sum_{q=-n}^{n-1} a_r^* a_q \left\{ \sum_{m=-n}^{n-1} \exp\left[i\frac{2\pi m}{N}r \right] \exp\left[-i\frac{2\pi m}{N}q \right] \right\}. \qquad (8.129)$$

From the orthogonality relation (equation 8.117), the expression inside the curly brackets is $N\delta_{rq}$, so equation 8.129 becomes

$$\sum_{m=-n}^{n-1} |F_m|^2 = \frac{1}{N^2} \sum_{r=-n}^{n-1} \sum_{q=-n}^{n-1} a_r^* a_q N\delta_{rq} = \frac{1}{N} \sum_{r=-n}^{n-1} |a_r|^2, \qquad (8.130)$$

which is Parseval's theorem.

If the a_r are all real, then $|F_m|^2 = |F_{-m}|^2 = A_m^2 + B_m^2$, and Parseval's theorem can be written as

$$A_0^2 + 2\sum_{m=1}^{n} \left(A_m^2 + B_m^2 \right) = \frac{1}{N} \sum_{r=-n}^{n-1} a_r^2. \qquad (8.131)$$

The power at frequency m in the one-sided spectrum is

$$P_m = \begin{cases} A_0^2, & m = 0 \\ 2(A_m^2 + B_m^2), & 1 \leq m \leq N/2 \end{cases}. \qquad (8.132)$$

If only frequency m contributes to the sequence, equation 8.126 becomes

$$a_r = 2A_m \cos\left(\frac{2\pi mr}{N} \right) + 2B_m \sin\left(\frac{2\pi mr}{N} \right), \qquad (8.133)$$

so the amplitude at frequency m is (compare to equation 8.35)

$$\alpha_m = \sqrt{4(A_m^2 + B_m^2)} = \sqrt{2P_m}. \qquad (8.134)$$

Equation 8.131 leads to an interesting alternative interpretation of Parseval's theorem. Note that

$$\frac{1}{N} \sum_{r=-n}^{n-1} a_r^2 = \langle a^2 \rangle, \qquad (8.135)$$

and also note that

$$A_0 = \frac{1}{N} \sum_{r=-n}^{n-1} a_r \cos(0) = \frac{1}{N} \sum_{r=-n}^{n-1} a_r = \langle a \rangle. \qquad (8.136)$$

Equation 8.131 can therefore be put in the form

$$2 \sum_{m=1}^{n} \left(A_m^2 + B_m^2 \right) = \langle a^2 \rangle - \langle a \rangle^2 \tag{8.137}$$

$$\sum_{m=1}^{n} P_m = \sigma_a^2. \tag{8.138}$$

Thus, the sum of all the power at frequencies greater than 0 is equal to the variance of the original signal. This result will be useful for interpreting power spectra of noise.

8.6 Convolution and the Convolution Theorem

8.6.1 Convolution

If $f(t)$ and $g(t)$ are two functions, their *convolution* $y(\tau)$ is defined to be

$$y(\tau) = \int_{-\infty}^{\infty} f(t)g(\tau - t)dt. \tag{8.139}$$

Note that the integral has $g(\tau - t)$, not $g(\tau + t)$, so $g(t)$ enters the convolution integral reversed in sign. At $\tau = 0$, it enters the integral mirrored about the origin. Also note that there is no complex conjugation, even if $f(t)$ and $g(t)$ are complex. If f_j and g_j are discrete sequences each sampled at the same uniform interval, their convolution is

$$y_n = \sum_{j=-\infty}^{\infty} f_j g_{n-j}. \tag{8.140}$$

If desired, this can be put into matrix notation. Let the components of the $\mathfrak{G}$ matrix be

$$(\mathfrak{G})_{nj} = \mathfrak{g}_{nj} = g_{n-j}. \tag{8.141}$$

In effect, each row of $\mathfrak{G}$ repeats g_j in reversed order and offset by n elements. Equation 8.140 can be written as $y_n = \sum_j \mathfrak{g}_{nj} f_j$ or, in matrix notation,

$$\mathbf{y} = \mathfrak{G}\mathbf{f}, \tag{8.142}$$

where $\mathbf{y}$ and $\mathbf{f}$ are vectors with elements y_n and f_j.

Despite appearances, convolution is symmetric between $f(t)$ and $g(t)$. To see this, perform a change of variables from t to $u = \tau - t$. Since $dt = -du$, we have

$$y(\tau) = \int_{t=-\infty}^{\infty} f(t)g(\tau - t)dt = -\int_{u=\infty}^{-\infty} f(\tau - u)g(u)du$$

$$= \int_{u=-\infty}^{\infty} f(\tau - u)g(u)du, \tag{8.143}$$

and the symmetry is manifest. It is the time reversal that produces the symmetry. The cross-covariance function, which is discussed in Chapter 10, is essentially a convolution without the time reversal. It is not symmetric between the two input functions.

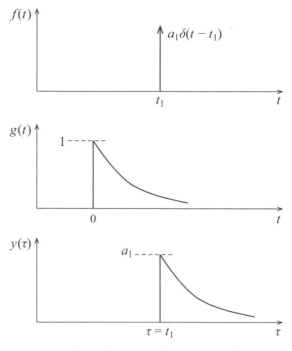

Figure 8.13: The delta function $a_1\delta(t-t_1)$ (top panel), the exponential function $\exp[-t]$, $t \geq 0$. (middle panel), and the convolution of the two (bottom panel). The exponential has been multiplied by a_1 and its origin has been shifted to $\tau = t_1$.

Let $f(t)$ be the linear combination of two other functions:

$$f(t) = a_1 f_1(t) + a_2 f_2(t). \tag{8.144}$$

The convolution of $f(t)$ with $g(t)$ is

$$y(\tau) = \int_{-\infty}^{\infty} \left[a_1 f_1(t) + a_2 f_2(t) \right] g(\tau - t) dt$$

$$= a_1 \int_{-\infty}^{\infty} f_1(t) g(\tau - t) dt + a_2 \int_{-\infty}^{\infty} f_1(t) g(\tau - t) dt. \tag{8.145}$$

Convolution is therefore a linear operation. Actually it is a bilinear operation, since it is linear in both $f(t)$ and $g(t)$.

There are several ways to understand convolution. If $f(t)$ is nonzero over only a small range, convolution approximately replicates $g(t)$, translating the replica to the place where $f(t)$ is nonzero. Figure 8.13 shows the extreme case when $f(t)$ is the delta function

$$f(t) = a_1 \delta(t - t_1), \tag{8.146}$$

and $g(t)$ is the exponential

$$g(t) = \begin{cases} 0, & t < 0 \\ \exp[-t], & t \geq 0 \end{cases}. \tag{8.147}$$

Their convolution is

$$y(\tau) = \int_{-\infty}^{\infty} f(t)g(\tau - t)dt = \int_{-\infty}^{\tau} a_1 \delta(t - t_1) \exp[-(\tau - t)]dt$$

$$= \begin{cases} 0, & \tau < t_1 \\ a_1 \exp[-(\tau - t_1)], & \tau \geq t_1 \end{cases}, \tag{8.148}$$

which is $g(t)$ multiplied by a_1 and translated in time to the position of the delta function. It is worth noting that even though $g(t)$ enters the convolution integral time reversed, it shows up in $y(\tau)$ unreversed.

If $f(t)$ is a continuous function, one can think of convolution as replicating $g(t)$ many times and then adding all the replicas together. To see this, divide t into small contiguous intervals with width Δt, as shown in Figure 8.14. The integral of $f(t)$ over Δt at t_i is approximately $f(t_i)\Delta t$. Each interval contributes an amount

$$y(\tau_i) \approx g(\tau - t_i)f(t_i)\Delta t \tag{8.149}$$

to the convolution integral, so the convolution is approximately

$$y(\tau) \approx \sum_i g(\tau - t_i)f(t_i)\Delta t. \tag{8.150}$$

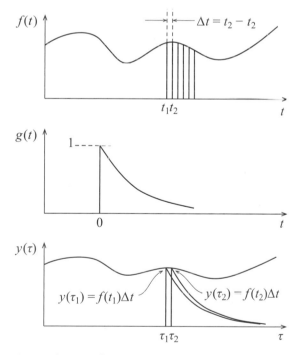

Figure 8.14: The continuous function $f(t)$ (top panel). The integral of $f(t)$ over a small interval Δt at t_i is approximately $f(t_i)\Delta t$. An exponential function (middle panel). The convolution of $f(t)$ with the exponential (bottom panel) is approximately the sum of many exponentials, each multiplied by $f(t_i)\Delta t$ and displaced to t_i.

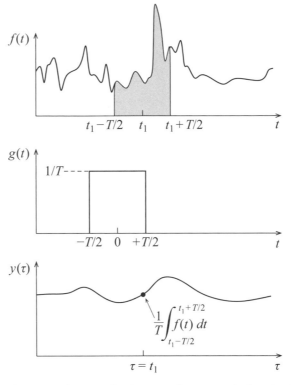

Figure 8.15: The rapidly varying function $f(t)$ (top panel), a rectangle function (middle panel), and the convolution of the two (bottom panel). The value of the convolution at $\tau = t_1$ is equal to the mean value of $f(t)$ over the interval centered on t_1 with width T.

Figure 8.14 shows how this works when $g(t)$ is the exponential function. In the limit as $\Delta t \rightarrow 0$ the sum becomes the convolution of $g(t)$ and $f(t)$.

Convolution can also be thought of as a smoothing process or a running average. Suppose $f(t)$ is a rapidly varying function, and $g(t)$ is the rectangle function with width T and height $1/T$ (see Figure 8.15). At the specific point $\tau = t_1$ the convolution of the two is

$$y(\tau = t_1) \; = \; \frac{1}{T} \int_{t_1 - T/2}^{t_1 + T/2} f(t) dt, \tag{8.151}$$

which is the average of $f(t)$ over an interval centered on t_1 with width T.

It is convenient to introduce a new notation. Let the symbol $\otimes$ stand for convolution, so that the convolution integral can be written compactly as

$$y(\tau) = \int_{-\infty}^{\infty} f(t) g(\tau - t) dt$$

$$= f(t) \otimes g(t). \tag{8.152}$$

Because convolution is symmetric between $f(t)$ and $g(t)$, there is no need to specify which function is reversed in time. The linearity of convolution can now be written as

$$[a_1 f_1(t) + a_2 f_2(t)] \otimes g(t) \; = \; a_1 f_1(t) \otimes g(t) + a_2 f_2(t) \otimes g(t). \tag{8.153}$$

Chapters 9 and 10 have many specific examples of convolutions, but several cases are worth mentioning now.

Convolution of Two Gaussians: Let $f(t)$ and $g(t)$ be two Gaussian functions with variances σ_1^2 and σ_2^2:

$$f(t) = \frac{1}{\sqrt{2\pi}\sigma_1} \exp\left[-\frac{1}{2}\frac{t^2}{\sigma_1^2}\right] \tag{8.154}$$

$$g(t) = \frac{1}{\sqrt{2\pi}\sigma_2} \exp\left[-\frac{1}{2}\frac{t^2}{\sigma_2^2}\right]. \tag{8.155}$$

Their convolution is also a Gaussian (see Section C.3 Appendix C), and its variance is $\sigma_1^2 + \sigma_2^2$:

$$y(\tau) = f(t) \otimes g(t) = \frac{1}{\sqrt{2\pi}(\sigma_1^2 + \sigma_2^2)^{1/2}} \exp\left[-\frac{1}{2}\frac{\tau^2}{\sigma_1^2 + \sigma_2^2}\right]. \tag{8.156}$$

Convolution of Two Rectangle Functions: Let $f(t)$ and $g(t)$ be the rectangle functions

$$f(t) = \begin{cases} 1, & -a \le t \le a \\ 0, & |t| > a \end{cases} \tag{8.157}$$

$$g(t) = \begin{cases} 1, & -b \le t \le b \\ 0, & |t| > b \end{cases}. \tag{8.158}$$

Their convolution is the trapezoid function shown in Figure 8.16. If the two rectangles are the same width ($a = b$), the trapezoid becomes a triangle with width $2a$.

Convolution of a Sine Curve with a Rectangle Function: Let $f(t)$ be the rectangle function

$$f(t) = \begin{cases} 1/2a, & -a \le t \le a \\ 0, & |t| > a \end{cases}, \tag{8.159}$$

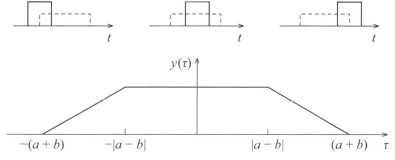

Figure 8.16: The convolution of two rectangle functions with widths a and b is the trapezoid function (bottom panel). The three figures (top panel) show the orientation of the two rectangle functions corresponding to the three parts of the trapezoid function. If $a = b$ the trapezoid becomes a triangle with width $2a$.

and let $g(t) = \sin(\omega t)$. Their convolution is

$$y(\tau) = \frac{1}{2a} \int_{-a}^{a} \sin \omega(\tau - t) dt$$

$$= \frac{1}{2a} \int_{-a}^{a} [\sin(\omega \tau) \cos(\omega t) - \cos(\omega \tau) \sin(\omega t)] \, dt. \tag{8.160}$$

The integral of the second term inside the square brackets is 0, because $\sin(\omega t)$ is antisymmetric about $t = 0$, so we have

$$y(\tau) = \frac{1}{2a} \sin(\omega \tau) \int_{-a}^{a} \cos(\omega t) dt = \frac{1}{2a} \sin(\omega \tau) \left. \frac{1}{\omega} \sin(\omega t) \right|_{-a}^{a}$$

$$= \frac{\sin(\omega a)}{\omega a} \sin(\omega \tau). \tag{8.161}$$

This is a sine curve with the same frequency as the original sine curve but with amplitude given by the sinc function $\sin(\omega a)/\omega a$.

8.6.2 Convolution Theorem

In Section 8.4 we saw that the Fourier transform of a rectangle times a sine curve is a sinc function (see equation 8.87); the previous example showed that the convolution of a rectangle with a sine curve gives a sine curve whose amplitude is a sinc function. This suggests a connection between the Fourier transform and convolution. The precise connection is given by the convolution theorem, which we now derive. Let two functions $f_1(t)$ and $f_2(t)$ have Fourier transforms $F_1(v)$ and $F_2(v)$, respectively. The product of the Fourier transforms is (no complex conjugation)

$$F_1(v)F_2(v) = \left[\int_{t=-\infty}^{\infty} f_1(t) \exp[-i2\pi vt] dt \right] \left[\int_{u=-\infty}^{\infty} f_2(u) \exp[-i2\pi vu] du \right]$$

$$= \int_{t=-\infty}^{\infty} \int_{u=-\infty}^{\infty} f_1(t) f_2(u) \exp[-i2\pi v(t + u)] dt du. \tag{8.162}$$

Perform a change of variables from (t, u) to (t, τ), where $\tau = t + u$. The transformation from $dt du$ to $dt d\tau$ is given by the Jacobian

$$dt du = \left| \frac{\partial(t, u)}{\partial(t, \tau)} \right| dt d\tau = \begin{vmatrix} 1 & -1 \\ 0 & 1 \end{vmatrix} dt d\tau = dt d\tau, \tag{8.163}$$

so the integral becomes

$$F_1(v)F_2(v) = \int_{\tau=-\infty}^{\infty} \left[\int_{t=-\infty}^{\infty} f_1(t) f_2(\tau - t) dt \right] \exp[-i2\pi v\tau] d\tau$$

$$= \int_{\tau=-\infty}^{\infty} [f_1(t) \otimes f_2(t)] \exp[-i2\pi v\tau] d\tau. \tag{8.164}$$

This remarkable equation says that the product of the Fourier transforms of two functions is equal to the Fourier transform of the convolution of the two functions. By similar logic one can derive the equally remarkable result:

$$F_1(v) \otimes F_2(v) = \int F_1(v')F_2(v-v')dv' = \int_{t=-\infty}^{\infty} f_1(t)f_2(t)\exp[-i2\pi vt]dt. \quad (8.165)$$

This equation says that the convolution of the Fourier transforms of two functions is equal to the Fourier transform of the product of the two functions. Equations 8.165 and 8.164 are alternative forms of the convolution theorem.

9

Analysis of Sequences: Power Spectra and Periodograms

9.1 Introduction

A sequence is an ordered set of objects. The spectrum of starlight is a sequence in which monochromatic flux is ordered by wavelength or frequency. The bases in a DNA molecule form a sequence, as do the exits on the Interstate 10 highway ordered by distance from its eastern terminus. One of the most common types of sequences is the time series, in which some quantity is arranged in order of increasing time. The variations of a star's brightness, the speed of a guided missile, and the record of the S&P 500 stock index are time series. These are all one-dimensional sequences, but multidimensional sequences are also common. A digital image is an obvious example: the pixels must be ordered correctly in two dimensions for the image to make sense. Multidimensional data can often be thought of as sequences of sequences. Thus a rectangular digital image can be thought of as an ordered sequence of lines, each of which is a sequence of intensities. A map of the cosmic microwave background across the sky is a set of sequences in which microwave fluxes are ordered along one angle (perhaps right ascension), and the sequences themselves are ordered along a second angle (perhaps declination). In this case the ends of the sequences join smoothly onto their beginnings.

This chapter introduces the analysis of sequences. The primary tools are power spectra and periodograms, both of which are useful for analyzing sequences containing periodic and nearly periodic signals. Chapter 10 moves on to convolution and then autocovariance and cross-covariance functions, which are more useful for analyzing sequences with nonperiodic or quasiperiodic signals.

9.2 Continuous Sequences: Data Windows, Spectral Windows, and Aliasing

Power spectra are clearly useful for analyzing sequences dominated by periodic signals. They are also useful for analyzing signals with periods that vary irregularly about some mean period. They are even useful for analyzing sequences that are nonperiodic if the sequences vary on some preferred time or length scale. As a result, power spectra are one of the principal tools for analyzing sequences. We begin our discussion of power spectra with the

Figure 9.1: The middle panel shows a sequence $s(t)$. The observed sequence $f(t)$ (left panel) can be thought of as the product of $s(t)$ with a rectangular data window $w(t)$ extending from t_1 to t_2 (right panel)

calculation of the power spectra of noise-free sequences. For convenience, we will generally use t and ν for the Fourier transform pairs, implying time and frequency, but as noted earlier, other interpretations are possible, most commonly length and wavenumber.

9.2.1 Data Windows and Spectral Windows

Let $s(t)$ be a continuous sequence whose power spectrum we wish to measure. In general only a portion of $s(t)$ has been sampled, not the entire sequence, and the way it has been sampled has a large effect on the measured power spectrum. Figure 9.1 shows a common situation. The sequence $s(t)$ is known only over the interval from t_1 to t_2. The observed sequence $f(t)$ can be thought of as the product of $s(t)$ with a rectangular function $w(t)$ extending from t_1 to t_2:

$$w(t) = \begin{cases} 0, & t < t_1 \\ 1, & t_1 \leq t \leq t_2 \\ 0, & t_2 < t \end{cases} . \tag{9.1}$$

The function $w(t)$ is called a *data window*. The Fourier transform of the observed sequence can be evaluated using the convolution theorem (see Section 8.6.2):

$$F(\nu) = \int f(t) \exp[-i2\pi \nu t] dt = \int s(t) w(t) \exp[-i2\pi \nu t] dt$$
$$= S(\nu) \otimes W(\nu), \tag{9.2}$$

where $S(\nu)$ and $W(\nu)$ are the Fourier transforms of $s(t)$ and $w(t)$.[1] The Fourier transform of the data window is called the *spectral window*. Equation 9.2 says that the Fourier transform of the observed sequence is the convolution of the spectral window with the Fourier transform of original, infinite-length sequence.

For a concrete example, suppose the cosine function $s(t) = \cos(2\pi \nu_0 t)$ has been observed over the interval $-b/2 \leq t \leq b/2$. The Fourier transform of the cosine function is the pair of delta functions located at ν_0 and $-\nu_0$ (see equation 8.89):

$$S(\nu) = \frac{1}{2}\delta(\nu - \nu_0) + \frac{1}{2}\delta(\nu + \nu_0), \tag{9.3}$$

[1] Equation 9.2 uses a shorthand notation for what is really a two-step process: first the convolution $F(\tau) = S(\nu) \otimes W(\nu)$ and then a remapping of $F(\tau) \to F(\nu)$ by setting $\tau = \nu$. The shorthand should cause no confusion and is used throughout this chapter.

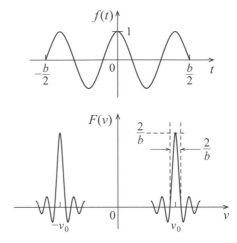

Figure 9.2: The function $f(t)$ is the product of the cosine function $\cos(2\pi \nu_0 t)$ with a rectangular data window extending from $t = -b/2$ to $t = b/2$ (top panel). The Fourier transform of the cosine function is two delta functions, one at $\nu = -\nu_0$ and one at $\nu = \nu_0$. The Fourier transform of the rectangular window function is a sinc function. The Fourier transform of $f(t)$ is the convolution of the two delta functions with the sinc function, producing two sinc functions, one centered at $\nu = -\nu_0$ and one at $\nu = \nu_0$ (bottom panel).

and the Fourier transform of the rectangle function is a sinc function (see equation 8.76). The Fourier transform of $f(t)$ is then

$$
\begin{aligned}
F(\nu) &= S(\nu) \otimes W(\nu) \\
&= [\frac{1}{2}\delta(\nu - \nu_0) + \frac{1}{2}\delta(\nu + \nu_0)] \otimes b\frac{\sin(\pi \nu b)}{\pi \nu b} \\
&= \frac{b}{2}\frac{\sin(\pi[\nu - \nu_0]b)}{\pi[\nu - \nu_0]b} + \frac{b}{2}\frac{\sin(\pi[\nu + \nu_0]b)}{\pi[\nu + \nu_0]b}.
\end{aligned}
\tag{9.4}
$$

Figure 9.2 shows $f(t)$ and $F(\nu)$. The effect of sampling a finite length of $s(t)$ is to broaden the two delta functions into two sinc functions whose full widths to the first zeros are $\Delta \nu = 2/b$. If $\nu_0 \gg \Delta \nu$ so that the two sinc functions are well separated, the power spectrum becomes two squared sinc functions. The one at positive frequency is

$$
P(\nu) = |F(\nu)|^2 = \frac{b^2}{4}\left\{\frac{\sin(\pi[\nu - \nu_0]b)}{\pi[\nu - \nu_0]b}\right\}^2
\tag{9.5}
$$

and is shown in Figure 9.3. The power spectrum has a central peak centered on ν_0 and a series of peaks called side lobes extending away symmetrically on both sides of the central peak. The heights of the first and second side lobes are 4.7% and 1.6% of the height of the central peak, respectively. The heights of the remaining peaks decrease roughly as $1/(\nu - \nu_0)^2$.

The rectangle data window has several deleterious effects on the power spectrum. First, the resolution of the transform is reduced. Signals at two different frequencies may no longer be distinguished as two distinct peaks in the power spectrum. Let $f(t)$ be the sum of two cosines with frequencies ν_1 and ν_2 multiplied by a rectangular data window $w(t)$:

$$
f(t) = w(t)[\cos(2\pi \nu_1 t) + \cos(2\pi \nu_2 t)].
\tag{9.6}
$$

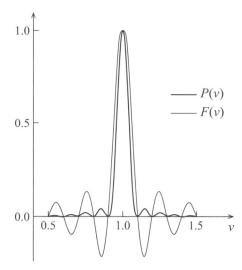

Figure 9.3: The thin line is the Fourier transform $F(\nu)$ of the cosine function $\cos(2\pi \nu t)$ with $\nu = 1$ times a rectangular data window extending from $t = -5$ to $t = 5$. The thick line is the power spectrum $P(\nu) = |F(\nu)|^2$. Both have been normalized to unity at $\nu = 1$.

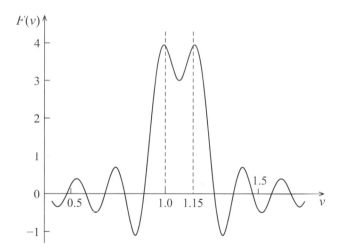

Figure 9.4: The solid line is the Fourier transform of the sum of two cosines $f(t) = \cos(2\pi \nu_1 t) + \cos(2\pi \nu_2 t)$ with $\nu_1 = 1.0$ and $\nu_2 = 1.15$, multiplied by a rectangular data window extending from $t = -5$ to $t = 5$. The two vertical dashed lines mark the frequencies of the two cosines. The two frequencies are barely resolved, and the peaks are not at the original frequencies.

The Fourier transform of $f(t)$ is the sum of two sinc functions, one centered on ν_1 and one on ν_2. Figure 9.4 shows the Fourier transform for the specific case $\nu_1 = 1.0$ and $\nu_2 = 1.15$, and for a rectangular data window extending from $t = -5$ to $t = 5$. The two sinc functions merge together with only a shallow dip between them. For $\nu_2 = 1.13$ there is no dip at all. The two vertical dashed lines in the figure mark ν_1 and ν_2. Note that the peaks are no longer at precisely the frequencies of the two constituent cosine functions.

A second effect of the data window is to introduce spectral leakage: power in a signal at one frequency spreads to other frequencies. Spectral leakage from a strong signal can distort

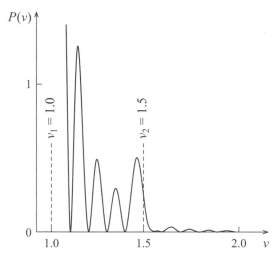

Figure 9.5: The power spectrum of the function $f(t) = \sin(2\pi v_1 t) + 0.1 \sin(2\pi v_2 t)$ with $v_1 = 1.0$, $v_2 = 1.5$, multiplied by a rectangular data window $w(t)$ extending from $t = -5$ to 5. The Fourier transform is the sum of two sinc functions, one centered at $v = 1.0$ and the other at $v = \pm 1.5$. The peak in the power spectrum at v_1 reaches $F(v_1) = 25$ but is truncated to save space. The peak at $v = 1.5$ is swamped by spectral leakage and betrays its presence only by the distortion of the side lobes of the first cosine.

or even obliterate weak signals at nearby frequencies. Figure 9.5 demonstrates this effect. It shows the power spectrum of the function

$$f(t) = w(t)[1.0\cos(2\pi v_1 t) + 0.1\cos(2\pi v_2 t)], \tag{9.7}$$

for $v_1 = 1.0$, $v_2 = 1.5$, and a rectangular data window extending from $t = -5$ to $t = 5$. The Fourier transform of $f(t)$ is the sum of sinc functions at $v = 1.0$ and $v = 1.5$. The sinc function at $v = 1.5$ is swamped by spectral leakage and betrays its presence only by the distortion of the side lobes of the sinc function at $v = 1.0$.

One can ameliorate the effect of the data window in a variety of ways. Perhaps the most important way is to redesign the experiment! For example, the width of the spectral window is inversely proportional to the length of the data sequence. Taking enough data to double the length of the sequence improves the spectral resolution by a factor of two.

Although spectral leakage is an inevitable result of a finite-length data window, the amount of leakage and the amplitude of the side lobes depends on the shape of the data window. The rectangular data window is a particularly bad offender because of the discontinuities at both of its ends. Data windows without discontinuities have less leakage and smaller side lobes. Figure 9.6 compares a cosine curve multiplied by a rectangular data window with one multiplied by the triangular window:

$$w(t) = \begin{cases} 1 - \dfrac{|t|}{b/2} & , |t| \le b/2 \\ 0 & , |t| > b/2 \end{cases}. \tag{9.8}$$

The amplitude of the cosine curve multiplied by the triangular data window tapers smoothly to 0, so there are no discontinuities. The figure also compares the power spectra of the two sequences. The power spectrum for the square window is a sinc function squared, but the

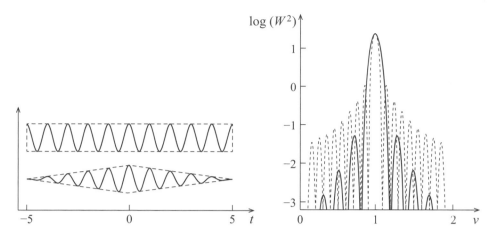

Figure 9.6: The left panel shows the cosine curve $s(t) = \cos(2\pi t)$ multiplied by the rectangular data window $w(t) = 1$, $|t| \leq 5$ and by the triangle data window $w(t) = 1 - |t|/5$. The right panel shows the power spectra of the two sequences, the narrow dashed curve corresponding to the square data window and the solid curve to the triangular data window.

power spectrum for the triangle window is the sinc function to the fourth power:

$$P(\nu) = |W(\nu)|^2 = \frac{b^2}{4}\left[\frac{\sin(\pi\nu b/2)}{\pi\nu b/2}\right]^4. \tag{9.9}$$

The side lobes in the power spectrum for the triangular data window fall off much more rapidly than the side lobes for the square window. Even the first side lobe has been reduced by more than an order of magnitude. But the factor of $b/2$ instead of b means that the resolution has been degraded by a factor of two: the central peak of the triangular spectral window is twice the width of the central peak in the square window.

Thus another way to reduce spectral leakage is to alter the shape of the data window. A square data window can easily be converted to a triangular data window merely by applying a linear taper to the data. A triangular data window is far from the best choice, though, because the discontinuities in its derivatives (the corners at the center and ends of the window) contribute to leakage and ringing. Many data windows have been proposed. One of the best is the "split cosine bell" data window, which smoothly tapers the end of the data window with a cosine function. The window, shown in Figure 9.7, is defined by

$$w(t) = \begin{cases} 0.5 - 0.5\cos\left(\pi\dfrac{b/2+t}{[1-\alpha]b/2}\right), & -b/2 \leq t < -\alpha b/2 \\[2mm] 1, & -\alpha b/2 \leq t \leq \alpha b/2 \\[2mm] 0.5 - 0.5\cos\left(\pi\dfrac{b/2-t}{[1-\alpha]b/2}\right), & \alpha b/2 < t \leq b/2 \end{cases}, \tag{9.10}$$

where α sets the width of the taper, $\alpha = 0.2$ giving a 10% taper at both ends. All derivatives of this window are continuous, minimizing ringing. The value of α can be chosen for the particular problem at hand, a smaller α yielding less leakage but poorer resolution.

A third way to deal with spectral leakage is to "pre-whiten" the spectrum. Pre-whitening is the process of subtracting a large signal from the original sequence, thus removing its leaked power from the power spectrum and allowing weaker signals to be detected.

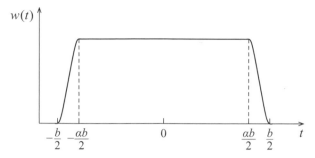

Figure 9.7: The split cosine data window (equation 9.10). The window shown here has $\alpha = 0.2$, giving a 10% taper at each end of the window.

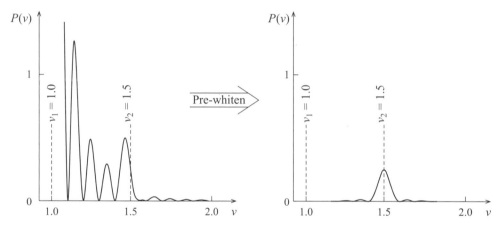

Figure 9.8: The left panel is a reproduction of Figure 9.5. It shows the power spectrum of the sum of two cosine curves, one at $\nu = 1.0$ and the other at $\nu = 1.5$ with an amplitude equal to 1/10 the amplitude of the first. The right panel shows the power spectrum after the cosine at $\nu = 1.0$ has been subtracted from the sequence. The lower-amplitude cosine is now easily visible.

Figure 9.5 is reproduced in the left panel of Figure 9.8. It shows the power spectrum of the sum of two cosine curves, one at $\nu = 1.0$ and the second at $\nu = 1.5$ with an amplitude that is 1/10 the amplitude of the first. The right panel of the figure shows the power spectrum after the cosine at $\nu = 1.0$ has been subtracted from the sequence. The lower-amplitude cosine is now easily visible. In practice, one often calculates the power spectrum to identify strong periodic signals. The light curve is pre-whitened by subtracting the strong signals, and then the power spectrum is recalculated to search for weaker signals.

Suppose that a sequence is the sum of a cosine and a constant,

$$s(t) = A + B\cos(2\pi \nu_0 t), \tag{9.11}$$

and the data window is a rectangle from $t = -b/2$ to $t = b/2$. The Fourier transform of the cosine is the sum of two sinc functions at $\pm \nu_0$ with peak amplitudes $Bb/2$ (see equation 9.4) plus a third sinc function at zero frequency:

$$F(\nu) = Ab\frac{\sin(\pi \nu b)}{\pi \nu b}. \tag{9.12}$$

The sinc function at zero frequency has spectral leakage with side lobes just like any other sinc function. In practice, A is often much larger than B, so the leakage from zero frequency can greatly alter the power spectrum. This zero-frequency signal must always be removed by pre-whitening, which in this case just means subtracting the mean value of the sequence. Slow trends and other slow variations in a sequence can be equally damaging, as they can introduce discontinuities at the ends of a sequence, introducing ringing in the transform. One often subtracts a low-order polynomial fit to the sequence.

In summary, the minimum preprocessing necessary to calculate a power spectrum of a sequence is to subtract means and trends from the sequence and to taper the sequence, in effect tapering the data window. Further processing by pre-whitening may also be useful if the signal is multiperiodic.

9.2.2 Aliasing

Aliasing is an extreme case of spectral leakage in which the power leaked to one or more other frequencies becomes comparable to the power at the true frequency. Figure 9.9 shows a typical way aliases can arise. The true sequence is the cosine function

$$s(t) = \cos(2\pi \nu_0 t). \tag{9.13}$$

The sequence is sampled twice at widely separated times, as shown by the solid curves labeled "Observed" between the vertical lines in the figure. The dashed lines correspond to the inferred sequence determined by extrapolating the observed sequences into the unobserved interval. Several different inferred sequences fit the data almost equally well, differing from one another by an integral number of cycles in the unobserved part of the

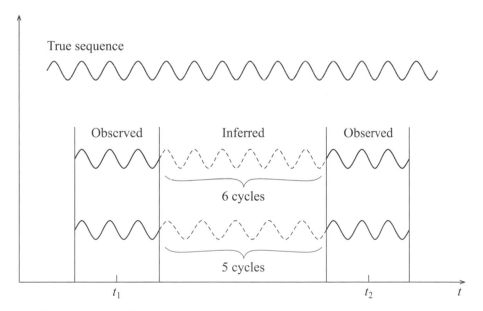

Figure 9.9: A sequence $s(t) = \cos(2\pi \nu_0 t)$ is sampled twice at widely separated times t_1 and t_2 shown by the solid curves labeled "Observed." The dashed lines correspond to the inferred sequence determined by extrapolating the observed sequences into the unobserved interval. Two possible inferred sequences are shown, one with five cycles in the unobserved part of the sequence, the other with six cycles. The two possible inferred periods are aliases of each other.

sequence. Two possible inferred sequences are shown in the figure, one with five cycles in the unobserved part of the sequence, the other with six cycles. These translate into different inferred periods for the cosine function. The two possible inferred periods are aliases of each other.

This case is simple enough to be analyzed analytically. The observed sequence $f(t)$ is the product of $s(t)$ with the data window $w(t)$:

$$f(t) = s(t)w(t), \tag{9.14}$$

where the data window consists of two rectangle functions centered at t_1 and t_2. We represent the window by a rectangle convolved with two delta functions:

$$w(t) = r(t) \otimes [\delta(t - t_1) + \delta(t - t_2)] \tag{9.15}$$

$$r(t) = \begin{cases} 1, & |t| \leq b/2 \\ 0, & |t| > b/2 \end{cases}. \tag{9.16}$$

From the convolution theorem, the Fourier transform of the observed sequence is

$$F(\nu) = S(\nu) \otimes W(\nu), \tag{9.17}$$

where $S(\nu) = \delta(\nu - \nu_0)$, and $W(\nu)$ is the the Fourier transform of $w(t)$. The Fourier transform of the rectangle function is a sinc function, so, again from the convolution theorem, the spectral window is

$$W(\nu) = b\frac{\sin(2\pi \nu b)}{2\pi \nu b}D(\nu), \tag{9.18}$$

where $D(\nu)$ is the Fourier transform of the delta function pair:

$$D(\nu) = \int_{-\infty}^{\infty} [\delta(t - t_1) + \delta(t - t_2)] \exp[-i2\pi \nu t]dt$$

$$= \exp[-i2\pi \nu t_1] + \exp[-i2\pi \nu t_2]. \tag{9.19}$$

It is easier to understand $D(\nu)$ by considering its absolute value squared:

$$|D(\nu)|^2 = |\exp[-i2\pi \nu t_1] + \exp[-i2\pi \nu t_2]|^2$$

$$= 2 + 2\cos[2\pi \nu(t_2 - t_1)]$$

$$= 4\cos^2[\pi \nu(t_2 - t_1)]. \tag{9.20}$$

If $t_2 - t_1$ is much greater than b, the power spectrum of $f(t)$ is

$$P(\nu) = |F(\nu)|^2 = \left| \delta(\nu - \nu_0) \otimes \left\{ b\frac{\sin(2\pi \nu b)}{2\pi \nu b}D(\nu) \right\} \right|^2$$

$$\approx \left\{ b\frac{\sin(2\pi[\nu - \nu_0]b)}{2\pi[\nu - \nu_0]b} \right\}^2 \cos^2(\pi[\nu - \nu_0][t_2 - t_1]). \tag{9.21}$$

Equation 9.21 is plotted in Figure 9.10. It can be understood as the rapidly oscillating function $\cos^2(\pi[\nu - \nu_0][t_2 - t_1])$ with a slowly varying amplitude given by the sinc function squared. The many narrow peaks separated by $\Delta\nu = 1/(t_2 - t_1)$ are called aliases. It can be

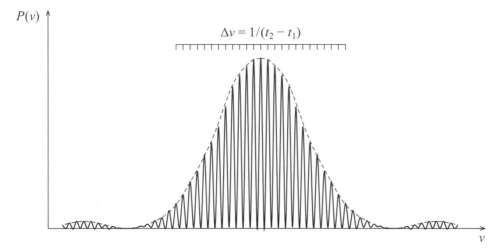

Figure 9.10: A plot of equation 9.21. It can be though of as the rapidly oscillating function $\cos^2(\pi v[t_2 - t_1])$ with a slowly varying amplitude given by the sinc function squared, shown as the dashed line. The many peaks are called aliases.

quite difficult to decide which of the many nearly equal peaks is the true period, particularly in noisy or multiperiodic sequences.

9.2.3 Arbitrary Data Windows

The foregoing discussion shows that it is crucial to know the data window and to calculate the corresponding spectral window. Data windows are often so complicated that calculating the corresponding spectral window analytically becomes impossible. Consider a realistic example from astronomy. A pulsating variable star is observed only at night, producing large gaps in its measured light curve every 24 hours. The observations are likely to be made only during particular phases of the moon, producing gaps every 29 days, and they may be made only at certain times of the year, introducing gaps at 1-year intervals. Cloudy weather introduces additional gaps. The spectral windows for complicated data windows must be calculated numerically. The procedure is:

1. Create an artificial sequence consisting of a pure cosine curve $s(t) = \cos(2\pi v_0 t)$.
2. Multiply the cosine curve by the known data window to create $f(t) = s(t)w(t)$.
3. Calculate the Fourier transform of $f(t)$ numerically to get

$$F(v) = S(v) \otimes W(v) = \delta(v - v_0) \otimes W(v) = W(v - v_0). \qquad (9.22)$$

Thus the Fourier transform of the artificial sequence is the spectral window displace to v_0. For all but the simplest spectral windows, it is a good idea to plot $|W(v - v_0)|^2$.

9.3 Discrete Sequences

In Section 9.2 the sequences were continuous, but in practice sequences of data are nearly always discrete, introducing additional complications. Let a_r be a sequence of $N = 2n$ values of a function $f(t)$ evaluated at an even spacing Δt. The Fourier transform of a_r and the inverse transform of the Fourier components are (see equations 8.108 and 8.109)

$$F_m = \frac{1}{N} \sum_{r=-n}^{n-1} a_r \exp\left[-i\frac{2\pi m}{N}r\right] \tag{9.23}$$

$$a_r = \sum_{m=-n}^{n-1} F_m \exp\left[i\frac{2\pi m}{N}r\right]. \tag{9.24}$$

Thus, the sequence a_r can be fully reconstructed from the N values of F_m.

9.3.1 The Need to Oversample F_m

While F_m contains all the information needed to reconstruct a_r, the properties of F_m may not be entirely clear, because it is discretely sampled. Consider the specific case $f(t) = \cos(2\pi \nu_0 t)$ sampled at $a_r = \cos(2\pi \nu_0 r\Delta t)$. The Fourier transform is equation 9.4 with $b = N\Delta t$ and $\nu = m/(N\Delta t)$. For simplicity, retain just the $\nu - \nu_0$ term:

$$F_m = F(\nu_m) = \frac{N\Delta t}{2} \frac{\sin(\pi[m/(N\Delta t) - \nu_0]N\Delta t)}{\pi[m/(N\Delta t) - \nu_0]N\Delta t}. \tag{9.25}$$

This is a sinc function centered on ν_0 and sampled at frequency interval $\Delta\nu = 1/(N\Delta t)$. If ν_0 is an integral multiple of $\Delta\nu$ so that $\nu_0 = n\Delta\nu = n/(N\Delta t)$, then

$$\sin(\pi[m/(N\Delta t) - \nu_0]N\Delta t) = \sin(\pi[m/(N\Delta t) - n/(N\Delta t)]N\Delta t)$$
$$= \sin(\pi[m - n]) = 0. \tag{9.26}$$

Thus, $F_m = 0$ at all the points where the Fourier transform is calculated except at ν_0 itself, where $F_m = 1/2$. (The other factor of 1/2 goes with the $\nu + \nu_0$ term that we have ignored.) The Fourier transform is shown in the right panel of Figure 9.11. It has a single high point at ν_0 and looks like it has no side lobes and no spectral leakage!

The sinc function is still present, though, and will manifest itself for all frequencies other than integer multiples of $\Delta\nu$. Suppose, for example, that ν_0 is the half-integer multiple $\nu_0 = (n+1/2)/(N\Delta t)$. Then

$$\sin(\pi[m/(N\Delta t) - \nu_0]N\Delta t) = \sin(\pi[m/(N\Delta t) - (n+1/2)/(N\Delta t)]N\Delta t)$$
$$= \sin(\pi[m - n - 1/2]) = \pm 1. \tag{9.27}$$

Now the Fourier transform is

$$F_m = \frac{\pm 1}{2\pi(n - m - 1/2)}. \tag{9.28}$$

This transform is shown in the left panel of Figure 9.11. Changing the frequency of the sine curve by just $\Delta\nu/2$ radically changed the appearance of the Fourier transform. Worse, none of the F_m gives the correct amplitude of the original cosine. This example shows that one should generally calculate the Fourier transform and the power spectrum at more frequencies than are necessary to reconstitute the original sequence. Oversampling by a factor of two is generally adequate. This can be easily done by calculating the Fourier transform at $F_{m+1/2}$ as well as at F_m. Only the F_m are needed to reconstruct the original sequence.

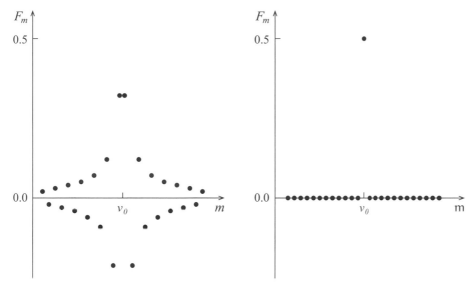

Figure 9.11: The Fourier transform F_m of a discretely sampled cosine given by equation 9.25. The left panel shows the Fourier transform when when v_0 falls halfway between two of the frequency points at which F_m is calculated. The right panel shows the Fourier transform when v_0 falls exactly on one of the frequency points.

9.3.2 Nyquist Frequency

Suppose a continuous sequence $s(t)$ is sampled at intervals Δt. Let $d(t)$ be

$$d(t) = \sum_{n=-\infty}^{\infty} \delta(t - n\Delta t), \tag{9.29}$$

which is an infinite string of delta functions separated by Δt. The sampled signal can be represented by

$$f(t) = s(t)d(t). \tag{9.30}$$

The Fourier transform of $f(t)$ is

$$F(v) = S(v) \otimes D(v), \tag{9.31}$$

where $D(v)$ is the Fourier transform of $d(t)$. This transform was worked out in Chapter 8 (see equation 8.99). The result is

$$D(v) = \sum_{k=-\infty}^{\infty} \delta(v - k/\Delta t). \tag{9.32}$$

Equation 9.31 becomes

$$F(v) = S(v') \otimes \sum_{k=-\infty}^{\infty} \delta(v' - k/\Delta t)$$

$$= \sum_{k=-\infty}^{\infty} S(v - k/\Delta t). \tag{9.33}$$

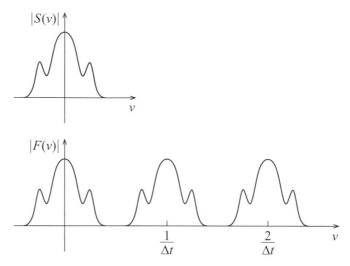

Figure 9.12: The Fourier transform of a sequence $s(t)$ is $S(v)$ (top panel). If $s(t)$ is sampled at intervals Δt, the Fourier transform of the discretely sampled function is $F(v)$ (bottom panel). The Fourier transform $S(v)$ is repeated at intervals $1/\Delta t$. For this case, $S(v)$ is narrow enough that the repeats do not overlap.

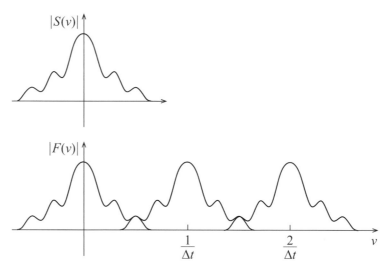

Figure 9.13: The Fourier transform of a sequence $s(t)$ is $S(v)$ (top panel). The Fourier transform of the discretely sampled function is $F(v)$ (bottom panel). In this case $S(v)$ is broader and the repeats do overlap, causing confusion about which frequencies are responsible for the signal at v.

This is the Fourier transform of $s(t)$ repeated infinitely many times, each time displaced by an amount $k/\Delta t$.

If $|S(v)|$ is a band-limited function equal to 0 at all frequencies greater than $1/2\Delta t$, as shown in Figure 9.12, there is no overlap between successive repeats of $S(v)$, so the repetition creates no problems. In contrast, if $|S(v)|$ is nonzero at frequencies greater than $1/2\Delta t$, successive repeats of $S(v)$ will overlap (Figure 9.13). The high-frequency signals in $s(t)$ appear at lower frequencies, causing confusion about which frequencies are responsible for

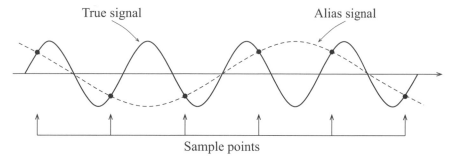

True signal

Alias signal

Sample points

Figure 9.14: The high-frequency sinusoidal signal shown by the solid curve is evenly sampled at the times indicated by the arrows. The value of the signal at the sample points is consistent with the lower-frequency signal shown by the dashed line. The lower-frequency signal is thus an alias.

the signal at ν. The reason for this is shown in Figure 9.14. The high-frequency sinusoidal signal shown by the solid curve is sampled at the evenly spaced times indicated by the arrows. The value of the signal at the sample points is consistent with the lower-frequency signal shown by the dashed line. The lower-frequency signal is thus an alias.

The frequency $\nu_N = 1/2\Delta t$ is called the *Nyquist frequency*. If a sine curve has a signal at a frequency in the range $\nu_N < \nu < 1/\Delta t$, the signal will also appear at the alias frequency ν_A, where

$$\nu_A = \nu_N - (\nu - \nu_N) = 2\nu_N - \nu. \tag{9.34}$$

To avoid this effect, the Nyquist frequency must be greater than the maximum frequency in the signal, or $\nu_N > \nu_{max}$. Equivalently, $1/\Delta t > 2\nu_{max}$, which leads to the *sampling theorem*: to measure a signal correctly, it is necessary to sample the signal at a frequency greater than twice the highest frequency in the signal.

Example: The discussions of the discretely sampled cosine at the beginning of this section, and of equation 9.25, were not quite correct, since they did not include aliasing. We now correct that deficiency. Suppose a cosine curve with frequency ν_0 has been sampled N times at intervals Δt over the interval $-N\Delta t/2 \leq t \leq N\Delta t/2$. The sequence can be represented by

$$f(t) = s(t)r(t)d(t),$$

where $s(t) = \cos(2\pi \nu_0 t)$, $r(t)$ is the rectangle function

$$r(t) = \begin{cases} 1, |t| \leq N\Delta t/2 \\ 0, |t| > N\Delta t/2 \end{cases},$$

and $d(t)$ is the sequence of delta functions

$$d(t) = \sum_{n=-\infty}^{\infty} \delta(t - n\Delta t).$$

Continued on page 270

The Fourier transform of $f(t)$ is

$$F(v) = \{S(v) \otimes R(v)\} \otimes D(v)$$

$$= \left\{ \frac{N\Delta t}{2} \frac{\sin(\pi[v - v_0]N\Delta t)}{\pi[v - v_0]N\Delta t} \right\} \otimes D(v)$$

$$F(v) = \frac{N\Delta t}{2} \frac{\sin(\pi[v - v_0 - k/\Delta t]N\Delta t)}{\pi[v - v_0 - k/\Delta t]N\Delta t}.$$

This is just the sinc function replicated at regular intervals $\Delta v = 1/\Delta t$. It replaces equation 9.4.

Now sample $F(v)$ at frequency points $v = m/(N\Delta t)$. Remembering that $F_m = F(v_m)$, we find

$$F_m = \sum_{k=-\infty}^{\infty} \frac{\Delta t}{2} \frac{\sin(\pi[m/(N\Delta t) - v_0 - k/\Delta t]N\Delta t)}{\pi[m/(N\Delta t) - v_0 - k/\Delta t]N\Delta t}.$$

For $k = 0$ this is just equation 9.25, the discretely sampled sinc function. The other values of k replicated the sinc function at higher frequencies.

9.3.3 Integration Sampling

Aliasing is somewhat ameliorated in real laboratory environments, because the samples of the original signal are generally not instantaneous but instead are averages over a finite time. Assume that the sampling process integrates the signal over the sampling interval as shown in Figure 9.15 instead of being an instantaneous sample. An integration sample of this sort is effectively a contiguous series of rectangular data windows

$$w(t) = r(t) \otimes \sum_{k=-\infty}^{\infty} \delta(t - k\Delta t), \tag{9.35}$$

where

$$r(t) = \begin{cases} 1, |t| \leq \Delta t/2 \\ 0, |t| > \Delta t/2 \end{cases}. \tag{9.36}$$

The spectral window is

$$W(v) = R(v) \sum_{k=-\infty}^{\infty} \delta\left(v - \frac{k}{\Delta t}\right), \tag{9.37}$$

where $R(v)$ is the Fourier transform of $r(t)$ and is the broad sinc function

$$R(v) = \Delta t \frac{\sin \pi v \Delta t}{\pi v \Delta t}. \tag{9.38}$$

Figure 9.15 shows a plot of $R(v)$. It drops off to 0.637 at the Nyquist frequency, goes to 0 at $v = 2v_N$, and thereafter oscillates about 0 with an amplitude that decreases with increasing frequency. Thus the low-frequency aliases of high frequency signals have lower amplitudes than the original signal. Note that the aliased frequencies are attenuated but not eliminated, and they are not attenuated by much if the original frequency lies between v_N and $2v_N$.

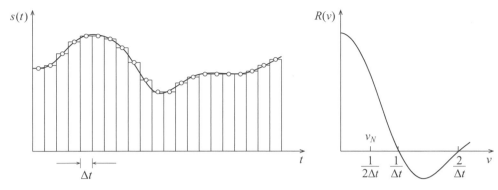

Figure 9.15: The signal $s(t)$ is sampled at intervals Δt, but the sampling process integrates the signal over the sampling interval (left panel). It is not an instantaneous sample. The open circles represent the resulting samples. The effect of the integration sampling is to introduce the broad sinc function spectral window (right panel). The spectral window attenuates high-frequency signals, somewhat ameliorating the effect of aliasing.

9.4 Effects of Noise

9.4.1 Deterministic and Stochastic Processes

It is useful to distinguish two types of processes that can produce sequences: deterministic processes and stochastic processes. A *deterministic process* is one that produces a single, predictable result for each member of the sequence. For example, the brightness of a pulsating variable star might be represented by the deterministic function

$$B(t) = A_0 + A_1 \sin(2\pi v t), \tag{9.39}$$

where v is the pulsation frequency, the independent variable is time t, and A_1 and A_2 are constants. There is only one possible value of B at each t.

If a sequence is produced by a stochastic process, each member of the sequence is a random number generated from a probability distribution. Suppose numbers ϵ_i are deviates generated from the Gaussian probability distribution

$$f(\epsilon) = \frac{1}{\sqrt{2\pi}\sigma} \exp\left\{-\frac{1}{2}\frac{(\epsilon-\mu)^2}{\sigma^2}\right\}. \tag{9.40}$$

Although the ϵ_i will cluster around the mean value μ, their precise values are not known before they are generated. The sequence formed by placing the ϵ_i in the order in which they were generated is a stochastic sequence. If the properties of a stochastic process do not change with time, it is called a *stationary stochastic process*. Some of the most interesting processes do change with time. An example is the sequence of random numbers chosen from a Gaussian probability distribution, the mean value of which varies with time:

$$f(\epsilon) = \frac{1}{\sqrt{2\pi}\sigma} \exp\left\{-\frac{1}{2}\frac{[\epsilon-\mu(t)]^2}{\sigma^2}\right\}. \tag{9.41}$$

A time series consisting of samples from this distribution has typical values that increase and decrease as $\mu(t)$ increases and decreases.

Sequences often have both a deterministic and a stochastic component. Suppose one measures the light curve of a pulsating star at times t_i and the measurements have random errors ϵ_i with a Gaussian distribution. The sequence is given by

$$B(t_i) = A_0 + A_1 \sin(2\pi \nu t_i) + \epsilon_i. \tag{9.42}$$

The behavior of the star is deterministic, but the noise is stochastic. It can be difficult to identify and separate the deterministic component from the stochastic component. Consider a stochastic sequence generated from equation 9.41 in which the mean is the sine curve $\mu(t) = A_0 + A_1 \sin(2\pi t/P + \phi)$:

$$f(\epsilon) = \frac{1}{\sqrt{2\pi}\sigma} \exp\left\{ -\frac{1}{2} \frac{[\epsilon - A_0 - A_1 \sin(2\pi \nu t)]^2}{\sigma^2} \right\}. \tag{9.43}$$

The time series is a noisy sine curve that looks essentially identical to the noisy sine curve produced by equation 9.42. In principle, the noise in equation 9.42 might be reduced by, for example, making more accurate measurements. The noise equation 9.43 is intrinsic, however, and cannot be eliminated by making more precise measurements.

9.4.2 Power Spectrum of White Noise

Up to now we have dealt with the Fourier transforms and power spectra of deterministic signals, and the main issue were the effects of the data window and of discrete sampling. This section discusses the simplest—but also the most important—of the stochastic signals: white noise. Suppose a signal has been discretely sampled at uniform intervals, producing the sequence $\{s_k\}$, $k = -n, \ldots, (n-1)$, where for convenience there are an even number of data points $N = 2n$. Let $s_k = \epsilon_k$, where the ϵ_k are samples from a probability distribution with properties

$$\langle \epsilon_k \rangle = 0 \tag{9.44}$$

$$\langle \epsilon_j \epsilon_k \rangle = \sigma^2 \delta_{jk}. \tag{9.45}$$

Thus, the samples have a 0 mean, they are uncorrelated, and their variance is σ^2. The Fourier transform of the sequence is

$$A_m = \frac{1}{N} \sum_{k=-n}^{n-1} \epsilon_k \cos\left(\frac{2\pi mk}{N}\right) \tag{9.46}$$

$$B_m = \frac{1}{N} \sum_{k=-n}^{n-1} \epsilon_k \sin\left(\frac{2\pi mk}{N}\right). \tag{9.47}$$

The mean values of A_m and B_m are

$$\langle A_m \rangle = \frac{1}{N} \sum_{k=-n}^{n-1} \langle \epsilon_k \rangle \cos\left(\frac{2\pi mk}{N}\right) = 0 \tag{9.48}$$

$$\langle B_m \rangle = \frac{1}{N} \sum_{k=-n}^{n-1} \langle \epsilon_k \rangle \sin\left(\frac{2\pi mk}{N}\right) = 0. \tag{9.49}$$

The mean value of A_m^2 is

$$
\begin{aligned}
\langle A_m^2 \rangle &= \left\langle \left\{ \frac{1}{N} \sum_{k=-n}^{n-1} \epsilon_k \cos\left(\frac{2\pi m k}{N}\right) \right\} \left\{ \frac{1}{N} \sum_{j=-n}^{n-1} \epsilon_j \cos\left(\frac{2\pi m j}{N}\right) \right\} \right\rangle \\
&= \frac{1}{N^2} \sum_{k=-n}^{n-1} \sum_{j=-n}^{n-1} \langle \epsilon_k \epsilon_j \rangle \cos\left(\frac{2\pi m k}{N}\right) \cos\left(\frac{2\pi m j}{N}\right) \\
&= \frac{1}{N^2} \sum_{k=-n}^{n-1} \sum_{j=-n}^{n-1} \sigma^2 \delta_{jk} \cos\left(\frac{2\pi m k}{N}\right) \cos\left(\frac{2\pi m j}{N}\right) \\
&= \frac{\sigma^2}{N^2} \sum_{k=-n}^{n-1} \cos^2\left(\frac{2\pi m k}{N}\right) = \frac{\sigma^2}{N^2} \frac{N}{2} \\
&= \frac{\sigma^2}{2N},
\end{aligned}
\tag{9.50}
$$

where the penultimate step used the orthogonality relation equation 8.120. Similarly, we also find

$$
\langle B_m^2 \rangle = \frac{\sigma^2}{2N}.
\tag{9.51}
$$

With the power spectrum defined as $P_m = A_m^2 + B_m^2$, the mean power becomes

$$
\langle P_m \rangle = \langle A_m^2 + B_m^2 \rangle = \frac{\sigma^2}{N}.
\tag{9.52}
$$

The mean power spectrum is independent of m and thus is flat. By analogy to the spectrum of light, the original sequence is called white noise. Note that this result depends explicitly on the assumption that the noise is uncorrelated: $\langle \epsilon_j \epsilon_k \rangle = \delta_{jk}$. Equation 9.52 is closely related to Parseval's theorem, one form of which states $\sum P_m = \sigma^2$ (see equation 8.138). Parseval's theorem always holds, but equation 9.52 holds only if the power is distributed uniformly over all frequencies—as it is for uncorrelated noise.

We have not yet specified the probability distribution from which the ϵ_k are drawn. First note that equations 9.48 and 9.49 are linear combinations of the ϵ_k. Let us assume that the ϵ_k have Gaussian distributions. One can show that a linear combination of random variables with Gaussian distributions also has a Gaussian distribution (see Section C.4 in Appendix C). Thus A_m and B_m have Gaussian distributions. Furthermore, from equations 9.48–9.51, we have $\sigma_{A_m}^2 = \langle A_m^2 \rangle = \sigma^2/2N$ and $\sigma_{B_m}^2 = \sigma^2/2N$, so the distributions of A_m and B_m have the same variance. However, this statement has a more general validity. The central limit theorem guarantees that A_m and B_m approach a Gaussian distribution for large N, the same distribution for both, no matter what distribution they came from. The only requirement is that the ϵ_k be independent and uncorrelated.

The joint probability distribution for A_m and B_m is therefore already a Gaussian distribution (if the ϵ_k have a Gaussian distribution) or approaches a Gaussian distribution as N becomes large (if they do not). It has the general form

$$
f(A_m, B_m) dA_m dB_m \propto \exp\left[-\frac{A_m^2 + B_m^2}{2c^2} \right] dA_m dB_m,
\tag{9.53}
$$

where c^2 has yet to be specified. Now change variables from (A_m, B_m) to $(P_m^{1/2}, \theta)$, where

$$A_m = P_m^{1/2} \cos \theta \tag{9.54}$$

$$B_m = P_m^{1/2} \sin \theta. \tag{9.55}$$

Then

$$dA_m dB_m = P_m^{1/2} d(P_m^{1/2}) d\theta = \frac{1}{2} d(P_m) d\theta, \tag{9.56}$$

and, after integrating over θ, the probability distribution function for P_m takes the form

$$\mathrm{f}(P_m) d(P_m) \propto \exp\left[-\frac{P_m}{2c^2}\right] d(P_m). \tag{9.57}$$

We recognize this as a χ^2 distribution with two degrees of freedom. From our previous work on exponential distributions, we know that the mean value is $\langle P_m \rangle = 2c^2$, and the normalization constant is $1/2c^2$, so we arrive at

$$\mathrm{f}(P_m) = \frac{1}{\langle P_m \rangle} \exp\left[-\frac{P_m}{\langle P_m \rangle}\right]. \tag{9.58}$$

For white noise, $\langle P_m \rangle$ is independent of m, so we can drop the subscript, and the distribution takes the simple form

$$\mathrm{f}(P) = \frac{1}{\langle P \rangle} \exp\left[-\frac{P}{\langle P \rangle}\right], \tag{9.59}$$

where, from equation 9.52, we have

$$\langle P \rangle = \langle P_m \rangle = \sigma^2 / N. \tag{9.60}$$

One sometimes plots the amplitude spectrum instead of the power spectrum. From equation 8.134, the amplitude at frequency m is $\alpha_m = (2P_m)^{1/2}$. Since $dP_m/d\alpha_m = \alpha_m$, the distribution of noise in the amplitude spectrum is

$$\mathrm{f}_\alpha(\alpha_m) = \mathrm{f}(2P_m = \alpha_m^2) \frac{dP_m}{d\alpha_m}$$

$$= \frac{2\alpha_m}{\langle \alpha_m^2 \rangle} \exp\left[-\frac{\alpha_m^2}{\langle \alpha_m^2 \rangle}\right], \tag{9.61}$$

where $\langle \alpha_m^2 \rangle = \langle P_m \rangle$.

The log-power spectrum is also sometimes useful. Set $y_m = \ln(P_m/\langle P_m \rangle)$. Then

$$\frac{dP_m}{dy_m} = \langle P_m \rangle \exp[y_m], \tag{9.62}$$

and the distribution of noise in the log-power spectrum is

$$\mathrm{f}_y(y_m) = \mathrm{f}(P_m = \langle P_m \rangle e^{y_m}) \frac{dP_m}{dy_m} = \frac{1}{\langle P_m \rangle} \exp[-e^{y_m}] \langle P_m \rangle \exp[y_m]$$

$$= \exp[y_m - e^{y_m}]. \tag{9.63}$$

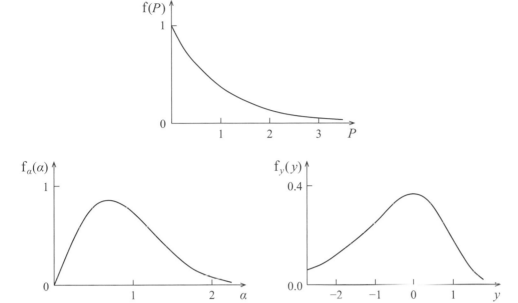

Figure 9.16: Probability distributions of noise. The top panel shows the distribution of noise in a power spectrum (equation 9.59). The bottom left panel shows the distribution of noise in an amplitude spectrum ($\alpha \propto P^{1/2}$, equation 9.61), and the bottom right panel shows the distribution of noise in a log-power spectrum ($y \propto \ln P$, equation 9.63).

If the noise is white noise, the subscript m can be dropped in equations 9.61 and 9.63. The noise distributions for the power spectrum, the amplitude spectrum, and the log-power spectrum are shown in Figure 9.16.

9.4.3 Deterministic Signals in the Presence of Noise

Suppose a sequence a_r consists of a deterministic signal $s(t)$ sampled at intervals Δt to give N values s_r, to which is added zero-mean uncorrelated noise ϵ_r with variance σ^2:

$$a_r = s_r + c_r. \tag{9.64}$$

The discrete Fourier transform of a_r is

$$F_m = S_m + E_m, \tag{9.65}$$

where S_m is the Fourier transform of just the s_r, and E_m is the Fourier transform of the ϵ_r. From equations 9.48 and 9.49, the mean value of E_m is $\langle E_m \rangle = 0$, and since the ϵ_r are uncorrelated with the s_r, the mean power spectrum is

$$\langle P_m \rangle = \langle |F_m|^2 \rangle = |S_m|^2 + \langle |E_m|^2 \rangle, \tag{9.66}$$

or, from equation 9.52,

$$\langle P_m \rangle = |S_m|^2 + \frac{\sigma^2}{N}. \tag{9.67}$$

Thus the effect of white noise is to increase P_m by an average amount σ^2/N. If the contribution from noise is large, it can dominate $\langle P_m \rangle$, drowning out the signal.

If the signal is periodic, the effect of noise can be ameliorated by obtaining more data, increasing N. Consider the specific case of a sine curve with amplitude α at frequency ν_0 with added noise. The Fourier transform of the sine curve is given by the discretely sampled sinc function centered on ν_0 with an amplitude proportional to α and independent of N. The power spectrum of the sine curve is the sinc function squared and is also independent of N. The noise level, however, decreases as N^{-1}. Thus, as N increases, the periodic signal stays at the same level, while the noise sinks. Furthermore, the ratio of the power in the deterministic signal to the power in noise increases linearly with N, not the more typical $\sqrt{N}$. This is why power spectra are so useful for detecting periodic signals hidden in noise.

Suppose the power spectrum of a sequence contains white noise with mean power $\langle P \rangle$ plus a high point at frequency ν_0 with power P_0. Is the high point due to a deterministic signal or merely a random fluctuation due to noise? One typically answers this question by calculating the probability that the high point could be due to a noise, sometimes called the "false-alarm probability." For a single, prespecified frequency, the false-alarm probability is just

$$P(\textit{false}) = \int_{P_0}^{\infty} \frac{1}{\langle P \rangle} \exp\left[-\frac{P}{\langle P \rangle}\right] dP = \exp\left[-\frac{P_0}{\langle P \rangle}\right]. \tag{9.68}$$

If the signal is known to fall within a range of frequencies but its precise frequency is unknown, any and all of the frequencies in the frequency range can produce high points. If there are M independent frequencies in the range, the false-alarm probability is

$$P(\textit{false}) = 1 - \left(1 - \exp\left[-\frac{P_0}{\langle P \rangle}\right]\right)^M$$

$$\approx M \exp\left[-\frac{P_0}{\langle P \rangle}\right]. \tag{9.69}$$

If the signal can have any frequency, then $M = N/2$, substantially increasing the false-alarm probability.

The very properties of power spectra that make them so useful for detecting periodic deterministic signals work against the detection and measurement of broad-band stochastic signals. We will see in Chapter 10 that the power in the power spectrum of a stochastic signal decreases as N^{-1}, where N is the length of the sequence. Since this is the same dependence as for white noise, the signal-to-noise ratio does not improve with the length of the sequence. However, the number of independent frequencies in the power spectrum does increase linearly with the length of the sequence, so a common practice when dealing with stochastic signals is to smooth the power spectrum, giving up spectral resolution to see the overall shape of the spectrum. Suppose the smoothing reduces the number of independent frequencies in the power spectrum by a factor k. The distribution of power in the power spectrum follows a χ^2 distribution with $\chi^2/2 = P/\langle P \rangle$ and $n = 2k$ degrees of freedom:

$$f_n(P) = \frac{1}{\langle P \rangle} \frac{(P/\langle P \rangle)^{(n-2)/2}}{(n/2 - 1)!} \exp\left[-\frac{P}{\langle P \rangle}\right]. \tag{9.70}$$

For no smoothing, $k = 1$ and $n = 2$, and equation 9.70 reduces to equation 9.49. The false-alarm probabilities are calculated analogously to equations 9.68 and 9.69. The results can be written uninformatively in terms of the incomplete gamma function but, in fact, are best handled numerically or from tables.

A final note: the definition of power in equation 8.132 was chosen so that the power due to a sine curve is easily related to the amplitude of the sine curve and is independent of the length of the sequence. This is a useful normalization when one is dealing with periodic signals. Other normalizations are possible. The power is often defined to be

$$P_m = N(A_m^2 + B_m^2),$$ (9.71)

which leads to

$$\sum P_m = N\sigma^2$$ (9.72)

and

$$\langle P_m \rangle = \sigma^2.$$ (9.73)

This normalization can be useful for analyzing stochastic signals.

9.4.4 Nonwhite, Non-Gaussian Noise

The power spectrum of noise is flat (equation 9.52) if the noise is uncorrelated (equation 9.45). Thus, uncorrelated noise is white noise. However, much of the noise encountered in nature is correlated, leading to nonflat, nonwhite power spectra. Noise that produces power spectra that rise toward lower frequencies is often called *red noise*; of which the most interesting cases are

$$\langle P \rangle \propto \frac{1}{\nu}, \qquad \text{pink noise}$$ (9.74)

$$\langle P \rangle \propto \frac{1}{\nu^2}, \qquad \text{brown noise.}$$ (9.75)

Because nonwhite noise is so ubiquitous, it is dangerous to thoughtlessly adopt equation 9.59 for the distribution of power, because $\langle P_m \rangle \neq \langle P \rangle$. One can describe noise with a nonwhite spectrum by allowing $\langle P_m \rangle$ to vary with frequency, retaining the subscript m in equation 9.58. If, for example, one wants to know the false-alarm probability for a peak with power P_0 at frequency m_0, one can measure the mean power in the spectrum at frequencies near m_0 and then replace $\langle P \rangle$ with $\langle P_{m_0} \rangle$ in equation 9.68 so that

$$P(\text{false}) = \exp\left[-\frac{P_0}{\langle P_{m_0} \rangle}\right].$$ (9.76)

In principle, noise that does not have a Gaussian distribution produces something other than an exponential distribution of noise in the power spectrum, but in practice equation 9.76 is almost always a good approximation. This is due in part to the dominating effect of the central limit theorem, especially for long sequences. In addition, the process of taking the squared absolute value of the Fourier components to make the power spectrum tends to reduce the effect of non-Gaussian noise. Suppose the non-Gaussian component of noise inserts an occasional point farther from the mean than expected from a Gaussian distribution. An outlier at time t_0 can be represented by $a_0\delta(t - t_0)$, where a_0 is the value of the outlier. The Fourier transform of the delta function is

$$F(\nu) = a_0 \exp[-i2\pi\nu t_0],$$ (9.77)

and its power spectrum is

$$P(\nu) = F(\nu)^* F(\nu) = a_0^2, \tag{9.78}$$

so the effect of the outlier is to increase the mean power uniformly at all frequencies. If a_0^2 is not too large, the increased power will masquerade as an increased variance of the Gaussian component of the noise. As a result, the power spectrum is a poor way to detect and measure a small departure from Gaussian noise.

Equation 9.78 also means that bad data points can have a disastrous effect on a power spectrum. A bad data point with value a_0 increases the level of the power spectrum at all frequencies by an amount a_0^2. If a_0 is large, the power level can be raised enough to drown out the features of interest in the spectrum. Accordingly, noise spikes in a sequence of data points should be identified and removed, generally by replacing them with the mean of a few surrounding points in the sequence.

9.5 Sequences with Uneven Spacing

The discrete Fourier transform explicitly assumes the data are evenly spaced and implicitly assumes the data points all have the same weight. In practice, both assumptions are often violated. A common way to handle unevenly spaced data is to interpolate between the data points to produce evenly spaced data. This is a good way to deal with small gaps in a sequence, but it seriously biases the power spectrum if the gaps are large. Furthermore, most interpolation schemes smooth the data, altering the noise properties of the resulting power spectrum. A linear interpolation, for example, is a low-pass filter. A better way to handle unevenly spaced data is to calculate a least squares periodogram. In this section we first discuss the relation between the discrete Fourier transform and least squares fits of sines and cosines. We next introduce the least squares periodogram for sequences in which the data have equal weights but unequal spacing, leading to the Lomb-Scargle periodogram. Then we extend these results to sequences in which the data have both unequal weights and unequal spacing.

9.5.1 Least Squares Periodogram

The discrete Fourier transform is intimately related to least squares fits of sines and cosines. To see this, fit the function

$$y = A_\omega \sin(\omega t) + B_\omega \cos(\omega t) \tag{9.79}$$

to a sequence of equally weighted data points (y_r, t_r), $r = -n, \ldots, (n-1)$, for which $\langle y_r \rangle = 0$. The sum of the squares of the residuals is

$$S = \sum_{r=-n}^{n-1} w_r [y_r - A_\omega \sin(\omega t_r) - B_\omega \cos(\omega t_r)]^2, \tag{9.80}$$

where w_r is the weight of data point r. Since the data are equally weighted,

$$w_r = w = 1/\sigma^2, \tag{9.81}$$

where σ^2 is the variance of the data. Minimize the sum in the usual way, by setting the derivatives of S with respect to A_ω and B_ω equal to zero:

$$\frac{\partial S}{\partial A_\omega} = 0 \tag{9.82}$$

$$\frac{\partial S}{\partial B_\omega} = 0. \tag{9.83}$$

This leads to the normal equations

$$\hat{A}_\omega \sum \sin^2(\omega t_r) + \hat{B}_\omega \sum \sin(\omega t_r)\cos(\omega t_r) = \sum y_r \sin(\omega t_r) \tag{9.84}$$

$$\hat{A}_\omega \sum \sin(\omega t_r)\cos(\omega t_r) + \hat{B}_\omega \sum \cos^2(\omega t_r) - \sum y_r \cos(\omega t_r), \tag{9.85}$$

where all sums are taken from $r = -n$ to $r = n - 1$. The weights cancel, because we have assumed equally weighted data. It is convenient to rewrite these equations in matrix notation:

$$\begin{pmatrix} \sum \sin^2(\omega t_r) & \sum \sin(\omega t_r)\cos(\omega t_r) \\ \sum \sin(\omega t_r)\cos(\omega t_r) & \sum \cos^2(\omega t_r) \end{pmatrix} \begin{pmatrix} \hat{A}_\omega \\ \hat{B}_\omega \end{pmatrix} = \begin{pmatrix} \sum y_r \sin(\omega t_r) \\ \sum y_r \cos(\omega t_r) \end{pmatrix}. \tag{9.86}$$

Now let the data points be evenly spaced at $t_r = r\Delta t$, and calculate the fit only for the frequencies $\omega = 2\pi m/(N\Delta t)$. The normal equations become

$$\begin{pmatrix} \sum \sin^2\left(\dfrac{2\pi mr}{N}\right) & \sum \sin\left(\dfrac{2\pi mr}{N}\right)\cos\left(\dfrac{2\pi mr}{N}\right) \\ \sum \sin\left(\dfrac{2\pi mr}{N}\right)\cos\left(\dfrac{2\pi mr}{N}\right) & \sum \cos^2\left(\dfrac{2\pi mr}{N}\right) \end{pmatrix} \begin{pmatrix} \hat{A}_m \\ \hat{B}_m \end{pmatrix}$$

$$= \begin{pmatrix} \sum y_r \sin\left(\dfrac{2\pi mr}{N}\right) \\ \sum y_r \cos\left(\dfrac{2\pi mr}{N}\right) \end{pmatrix}. \tag{9.87}$$

The elements of the matrix satisfy the orthogonality relations equations 8.118, 8.119, and 8.120, so the normal equations collapse to

$$\begin{pmatrix} \dfrac{N}{2} & 0 \\ 0 & \dfrac{N}{2} \end{pmatrix} \begin{pmatrix} \hat{A}_m \\ \hat{B}_m \end{pmatrix} = \begin{pmatrix} \sum y_r \sin\left(\dfrac{2\pi mr}{N}\right) \\ \sum y_r \cos\left(\dfrac{2\pi mr}{N}\right) \end{pmatrix}. \tag{9.88}$$

By inspection, we see that the solutions for $\hat{A}_m$ and $\hat{B}_m$ are

$$\hat{A}_m = \frac{2}{N} \sum y_r \sin\left(\frac{2\pi mr}{N}\right) \tag{9.89}$$

$$\hat{B}_m = \frac{2}{N} \sum y_r \cos\left(\frac{2\pi mr}{N}\right). \tag{9.90}$$

Except for the factor of 2, these are identical to equations 8.127 and 8.128 for the coefficients of the terms in the discrete Fourier series. The factor of 2 is also present in the discrete Fourier transform but is placed in equation 8.126, which reconstructs the original sequence from the Fourier series. Thus for a sequence with evenly spaced and equally weighted data points, the discrete Fourier transform is identical to least squares fits of sine curves.

9.5.2 Lomb-Scargle Periodogram

The orthogonality relations do not hold when the data points are not evenly spaced, so the connection between least squares and the discrete Fourier transform is no longer obvious; but they are connected nevertheless. To see this, offset the zero point in t by an amount τ. We can do this without affecting the amplitude $(\hat{A}_\omega^2 + \hat{B}_\omega^2)^{1/2}$. The normal equations become

$$\begin{pmatrix} \sum \sin^2(\omega[t_r - \tau]) & \sum \sin(\omega[t_r - \tau])\cos(\omega[t_r - \tau]) \\ \sum \sin(\omega[t_r - \tau])\cos(\omega[t_r - \tau]) & \sum \cos^2(\omega[t_r - \tau]) \end{pmatrix} \begin{pmatrix} \hat{A}_\omega \\ \hat{B}_\omega \end{pmatrix}$$

$$= \begin{pmatrix} \sum y_r \sin(\omega[t_r - \tau]) \\ \sum y_r \cos(\omega(t_r - \tau]) \end{pmatrix}, \quad (9.91)$$

where all sums are taken over the $2n$ data points. Now choose τ so that the off-diagonal terms in the matrix disappear. This can always be done by setting

$$0 = \sum \sin(\omega[t_r - \tau])\cos(\omega[t_r - \tau]). \quad (9.92)$$

Expanding the sines and cosines and collecting terms, we find

$$0 = \sum \left\{ \sin(\omega t_r)\cos(\omega t_r)\left[\cos^2(\omega\tau) - \sin^2(\omega\tau)\right] - \left[\cos^2(\omega t_r) - \sin^2(\omega t_r)\right]\cos(\omega\tau)\sin(\omega\tau) \right\}$$

$$= \frac{1}{2}\sum \left\{ \cos(2\omega\tau)\sin(2\omega t_r) - \sin(2\omega\tau)\cos(2\omega t_r) \right\}, \quad (9.93)$$

so τ is given by

$$\tan(2\omega\tau) = \frac{\sum \sin(2\omega t_r)}{\sum \cos(2\omega t_r)}. \quad (9.94)$$

Since τ is a function of ω, we will denote it by τ_ω from now on. With this choice of τ_ω, the normal equations become

$$\begin{pmatrix} \sum \sin^2(\omega[t_r - \tau_\omega]) & 0 \\ 0 & \sum \cos^2(\omega[t_r - \tau_\omega]) \end{pmatrix} \begin{pmatrix} \hat{A}_\omega \\ \hat{B}_\omega \end{pmatrix} = \begin{pmatrix} \sum y_r \sin(\omega[t_r - \tau_\omega]) \\ \sum y_r \cos(\omega[t_r - \tau_\omega]) \end{pmatrix}. \quad (9.95)$$

The solutions for $\hat{A}_\omega$ and $\hat{B}_\omega$ are

$$\hat{A}_\omega = \frac{\sum y_r \sin(\omega[t_r - \tau_\omega])}{\sum \sin^2(\omega[t_r - \tau_\omega])} \quad (9.96)$$

$$\hat{B}_\omega = \frac{\sum y_r \cos(\omega[t_r - \tau_\omega])}{\sum \cos^2(\omega[t_r - \tau_\omega])}. \quad (9.97)$$

The covariance matrix is

$$\mathbf{C} = \sigma^2 \mathbf{N}^{-1} = \sigma^2 \begin{pmatrix} \sum \sin^2(\omega[t_r - \tau_\omega]) & 0 \\ 0 & \sum \cos^2(\omega[t_r - \tau_\omega]) \end{pmatrix}^{-1}, \qquad (9.98)$$

where σ^2 is the variance of a single data point (see equation 5.69 with $\mathbf{W}^{-1} = \sigma^2 \mathbf{I}$, where $\mathbf{I}$ is the unit matrix), so the variances of $\hat{A}_\omega$ and $\hat{B}_\omega$ are

$$\sigma^2_{\hat{A}_\omega} = \sigma^2 \left[\sum \sin^2(\omega[t_r - \tau_\omega]) \right]^{-1} \qquad (9.99)$$

$$\sigma^2_{\hat{B}_\omega} = \sigma^2 \left[\sum \cos^2(\omega[t_r - \tau_\omega]) \right]^{-1}. \qquad (9.100)$$

If there is some suspicion that the variance of the data points has not been reported correctly, one should replace σ^2 with the estimated variance

$$\hat{\sigma}^2 = \frac{1}{N-1} \sum (y_r - \langle y \rangle)^2, \qquad (9.101)$$

where $N = 2n$ is the number of data points.

It is tempting at this point to calculate the power spectrum, $P_\omega = \hat{A}_\omega^2 + \hat{B}_\omega^2$, but the power spectrum now has poor noise properties. While $\hat{A}_\omega$ and $\hat{B}_\omega$ still have Gaussian distributions, their variances are no longer the same, because $\sum \sin^2(\omega[t_r - \tau_\omega])$ is not equal to $\sum \cos^2(\omega[t_r - \tau_\omega])$. If the variances are not the same, the equivalent of equation 9.53 is no longer valid, and the noise in the spectrum no longer has a simple distribution. Worse, the variances of $\hat{A}_\omega$ and $\hat{B}_\omega$ are functions of ω, which means that every point in the periodogram has a different noise distribution.

The poor noise properties of the power spectrum can be avoided by switching to the Lomb-Scargle significance, P_{LS}, which, following Horne and Baliunas,[2] we define as

$$P_{LS} = \frac{1}{2} \frac{\hat{A}_\omega^2}{\sigma^2_{\hat{A}_\omega}} + \frac{1}{2} \frac{\hat{B}_\omega^2}{\sigma^2_{\hat{B}_\omega}}. \qquad (9.102)$$

This quantity measures the amount by which $\hat{A}_\omega^2$ and $\hat{B}_\omega^2$ exceed their variances. If, for example, both $\hat{A}_\omega^2$ and $\hat{B}_\omega^2$ exceed their variances by a factor of 10, we would find $P_{LS} = 10$ and would conclude that the signal at ω is highly significant. A plot of P_{LS} against ω is usually called the *Lomb-Scargle periodogram* after its two inventors, Lomb[3] and Scargle.[4] From equations 9.96–9.100 P_{LS} can be written as

$$P_{LS} = \frac{1}{2} \frac{\left\{ \sum y_r \sin(\omega[t_r - \tau_\omega]) \right\}^2}{\sigma^2 \sum \sin^2(\omega[t_r - \tau_\omega])} + \frac{1}{2} \frac{\left\{ \sum y_r \cos(\omega[t_r - \tau_\omega]) \right\}^2}{\sigma^2 \sum \cos^2(\omega[t_r - \tau_\omega])}. \qquad (9.103)$$

[2] J. H. Horne and S. L. Baliunas. 1986. "A Prescription for Period Analysis of Unevenly Sampled Time Series." *Astrophysical Journal* vol. 302, p. 757.

[3] N. R. Lomb. 1976. "Least-Squares Frequency Analysis of Unequally Spaced Data." *Astrophysics and Space Science* vol. 39, p. 447.

[4] J. D. Scargle. 1982. "Studies in Astronomical Time Series Analysis. II—Statistical Aspects of Spectral Analysis of Unevenly Spaced Data." *Astrophysical Journal* vol. 263, p. 835.

For evenly spaced data, the orthogonality relations hold, and this reduces to $P_{LS} = (N/4)(A_\omega^2 + \hat{B}_\omega^2)$, which is the standard Fourier transform power spectrum (equation 8.132) except for a constant factor. One often sees P_{LS} written without the factor of σ^2 that appears in equation 9.103. As we will show below, the factor of σ^2 is necessary to normalize the spectrum to $\langle P_{LS} \rangle = 1$ (equation 9.111) and to give the noise in the spectrum a normalized χ^2 distribution (equation 9.113).

To determine the noise properties of the Lomb-Scargle periodogram, let the $y_r = \epsilon_r$, where ϵ_r is uncorrelated noise with 0 mean and variance σ^2:

$$\langle \epsilon_r \rangle = 0 \tag{9.104}$$

$$\langle \epsilon_r \epsilon_q \rangle = \sigma^2 \delta_{rq}. \tag{9.105}$$

To be explicit, this σ^2 is the same as the σ^2 in equation 9.81 and in equations 9.99 and 9.100. For convenience, let $P_{LS} = X_\omega^2 + Y_\omega^2$, where

$$X_\omega = \frac{1}{\sqrt{2}} \frac{\sum \epsilon_r \sin(\omega[t_r - \tau_\omega])}{\{\sigma^2 \sum \sin^2(\omega[t_r - \tau_\omega])\}^{1/2}} \tag{9.106}$$

$$Y_\omega = \frac{1}{\sqrt{2}} \frac{\sum \epsilon_r \cos(\omega[t_r - \tau_\omega])}{\{\sigma^2 \sum \cos^2(\omega[t_r - \tau_\omega])\}^{1/2}}. \tag{9.107}$$

Following the same logic used to derive equations 9.48 and 9.49, the mean values of these quantities are

$$\langle X_\omega \rangle = \langle Y_\omega \rangle = 0. \tag{9.108}$$

The variance of X_ω is

$$\begin{aligned}
\sigma_{X_\omega}^2 = \langle X_\omega^2 \rangle &= \frac{1}{2} \frac{1}{\sigma^2 \sum \sin^2(\omega[t_r - \tau_\omega])} \left\langle \left\{ \sum_r \epsilon_r \sin(\omega[t_r - \tau_\omega]) \right\} \left\{ \sum_q \epsilon_q \sin(\omega[t_q - \tau_\omega]) \right\} \right\rangle \\
&= \frac{1}{2} \frac{1}{\sigma^2 \sum \sin^2(\omega[t_r - \tau_\omega])} \sum_r \sum_q \langle \epsilon_q \epsilon_r \rangle \sin(\omega[t_r - \tau_\omega]) \sin(\omega[t_q - \tau_\omega]) \\
&= \frac{1}{2} \frac{1}{\sigma^2 \sum \sin^2(\omega[t_r - \tau_\omega])} \sum_r \sum_q \sigma^2 \delta_{rq} \sin(\omega[t_r - \tau_\omega]) \sin(\omega[t_q - \tau_\omega]) \\
&= \frac{1}{2} \frac{1}{\sum \sin^2(\omega[t_r - \tau_\omega])} \sum_r \sin^2(\omega[t_r - \tau_\omega]) \\
&= \frac{1}{2}. \tag{9.109}
\end{aligned}$$

Similarly, the variance of Y_ω is

$$\sigma_{Y_\omega}^2 = \langle Y_\omega^2 \rangle = \frac{1}{2}. \tag{9.110}$$

Together equations 9.109 and 9.110 give

$$\langle P_{LS} \rangle = \langle X_\omega^2 \rangle + \langle Y_\omega^2 \rangle = 1. \tag{9.111}$$

We are now in position to derive the distribution of noise in the Lomb-Scargle periodogram. The logic is identical to that used to derive the noise in an ordinary power spectrum (equation 9.59), so we will skip most of the steps. If the ϵ_r have a Gaussian distribution, or if there are enough of the ϵ_r that the central limit theorem applies, then both X_ω and Y_ω have Gaussian distributions. Since their means are both equal to 0 and their variances both equal to 1/2, their joint probability distribution is

$$f(X_\omega, Y_\omega)dX_\omega dY_\omega \propto \exp\left[-\frac{1}{2}\left(\frac{X_\omega^2}{1/2}\right) - \frac{1}{2}\left(\frac{Y_\omega^2}{1/2}\right)\right] dX_\omega dY_\omega. \qquad (9.112)$$

Remembering that $P_{LS} = X_\omega^2 + Y_\omega^2$, perform a change of variables from (X_ω, Y_ω) to $(P_m^{1/2}, \theta)$ and then integrate over θ to find

$$f(P_{LS}) = \exp[-P_{LS}], \qquad (9.113)$$

which is already correctly normalized.

Unlike the ordinary power spectrum, the Lomb-Scargle periodogram is not simply related to the amplitudes of the Fourier components. This is much to give up, but equation 9.113 justifies the sacrifice. The noise properties of the periodogram are simple, making it easy to determine the significance of peaks in the periodogram. If a significant period has been identified and if the signal is a sine curve, its amplitude can be calculated from equations 9.96 and 9.97; or, if the function is more complicated, one can fit the function directly to the data by standard methods.

One typically shows false-alarm probabilities along with the Lomb-Scargle periodogram. As for the standard power spectrum, the probability that a frequency in the Lomb-Scargle periodogram for which $P_{LS} = P_0$ is a false alarm is

$$P(\textit{false}) = 1 - (1 - \exp[-P_0])^M \approx M \exp[-P_0], \qquad (9.114)$$

where M is the number of independent frequencies in the periodogram. Because the spectral window of a Lomb-Scargle periodogram is usually complicated (see below), it is not at all obvious what to use for M. Horne and Baliunas[5] suggest that if the points in the sequence are not too strongly clumped, one can take $M \approx n$, but if the data points are strongly clumped, M can be much less than n. Cumming[6] has suggested that M is roughly equal to the number of peaks in the periodogram or, if one is examining a restricted range of frequencies, the number of peaks in the restricted range.

All things considered, though, it is difficult or even impossible to choose a defensible value for M, often forcing one to resort to Monte Carlo techniques to determine false-alarm probabilities. There are a variety of possible approaches to the Monte Carlo calculation. Perhaps the most common is Fisher randomization. Starting with the original data set (y_r, t_r), one creates an artificial sequence by leaving the values of the y_r and t_r individually unchanged but randomly shuffling the y_r among the t_r. Because the t_r are unchanged, the spectral window of the artificial sequence is the same as for the original sequence; and because the y_r are unchanged, the σ^2 of the artificial sequence is the same as for the original sequence. But because the y_r have been shuffled to different times, all signals and correlations in the original sequence have been destroyed. One calculates Lomb-Scargle periodograms for a large number of these artificial sequences. The distribution of the power

[5] J. H. Horne and S. L. Baliunas. 1986. "A Prescription for Period Analysis of Unevenly Sampled Time Series." *Astrophysical Journal* vol. 302, p. 757.

[6] A. Cumming. 2004. "Detectability of Extrasolar Planets in Radial Velocity Surveys." *Monthly Notices of the Royal Astronomical Society* vol. 354, p. 1165.

in all the periodograms yields an empirical $f(P_{LS})$, from which the false-alarm probabilities can be deduced. Fisher randomization is a powerful tool, but it must not be used blindly. Because shuffling the y_r destroys any correlations in the original data sequence, Fisher randomization gives incorrect results for nonwhite noise.

Finally, a major issue for unevenly spaced data is that the spectral window generally has many large side lobes and aliases. To understand a periodogram, it is crucial to calculate the spectral window. To calculate the spectral window:

1. Create an artificial sequence (s_r, t_r) by sampling a pure sine curve at the same times t_r as the original data sequence, $s_r = \sin(2\pi v_0 t_r)$.
2. Calculate the Lomb-Scargle periodogram of the artificial sequence.

The periodogram is the spectral window displaced to v_0. The spectral window is usually plotted along with the periodogram of the original data sequence.

9.5.3 Generalized Lomb-Scargle Periodogram

Equations 9.79–9.81 lie at the heart of the ordinary Lomb-Scargle periodogram. They explicitly assume that the data points have equal weights and implicitly assume that a single mean value of the data is valid for all frequencies. The first assumption is usually unjustified for real data, and the second assumption fails when a frequency is poorly sampled (think of a sine curve that is only sampled at phases between 0 and π). The generalized Lomb-Scargle periodogram developed by Zechmeister and Kürster rectifies the deficiencies in the ordinary Lomb-Scargle periodogram.[7]

Suppose we are given a sequence of $2n$ data points with unequal spacing and unequal weights, (y_r, t_r, σ_r), and we wish to fit the function

$$y(t_r) = A_\omega \sin(\omega t_r) + B_\omega \cos(\omega t_r) + c_\omega \tag{9.115}$$

to the data to determine values and variances for A_ω and B_ω. The least squares estimates are given by minimizing the sum

$$S = \sum_{r=1}^{2n} w_r \left[y_r - A_\omega \sin(\omega t_r) - B_\omega \cos(\omega t_r) - c_\omega \right]^2, \tag{9.116}$$

where $w_r = 1/\sigma_r^2$. Setting the derivatives of S with respect to A_ω, B_ω, and c_ω equal to zero yields the normal equations:

$$\hat{A}_\omega \sum w_r \sin^2(\omega t_r) + \hat{B}_\omega \sum w_r \sin(\omega t_r) \cos(\omega t_r) + \hat{c}_\omega \sum w_r \sin(\omega t_r) = \sum w_r y_r \sin(\omega t_r) \tag{9.117}$$

$$\hat{A}_\omega \sum w_r \sin(\omega t_r) \cos(\omega t_r) + \hat{B}_\omega \sum w_r \cos^2(\omega t_r) + \hat{c}_\omega \sum w_r \cos(\omega t_r) = \sum w_r y_r \cos(\omega t_r) \tag{9.118}$$

$$\hat{A}_\omega \sum w_r \sin(\omega t_r) + \hat{B}_\omega \sum w_r \cos(\omega t_r) + \hat{c}_\omega \sum w_r = \sum w_r y_r, \tag{9.119}$$

[7] M. Zechmeister and M. Kürster. 2009. "The Generalised Lomb-Scargle Periodogram: A New Formalism for the Floating-Mean and Keplerian Periodograms." *Astronomy and Astrophysics* vol. 496, p. 577.

where all sums are taken over r. For convenience, we use a notation similar (but *not* identical) to that of Zechmeister and Kürster:

$$
\begin{aligned}
I &= \sum w_r & \overline{YY} &= \sum w_r y_r^2 \\
Y &= \sum w_r y_r & \overline{YS} &= \sum w_r y_r \sin(\omega t_r) \\
S &= \sum w_r \sin(\omega t_r) & \overline{YC} &= \sum w_r y_r \cos(\omega t_r) \\
C &= \sum w_r \cos(\omega t_r) & \overline{SS} &= \sum w_r \sin^2(\omega t_r) \\
& & \overline{CC} &= \sum w_r \cos^2(\omega t_r) \\
& & \overline{SC} &= \sum w_r \sin(\omega t_r) \cos(\omega t_r).
\end{aligned}
\tag{9.120}
$$

Zechmeister and Kürster assume that the weights have been normalized to $I = \sum w_r = 1$. We do not make this assumption. As a result, the equations we derive will differ from theirs by a factor of I in many of the terms.

To solve the normal equations, first use equation 9.119 to eliminate $\hat{c}_\omega$ from equations 9.117 and 9.118, arriving at

$$
\hat{A}_\omega(\overline{SS} - S^2/I) + \hat{B}_\omega(\overline{SC} - SC/I) = \overline{YS} - YS/I \tag{9.121}
$$

$$
\hat{A}_\omega(\overline{SC} - SC/I) + \hat{B}_\omega(\overline{CC} - C^2/I) = \overline{YC} - YC/I, \tag{9.122}
$$

or, in matrix form,

$$
\begin{pmatrix} \overline{SS} - S^2/I & \overline{SC} - SC/I \\ \overline{SC} - SC/I & \overline{CC} - C^2/I \end{pmatrix} \begin{pmatrix} \hat{A}_\omega \\ \hat{B}_\omega \end{pmatrix} = \begin{pmatrix} \overline{YS} - YS/I \\ \overline{YC} - YC/I \end{pmatrix}. \tag{9.123}
$$

Choose a displacement τ_ω of the zero point in time that makes the off-diagonal terms in the matrix equal to 0,

$$
0 = \overline{SC} - \frac{SC}{I}, \tag{9.124}
$$

or, written out explicitly,

$$
\begin{aligned}
0 = &\sum w_r \sin(\omega[t_r - \tau_\omega]) \cos(\omega[t_r - \tau_\omega]) \\
&- \frac{1}{\sum w_r} \left\{ \sum w_r \sin(\omega[t_r - \tau_\omega]) \right\} \left\{ \sum w_r \cos(\omega[t_r - \tau_\omega]) \right\}.
\end{aligned}
\tag{9.125}
$$

After some exercise with trigonometric identities, we find

$$
\tan(2\omega\tau_\omega) = \frac{\sum w_r \sin(2\omega t_r) - 2\left[\sum w_r \sin(\omega t_r)\right]\left[\sum w_r \cos(\omega t_r)\right]/\sum w_r}{\sum w_r \cos(2\omega t_r) - \left\{\left[\sum w_r \cos(\omega t_r)\right]^2 - \left[\sum w_r \sin(\omega t_r)\right]^2\right\}/\sum w_r}. \tag{9.126}
$$

The normal equations now become

$$
\begin{pmatrix} \overline{CC} - C^2/I & 0 \\ 0 & \overline{SS} - S^2/I \end{pmatrix} \begin{pmatrix} \hat{A}_\omega \\ \hat{B}_\omega \end{pmatrix} = \begin{pmatrix} \overline{YC} - YC/I \\ \overline{YS} - YS/I \end{pmatrix}, \tag{9.127}
$$

where it is understood that all times have been displaced by τ_ω, so that t_r is everywhere replaced by $t_r - \tau_\omega$. The solution to the normal equations can now be written down by inspection:

$$\hat{A}_\omega = \frac{\overline{YC} - YC/I}{\overline{CC} - C^2/I} \tag{9.128}$$

$$\hat{B}_\omega = \frac{\overline{YS} - YS/I}{\overline{SS} - S^2/I}. \tag{9.129}$$

The estimated covariance matrix for a weighted least squares fit is

$$\hat{\mathbf{C}} = \hat{\sigma}^2(\mathbf{N})^{-1} = \sigma^2 \begin{pmatrix} \overline{CC} - C^2/I & 0 \\ 0 & \overline{SS} - S^2/I \end{pmatrix}^{-1} \tag{9.130}$$

(see Sections 5.3 and 5.5), where $\mathbf{N}$ is the normal matrix. Looking ahead, we note that when calculating false-alarm probabilities in the Lomb-Scargle periodogram, the null hypothesis is that there are no signals present in the data. For this case

$$\sigma^2 = \frac{1}{n-1} \sum w_r (y_r - \langle y \rangle)^2 \tag{9.131}$$

and

$$\langle y \rangle = \frac{\sum w_r y_r}{\sum w_r}. \tag{9.132}$$

Because we have forced the normal matrix to be diagonal, the expressions for the variances reduce to

$$\sigma_{\hat{A}_\omega}^2 = \frac{\sigma^2}{\overline{CC} - C^2/I} \tag{9.133}$$

$$\sigma_{\hat{B}_\omega}^2 = \frac{\sigma^2}{\overline{SS} - S^2/I}. \tag{9.134}$$

Since $\sigma_{\hat{A}_\omega}^2 \neq \sigma_{\hat{B}_\omega}^2$, the quantity $\hat{A}_\omega^2 + \hat{B}_\omega^2$ has poor noises properties and is a poor choice for a periodogram. Instead, we define the generalized Lomb-Scargle significance to be

$$P_{GLS} = \frac{1}{2} \frac{\hat{A}_\omega^2}{\sigma_{\hat{A}_\omega}^2} + \frac{1}{2} \frac{\hat{B}_\omega^2}{\sigma_{\hat{B}_\omega}^2} \tag{9.135}$$

$$= \frac{1}{2\sigma^2} \frac{(\overline{YC} - YC/I)^2}{\overline{CC} - C^2/I} + \frac{1}{2\sigma^2} \frac{(\overline{YS} - YS/I)^2}{\overline{SS} - S^2/I}. \tag{9.136}$$

The generalized Lomb-Scargle periodogram is a plot of P_{GLS} against frequency or period.

To investigate the noise properties of P_{GLS}, let the $y_r = \epsilon_r$, where ϵ_r is uncorrelated noise with 0 mean and variance σ_r^2:

$$\langle \epsilon_r \rangle = 0 \tag{9.137}$$

$$\langle \epsilon_r \epsilon_q \rangle = \sigma_r^2 \delta_{rq}. \tag{9.138}$$

Following logic similar to that used to derive equations 9.48–9.51 (or equivalently, equations 9.108–9.110), one can show that $\langle \hat{A}_\omega \rangle = \langle \hat{B}_\omega \rangle = 0$ and that

$$\langle \hat{A}_\omega^2 \rangle = \sigma_{\hat{A}_\omega}^2 \qquad (9.139)$$

$$\langle \hat{B}_\omega^2 \rangle = \sigma_{\hat{B}_\omega}^2. \qquad (9.140)$$

Therefore,

$$\frac{\langle \hat{A}_\omega^2 \rangle}{\sigma_{\hat{A}_\omega}^2} = \frac{\langle \hat{B}_\omega^2 \rangle}{\sigma_{\hat{B}_\omega}^2} = 1, \qquad (9.141)$$

and furthermore, the mean value of P_{GLS} is

$$\langle P_{GLS} \rangle = \frac{1}{2} \frac{\langle \hat{A}_\omega^2 \rangle}{\sigma_{\hat{A}_\omega}^2} + \frac{1}{2} \frac{\langle \hat{B}_\omega^2 \rangle}{\sigma_{\hat{B}_\omega}^2} = 1. \qquad (9.142)$$

To derive the probability distribution for P_{GLS}, we invoke yet again the logic used to derive equations 9.59 and 9.113. If the y_r have a Gaussian distribution, or if there are enough of the y_r that the central limit theorem applies, then P_{GLS} has a χ^2 distribution with two degrees of freedom. The distribution of noise in the periodogram is

$$\mathrm{f}(P_{GLS}) = \frac{1}{\langle P_{GLS} \rangle} \exp\left[-\frac{P_{GLS}}{\langle P_{GLS} \rangle} \right] = \exp[-P_{GLS}]. \qquad (9.143)$$

The probability of getting a false alarm with $P_{GLS} > P_0$ is

$$P(\textit{false}) = 1 - (1 - \exp[-P_0])^M \approx M \exp[-P_0], \qquad (9.144)$$

where M is the number of independent frequencies in the periodogram. As for the ordinary Lomb-Scargle periodogram, the value of M is not well defined. The reader is referred to the extended discussion of M and of spectral windows for ordinary Lomb-Scargle periodograms at the end of Section 9.5.2, as that discussion applies equally to the generalized Lomb-Scargle periodogram.

9.6 Signals with Variable Periods: The O−C Diagram

Techniques based directly or indirectly on Fourier analysis make the fundamental assumption that a sequence is composed of sines and cosines with constant periods, amplitudes, and phases. Yet nature rarely, perhaps never, exhibits phenomena with such perfectly constant properties. The Fourier transform of even the simplest of functions with a changing period can be truly inscrutable. Consider a cosine defined between times t_0 and t_1. Let its angular frequency be ω_0 at t_0 and let its frequency change linearly with time at rate $\dot{\omega}_0 = d\omega/dt|_{t_0}$, so that

$$f(t) = \cos(\omega_0[t - t_0] + \dot{\omega}_0[t - t_0]^2/2), \quad t_0 \le t \le t_1. \qquad (9.145)$$

The Fourier transform of $f(t)$ is the convolution of a sinc function, a delta function at ω_0, and the Fourier transform of $\cos(\dot{\omega}_0 t^2/2)$, which, from standard tables of Fourier

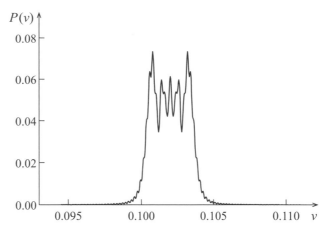

Figure 9.17: Power spectrum of a cosine curve with a period that increases linearly with time (equation 9.145). The initial period is 10 seconds, Δt is 5000 seconds, and $\dot{P} = 8 \times 10^{-5}$. The cosine has 10 fewer cycles over the length of the sequence compared to a cosine with a constant period.

transforms, is $\cos(\omega^2/2\dot{\omega}_0 - \pi/4)$.[8] The resulting Fourier transform and its corresponding power spectrum can be complex, with many nearly equal peaks. Figure 9.17 shows the power spectrum of a particular sequence generated from equation 9.145. It would be easy to misinterpret the power spectrum and conclude, for example, that the sequence consists of five superimposed sine curves instead of a single sine curve with a changing period.

The $O - C$ ("Observed minus Calculated") diagram is a useful tool for investigation phenomena with variable periods. Suppose that recurring events in a time sequence can be individually distinguished and their times measured. Good examples would be the times of the eclipses of an eclipsing binary star or the times of maximum brightness of a pulsating variable star. Suppose further that each event can be numbered with an integer E that gives its order in the sequence, and let $O(E)$ be the observed time at which event E occurs. Now suppose that there is a model from which the expected times of the events $C(E)$ can be calculated. The quantity $O(E) - C(E)$ measures the departure of the observed times of the events from their expected times. The $O - C$ diagram is a plot of $O - C$ against either E or time.

If the period is not changing too fast, the $O - C$ diagram often gives a more directly intuitive understanding of the properties of a sequence. To see this, return to equation 9.145. Let the events be the maxima of the cosine curve, and compare the times of the maxima, $O(E)$, to the times predicted by a model in which the period of the cosine curve is constant:

$$C(E) = t_0 + P_0 E. \tag{9.146}$$

Figure 9.18 shows the $O - C$ diagram for the sequence from which the power spectrum in Figure 9.17 was calculated. The observed times depart quadratically from the predicted times. To see why this is true, note that the observed peaks of the cosine curve occur at times

$$2\pi E = \omega_0(t - t_0) + \dot{\omega}_0(t - t_0)^2/2. \tag{9.147}$$

---

[8] The Fourier transform of equation 9.145 can be calculated by converting the function to complex notation and completing the square.

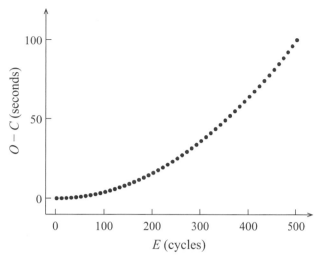

Figure 9.18: The $O - C$ diagram for the peaks of a cosine curve with a period that changes linearly with time. The initial period is 10 seconds, Δt is 5000 seconds, and $\dot{P} = 8 \times 10^{-5}$. The cosine has 10 fewer cycles over the length of the sequence compared to a cosine with a constant period. This is the same sequence used to calculate the power spectrum shown in Figure 9.17.

Recognizing that $\omega = 2\pi/P$ and $\dot{\omega} = -2\pi\dot{P}/P^2$, this is

$$E = \frac{1}{P_0}(t - t_0) - \frac{1}{2}\frac{\dot{P}_0}{P_0^2}(t - t_0)^2. \tag{9.148}$$

Solving for t and expanding in a power series about t_0, we find

$$O(E) = t = t_0 + P_0 E + \frac{1}{2}P_0\dot{P}_0 E^2 + \cdots. \tag{9.149}$$

Now subtract $C(E)$ to get

$$O - C = \frac{1}{2}P_0\dot{P}_0 E^2 + \cdots, \tag{9.150}$$

which is the expected quadratic dependence.

The most common comparison models in $O - C$ diagrams are low-order polynomials. One generally expands $C(E)$ directly in terms of E to get

$$C(E) = t_0 + \left.\frac{dt}{dE}\right|_{t_0} E + \frac{1}{2}\left.\frac{d^2t}{dE^2}\right|_{t_0} E^2 + \frac{1}{6}\left.\frac{d^3t}{dE^3}\right|_{t_0} E^3 + \cdots. \tag{9.151}$$

Recognizing that

$$\frac{dt}{dE} = \text{time per cycle} = P \tag{9.152}$$

and

$$\frac{d^2t}{dE^2} = \frac{dP}{dE} = \frac{dt}{dE}\frac{dP}{dt} = P\dot{P}, \tag{9.153}$$

we arrive at

$$C(E) = t_0 + P_0 E + \frac{1}{2}P_0\dot{P}_0 E^2 + \frac{1}{6}\left[P_0\dot{P}_0^2 + P_0^2\ddot{P}\right]E^3 + \cdots. \tag{9.154}$$

Equation 9.154 can be expanded to higher order if desired.

While one can fit a model directly to observed times of events, it is often more convenient to fit the residuals in the $O - C$ diagram. Suppose, for example, that the calculated event times come from a quadratic model (equation 9.154 truncated after the quadratic term). If the model agrees with the observations perfectly, $O - C = 0$ for all E; but if the model is imperfect, $O - C$ will drift upward or downward in the $O - C$ diagram, as it does in Figure 9.18. To correct the parameters of the model, fit the quadratic equation

$$O - C = a + bE + cE^2 \tag{9.155}$$

to the drifting $O - C$ values, perhaps by χ^2 minimization. The revised quadratic model is

$$C = t'_0 + P'_0 E + \frac{1}{2} P'_0 \dot{P}'_0 E^2, \tag{9.156}$$

where

$$t'_0 = t_0 + a \tag{9.157}$$

$$P'_0 = P_0 + b \tag{9.158}$$

$$\dot{P}'_0 = \frac{P_0 \dot{P}_0 + 2c}{P_0 + b}. \tag{9.159}$$

One special case deserves comment. For some phenomena, the $O - C$ residuals can themselves be periodic. Consider events whose times are given by

$$O(E) = t_0 + P_0 E + f(\Pi, E), \tag{9.160}$$

where $f(\Pi, E)$ is periodic with period Π. This can happen if, for example, one observes a pulsating variable star that is in orbit around another star. The distance from the pulsating star to the Earth increases and decreases as the star moves around its orbit, causing the observed times of the pulsations to be late or early by the length of time it takes light to travel

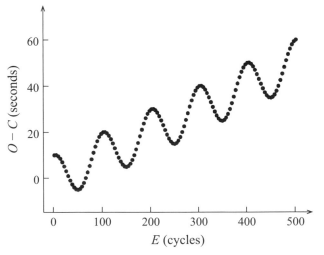

Figure 9.19: The $O - C$ diagram for the peaks of a cosine curve with a period that changes sinusoidally with time, as in equation 9.161. The overall slope in the $O - C$ diagram is nonzero, because C was calculated with a the reference period that is too short. The sinusoid superimposed on the overall slope has a period $\Pi = 100P$ and an amplitude of 10 seconds.

the increased or decreased distance, respectively. If the orbit is circular and the maximum delay is A at orbital phase ϕ_0, the observed times of the pulsations occur at

$$O(E) \;=\; t_0 + P_0 E + A \cos(2\pi \frac{P_0}{\Pi} E - \phi_0). \tag{9.161}$$

If the calculated times are $C = t_0 + P_0 E$, then $O - C$ oscillates sinusoidally about 0. Figure 9.19 shows the $O - C$ diagram for a specific example. The figure is easily interpretable as a single sine curve whose frequency is sinusoidally modulated.

10

Analysis of Sequences: Convolution and Covariance

Power spectra and periodograms are powerful tools for analyzing deterministic signals that are periodic or whose periods vary slowly. If the signal is not periodic, if its period changes rapidly, or if it is a stochastic signal, convolution functions, covariance functions, and related tools are often more useful. Section 10.1 introduces impulse response functions and frequency response functions as alternate ways to understand convolution. Section 10.2 discusses deconvolution and the reconstruction of sequences that have been degraded by noise and smearing. Most of the section is devoted to Wiener deconvolution and the Richardson-Lucy algorithm, which are bases for many image reconstruction techniques. Section 10.3 discusses autocovariance functions, especially autocovariance functions of stochastic processes, culminating with a lengthy analysis of quasiperiodic oscillations. Section 10.4 is devoted to cross-covariance functions and their use for finding weak, nonperiodic signals buried in noise.

10.1 Convolution Revisited

In Chapter 9 convolution was used to analyze spectral leakage and aliasing in power spectra and periodograms: the product of a sequence with a data window becomes the convolution of their Fourier transforms. In this chapter we examine the effect of convolving a function directly with the data sequence. Impulse response functions and frequency response functions are more appropriate for understanding direct convolution than are data windows and spectral windows.

10.1.1 Impulse Response Function

We begin by reexamining an example from Section 8.6.1. Let $f(t)$ be the delta function

$$f(t) = a_1 \delta(t - t_1) \tag{10.1}$$

and $h(t)$ be the exponential function

$$h(t) = \begin{cases} 0, & t < 0 \\ \exp[-t], & t \geq 0 \end{cases} . \tag{10.2}$$

Their convolution is

$$g(\tau) = \int_{-\infty}^{\infty} f(t)h(\tau - t)dt = \int_{-\infty}^{\infty} a_1\delta(t - t_1)h(\tau - t)dt$$

$$= \begin{cases} 0, & \tau < t_1 \\ a_1\exp[-(\tau - t_1)], & \tau \geq t_1 \end{cases}. \tag{10.3}$$

This is $h(t)$ multiplied by a_1 and translated in time to the position of the delta function. If $f(t)$ is the sum of n delta functions,

$$f(t) = \sum_{j=1}^{n} a_j\delta(t - t_j), \tag{10.4}$$

its convolution with $h(t)$ is

$$g(\tau) = \left[\sum_{j=1}^{n} a_j\delta(t - t_j)\right] \otimes h(t) = \sum_{j=1}^{n} a_j\left[\delta(t - t_j) \otimes h(t)\right] = \sum_{i=1}^{n} a_j h(\tau - t_j). \tag{10.5}$$

Thus $g(\tau)$ is the sum of many exponentials, each replicated at the position of a delta function. Figure 10.1 shows an example in which three delta functions are convolved with an exponential. One can think of the delta functions as producing the exponentials.

This leads to a different way of interpreting convolution. Consider a physical system that is described by a differential equation. For example, an object whose velocity v is slowed by a damping force proportional to its velocity can be described by the differential equation

$$\frac{dv}{dt} + \alpha v = 0, \tag{10.6}$$

where α is the damping coefficient. The solution to the differential equation is

$$v(t) = v_0\exp[-\alpha t], \tag{10.7}$$

which we recognize as an exponentially decreasing velocity. The response of this system to an arbitrary driving force $f(t)$ is given by the inhomogeneous differential equation

$$\frac{dv}{dt} + \alpha v = f(t). \tag{10.8}$$

Suppose that the velocity of the object is initially zero but is abruptly increased by one unit at time $t = 0$. Denote the resulting velocity by $h(t)$. One can form a differential equation for $h(t)$ by replacing v by h in equation 10.8 and replacing $f(t)$ by the delta function $\delta(t)$ to give

$$\frac{dh}{dt} + \alpha h = \delta(t). \tag{10.9}$$

For this case, we can write down $h(t)$ by inspection:

$$h(t) = \begin{cases} 0, & t < 0 \\ \exp[-\alpha t], & t \geq 0 \end{cases}. \tag{10.10}$$

If desired, equation 10.9 can be solved for $h(t)$ formally using Green's functions (see Appendix G).

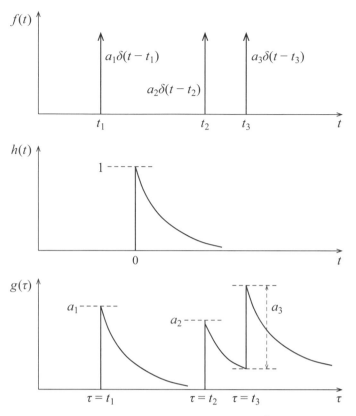

Figure 10.1: (Top panel) The sum of three delta functions $f(t) = \sum_{j=1}^{3} a_j \delta(t - t_j)$. (Middle panel) An exponential function $h(t) = \exp[-t]$, $t \geq 0$. (Bottom panel) The convolution of the two, $g(\tau) = f(t) \otimes g(t)$. It is the sum of three replicated exponentials, one for each delta function.

Once $h(t)$ is known, the response of the system to *any* arbitrary $f(t)$ is given by the convolution

$$v(t) = \int h(t - u) f(u) du. \qquad (10.11)$$

To see that $v(t)$ does indeed satisfy the original differential equation, take

$$\frac{dv}{dt} + bv = \frac{d}{dt}\left[\int h(t-u)f(u)du\right] + b\left[\int h(t-u)f(u)du\right]$$

$$= \int \left[\frac{dh(t-u)}{dt} + bh(t-u)\right]f(u)du = \int \delta(t-u)f(u)du$$

$$= f(t). \qquad (10.12)$$

The delta function is an impulse to the system, so $h(t)$ is called the *impulse response function*. It is the residual effect of the impulse at time t after the impulse. The product $h(t - u)f(u)$ is the residual effect at time $t - u$ of a perturbation $f(t)$ to the system at time t. The convolution integral sums the residual effects at t of all the previous perturbations.

We can now reinterpret the example at the beginning of this section. The solution to equation 10.9 is $h(t) = \exp[-\alpha t]$, which is the same as equation 10.2 if the damping coefficient is $\alpha = 1$. Thus equation 10.2 is the impulse response function of an object whose velocity is slowed by friction. Equation 10.4 is the forcing function, a series of sharp perturbations to the velocity; equation 10.5 is the resulting velocity of the object.

Similar results hold for any inhomogeneous linear ordinary differential equation. The general inhomogeneous linear ordinary differential equation can be written as

$$z_n(t)\frac{d^n g}{dt^n} + z_{n-1}(t)\frac{d^{n-1}g}{dt^{n-1}} + \cdots + z_1(t)\frac{dg}{dt} + z_0(t)g = f(t), \tag{10.13}$$

where the $z_j(t)$ are functions of t. The left-hand side of this equation can be thought of as a description of a complicated system, and $f(t)$ as a force driving the system. One can replace $f(t)$ by $\delta(t)$ and, in principle, solve the differential equation for the impulse response function $h(t)$. The response of the system to the driving force is given by the convolution

$$g(t) = h(t) \otimes f(t). \tag{10.14}$$

One usually normalizes the impulse response function by setting

$$\int h(t)dt = 1. \tag{10.15}$$

If $h(t)$ is normalized, then

$$\int g(\tau)d\tau = \int_\tau \int_t h(\tau - t)f(t)dtd\tau = \int_t f(t)\left[\int_\tau h(\tau - t)d\tau\right]dt = \int f(t)dt, \tag{10.16}$$

which means that the impulse response function conserves flux (see the following example).

Why use impulse response functions? Since the differential equation is a model of the system, the impulse response function is also a model of the system. In many cases the impulse response function is known or easily measured, but the corresponding differential equation is not. Even if the differential equation is known, it is often easier to calculate the response from the convolution integral rather than by solving an inhomogeneous differential equation. This property makes the impulse response function interesting and useful.

In the following two examples, the impulse response functions can be directly measured and provide a natural way to understand the properties of the experimental data. There is no need to know the differential equation corresponding to the impulse response function. The first example also shows one way to extend the results for a one-dimensional sequence to a two-dimensional image.

Example: Stars are so far away that, with few exceptions, they are point sources of light. However, images of stars obtained with cameras on telescopes are always extended. The images are smeared out by the optics of the telescope and camera and, if the telescope is on the Earth, they are also smeared by passage through the Earth's atmosphere.

The undistorted image of a single star can be represented by the two-dimensional delta function $a_1\delta(x - x_1, y - y_1) = a_1\delta(x - x_1)\delta(x - x_2)$, where a_1 is the flux of light

Continued on page 296

from the star, and (x_1, y_1) is the location of the star image. An image containing n stars can be represented by the sum of n delta functions

$$f(x,y) = \sum_{j=1}^{n} a_j \delta(x - x_j, y - y_j).$$

The smearing of this ideal image is often represented by the point spread function $p(x,y)$, such that the measured image $I(x,y)$ is the convolution of the point spread function with the ideal image:

$$I(x,y) = \int_u \int_v \left\{ \sum_{j=1}^{n} a_j \delta(u - x_j, v - y_j) \right\} p(x - u, y - v) du\, dv$$

$$= \sum_{j=1}^{n} a_j p(x - x_j, y - y_j).$$

The two-dimensional impulse response function $h(\sigma, \tau)$ for the telescope/camera/atmosphere is the response to a delta function at the origin:

$$h(\sigma, \tau) = \int_u \int_v \delta(u, v) p(\sigma - u, \tau - v) du\, dv = p(\sigma, \tau).$$

We now recognize that the point spread function is the same as the impulse response function.

This also suggests a way to determine the impulse response function empirically: Take the weighted mean of observed images of well separated stars. Once the impulse response function is known, the response of the system to any image, not just stars, can be calculated by convolving the impulse response function with the image.

Example: One of the most important ways to describe the light from a star is by its spectrum, one version of which is a plot of the intensity of the light as a function of wavelength. The measured spectrum $I(\lambda)$ is not the same as the true spectrum $I_0(\lambda)$, because the spectrometer has limited resolution and optical imperfections. The effect of the spectrograph can be described by a convolution of the instrumental profile $p(\lambda)$ with the true spectrum:

$$I(\lambda) = \int_\infty^\infty I_0(u) p(\lambda - u) du.$$

Suppose one shines monochromatic light at wavelength λ_0 into the spectrograph, perhaps from a laser. The spectrum of the monochromatic light can be described by a delta function $\delta(\lambda - \lambda_0)$. This will be degraded by the spectrograph to

$$I(\lambda) = \int_\infty^\infty \delta(u - \lambda_0) p(\lambda - u) du = p(\lambda - \lambda_0).$$

We recognize this as the impulse response function recentered to λ_0. Thus the instrumental profile of the spectrograph is the same as its impulse response function.

10.1.2 Frequency Response Function

Suppose that a system described by the impulse response function $h(t)$ is subjected to a sinusoidal driving force

$$f(t) = \sin(2\pi \nu t). \tag{10.17}$$

From equation 10.14, the response of the system is

$$g(t) = h(t) \otimes f(t) = \int_{-\infty}^{\infty} h(u) \sin[2\pi \nu (t - u)] du. \tag{10.18}$$

Expanding the sine function, we find

$$g(t) = \int_{-\infty}^{\infty} h(u) [\sin(2\pi \nu t) \cos(2\pi \nu u) - \cos(2\pi \nu t) \sin(2\pi \nu u)] \, du$$
$$= A(\nu) \sin(2\pi \nu t) + B(\nu) \cos(2\pi \nu t), \tag{10.19}$$

where

$$A(\nu) = \int_{-\infty}^{\infty} h(u) \cos(2\pi \nu u) du \tag{10.20}$$

$$B(\nu) = -\int_{-\infty}^{\infty} h(u) \sin(2\pi \nu u) du. \tag{10.21}$$

It is more illuminating to put $g(t)$ in the form

$$g(t) = Z(\nu) \sin[2\pi \nu t + \phi(\nu)], \tag{10.22}$$

where

$$Z(\nu) = \left[A^2(\nu) + B^2(\nu) \right]^{1/2} \tag{10.23}$$

$$\tan[\phi(\nu)] = \frac{B(\nu)}{A(\nu)}. \tag{10.24}$$

Thus the response of the system to a sinusoidal driving function is a sinusoid with a frequency equal to the driving frequency. The amplitude of the response is $Z(\nu)$, and the response is shifted in phase by the amount $\phi(\nu)$. The function $Z(\nu)$ is called the *gain* and $\phi(\nu)$ is the *phase shift*.

It is often convenient to write the driving force in complex form:

$$f(t) = \exp[i2\pi \nu t]. \tag{10.25}$$

The response is then

$$g(t) = \int_{-\infty}^{\infty} h(u) \exp[i2\pi \nu (t - u)] du$$
$$= \exp[i2\pi \nu t] \int_{-\infty}^{\infty} h(u) \exp[-i2\pi \nu u] du. \tag{10.26}$$

This can be written as

$$g(t) = H(\nu) \exp[i2\pi \nu t], \tag{10.27}$$

where $H(\nu)$ is called the *frequency response function* and is given by

$$H(\nu) = \int_{-\infty}^{\infty} h(u) \exp[-i2\pi \nu u] du. \tag{10.28}$$

The frequency response function is the Fourier transform of the impulse response function. The frequency response function is related to the gain and phase shift through

$$H(\nu) = Z(\nu) \exp[i\phi(\nu)], \tag{10.29}$$

where

$$Z(\nu) = \left[H(\nu)H^*(\nu)\right]^{1/2} \tag{10.30}$$

$$\tan[\phi(\nu)] = \frac{\sin[\phi(\nu)]}{\cos[\phi(\nu)]} = \frac{\mathrm{Im}[H(\nu)]}{\mathrm{Re}[H(\nu)]}. \tag{10.31}$$

The following example shows how the frequency response function can yield information about the behavior of a system that is not immediately apparent from the impulse response function.

Example: Let the impulse response function of a system be an exponential with a decay time T:

$$h(t) = \begin{cases} 0, & t < 0 \\ \dfrac{1}{T} \exp[-t/T], & t \geq 0 \end{cases}. \tag{10.32}$$

The corresponding frequency response function is

$$H(\nu) = \int_{-\infty}^{\infty} h(u) \exp[-i2\pi \nu u] du = \int_{0}^{\infty} \frac{1}{T} \exp[-u/T] \exp[-i2\pi \nu u] du.$$

To evaluate this integral, change variables to $\zeta = u/T$. Then

$$H(\nu) = \int_{0}^{\infty} \exp[-\zeta] \exp[-i2\pi \nu T \zeta] \, d\zeta = \int_{0}^{\infty} \exp[-(1 + i2\pi \nu T)\zeta] \, d\zeta$$

$$= \frac{1}{1 + i2\pi \nu T} = \frac{1 - i2\pi \nu T}{1 + (2\pi \nu T)^2}. \tag{10.33}$$

From equation 10.30, the gain is

$$Z(\nu) = \frac{1}{\left[1 + (2\pi \nu T)^2\right]^{1/2}}, \tag{10.34}$$

and from equation 10.31, the phase shift is

$$\tan[\phi(\nu)] = -2\pi \nu T. \tag{10.35}$$

Figure 10.2 shows plots of $h(t)$, $Z(\nu)$, and $\phi(\nu)$. At low driving frequencies, the gain is near 1 and the phase shift is near 0, so the system follows the driving force; but as the frequency increases, the response of the system decreases in amplitude and lags in phase.

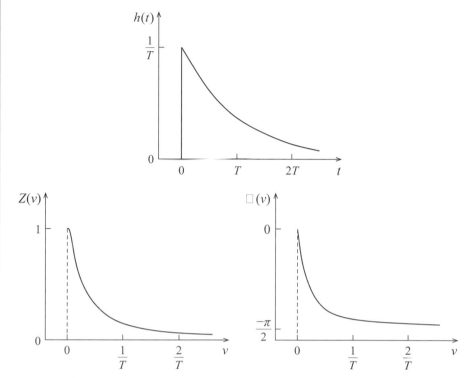

Figure 10.2: (Top panel) The exponential impulse response function $h(t) = (1/T)\exp[-t/T]$. The gain $Z(v)$ (bottom left panel) and phase shift $\phi(v)$ (bottom right panel) of the system are shown in response to a sinusoidal driving force with frequency v (equations 10.34 and 10.35).

Suppose that a system with impulse response function $h(t)$ is subjected to an arbitrary driving force $f(t)$, yielding the response $g(t) = h(t) \otimes f(t)$. From the convolution theorem, the Fourier transform of the response is

$$G(v) = H(v)F(v), \tag{10.36}$$

where $F(v)$ is the Fourier transform of the input function.[1] The inverse Fourier transform of $G(v)$ returns the original response, so

$$g(t) = \int_{-\infty}^{\infty} G(v)\exp[+i2\pi vt]dv = \int_{-\infty}^{\infty} H(v)F(v)\exp[+i2\pi vt]dv. \tag{10.37}$$

The response of a system to an arbitrary input can therefore be calculated from the frequency response function, albeit through two Fourier transforms, the first deriving $F(v)$ from $f(t)$ and the second deriving $g(t)$ from $G(v)$.

---

[1] It is worth a warning here and in connection with equation 10.39: do not confuse the frequency response function with a closely related function called the *transfer function*, $H(s)$. The transfer function is typically defined to be $H(s) = G(s)/F(s)$, where $G(s)$ and $F(s)$ are the Laplace transforms (not Fourier transforms) of $g(t)$ and $f(t)$, respectively.

Example: The top panel of Figure 10.3 shows the sequence described by the function

$$f(t) = \sin(2\pi \nu_0 t) + \sin(4\pi \nu_0 t),$$

with $\nu_0 = 0.001$. We wish to convolve $f(t)$ with the exponential impulse response function

$$h(t) = 4\nu_0 \exp[-4\nu_0 T].$$

Noting that $h(t)$ is the same as $h(t)$ in equation 10.32 with $T = 1/4\nu_0$, we can greatly speed the convolution by using the gain and phase shift functions corresponding to $h(t)$. Equation 10.34 becomes

$$Z(\nu) = \frac{1}{[1 + (\pi/2)^2 (\nu/\nu_0)^2]^{1/2}},$$

and the gains for the two sine curves that make up $f(t)$ are 0.537 and 0.303. Equation 10.35 becomes

$$\tan[\phi(\nu)] = -\frac{\pi}{2}\frac{\nu}{\nu_0},$$

and the phase shifts are -1.00 and -1.26. The convolved function is therefore

$$g(t) = 0.537 \sin(2\pi \nu_0 t - 1.00) + 0.303 \sin(4\pi \nu_0 t - 1.26). \tag{10.38}$$

Equation 10.38 is plotted in the bottom panel of Figure 10.3.

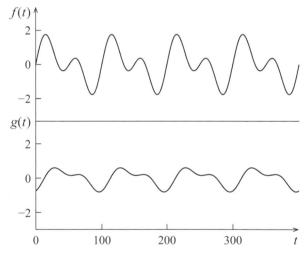

Figure 10.3: (Top panel) A plot of the function $f(t) = \sin(2\pi \nu_0 t) + \sin(4\pi \nu_0 t)$ with $\nu_0 = 0.01$. (Bottom panel) A plot of $g(t) = f(t) \otimes h(t)$, where $h(t)$ is the exponential impulse response function, $h(t) = 4\nu_0 \exp[-4\nu_0 t/T]$. The plot was made not by direct numerical convolution, but instead by plotting the functional form of $g(t)$ (equation 10.38), which was determined by calculating the gains and phase shifts of the two sine curves that make up $f(t)$.

The amplitudes and phases of the two sine curves are changed, and they are changed by different amounts. As a result, the shape of the sequence is altered greatly by the convolution.

Equations 10.36 and 10.37 provide an alternate way to measure the impulse response function. Suppose that the impulse and frequency response functions for a system are unknown. They can be measured by driving the system with a known forcing function $f(t)$ and measuring the response $g(t)$. Calculate the Fourier transforms of the known driving force and the measured response to get $F(v)$ and $G(v)$. From equation 10.37, the frequency response function is

$$H(v) = \frac{G(v)}{F(v)} = \frac{F^*(v)G(v)}{|F(v)|^2}. \tag{10.39}$$

The impulse response function is given by the inverse Fourier transform of the frequency response function

$$h(t) = \int_{-\infty}^{\infty} H(v)\exp[+i2\pi vt]dv = \int_{-\infty}^{\infty} \frac{F^*(v)G(v)}{|F(v)|^2}\exp[+i2\pi vt]dv. \tag{10.40}$$

In a laboratory, this can easily be done by driving the system with a sine curve of constant amplitude, $f(t) = \sin(2\pi vt)$, sweeping the sine curve through frequency, and measuring the gain $Z(v)$ and phase shift $\phi(v)$ at each frequency. Then $F(v) = $ constant, and the impulse response function is calculated from

$$h(t) = \int_{\infty}^{\infty} G(v)\exp[+i2\pi vt]dv = \int_{\infty}^{\infty} Z(v)\exp[i\phi(v)]\exp[+i2\pi vt]dv. \tag{10.41}$$

If necessary, $h(t)$ can then be normalized.

10.2 Deconvolution and Data Reconstruction

10.2.1 Effect of Noise on Deconvolution

Suppose one measures a sequence $g(t)$ produced by convolving a known impulse response function $h(t)$ with an input sequence $f(t)$:

$$g(t) = h(t) \otimes f(t). \tag{10.42}$$

One sometimes wants to deconvolve $g(t)$ to retrieve the original input sequence. The image of a cluster of stars is an example. The image of the cluster is degraded by passage through the Earth's atmosphere and then through the telescope and camera optics. The degradation can be described as a convolution of the original image with a point spread function that models the effect of the atmosphere and optics. We may want to deconvolve the measured image to retrieve the original image of the cluster, perhaps to reveal close pairs of stars.

Superficially, deconvolution appears to be a straightforward process if the impulse response function is known. Suppose g_j, h_j, and f_j are all discrete sequences with the same uniform intervals. The measured sequence g_j is the convolution of the impulse response function h_j with the unknown input sequence f_j (see equation 8.140):

$$g_j = \sum_{k=1}^{n} h_{j+1-k}f_k, \quad j=1,\ldots,m. \tag{10.43}$$

Here n is the length of the original sequence, m is the length of the measured sequence, and we assume that h_j is known over the range $j = -n$ to $j = n$. We will assume throughout that the impulse response function is normalized, so that

$$\sum_j h_j = 1. \tag{10.44}$$

Equation 10.43 can be written as the matrix equation (see equation 8.142)

$$\mathbf{g} = \mathfrak{H}\mathbf{f}, \tag{10.45}$$

where the elements of the vectors and matrix are

$$(\mathbf{g})_j = g_j \tag{10.46}$$

$$(\mathbf{f})_j = f_j \tag{10.47}$$

$$(\mathfrak{H})_{jk} = \mathfrak{h}_{jk} = h_{j+1-k}. \tag{10.48}$$

Note that the rows of the $\mathfrak{H}$ matrix are equal to h_j reversed and shifted by an amount equal to the row number k. If $n = m$ and if the matrix is nonsingular, the original sequence can be retrieved by inverting the matrix and then multiplying the inverse into $\mathbf{g}$:

$$\mathfrak{H}^{-1}\mathbf{g} = \mathfrak{H}^{-1}\mathfrak{H}\mathbf{f} = \mathbf{f}. \tag{10.49}$$

Alternatively, $g(t)$ can be deconvolved in Fourier space. The Fourier transform of equation 10.42 is

$$G(\nu) = H(\nu)F(\nu), \tag{10.50}$$

where $G(\nu)$, $H(\nu)$, and $F(\nu)$ are the Fourier transforms of $g(t)$, $h(t)$, and $f(t)$, respectively. Note that $H(\nu)$ is the frequency response function (see equation 10.28). To retrieve $F(\nu)$, divide $G(\nu)$ by the frequency response function:

$$F(\nu) = \frac{G(\nu)}{H(\nu)} = \frac{H^*(\nu)G(\nu)}{|H(\nu)|^2} = \frac{H^*(\nu)G(\nu)}{Z(\nu)^2}, \tag{10.51}$$

where $Z(\nu)$ is the gain (see equations 10.23 and 10.32). The original sequence can be reconstructed by taking the inverse Fourier transform of $F(\nu)$:

$$f(t) = \int_\infty^\infty F(\nu)\exp[i2\pi\nu t]d\nu = \int_\infty^\infty \frac{H^*(\nu)G(\nu)}{Z(\nu)^2}\exp[i2\pi\nu t]d\nu. \tag{10.52}$$

Equation 10.49 will fail if $\mathfrak{H}$ is singular, and equation 10.52 will fail if $Z(\nu)$ equals 0 at any ν. A more serious problem is that the measured sequence is nearly always contaminated by noise, so equation 10.42 is not an accurate description of how $g(t)$ was produced. If the noise is additive, the correct equation is

$$g(t) = h(t) \otimes f(t) + \epsilon(t), \tag{10.53}$$

where $\epsilon(t)$ is the noise. Equation 10.51 must be replaced by

$$\frac{G(\nu)}{H(\nu)} = \frac{G(\nu) + E(\nu)}{H(\nu)} = F(\nu) + \frac{H^*(\nu)E(\nu)}{Z(\nu)^2}, \tag{10.54}$$

where $E(\nu)$ is the Fourier transform of $\epsilon(t)$. Thus simple deconvolution will amplify the noise by an amount $H^*(\nu)/Z(\nu)^2 \approx 1/Z(\nu)$. The following example shows that the noise amplification can be large.

Example: Let us return to the example in Figure 10.3 The top panel of Figure 10.3 shows the function $f(t) = \sin(2\pi \nu_0 t) + \sin(4\pi \nu_0 t)$ with $\nu_0 = 0.01$, and the bottom panel shows the convolution $g(t) = h(t) \otimes f(t)$, where the impulse response function is $h(t) = 4\nu_0 \exp[-4\nu_0 t/T]$. We will attempt to reconstruct $f(t)$ from $g(t)$ twice, first when $g(t)$ is free of noise and second, when a small amount of noise has been added to $g(t)$.

Deconvolution in the Absence of Noise: Figure 10.4 shows the result of deconvolving $g(t)$ using using the Fourier transform method. The convolved function $g(t)$ (equation 10.38) is reproduced in the top left panel of the figure. Its Fourier transform $G(\nu)$ was calculated from equations 8.127 and 8.128. The sequence in the bottom left panel was constructed by calculating $F(\nu)$ from

$$F(\nu) \;=\; G(\nu)/H(\nu) \;=\; G(\nu)(1+i2\pi\nu T) \;=\; G(\nu)\left(1+i\frac{\pi}{2}\frac{\nu}{\nu_0}\right) \qquad (10.55)$$

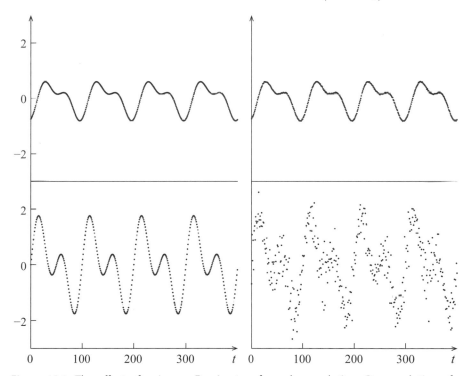

Figure 10.4: The effect of noise on Fourier transform deconvolution. Deconvolution of a noise-free sequence (left panels). The top left panel shows $g(t) = f(t) \otimes h(t)$, where $f(t)$ is the sum of two sine curves, and $h(t) = 4\nu_0 \exp[-4\nu_0/T]$ (equation 10.38). In the bottom left panel the original sequence has been successfully reconstructed using Fourier deconvolution (equation 10.55). Deconvolution of a noisy sequence (right panels). The top right panel shows the convolved sequence again, but now uncorrelated white noise has been added. The amplitude of the noise is so small that the noise is almost invisible in the figure. The bottom right panel shows the deconvolved sequence. While the original sequence has been reconstructed correctly in the mean, the noise is greatly magnified.

Continued on page 304

(see equation 10.33) and then calculating $f(t)$ from the inverse Fourier transform of $F(v)$ (equation 8.126). The original function has been reconstructed almost perfectly.

Deconvolution When Noise Is Present: The top right panel of Figure 10.4 replots $g(t)$ with uncorrelated white noise added. The noise has a Gaussian distribution with a 0 mean and a standard deviation of $\sigma = 0.01$. The amplitude of the noise is so small that it is almost invisible in the figure. The bottom panel shows the result of deconvolving this noisy sequence in exactly the same way as the noise-free sequence. The original sequence has been reconstructed correctly in the mean, but the noise is greatly magnified.

The reason the noise was amplified so much in this example is that the mean power spectrum of uncorrelated noise is flat and extends all the way to the Nyquist frequency. The Fourier transform deconvolution multiplies the Fourier components by $[1 + i(\pi/2)(v/v_0)]$. For this case, the Nyquist frequency is $50v_0$, so the high-frequency components of $G(v)$—which are due entirely to the noise, not signal—are amplified by up to $\sim 25\pi$, a huge amount.

Noise amplification causes problems for any deconvolution method, not just Fourier deconvolution. Most convolutions smooth the original sequence. If noise is introduced into the observed sequence, matrix-inversion deconvolution (equation 10.49) will "unsmooth" the abrupt point-to-point variations introduced by the noise as if they were real variations, magnifying the amplitude of the variations in the deconvolved sequence.

If prior knowledge about the noise and the original sequence is available, the effect of noise on the deconvolution can be ameliorated. One way to do this is by a constrained maximum likelihood or χ^2 minimization solution. Suppose a measured sequence g_j with m points is contaminated by uncorrelated Gaussian noise with a variances σ_j^2. Denote the reconstructed image by $\hat{f}_k$, and allow $\hat{f}_k$ to differ from the original sequence f_k by an amount consistent with the noise. If f_k is replaced by $\hat{f}_k$ in equation 10.43, the χ^2 statistic is

$$\chi^2 = \sum_{j=1}^{m} \frac{(g_j - \sum_{k=1}^{n} h_{j+1-k}\hat{f}_k)^2}{\sigma_j^2}. \tag{10.56}$$

One adjusts the $\hat{f}_k$ so as to minimize χ^2. This is essentially a linear least squares fit. However, there are as many values of the $\hat{f}_k$ as there are of the f_k, and equation 10.56 adds only one constraint. The reconstructed sequence is just as likely to have unsatisfactory artifacts as the sequence reconstructed by a naive matrix deconvolution or Fourier transform deconvolution. The artifacts can be ameliorated by adding additional constraints, usually many additional constraints. One might, for example, attempt to limit the size of point-to-point variations in the reconstructed sequence. The penalty is considerably greater complexity and often a lack of good motivation for the extra constraints.

Maximum likelihood is nevertheless a promising approach, because it can serve as a basis for maximum entropy and Bayesian image reconstructions. But the deconvolution and image reconstruction techniques more commonly employed are generally based on variants of Wiener deconvolution or the Richardson-Lucy algorithm, in part because they are easy to implement.

10.2.2 Wiener Deconvolution

Suppose that a measured function $g(t)$ is the convolution of an unknown function $f(t)$ with a known impulse response function $h(t)$ to which noise $\epsilon(t)$ has been added:

$$g(t) = f(t) \otimes h(t) + \epsilon(t). \tag{10.57}$$

The Fourier transform of this equation is

$$G(\nu) = F(\nu)H(\nu) + E(\nu), \tag{10.58}$$

where $G(\nu)$, $F(\nu)$, $H(\nu)$, and $E(\nu)$ are the Fourier transforms of $g(t), f(t), h(t)$, and $\epsilon(t)$, respectively. Wiener deconvolution constructs a function $W(\nu)$ such that the product of $W(\nu)$ and $G(\nu)$ yields a good approximation to $F(\nu)$ even when noise is present. The inverse Fourier transform of the approximation is the desired approximation to $f(t)$.

Denoting the approximate Fourier transform by $\hat{F}(\nu)$, so that

$$\hat{F}(\nu) = W(\nu)G(\nu), \tag{10.59}$$

we wish to find a $W(\nu)$ that minimizes the difference between $\hat{F}(\nu)$ and $F(\nu)$. We avoid using $[F(\nu) - \hat{F}(\nu)]^2$ to measure the difference, because this quantity is complex, and instead use the absolute value squared: $\Delta(\nu) = |F(\nu) - \hat{F}(\nu)|^2$. Because the difference is caused by noise, whose properties will be specified only in the mean, we characterize the difference between $\hat{F}(\nu)$ and $F(\nu)$ in the mean:

$$\langle \Delta(\nu) \rangle = \langle |F(\nu) - \hat{F}(\nu)|^2 \rangle. \tag{10.60}$$

The problem becomes: find the complex function $W(\nu)$ that minimizes $\langle \Delta(\nu) \rangle$. Expanding $\Delta(\nu)$ using equations 10.58 and 10.59, we find

$$\begin{aligned}
\Delta(\nu) &= |F(\nu) - W(\nu)F(\nu)H(\nu) - W(\nu)E(\nu)|^2 \\
&= |[1 - W(\nu)H(\nu)]F(\nu) - W(\nu)E(\nu)|^2 \\
&= [1 - W(\nu)H(\nu)]^* [1 - W(\nu)H(\nu)]F^*(\nu)F(\nu) \\
&\quad - [1 - W(\nu)H(\nu)]^* F^*(\nu)W(\nu)E(\nu) \\
&\quad - [1 - W(\nu)H(\nu)]F(\nu)W^*(\nu)E^*(\nu) \\
&\quad + W^*(\nu)E^*(\nu)W(\nu)E(\nu).
\end{aligned} \tag{10.61}$$

Notice that $F^*(\nu)F(\nu)$ and $E^*(\nu)E(\nu)$ are the power spectra of the signal and the noise, respectively. Denote these by

$$S(\nu) = F^*(\nu)F(\nu) \tag{10.62}$$

$$N(\nu) = E^*(\nu)E(\nu). \tag{10.63}$$

If the noise is uncorrelated with the signal, then

$$\langle F^*(\nu)E(\nu) \rangle = \langle F(\nu)E^*(\nu) \rangle = 0. \tag{10.64}$$

The mean value of equation 10.61 becomes

$$\langle \Delta(\nu) \rangle = [1 - W(\nu)H(\nu)]^* [1 - W(\nu)H(\nu)]S(\nu) + W^*(\nu)W(\nu)N(\nu). \tag{10.65}$$

Minimize the mean difference by setting the functional derivative of $\langle \Delta(\nu) \rangle$ with respect to $W(\nu)$ equal to 0:

$$\frac{\partial \langle \Delta(\nu) \rangle}{\partial W(\nu)} = -2[1 + W^*(\nu)H^*(\nu)]H(\nu)S(\nu) + 2W^*(\nu)N(\nu) = 0, \qquad (10.66)$$

from which we find

$$W(\nu) = \frac{H^*(\nu)S(\nu)}{N(\nu) + |H(\nu)|^2 S(\nu)} = \frac{1}{H(\nu)} \frac{|H(\nu)|^2}{|H(\nu)|^2 + N(\nu)/S(\nu)}. \qquad (10.67)$$

This is the Wiener deconvolution filter. In effect, it attenuates Fourier components at frequencies where mean noise is large compared to the signal. To use Wiener deconvolution, one must have some prior knowledge about the power spectra of the signal and the noise.

In the following extended example, Wiener deconvolution is used with good success to deconvolve a noisy sequence.

Example: Let us apply Wiener deconvolution to the example shown in Figure 10.4, for which simple Fourier deconvolution gave such a poor result.

To reprise: the original sequence $f(t)$ was the sum of two sine curves

$$f(t) = \sin(2\pi \nu_0 t) + \sin(4\pi \nu_0 t),$$

with $\nu_0 = 0.01$. This was convolved with the exponential impulse response function

$$h(t) = 4\nu_0 \exp[-4\nu_0 t], \quad t \geq 0,$$

and then noise $\epsilon(t)$ was added to give

$$g(t) = f(t) \otimes h(t) + \epsilon(t),$$

where $\epsilon(t)$ was uncorrelated noise with a Gaussian distribution, 0 mean, and a standard deviation of only $\sigma = 0.01$. Figure 10.4 shows plots of $g(t)$ before and after the addition of noise. We first attempted to reconstruct the original sequence by calculating

$$F(\nu) = G(\nu)/H(\nu),$$

where $F(\nu)$, $G(\nu)$, and $H(\nu)$ are the Fourier transforms of $f(t)$, $g(t)$, and $h(t)$, respectively, and

$$H(\nu) = \frac{1}{1 + i(\pi/2)(\nu/\nu_0)}.$$

When noise was present, the reconstructed sequence, shown in the bottom right panel of Figure 10.4, was unsatisfactory, because the noise was greatly amplified.

We now try Wiener deconvolution. A common way to determine the noise and signal in a sequence is to examine the power spectrum of the sequence. The power spectrum of $g(t)$ is shown in the top panel of Figure 10.5. Noting that the signal appears to be concentrated at low frequencies and that the noise at high frequencies is white (and aided by knowledge of the way the sequence was constructed!), we set the noise power to

$$N(\nu) = \text{constant} \approx 2.5 \times 10^{-7}.$$

In a real-world example, we would not know that the signal is confined to just two frequencies and would prefer to allow the signal to occur anywhere in a broad low-frequency band. Accordingly, we adopt a Gaussian for the distribution of signal power:

$$S(v) \propto A_G \exp\left[-\frac{1}{2}\frac{v^2}{\sigma_G^2}\right],$$

with the amplitude A_G set to 5 and σ_G set to 0.01 to make $S(v)$ outline the likely band of frequencies containing power. Figure 10.5 shows plots of $S(v)$ and $N(v)$ superimposed on the power spectrum of $g(t)$. Armed with $G(v)$, $H(v)$, $S(v)$, and $N(v)$, we reconstruct $F(v)$ from

$$F(v) = W(v)G(v),$$

where $W(v)$ is given by equation 10.67. The inverse Fourier transform of $F(v)$ yields the reconstructed $f(t)$. The reconstructed sequence is shown in the bottom panel of Figure 10.5.

The reconstructed signal is close enough to the original signal that the differences are not readily apparent, and the noise amplification has been ameliorated. The result is much more satisfactory than that form simple Fourier transform deconvolution. Several properties of $g(t)$ contributed to this good outcome: the signal was confined to a narrow band of frequencies, the noise had a relatively low amplitude, and the noise was spread over a much wider band of frequencies than the signal. Unfortunately, real data often lack one or more of these properties, so Wiener deconvolution often yields less impressive results for real data than it does for this example.

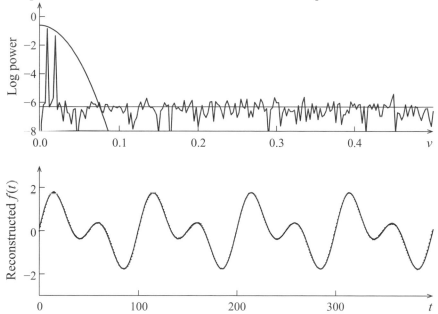

Figure 10.5: An example of Wiener deconvolution. The top panel shows the log power spectrum of the noisy sequence in the top right panel of Figure 10.4. The horizontal straight line is the mean power level introduced by the noise. (The location of the line looks odd to the eye because the vertical scale is log(power).) The inverted parabola is the Gaussian "signal" assumed for the Wiener deconvolution filter. The bottom panel shows the sequence reconstructed using Wiener deconvolution. It should be compared to the bottom right panel of Figure 10.4. The solid line is the original, noise-free sequence.

10.2.3 Richardson-Lucy Algorithm

The Richardson-Lucy algorithm avoids Fourier transforms altogether. Suppose that the original sequence f_j has been degraded to an observed sequence g_j by convolving it with an impulse response function h_j and then adding noise ϵ_j:

$$g_j = \sum_k h_{j+1-k} f_k + \epsilon_j = \sum_k \mathfrak{h}_{jk} f_k + \epsilon_j, \tag{10.68}$$

where $\mathfrak{h}_{jk} = h_{j+1-k}$. We search for a linear operator Q_{kj} that, when operating on g_j, produces an approximation $\hat{f}_j$ to f_j:

$$\hat{f}_k = \sum_j Q_{kj} g_j. \tag{10.69}$$

In the absence of noise we can solve for Q_{kj} exactly, since this equation is just equation 10.49 in disguise, so Q_{kj} is the same matrix as $\mathfrak{H}^{-1}$.

When noise is present, $\mathfrak{H}^{-1}$ is not a good choice for Q_{kj}, because it can greatly amplify the noise. Richardson and Lucy were led to a better choice for Q_{kj} by considering Bayes's theorem. Suppose we were to think of f_j and g_j as random deviates drawn from probability distributions. Bayes's theorem for the distributions would be written

$$P_1(f_k|g_j) = \frac{L(g_j|f_k)P_0(f_k)}{\sum_k L(g_j|f_k)P_0(f_k)}, \tag{10.70}$$

where P_0 and P_1 are the prior and posterior probability distributions, respectively, and L is the likelihood function (see Section 7.1). Comparison of the numerator of equation 10.70 to equation 10.68 suggests that (by analogy!) we should take

$$Q_{kj} = \frac{\mathfrak{h}_{jk} f_k}{\sum_k \mathfrak{h}_{jk} f_k}. \tag{10.71}$$

This is, in the absence of noise, an exact solution for Q_{kj}, since

$$\sum_j Q_{kj} g_j = \sum_j Q_{kj} \left[\sum_k \mathfrak{h}_{jk} f_k \right] = \sum_j \frac{\mathfrak{h}_{jk} f_k}{\sum_k \mathfrak{h}_{jk} f_k} \left[\sum_k \mathfrak{h}_{jk} f_k \right] = \sum_j \mathfrak{h}_{jk} f_k = f_k, \tag{10.72}$$

where the last step holds because h_j is normalized (equation 10.44). But as it stands, equation 10.71 is not a useful expression for Q_{kj}. It has two problems. First, when noise is present, the first equality in equation 10.72 is no longer true, so $\sum_j Q_{kj} g_j \neq f_k$; and second, we are trying to solve for f_k, but the expression requires that we already know f_k before we can calculate Q_{jk}.

The Richardson-Lucy algorithm replaces equation 10.71 for Q_{kj} by the iterative solution

$$Q_{kj}^{(r)} = \frac{\mathfrak{h}_{jk} \hat{f}_k^{(r)}}{\sum_k \mathfrak{h}_{jk} \hat{f}_k^{(r)}}, \tag{10.73}$$

and

$$\hat{f}_k^{(r+1)} = \sum_j Q_{kj}^{(r)} g_j, \tag{10.74}$$

where the superscript (r) means the rth iteration. The approximate sequence produced at the end of one iteration is used to calculate Q_{kj} for the next iteration. If desired, equations 10.73

and 10.74 can be combined:

$$\hat{f}_k^{(r+1)} = \hat{f}_k^{(r)} \frac{\sum_j \mathfrak{h}_{jk} g_j}{\sum_k \mathfrak{h}_{jk} \hat{f}_k^{(r)}}. \tag{10.75}$$

The Richardson-Lucy algorithm is insensitive to the beginning point of the iteration, so one typically just chooses $\hat{f}_k^{(0)} = 1$, or equivalently, $Q_{kj}^{(0)} = \mathfrak{h}_{jk}$.

The algorithm tends to converge slowly, requiring dozens or even hundreds of iterations. It is known to converge under most conditions and when it does, it converges to $\mathfrak{H}^{-1}\mathbf{g}$, that is, it deconvolves g_j as if it were a noise-free signal. This is *not* the desired result! The iteration must be terminated long before it comes close to its converged limit to avoid unwanted noise amplification.

When should the iteration be terminated? There is no one answer. Some users may stop when the reconstructed sequence (or image) "looks good." Others may stop when the changes in the reconstructed sequence from iteration to iteration slow down. The justification is that the changes slow down when the algorithm is attempting to reconstruct the rapid point-to-point variations in the sequence introduced by the noise. Lucy himself suggested iterating until the χ^2 defined by equation 10.56 has been reduced to

$$\chi^2 = \sum_{k=1}^{m} \frac{\left(g_k - \sum_{j=1}^{n} h_{k+1-j}\hat{f}_j\right)^2}{\sigma_k^2} \approx m, \tag{10.76}$$

where σ_k^2 is the variance of g_k due to noise. If the noise has a Poisson distribution, as it would if the signal is detected photons, then one should set $\sigma_k^2 = \hat{f}_j$.

10.3 Autocovariance Functions

10.3.1 Basic Properties of Autocovariance Functions

Autocovariance Function: The autocovariance function is a measure of the similarity between a sequence and the same sequence at a later time. If $f(t)$ is a continuous sequence whose mean value at t is $\mu(t)$, the autocovariance of $f(t)$ is

$$\gamma_{ff}(\tau) = \langle [f(t) - \mu(t)][f(t+\tau) - \mu(t+\tau)] \rangle, \tag{10.77}$$

In effect, $f(t)$ is duplicated and then shifted by an amount τ called the lag. After their local mean values have been subtracted, the function and its shifted duplicate are multiplied together. The autocovariance is the mean value of the product. The subscript ff is inserted to distinguish the autocovariance function from the cross-covariance function discussed in Section 10.4. The autocovariance function is symmetric about $\tau = 0$. To see this, begin with

$$\gamma_{ff}(-\tau) = \langle [f(t) - \mu(t)][f(t-\tau) - \mu(t-\tau)] \rangle. \tag{10.78}$$

Change variables to $\zeta = t - \tau$ to get

$$\gamma_{ff}(-\tau) = \langle [f(\zeta+\tau) - \mu(\zeta+\tau)][f(\zeta) - \mu(t+\zeta)] \rangle = \gamma_{ff}(\tau). \tag{10.79}$$

As a result, one typically calculates the autocovariance function only for positive lag.

If the mean value is independent of time, the autocovariance function simplifies somewhat to

$$\gamma_{ff}(\tau) = \langle [f(t) - \mu][f(t+\tau) - \mu] \rangle = \langle f(t)f(t+\tau) \rangle - \mu^2. \tag{10.80}$$

The autocovariance function at zero lag is equal to the covariance of $f(t)$. This is true whether or not the mean is independent of time, but it is manifestly true if μ is a constant, since

$$\gamma_{ff}(0) = \langle [f(t) - \mu]^2 \rangle \equiv \sigma_f^2. \tag{10.81}$$

From now on we will assume that μ has already been subtracted from $f(t)$. The autocovariance function then takes the simple form

$$\gamma_{ff}(\tau) = \langle f(t)f(t+\tau) \rangle. \tag{10.82}$$

This assumption simplifies the presentation with little loss of generality.

How the mean value is calculated depends somewhat on context. If $f(t)$ is defined only for $0 \le t \le T$ or is a measured sequence with length T, the autocovariance function is usually taken to be

$$\gamma_{ff}(\tau) = \frac{1}{T} \int_0^{T-\tau} f(t)f(t+\tau)dt, \quad 0 \le \tau < T. \tag{10.83}$$

This is sometimes called the *sample autocovariance function*. Infinitely long sequences can often be handled by calculating the autocovariance function for some finite length T and then taking the limit as $T \to \infty$. Some authors use a factor of $1/(T - \tau)$ instead of $1/T$ in equation 10.83. There is no clear advantage to either usage: at large lags $1/(T - \tau)$ yields an unbiased but noisy estimate of the autocovariance function, while $1/T$ yields a biased but less noisy estimate. If f_j is a discrete sequence with n elements at a uniform interval Δt, its autocovariance function is usually taken to be

$$\gamma_{ff}(\tau_k) = \frac{1}{n} \sum_{j=1}^{n-k} f_j f_{j+k}, \tag{10.84}$$

where $\tau_k = k\Delta t$. Again, some authors use a factor $1/(n - k)$ instead of $1/n$.

If $f(t)$ is a periodic sequence with period P,

$$f(t + mP) = f(t), \tag{10.85}$$

where m is an integer, then the autocovariance function of $f(t)$ is also periodic at P, since

$$\gamma_{ff}(\tau + mP) = \langle f(t)f(t+\tau+mP) \rangle = \langle f(t)f(t+\tau) \rangle$$
$$= \gamma_{ff}(\tau). \tag{10.86}$$

This allows equation 10.83 to be written as

$$\gamma_{ff}(\tau) = \frac{1}{P} \int_{t_0}^{t_0+P} f(t)f(t+\tau)dt, \tag{10.87}$$

where t_0 is an arbitrary starting point for the integration.

Equation 10.83 makes the autocovariance function look somewhat like the convolution of a sequence with itself, but there are two important differences between autocovariance and convolution. First, when convolving two functions, time is reversed in one of them.

There is no time reversal when calculating the autocovariance function. Second, the autocovariance function is the mean value of the lagged products, whereas convolution does not involve taking a mean value. The mean value makes the autocovariance function well suited for discussions of noisy sequences and noise-driven processes.

The following example calculates the autocovariance functions for the rectangle function and then a square wave with period P.

Example: The rectangle function with width T is

$$f(t) = \begin{cases} 1, & 0 \leq t \leq T \\ 0, & \text{otherwise} \end{cases}.$$

From equation 10.83, the autocovariance function for positive lag is

$$\gamma_{ff}(\tau \geq 0) = \frac{1}{T} \int_{t=0}^{T-\tau} dt = 1 - \tau/T, \quad 0 \geq \tau/T \geq 1.$$

Since the autocovariance function is symmetric about 0, the full autocovariance function is

$$\gamma_{ff}(\tau) = \begin{cases} 1 - |\tau|/T, & |\tau|/T \leq 1 \\ 0, & \text{otherwise} \end{cases}.$$

This is the triangle function shown in the top right panel of Figure 10.6.

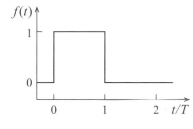
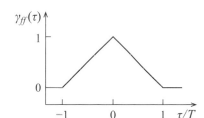
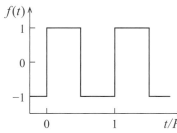
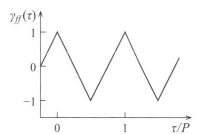

Figure 10.6: A rectangle with width T (top left panel) and its autocovariance function (top right panel). A square wave with period P (bottom left panel) and its autocovariance function (bottom right panel).

The square wave can be expressed as

$$f(t) = \text{sgn}\left[\sin(2\pi t/P)\right],$$

where $\text{sgn}(z)$ equals 0 if $z = 0$ and $z/|z|$ otherwise. By inspection the value of the autocovariance function at $\tau = 0$ is 1, and its value at $\tau = P/2$ is -1. From the example of the rectangle function, we infer that the autocovariance varies linearly between these two values and is periodic with period P. We can therefore write down the full autocovariance function without evaluating equation 10.87 explicitly:

$$\gamma_{ff}(\tau + mP) = \gamma_{ff}(\tau) = \begin{cases} 1 - 4\tau/P, & 0 \leq \tau \leq P/2 \\ -1 + 4\tau/P, & P/2 < \tau \leq P \end{cases},$$

where m is an integer. The autocovariance function is shown in the bottom right panel of Figure 10.6.

In the following example, we calculate the autocovariance function of a sine curve of infinite length.

Example: Let a sequence be the sine curve $f(t) = \sin(2\pi\nu t + \phi)$. It is straightforward to calculate the autocovariance function of the sine curve from equation 10.87, since the sine curve is periodic. Instead, for illustrative purposes, let us calculate its autocovariance function from equation 10.83, initially restricting the integral to the finite interval $0 \leq t \leq T - \tau$, but then allowing T to become large. The autocovariance function of the sine curve is

$$\gamma_{ff}(\tau) = \frac{1}{T} \int_0^{T-\tau} \sin(2\pi\nu t + \phi) \sin(2\pi\nu[t + \tau] + \phi) dt$$

$$= \frac{1}{T} \int_0^{T-\tau} \sin(2\pi\nu t + \phi)$$

$$[\sin(2\pi\nu t + \phi)\cos(2\pi\nu\tau) + \cos(2\pi\nu t + \phi)\sin(2\pi\nu\tau)] dt$$

$$= \frac{\cos(2\pi\nu\tau)}{T} \int_0^{T-\tau} \sin^2(2\pi\nu t + \phi) dt$$

$$+ \frac{\sin(2\pi\nu\tau)}{T} \int_0^{T-\tau} \sin(2\pi\nu t + \phi)\cos(2\pi\nu t + \phi) dt.$$

The integral in the second term oscillates back and forth between -1 and 1, so as T becomes large, the second term goes to 0, and the autocovariance function becomes

$$\gamma_{ff}(\tau) = \frac{\cos(2\pi\nu\tau)}{T} \int_0^{T-\tau} \sin^2(2\pi\nu t + \phi) dt.$$

Using the trigonometric half-angle formula, this can be written as

$$\gamma_{ff}(\tau) = \frac{\cos(2\pi\nu\tau)}{T} \int_0^{T-\tau} \frac{1}{2}[1 - \cos(4\pi\nu t + 2\phi)] dt.$$

Once again, $\cos(4\pi\nu t + 2\phi)$ oscillates back and forth between -1 and 1, so as T becomes large, its integral goes to 0, and the autocovariance function becomes

$$\gamma_{ff}(\tau) = \frac{\cos(2\pi\nu\tau)}{T} \int_0^{T-\tau} \frac{1}{2} dt = \cos(2\pi\nu\tau)\left(\frac{T-\tau}{2T}\right).$$

In the limit as $T \to \infty$, we arrive at

$$\gamma_{ff}(\tau) = \frac{1}{2}\cos(2\pi\nu\tau).$$

Thus the autocovariance function of a sine curve with frequency ν is a cosine curve that oscillates back and forth between $1/2$ and $-1/2$ with the same frequency.

Note that the autocorrelation function is independent of the phase of the original sequence. This is a specific example of a more general result: phase information is lost when calculating the autocovariance function.

Autocorrelation Function: The autocorrelation function is defined to be

$$\rho_{ff}(\tau) = \frac{\gamma_{ff}(\tau)}{\gamma_{ff}(0)} = \frac{\gamma_{ff}(\tau)}{\sigma_f^2}, \tag{10.88}$$

which is the same as the autocovariance function, except that it has been normalized to $\rho_{ff}(0) = 1$. The absolute value of the autocorrelation function is always less than or equal to 1. To see this, take

$$\left|\rho_{ff}(\tau)\right|^2 = \left|\frac{\gamma_{ff}(\tau)}{\gamma_{ff}(0)}\right|^2 = \frac{1}{\gamma_{ff}^2(0)}\left|\gamma_{ff}(\tau)\right|^2. \tag{10.89}$$

One version of the Cauchy-Schwarz inequality is $\left|\langle xy \rangle\right|^2 \leq \langle x^2 \rangle \langle y^2 \rangle$. Write out $\gamma_{ff}(\tau)$ in equation 10.89 explicitly, and apply the Cauchy-Schwarz inequality to get

$$\left|\rho_{ff}(\tau)\right|^2 = \frac{1}{\gamma_{ff}^2(0)}\left|\langle f(t)f(t+\tau)\rangle\right|^2$$

$$\leq \frac{1}{\gamma_{ff}^2(0)}\langle f^2(t)\rangle\langle f^2(t+\tau)\rangle$$

$$\leq \frac{1}{\gamma_{ff}^2(0)}\gamma_{ff}(0)\gamma_{ff}(0)$$

$$\leq 1. \tag{10.90}$$

The equality holds if there is any period P for which $f(t+P) = \pm f(t)$, since

$$\rho_{ff}(\tau = P) = \frac{\langle f(t)f(t+P)\rangle}{\gamma_{ff}(0)} = \pm\frac{\langle f(t)f(t)\rangle}{\gamma_{ff}(0)} = \pm\frac{\gamma_{ff}(0)}{\gamma_{ff}(0)} = \pm 1. \tag{10.91}$$

10.3.2 Relation to the Power Spectrum

In this section we show that the power spectrum is the Fourier transform of the sample autocovariance function. Although closely related to the convolution theorem, this result is more cumbersome to derive because of the finite limits on the integrals; but the result is worth the extra effort.

Let $f(t)$ be a continuous function that, for convenience, is defined between $-T/2$ and $T/2$ instead of 0 and T. For these limits, the sample autocovariance function must be rewritten as

$$\gamma_{ff}(\tau) = \begin{cases} \dfrac{1}{T} \displaystyle\int_{-T/2}^{T/2-\tau} f(t)f(t+\tau)dt, & 0 \leq \tau \leq T \\[3mm] \dfrac{1}{T} \displaystyle\int_{-T/2-\tau}^{T/2} f(t)f(t+\tau)dt, & -T \leq \tau < 0 \end{cases} \tag{10.92}$$

(compare to equation 10.83), where we have explicitly allowed for negative lags. If $F(\nu_n)$ is the Fourier transform of $f(t)$, the power spectrum of $f(t)$ is

$$P(\nu_n) = F(\nu_n)F^*(\nu_n), \tag{10.93}$$

where $\nu_n = n/T$, and n is a positive integer. Using equation 8.57 to write out the Fourier transforms explicitly, we have

$$P(\nu_n) = \left[\frac{2\sqrt{\pi}}{T} \int_{-T/2}^{T/2} f(t) \exp[-i2\pi \nu_n t]dt \right] \left[\frac{2\sqrt{\pi}}{T} \int_{-T/2}^{T/2} f(t) \exp[+i2\pi \nu_n t]dt \right],$$

$$= \frac{4\pi}{T^2} \int_{-T/2}^{T/2} \int_{-T/2}^{T/2} f(t)f(t') \exp\left[-i2\pi \nu_n (t-t')\right] dt\, dt'. \tag{10.94}$$

Note that complex conjugation reverses the sign on the exponent in the second integral.

Now perform a change of variables to

$$\tau = t - t' \tag{10.95}$$

$$\nu = t'. \tag{10.96}$$

The Jacobian of the transformation is

$$\left| \frac{\partial(t,t')}{\partial(\tau,\nu)} \right| = \begin{vmatrix} 1 & 1 \\ 0 & 1 \end{vmatrix} = 1, \tag{10.97}$$

and the integration limits in the (τ,ν) plane are shown in Figure 10.7. Because the limits on ν are functions of τ, the integration must be broken into two parts, one for τ less than 0 and

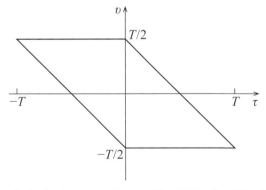

Figure 10.7: Integration limits for the integral in equation 10.94 after it has been transformed from the (t,t') plane to the (τ,ν) plane. Because the limits on ν are functions of τ, it is easier to perform the integration by breaking it into two parts, one for τ less than 0 and one for τ greater than 0.

one for τ greater than 0. Equation 10.94 transforms to

$$P(v_n) = \frac{4\pi}{T^2} \int_{\tau=0}^{T} \int_{v=-T/2}^{T/2-\tau} f(v+\tau)f(v) \exp[-i2\pi v_n \tau] d\tau \, dv$$

$$+ \frac{4\pi}{T^2} \int_{\tau=-T}^{0} \int_{v=-T/2-\tau}^{T/2} f(v+\tau)f(v) \exp[-i2\pi v_n \tau] d\tau \, dv$$

$$= \frac{4\pi}{T} \int_{\tau=0}^{T} \left[\frac{1}{T} \int_{v=-T/2}^{T/2-\tau} f(v+\tau)f(v)dv \right] \exp[-i2\pi v_n \tau] d\tau$$

$$+ \frac{4\pi}{T} \int_{\tau=-T}^{0} \left[\frac{1}{T} \int_{v=-T/2-\tau}^{T/2} f(v+\tau)f(v)dv \right] \exp[-i2\pi v_n \tau] d\tau. \quad (10.98)$$

The integrals inside the square brackets are the two halves of the autocovariance function as written in equation 10.92, the first for $\tau \geq 0$, the second for $\tau < 0$. Equation 10.98 is therefore

$$P(v_n) = \frac{4\pi}{T} \int_{\tau=0}^{T} \gamma_{ff}(\tau) \exp[-i2\pi v_n \tau] d\tau + \frac{4\pi}{T} \int_{\tau=-T}^{0} \gamma_{ff}(\tau) \exp[-i2\pi v_n \tau] d\tau$$

$$= \frac{4\pi}{T} \int_{-T}^{T} \gamma_{ff}(\tau) \exp[-i2\pi v_n \tau] d\tau, \quad (10.99)$$

which is the desired result: the power spectrum is the Fourier transform of the autocovariance function. Note that the integral now runs from $-T$ to T. The extra factor of 4π traces back to the decision to impose symmetry on the normalizing constants in the equations for the forward and reverse Fourier series (equations 8.56 and 8.57).

In fact, the relation between the autocovariance function and the power spectrum is even simpler, because the autocovariance function is real and symmetric. Taking advantage of the symmetry, we can write equation 10.99 as

$$P(v_n) = \frac{4\pi}{T} \int_{-T}^{T} \gamma_{ff}(\tau) \exp[-i2\pi v_n \tau] d\tau$$

$$= \frac{4\pi}{T} \int_{0}^{T} \gamma_{ff}(\tau) \exp[-i2\pi v_n \tau] d\tau + \frac{4\pi}{T} \int_{0}^{T} \gamma_{ff}(-\tau) \exp[+i2\pi v_n \tau] d\tau$$

$$= \frac{4\pi}{T} \int_{0}^{T} \gamma_{ff}(\tau) \left(\exp[-i2\pi v_n \tau] + \exp[+i2\pi v_n \tau] \right) d\tau$$

$$= \frac{8\pi}{T} \int_{0}^{T} \gamma_{ff}(\tau) \cos(2\pi v_n \tau) d\tau. \quad (10.100)$$

Integrals like that in equation 10.100 are called *cosine transforms*. Thus, the power spectrum is the cosine transform of the autocovariance function.

The autocovariance function can be reconstructed from the power spectrum. From equation 8.56, we have

$$\gamma_{ff}(\tau) = \frac{1}{2\sqrt{\pi}} \sum_{n=-\infty}^{\infty} P(v_n) \exp[i2\pi v_n \tau]. \quad (10.101)$$

Since $P(\nu_n)$ is symmetric, this can be written as

$$\gamma_{ff}(\tau) \;=\; \frac{1}{\sqrt{\pi}} \sum_{n=0}^{\infty} P(\nu_n)\cos(2\pi\nu_n\tau). \qquad (10.102)$$

It is sometimes more efficient to calculate power spectra by first calculating the autocovariance function and then taking the cosine transform instead of calculating the power spectrum from the full Fourier transform. The advantage of this indirect method has been rendered moot, at least for evenly spaced sequences, by the invention of the fast Fourier transform algorithm and the development of fast computers.

10.3.3 Application to Stochastic Processes

Autocovariance and autocorrelation functions are particularly useful for evaluating the properties of stochastic processes. In this section we derive autocovariance functions for a few of the more important of these processes.

Additive Noise: Suppose that noise $\epsilon(t)$ with zero mean, $\langle\epsilon(t)\rangle = 0$, is added to a sequence $g(t)$ to give $f(t) = g(t) + \epsilon(t)$. The autocovariance function of $f(t)$ is

$$\gamma_{ff}(\tau) \;=\; \langle f(t)f(t+\tau)\rangle \;=\; \frac{1}{T}\int_0^{T-\tau} [g(t)+\epsilon(t)][g(t+\tau)+\epsilon(t+\tau)]dt. \qquad (10.103)$$

Expanding the integral and taking the mean value of the autocovariance function, we find

$$\langle\gamma_{ff}(\tau)\rangle = \frac{1}{T}\int_0^{T-\tau} g(t)g(t+\tau)dt + \frac{1}{T}\int_0^{T-\tau} g(t)\langle\epsilon(t+\tau)\rangle dt$$
$$+ \frac{1}{T}\int_0^{T-\tau} \langle\epsilon(t)\rangle g(t+\tau)dt + \frac{1}{T}\left\langle\int_0^{T-\tau} \epsilon(t)\epsilon(t+\tau)dt\right\rangle. \qquad (10.104)$$

Since $\langle\epsilon(t)\rangle = 0$, this becomes

$$\langle\gamma_{ff}(\tau)\rangle \;=\; \gamma_{gg}(\tau) + \langle\gamma_{\epsilon\epsilon}(\tau)\rangle, \qquad (10.105)$$

where $\langle\gamma_{\epsilon\epsilon}(\tau)\rangle$ is the mean autocovariance function of the noise. Thus, in the mean, the autocovariance function of the noise is added to the autocovariance function of the noise-free sequence.

White Noise: Suppose the elements of a discrete sequence are evenly spaced at separations Δt and consist of noise ϵ_j that is uncorrelated and has zero mean:

$$\langle\epsilon_j\rangle = 0 \qquad (10.106)$$
$$\langle\epsilon_j\epsilon_k\rangle = \sigma_\epsilon^2\delta_{jk}. \qquad (10.107)$$

From our discussion of power spectra (Section 9.4.2), we know that noise with these properties produces a flat power spectrum. Thus the designation "white noise."

The autocovariance function of the sequence is

$$\gamma_{\epsilon\epsilon}(\tau_k) \;=\; \frac{1}{n}\sum_{j=1}^{n-k} \epsilon_j\epsilon_{j+k}, \qquad (10.108)$$

where $\tau_k = k\Delta t$. In the mean this is

$$\langle \gamma_{\epsilon\epsilon}(\tau_k) \rangle = \frac{1}{n} \sum_{j=1}^{n-k} \langle \epsilon_j \epsilon_{j+k} \rangle = \frac{1}{n} \sum_{j=1}^{n-k} \sigma_\epsilon^2 \delta_{j,j+k}. \tag{10.109}$$

If $k \neq 0$, then $\delta_{j,j+k} = 0$ and the sum is equal to 0. If $k = 0$, then $\delta_{j,j+k} = 1$ and the mean autocovariance function is

$$\langle \gamma_{\epsilon\epsilon}(\tau_k) \rangle = \frac{1}{n} \sum_{j=1}^{n} \sigma_\epsilon^2 = \sigma_\epsilon^2. \tag{10.110}$$

These properties can be combined into one equation:

$$\langle \gamma_{\epsilon\epsilon}(\tau_k) \rangle = \sigma_\epsilon^2 \delta_{0k}. \tag{10.111}$$

Thus, in the mean, the autocovariance function of uncorrelated noise is 0 everywhere except at $\tau = 0$, where it is σ_ϵ^2.

We can combine this result with the previous result for arbitrary additive noise: If a discrete sequence f_j is the sum of a sequence g_j and white noise ϵ_j, the mean autocovariance function of f_j is

$$\langle \gamma_{ff}(\tau_k) \rangle = \gamma_{gg}(\tau_k) + \sigma_\epsilon^2 \delta_{0k}. \tag{10.112}$$

The only effect of white noise on the mean autocovariance function is to add a spike at zero lag.

Equations 10.111 and 10.112 are true only in the mean. There is substantial scatter in measured autocovariance functions. For large n, the central limit theorem holds, and the scatter in $\gamma_{\epsilon\epsilon}(\tau_k)$ approaches a Gaussian distribution with variance σ_ϵ^2/n. In an autocorrelation function, the factor σ_ϵ^2 divides out, and the noise approaches a variance of $1/n$.

Shot Noise: Once again let f_j be a discrete sequence with elements that are evenly spaced at separations Δt, but for this case we let the sequence comprise a series of shots with random amplitudes. The difference between a shot and, say, white noise is that a shot extends over time. The profile of the shot is given by h_k, where we use the symbol h because the shot profile is essentially an impulse response function. If a new shot is added to the sequence at each position j and if the amplitudes of the shots are random numbers ϵ_j, the sequence is given by the convolution

$$f_j = \sum_{\zeta=1}^{n} \epsilon_\zeta h_{j+1-\zeta}. \tag{10.113}$$

If the ϵ_j are uncorrelated and have 0 mean (equations 10.106 and 10.107), the autocovariance function of f_j is

$$\gamma_{ff}(\tau_k) = \frac{1}{n} \sum_{j=1}^{n-k} f_j f_{j+k} = \frac{1}{n} \sum_{j=1}^{n-k} \left[\sum_{\zeta=1}^{n} \epsilon_\zeta h_{j+1-\zeta} \right] \left[\sum_{\xi=1}^{n} \epsilon_\xi h_{j+1+k-\xi} \right]. \tag{10.114}$$

Rearranging the summations and taking the mean value, we find

$$\langle \gamma_{ff}(\tau_k) \rangle = \frac{1}{n} \sum_{j=1}^{n-k} \sum_{\zeta=1}^{n} \sum_{\xi=1}^{n} h_{j+1+k-\xi} h_{j+1-\zeta} \langle \epsilon_\zeta \epsilon_\xi \rangle$$

$$= \frac{1}{n} \sum_{j=1}^{n-k} \sum_{\zeta=1}^{n} h_{j+1+k-\zeta} h_{j+1-\zeta} \sigma_\epsilon^2$$

$$= \sigma_\epsilon^2 \sum_{\zeta=1}^{n} \left[\frac{1}{n} \sum_{j=1}^{n-k} h_{j+1+k-\zeta} h_{j+1-\zeta} \right], \quad (10.115)$$

where the penultimate step made use of equation 10.107. The term inside the square brackets is the autocovariance function of h_j,

$$\frac{1}{n} \sum_{j=1}^{n-k} h_{j+1+k-\zeta} h_{j+1-\zeta} = \gamma_{hh}(\tau_k), \quad (10.116)$$

so we arrive at

$$\langle \gamma_{ff}(\tau_k) \rangle = \sigma_\epsilon^2 \sum_{\zeta=1}^{n} \gamma_{hh}(\tau_k) = n\sigma_\epsilon^2 \gamma_{hh}(\tau_k). \quad (10.117)$$

This is remarkable. Given enough data, we can extract the autocovariance function of the shots even when the shots grossly overlap. While the derivation of equation 10.117 assumed that a new shot started at each point in the sequence, this assumption is not really necessary. An arbitrary selection of the ϵ_j can be set to 0 and the only change is that the equality in equation 10.117 becomes a proportionality: $\langle \gamma_{ff}(\tau_k) \rangle \propto \gamma_{hh}(\tau_k)$. The autocorrelation function does not change at all.

The following example shows that the autocorrelation function can be a powerful tool for recognizing shot noise and measuring its properties.

Example: Let us calculate the autocorrelation function of a sequence composed of random shots. The profile of the shots is $h_j = 0$ except for

$$h_1 = 1$$
$$h_{10} = -1$$
$$h_{12} = 1$$

The shot profile is plotted in the top left panel of Figure 10.8. The autocorrelation function of h_j is shown in the bottom left panel of the figure.

To construct the sequence f_j, first create a sequence of n uncorrelated random deviates ϵ_j drawn from a Gaussian distribution. The deviates have 0 mean and variance σ_ϵ^2. Then insert a new shot with amplitude ϵ_j beginning at each location j in f_j. When done, add a constant to make $\langle f_j \rangle = 1.0$. Thus, f_j is the convolution of ϵ_j with h_j plus the constant:

$$f_j = \sum_{k=1}^{n} \epsilon_k h_{j+1-k} + 1.0.$$

For this example, $\sigma_\epsilon = 0.01$ and $n = 10,000$. A portion of f_j is shown in the top right panel of Figure 10.8. The underlying pulse profile is invisible to the eye.

The autocorrelation function of f_j is shown in the bottom right panel of the figure. It is a close match to the autocorrelation function of h_j. The standard deviation of the noise in the autocorrelation function is $n^{-1/2} = 0.01$.

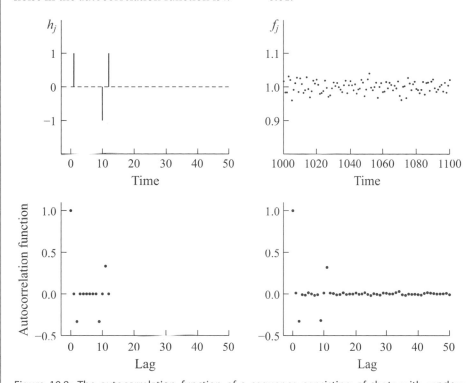

Figure 10.8: The autocorrelation function of a sequence consisting of shots with random amplitudes. The shot profile is shown in the top left panel, and the autocorrelation function of the shot profile is shown in the bottom left panel. The sequence plotted in the top right panel is composed of a series of these shots. The amplitudes ϵ_j of the shots are random deviates drawn from a Gaussian distribution with $\sigma_\epsilon^2 = 10^{-4}$. The autocorrelation function of f_j is shown in the bottom right panel. Even though the shots cannot be distinguished in the the sequence, the autocorrelation function of the sequence retrieves the autocorrelation function of the shots.

Quasiperiodic Oscillations: There is no unique definition of quasiperiodic oscillations, but one of the most useful definitions is behavior that can be described by a damped simple harmonic oscillator driven by noise. The oscillator is periodic, but the noise alters the amplitude and phase of the oscillator so that it eventually loses memory of its initial conditions and is no longer strictly periodic. Many physical systems fit this description. Among musical instruments, one might think of a trumpet or an organ pipe driven by the random pressure variations in the stream of air passing through the instrument. An astronomical example is the pulsations of a star driven by random turbulence in the star's convective layers.

Since driven damped simple harmonic oscillators are discussed extensively in elementary textbooks on mechanics and are familiar to most readers, we will forego extensive derivations and merely quote their properties. The differential equation for a driven damped simple harmonic oscillator can be written as

$$m\frac{d^2f}{dt^2} + b\frac{df}{dt} + kf = \epsilon(t). \tag{10.118}$$

If the physical model is an object on a spring, then f is the position of the object, m is its mass, k is the spring constant, $b(df/dt)$ is velocity-dependent damping, and $\epsilon(t)$ is the driving force. If there are no driving forces, $\epsilon(t) = 0$, and the solution to the homogeneous differential equation is

$$f(t) = A\exp[-\alpha t]\sin(\omega_1 t + \phi), \tag{10.119}$$

where $\omega_0 = \sqrt{k/m}$ is the angular frequency of the undamped oscillator, $\alpha = b/2m$ is the damping coefficient, and $\omega_1^2 = \omega_0^2 - \alpha^2$; A and ϕ are integration constants set by initial conditions. Since we are interested in quasiperiodic oscillations, assume that the system is underdamped: $\omega_1^2 > 0$.

The frequency response function of the system is derived by setting $\epsilon(t) = \exp[i\omega t]$, which yields

$$H(\omega) = \frac{1/m}{\omega_0^2 - \omega^2 + 2i\alpha\omega}. \tag{10.120}$$

The corresponding gain and phase shift are

$$Z(\omega) = \frac{1/m}{[(\omega_0^2 - \omega^2)^2 + 4\alpha^2\omega^2]^{1/2}} \tag{10.121}$$

$$\tan[\phi(\omega)] = \frac{2\alpha\omega}{\omega_0^2 - \omega^2}. \tag{10.122}$$

The maximum gain and maximum phase shift occur when $\omega = \omega_0$, so ω_0 is often called the resonant frequency. The strength of the resonance is can be characterized by the Q-factor, defined as $Q = \omega_0/\Delta\omega$, where $\Delta\omega$ is the full width at half maximum of $Z(\omega)^2$.

The properties of quasiperiodic oscillations in a sequence can be and sometimes are determined from the power spectrum of the sequence. One way to do this is to fit $Z(\omega)^2$ to the power spectrum with ω_0 and α as parameters of fit. However, this approach can be less than satisfactory, because power spectra of quasiperiodic oscillations are noisy, and the noise does not decrease as the length of the sequence increases (see Figure 10.9).

The properties of the quasiperiodic oscillations can also be extracted from the autocovariance function of the sequence. To show this, we need the impulse response function of a damped simple harmonic oscillator and the autocovariance function of the impulse response function. The impulse response function can be found either by calculating the Fourier transform of the frequency response function or by calculating the Green's function for the original differential equation (see Appendix G for the Green's function derivation). The impulse response function is

$$h(t) = \begin{cases} \dfrac{1}{\omega_1}\exp[-\alpha t]\sin(\omega_1 t), & t \geq 0 \\ 0, & t < 0 \end{cases}. \tag{10.123}$$

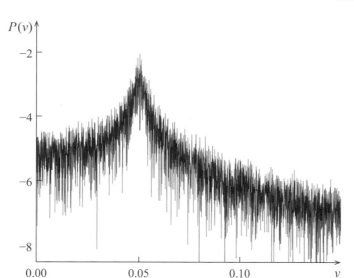

Figure 10.9: Power spectrum of the sequence of quasiperiodic oscillations shown in the top panel of Figure 10.10. The oscillations have a frequency of 0.05 Hz and a damping constant of 0.01 second^{-1}.

If $T \gg \tau$, the autocovariance of the impulse response function is

$$\gamma_{hh}(\tau) = \frac{1}{T} \int_0^T h(t)h(t+\tau)dt$$

$$= \frac{1}{\omega_1^2 T} \int_0^T \exp[-\alpha t] \sin(\omega_1 t) \exp[-\alpha(t+\tau)] \sin(\omega_1[t+\tau])dt. \quad (10.124)$$

For $2\alpha T \gg 1$, and after much algebra (which is greatly simplified by moving into the complex plane), we arrive at

$$\gamma_{hh}(\tau) = \frac{1}{2\omega_1^2 T} \exp[-\alpha\tau]\{\omega_1 \cos(\omega_1\tau)\alpha \sin(\omega_1\tau)\} \quad (10.125)$$

or

$$\gamma_{hh}(\tau) \propto \exp[-\alpha\tau]\cos(\omega_1\tau + \phi), \quad (10.126)$$

where $\tan(\phi) = -\alpha/\omega_1$. For many real physical systems, $\alpha/\omega_1 \ll 1$, so ϕ differs from 0 by only a few degrees. In summary, one can determine the properties of a driven, damped harmonic oscillator from the autocovariance function of its impulse response function: the frequency and damping of the autocovariance function are the same as the frequency and damping of the underlying oscillator.

We are now prepared to calculate the autocovariance function of the noise-driven oscillator. The response of the oscillator to the driving force is the convolution of $h(t)$ with $\epsilon(t)$:

$$f(t) = h(t) \otimes \epsilon(t) = \int_{u=-\infty}^{\infty} h(u)\epsilon(t-u)du. \quad (10.127)$$

For $T \gg \tau$, the autocovariance function of $f(t)$ is

$$
\begin{aligned}
\gamma_{ff}(\tau) &= \frac{1}{T} \int_{t=0}^{T} f(t)f(t+\tau)dt \\
&= \frac{1}{T} \int_t \left[\int_u h(u)\epsilon(t-u)du \right] \left[\int_v h(v)\epsilon(t+\tau-v)dv \right] dt. \quad (10.128)
\end{aligned}
$$

Let us assume that the driving force is zero-mean, uncorrelated noise with variance σ_ϵ^2, so that

$$
\langle \epsilon(t_1)\epsilon(t_2) \rangle = \sigma^2 \delta(t_2 - t_1). \quad (10.129)
$$

The mean value of the autocovariance function of $f(t)$ becomes

$$
\langle \gamma_{ff}(\tau) \rangle = \frac{1}{T} \int_t \int_u \int_v h(u)h(v)\langle \epsilon(t-u)\epsilon(t+\tau-v) \rangle dv\,du\,dt. \quad (10.130)
$$

The mean value of the noise term is equal to 0 unless $t - u = t + \tau - v$, or equivalently, $v = u + \tau$, so equation 10.130 becomes

$$
\begin{aligned}
\langle \gamma_{ff}(\tau) \rangle &= \frac{1}{T} \int_t \int_u \int_v h(u)h(v)\sigma_\epsilon^2 \delta(v-u-\tau)dv\,du\,dt \\
&= \frac{\sigma_\epsilon^2}{T} \int_t \int_u h(u)h(u+\tau)du\,dt = \frac{\sigma_\epsilon^2}{T} \int_t \gamma_{hh}(\tau)dt = \gamma_{hh}(\tau)\frac{\sigma_\epsilon^2}{T} \int_t dt \\
&= \sigma_\epsilon^2 \gamma_{hh}(\tau), \quad (10.131)
\end{aligned}
$$

and therefore

$$
\langle \gamma_{ff}(\tau) \rangle \propto \exp[-\alpha\tau]\cos(\omega_1 \tau + \phi). \quad (10.132)
$$

This is a remarkable result. The frequency and decay time of the underlying system can be extracted from the autocovariance function of a sequence of quasiperiodic oscillations produced by the system. As noted in the discussion following equation 10.112, the variance of the scatter in an autocovariance function calculated from a measured sequence decreases as $1/n$, or as $1/T$ in this case. Thus, unlike the noise in a power spectrum of quasiperiodic oscillations, the noise in the autocovariance function of the oscillations decreases as the length of the measured sequence increases.

In the following example we generate a sequence of quasiperiodic oscillations and then calculate the power spectrum and autocovariance function of the sequence.

Example: A sequence of discretely sampled quasiperiodic oscillations with a frequency ω_0 and damping coefficient α can be generated using the recurrence relation

$$
x_j = a_1 x_{j-1} + a_2 x_{j-2} + \epsilon_j, \quad (10.133)
$$

where

$$
a_2 = -\exp[-2\alpha] \quad (10.134)
$$

$$
a_1 = -\left(\frac{4a_2}{1-a_2} \right) \cos \omega_0, \quad (10.135)
$$

and the ϵ_j are uncorrelated Gaussian random deviates with zero mean. The recurrence relation can be viewed as the discrete equivalent of the differential equation for the damped, driven harmonic oscillator (equation 10.118) or as a way (a poor way!) to integrate that equation numerically. The equation can also be viewed as a second-order autoregressive process. While beyond the scope of this book, autoregressive models are well suited for representing stochastic processes. A discussion of the second-order autoregressive process and derivations of equations 10.134 and 10.135 can be found in Appendix H.

A portion of the sequence is shown in the top panel of Figure 10.10. The frequency of the oscillations is $\nu_0 = \omega_0/2\pi = 0.05$ second^{-1}, and the damping coefficient is $\alpha = 0.01$ second^{-1}. The full sequence is 20,000 seconds long.

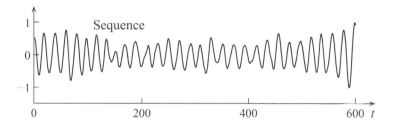

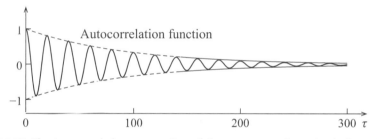

Figure 10.10: The top panel shows a portion of the sequence of quasiperiodic oscillations created using the recurrence relation 10.133. The frequency of the oscillations is 0.05 Hz, and the damping constant is 0.01 second^{-1}. The full sequence is 20,000 seconds long. The bottom panel shows the autocorrelation function of the sequence with the expected exp[−0.01t] damping overplotted as the dashed lines.

The power spectrum of the sequence is shown in Figure 10.9. Although the power spectrum does conform to the square of the gain function, it does so only in the mean. The power spectrum is noisy even though the sequence of oscillations is long. (Because the sequence is discrete, the gain function is actually equation H.20, not equation 10.121, but the two are almost identical near the resonant frequency.)

The autocorrelation function of the sequence is shown in the bottom panel of Figure 10.10. The autocorrelation function agrees well with the exponentially damped sine curve of equation 10.132. In this case $\tan(\phi) = -\alpha/\omega_1 = 0.2$, so $\phi = 11°$, and the phase shift is not noticeable to the eye. In contrast to the power spectrum, the long sequence reduces the standard deviation of the noise in the autocovariance function to less than 1%.

10.4 Cross-Covariance Functions

10.4.1 Basic Properties of Cross-Covariance Functions

Let $f(t)$ and $g(t)$ be continuous sequences. The cross-covariance function at lag τ is defined to be

$$\gamma_{fg}(\tau) = \langle [f(t) - \mu_f(t)][g(t+\tau) - \mu_g(t+\tau)] \rangle, \tag{10.136}$$

where $\mu_f(t)$ and $\mu_g(t)$ are the local mean values of the two sequences. If $f(t) = g(t)$, then the cross-covariance function reduces to

$$\gamma_{fg}(\tau) = \langle [f(t) - \mu_f(t)][f(t+\tau) - \mu_f(t+\tau)] \rangle = \gamma_{ff}(\tau). \tag{10.137}$$

The autocovariance function is therefore a special case of the cross-covariance function (see equation 10.77). As for the autocovariance function, assume that the means have already been removed from the sequences, so the cross-covariance function becomes

$$\gamma_{fg}(\tau) = \langle f(t)g(t+\tau) \rangle. \tag{10.138}$$

Calculating the autocovariance function is a linear operation. Set $f(t)$ equal to a linear combinations of two other functions:

$$f(t) = a_1 f_1(t) + a_2 f_2(t). \tag{10.139}$$

Since taking a mean value is a linear operation, the cross-covariance becomes

$$\begin{aligned} \gamma_{fg}(\tau) &= \langle [a_1 f_1(t) + a_2 f_2(t)]g(t+\tau) \rangle \\ &= \langle a_1 f_1(t)g(t+\tau) \rangle + \langle a_2 f_2(t)g(t+\tau) \rangle \\ &= a_1 \gamma_{f_1 g}(\tau) + a_2 \gamma_{f_2 g}(\tau). \end{aligned} \tag{10.140}$$

Thus, the cross-covariance function is the same linear combination of the individual cross-correlation functions. We could also have replaced $g(t)$ by a linear combination of functions and achieved the same result, so the cross-covariance function is bilinear.

Unlike the autocovariance function, the cross-covariance function is not symmetric about $\tau = 0$. Evaluate $\gamma_{fg}(\tau)$ at $-\tau$,

$$\gamma_{fg}(-\tau) = \langle f(t)g(t-\tau) \rangle, \tag{10.141}$$

and then change variables to $\zeta = -t$ to find

$$\gamma_{fg}(-\tau) = \langle f(-\zeta)g(-\zeta-\tau) \rangle. \tag{10.142}$$

This is equation 10.138 with the signs of both t and τ reversed in f and g, which produces a mirror image, not equality, around $\tau = 0$. In general, therefore, $\gamma_{fg}(-\tau) \neq \gamma_{fg}(\tau)$. As result, plots must show $\gamma_{fg}(\tau)$ for both positive and negative values of τ. There is nevertheless a symmetry to the cross-covariance function. Change variables in equation 10.136 to $\zeta = t + \tau$. Then

$$\begin{aligned} \gamma_{fg}(\tau) &= \langle f(t)g(t+\tau) \rangle = \langle f(\zeta-\tau)g(\zeta) \rangle \\ &= \gamma_{gf}(-\tau). \end{aligned} \tag{10.143}$$

Equality holds when the roles of $f(t)$ and $g(t)$ are exchanged *and* τ changes to $-\tau$. Changing the sign of τ reflects $\gamma_{fg}(\tau)$ about $\tau = 0$, but exchanging the roles of $f(t)$ and $g(t)$ also reflects

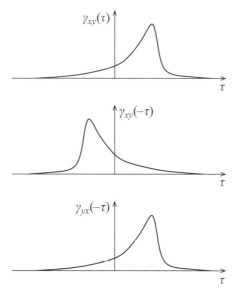

Figure 10.11: Symmetries of the cross-covariance function. Changing the sign of τ reflects $\gamma_{xy}(\tau)$ about $\tau = 0$ (top panel). Reversing the roles of $x(t)$ and $y(t)$ also reflects $\gamma_{xy}(\tau)$ about the origin (middle panel). Applying both operations retrieves the original $\gamma_{xy}(\tau)$ (bottom panel).

$\gamma_{fg}(\tau)$ about the origin. Applying both operations returns the original cross-covariance function. Figure 10.11 shows the effect of applying both operations to the cross-covariance function. Because of the symmetry, a plot of either $\gamma_{fg}(\tau)$ or $\gamma_{gf}(\tau)$ at both positive and negative τ gives full information about the autocovariance function. One need not plot both.

The way the average is calculated depends on the properties of $f(t)$ and $g(t)$. If they are both defined for $0 \le t \le T$, the sample cross-covariance function is usually taken to be

$$\gamma_{fg}(\tau) = \begin{cases} \dfrac{1}{T} \displaystyle\int_0^{T-\tau} f(t)g(t+\tau)dt, & 0 \le \tau \le T \\[4mm] \dfrac{1}{T} \displaystyle\int_{-\tau}^{T} f(t)g(t+\tau)dt, & -T \le \tau < 0 \end{cases}. \tag{10.144}$$

As for the autocovariance function, some authors use a factor of $1/(T - |\tau|)$ instead of $1/T$. If f_j and g_j are discrete sequences, each with n elements spaced at a uniform interval Δt, the sample cross-covariance function is

$$\gamma_{fg}(\tau_k) = \begin{cases} \dfrac{1}{n} \displaystyle\sum_{j=1}^{n-k} f_j g_{j+k}, & 0 \le k \le (n-1) \\[4mm] \dfrac{1}{n} \displaystyle\sum_{j=1-k}^{n} f_j g_{j+k}, & -(n-1) \le k \le -1 \end{cases}, \tag{10.145}$$

where $\tau_k = k\Delta t$. Again, some authors use a factor of $1/(n - |k|)$ instead of $1/n$.

The inelegance of equations 10.144 and 10.145 arises from edge effects. As the sequences are shifted, they fall off each other's ends, forcing special handling. In practice, however, one of the sequences is often much longer than the other. For example, one may be searching for a short pulse train f buried in a sequence g, whose length is much greater than the length of f.

If so, one can ignore edge effects, and the equations become much simpler. Equation 10.144 simplifies to

$$\gamma_{fg}(\tau) = \frac{1}{T}\int_0^T f(t)g(t+\tau)dt, \tag{10.146}$$

where $f(t)$ has length T, and equation 10.145 to

$$\gamma_{fg}(\tau_k) = \frac{1}{n}\sum_{j=1}^{n} f_j g_{j+k}, \tag{10.147}$$

where f_j has n elements.

The cross-correlation function is defined to be

$$\rho_{fg}(\tau) = \frac{\gamma_{fg}(\tau)}{\sqrt{\gamma_{ff}(0)\gamma_{gg}(0)}} = \frac{\gamma_{fg}(\tau)}{\sigma_f \sigma_g}, \tag{10.148}$$

where the last step made use of equation 10.81. Since the cross-correlation function is identical to the cross-covariance function except for normalizing constants, the cross-correlation function has the same symmetry:

$$\rho_{fg}(\tau) = \rho_{gf}(-\tau). \tag{10.149}$$

It can be shown that

$$|\rho_{fg}(\tau)| \leq 1, \tag{10.150}$$

and that $|\rho_{fg}(\tau)| = 1$ only if $f(t)$ and $g(t)$ are identical when shifted by an amount τ. The proofs of these two statements are nearly identical to the proofs of similar statements about $\rho_{ff}(\tau)$ (see equations 10.90 and 10.91) and are omitted here.

10.4.2 Relation to χ^2 and to the Cross Spectrum

Relation to χ^2: One can gain some insight into the meaning of the the cross-covariance function by comparing it to χ^2. Suppose that f_j and g_j are discrete sequences with elements spaced at a constant interval Δt, and suppose that f_j has n elements and that g_j is much longer than f_j. Let the values of their elements be randomly distributed with variances $\sigma_{f_j}^2 = \langle f_j^2 \rangle$ and $\sigma_{g_j}^2 = \langle g_j^2 \rangle$. If sequence g_j is shifted by $\tau_k = k\Delta t$ with respect to sequence f_j, the difference between the two sequences at position j is

$$\epsilon_j(\tau_k) = f_j - g_{j+k}. \tag{10.151}$$

The variance of the difference is

$$\sigma_j^2 = \sigma_{f_j}^2 + \sigma_{g_{j+k}}^2. \tag{10.152}$$

We can characterize the goodness of the match between f_j and g_{j+k} by χ^2, where

$$\chi^2(\tau_k) = \sum_j \frac{\epsilon_j^2}{\sigma_j^2} = \sum_{j=1}^{n} \frac{(f_j - g_{j+k})^2}{\sigma_j^2} = \sum_{j=1}^{n} \frac{f_j^2 + g_{j+k}^2}{\sigma_j^2} - 2\sum_{j=1}^{n} \frac{f_j g_{j+k}}{\sigma_j^2}, \tag{10.153}$$

and we have explicitly recognized the dependence of χ^2 on the shift.

Now let us assume that the variances are independent of j, so that

$$\sigma_{f_j}^2 = \langle f_j^2 \rangle = \sigma_f^2 \tag{10.154}$$

$$\sigma_{g_j}^2 = \langle g_j^2 \rangle = \sigma_g^2. \tag{10.155}$$

Equation 10.153 becomes

$$\chi^2(\tau_k) = \frac{1}{\sigma_f^2 + \sigma_g^2} \sum_{j=1}^{n} (f_j^2 + g_{j+k}^2) - \frac{2}{\sigma_f^2 + \sigma_g^2} \sum_{j=1}^{n} f_j g_{j+k}. \tag{10.156}$$

The second term on the right-hand side of this equation is proportional to $\gamma_{fg}(\tau_k)$ as given by equation 10.147. For large n the first term on the right-hand side approaches

$$\frac{1}{\sigma_f^2 + \sigma_g^2} \sum_{j=1}^{n} (f_j^2 + g_{j+k}^2) = \frac{1}{\sigma_f^2 + \sigma_g^2} n(\sigma_f^2 + \sigma_g^2) = n. \tag{10.157}$$

Thus we find

$$\chi^2(\tau_k) = n - \frac{2n}{\sigma_f^2 + \sigma_j^2} \gamma_{fg}(\tau_k). \tag{10.158}$$

This result is not entirely obvious. The cross-covariance function is based on the product of two sequences, but χ^2 is based on the difference between the two sequences. Yet for long sequences of equally weighted data (or equivalently, unweighted data), there is a one-to-one linear correspondence between them. The lag at which $\gamma_{fg}(\tau_k)$ is maximized is the same as the lag at which $\chi^2(\tau_k)$ is minimized.

The standard definition of the cross-covariance function gives equal weight to all the data points in the two sequences. It can fail to give meaningful results when applied to sequences in which the data points have unequal weights. Equation 10.153 suggests a way to modify the cross-covariance function to allow for unequal weights. If the weighted cross-covariance function is defined to be

$$\gamma_{fg}'(\tau_k) = \frac{1}{n} \sum_{j=1}^{n-k} \frac{f_j g_{j+k}}{\sigma_{f_j}^2 + \sigma_{g_{j+k}}^2}, \tag{10.159}$$

we retain the one-to-one linear correspondence to $\chi^2(\tau_k)$, at least in the mean. The meaning of this weighted version of the cross-covariance function differs from the meaning of the standard version, though. The difference is similar to the difference between the meanings of the Lomb-Scargle periodogram and a standard power spectrum: $\gamma_{fg}'(\tau_k)$ is a measure of the significance of a correlation at lag τ_k, not its strength.

Relation to the Cross Spectrum: If $f(t)$ and $g(t)$ are continuous functions defined between $-T/2$ and $T/2$, and if $F(\nu_n)$ and $G(\nu_n)$ are their Fourier transforms, their cross spectrum is

$$C_{FG}(\nu_n) = F^*(\nu_n)G(\nu_n), \tag{10.160}$$

where $\nu_n = n/T$, and n is a positive integer. The meaning of this rather opaque quantity can be clarified if we write it in terms of the cross-amplitude and the phase spectra. Remembering that any complex number z can be written as $z = \rho \exp[i\phi]$, we can rewrite the Fourier transforms as

$$F(\nu_n) = \rho_F(\nu_n)\exp[i\phi_F(\nu_n)] \tag{10.161}$$

$$G(\nu_n) = \rho_G(\nu_n)\exp[i\phi_G(\nu_n)] \tag{10.162}$$

$$C_{FG}(\nu_n) = \rho_{FG}(\nu_n)\exp[i\phi_{FG}(\nu_n)]. \tag{10.163}$$

Equation 10.160 can therefore be written as

$$\rho_{FG}(\nu_n)\exp[i\phi_{FG}(\nu_n)] = \{\rho_F(\nu_n)\exp[-i\phi_F(\nu_n)]\}\{\rho_G(\nu_n)\exp[i\phi_G(\nu_n)]\}$$
$$= \rho_F(\nu_n)\rho_G(\nu_n)\exp[i\{\phi_G(\nu_n) - \phi_F(\nu_n)\}]. \tag{10.164}$$

The cross-amplitude spectrum is defined to be

$$\rho_{FG}(\nu_n) = \rho_F(\nu_n)\rho_G(\nu_n) \tag{10.165}$$

and the phase spectrum to be

$$\phi_{FG}(\nu_n) = \phi_G(\nu_n) - \phi_F(\nu_n). \tag{10.166}$$

The cross amplitude at frequency ν_n is large if the amplitudes of the Fourier components of $f(t)$ and $g(t)$ are both large at ν_n. The phase spectrum gives the phase lag at ν_n.

In Section 10.3.2 we showed that the Fourier transform of the autocovariance function yields the power spectrum of the sequence. Here we show that the Fourier transform of the cross-covariance function yields the cross spectrum. We proceed in much the same way as we did in Section 10.3.2. First, since $f(t)$ and $g(t)$ are defined between $-T/2$ and $T/2$ instead of 0 to T, equation 10.144 for the cross-covariance function must be rewritten as

$$\gamma_{fg}(\tau) = \begin{cases} \dfrac{1}{T}\displaystyle\int_{-T/2}^{T/2-\tau} f(t)g(t+\tau)dt, & 0 \leq \tau \leq T \\[4mm] \dfrac{1}{T}\displaystyle\int_{-T/2-\tau}^{T/2} f(t)g(t+\tau)dt, & -T \leq \tau < 0 \end{cases}. \tag{10.167}$$

Using equation 8.57 to write out the Fourier transforms explicitly, equation 10.160 becomes

$$C_{FG}(\nu_n) = \left[\frac{2\sqrt{\pi}}{T}\int_{-T/2}^{T/2} f(t)\exp[+i2\pi\nu_n t]dt\right]\left[\frac{2\sqrt{\pi}}{T}\int_{-T/2}^{T/2} g(t)\exp[-i2\pi\nu_n t]dt\right],$$
$$= \frac{4\pi}{T^2}\int_{-T/2}^{T/2}\int_{-T/2}^{T/2} f(t)g(t')\exp\left[-i2\pi\nu_n(t'-t)\right]dt dt'. \tag{10.168}$$

Now change variables to $\tau = t' - t$ and $v = t$. The limits of the integration over τ and v are shown in Figure 10.7. Because the limits on v are functions of τ, the integration must be broken into two parts, one for τ less than 0 and one for τ greater than 0. Equation 10.168 becomes

$$C_{FG}(v_n) = \frac{4\pi}{T^2} \int_{\tau=0}^{T} \int_{v=-T/2}^{T/2-\tau} f(v)g(v+\tau)\exp[-i2\pi v_n \tau]d\tau\,dv$$

$$+ \frac{4\pi}{T^2} \int_{\tau=-T}^{0} \int_{v=-T/2-\tau}^{T/2} f(v)g(v+\tau)\exp[-i2\pi v_n \tau]d\tau\,dv$$

$$= \frac{4\pi}{T} \int_{\tau=0}^{T} \left[\frac{1}{T} \int_{v=-T/2}^{T/2-\tau} f(v)g(v+\tau)dv \right] \exp[-i2\pi v_n \tau]d\tau$$

$$+ \frac{4\pi}{T} \int_{\tau=-T}^{0} \left[\frac{1}{T} \int_{v=-T/2-\tau}^{T/2} f(v)g(v+\tau)dv \right] \exp[-i2\pi v_n \tau]d\tau. \quad (10.169)$$

The integrals inside the square brackets are the cross-covariance function $\gamma_{fg}(\tau)$, the first for $\tau \geq 0$ and the second for $\tau < 0$. Equation 10.169 is therefore

$$C_{FG}(v_n) = \frac{4\pi}{T} \int_{\tau=0}^{T} \gamma_{fg}(\tau)\exp[-i2\pi v_n \tau]d\tau + \frac{4\pi}{T} \int_{\tau=-T}^{0} \gamma_{fg}(\tau)\exp[-i2\pi v_n \tau]d\tau$$

$$= \frac{4\pi}{T} \int_{-T}^{T} \gamma_{fg}(\tau)\exp[-i2\pi v_n \tau]d\tau. \quad (10.170)$$

Thus, the cross spectrum is the Fourier transform of the cross-covariance function. Equation 10.170 cannot be converted to a cosine transform as we did for the autocovariance function, because the cross-covariance function lacks the symmetry of the autocovariance function.

10.4.3 Detection of Pulsed Signals in Noise

Cross-covariance functions are commonly used to detect pulsed signals buried in noise. We consider two examples. In the first the pulse profile is known beforehand, and one wants to determine where in a noisy sequence the pulses occur. This example corresponds to many kinds of radar signals. In the second, the properties of the pulses are unknown. This example corresponds to several kinds of astronomical applications, notably echo mapping of the gas around the central black holes in quasars and X-ray binary stars.

Example: The top left panel of Figure 10.12 shows shows a (known) pulse $f(t)$ consisting of five cycles of a sine curve with a period of 10 seconds. The bottom left panel shows the cross covariance of the pulse with itself. The top right panel of the figure shows a sequence $g(t)$ containing two of the pulses, one at 150 seconds and the other at 650 seconds. Uncorrelated Gaussian noise has been added to $g(t)$, making the pulses invisible to the eye, although, with the benefit of hindsight, one might notice increased scatter in $g(t)$ at the positions of the pulses. The bottom right panel is the cross covariance $\gamma_{fg}(\tau)$. The two pulses are clearly revealed in the cross-covariance function.

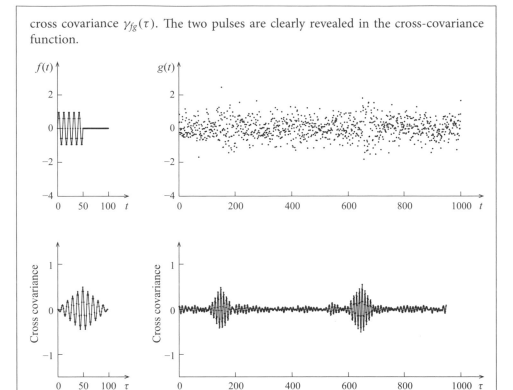

Figure 10.12: Detection of pulsed signals in noise when the pulse shape is known. The top left panel shows a pulse $f(t)$ consisting of five cycles of a sine curve. The bottom left panel shows the cross covariance of the pulse with itself. The top right panel shows a sequence $g(t)$ containing two of the pulses. The noise makes the pulses difficult to discern. The bottom right panel is the cross covariance $\gamma_{fg}(\tau) = \langle f(t)g(t+\tau)\rangle$. The two pulses are clearly revealed.

Example: The top panel in Figure 10.13 displays the beginning of a sequence $f(t)$ that contains a series of pulses at irregular intervals. Each pulse is one cycle of a sine curve with a period of 8 seconds, and there are 16 pulses in the sequence. Enough uncorrelated Gaussian noise has been added to the sequence to hide the individual pulses. The middle panel displays a sequence $g(t)$ that contains the same series of pulses as $f(t)$ but delayed in time by 130 seconds. The pulses in $g(t)$ are also buried in noise and are invisible. The bottom panel shows the cross covariance $\gamma_{fg}(\tau)$ between the two sequences. The cross-covariance function clearly picks out the 130-second delay between the pulses in the two sequences. This is surely one of the more interesting examples in this book. One can extract useful information—the delay time between two sequences—without having any information about the delayed signal itself.

Continued on page 332

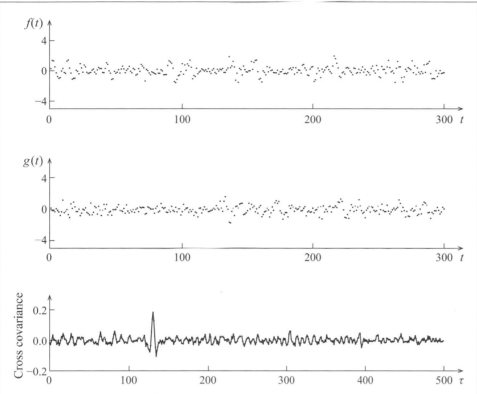

Figure 10.13: Measurement of an echo delay when the profile of the echoed pulse is unknown. The sequence $f(t)$ contains a series of pulses at irregular intervals to which uncorrelated Gaussian noise has been added (top panel). The sequence $g(t)$ is a noisy "echo" of $f(t)$. It contains the same series of pulses delayed in time also with noise added (middle panel). The cross-covariance function picks out the 130-second delay, even though the pulse shape is not known and the pulses cannot be distinguished in the two sequences (bottom panel).

Haec sit finis.

Some Useful Definite Integrals

A.1 Integral of sin(x)/x

The integral

$$I = \int_0^\infty \frac{\sin(x)}{x} dx \tag{A.1}$$

can be evaluated by contour integration, but since the singularity at $x = 0$ is not real, it can also be evaluated by more elementary methods—plus a bit of trickery. Introduce the dummy variable a

$$I(a) = \int_0^\infty \exp[-ax] \frac{\sin x}{x} dx, \tag{A.2}$$

so that

$$I = I(0). \tag{A.3}$$

Now differentiate, and then integrate $I(a)$ with respect to a:

$$\frac{dI(a)}{da} = -\int_0^\infty \exp[-ax] \sin(x) dx$$

$$= -\frac{1}{2i} \int_0^\infty \exp[-ax] \{\exp[ix] - \exp[-ix]\} dx$$

$$= -\frac{1}{a^2 + 1} \tag{A.4}$$

$$I(a) = -\int \frac{1}{a^2 + 1} da$$

$$= -\tan^{-1} a + c. \tag{A.5}$$

From equation A.2, $I(\infty) = 0$, so the integration constant c must equal $\pi/2$. Equation A.5 becomes

$$I(a) = \frac{\pi}{2} - \tan^{-1} a, \tag{A.6}$$

and therefore

$$\int_0^\infty \frac{\sin(x)}{x} dx = I(0) = \frac{\pi}{2}. \tag{A.7}$$

A.2 Integrals of $x^n \exp[-ax^2]$

We first derive the definite integral of $\exp[ax^2]$ and then give two different derivations for the higher moments, one based on a recursion relation and one on the moment generating function.

Integral of $\exp[-ax^2]$: The definite integral

$$I = \int_{-\infty}^\infty \exp[-x^2] dx \tag{A.8}$$

is evaluated by calculating I^2, not I:

$$I^2 = \left\{ \int_{-\infty}^\infty \exp[-x^2] dx \right\} \left\{ \int_{-\infty}^\infty \exp[-y^2] dy \right\} = \int_{-\infty}^\infty \int_{-\infty}^\infty \exp\left[-(x^2 + y^2)\right] dx dy. \tag{A.9}$$

Convert to polar coordinates:

$$x = r\cos\theta \tag{A.10}$$

$$y = r\sin\theta, \tag{A.11}$$

so that equation A.9 becomes

$$I^2 = \int_{\theta=0}^{2\pi} \int_{r=0}^\infty \exp[-r^2] r dr d\theta = 2\pi \int_{r=0}^\infty \exp[-r^2] r dr. \tag{A.12}$$

After a second coordinate transformation, $z = r^2$, equation A.12 becomes

$$I^2 = \pi \int_{z=0}^\infty \exp[-z] dz = \pi, \tag{A.13}$$

and thus $I^2 = \pi$, and

$$\int_{-\infty}^\infty \exp[-x^2] dx = \sqrt{\pi}. \tag{A.14}$$

Setting $x = \sqrt{a} z$, one also has

$$\int_{-\infty}^\infty \exp[-az^2] dz = \sqrt{\frac{\pi}{a}}. \tag{A.15}$$

Integral of $x^n \exp[-ax^2]$: Now let the function $I(n)$ be defined by

$$I(n) = \int_{-\infty}^\infty x^n \exp[-ax^2] dx. \tag{A.16}$$

First note that the derivative of $I(n)$ with respect to x is 0, because there is no dependence on x remaining after integrating over x. One then has

$$0 = \frac{\partial I(n)}{\partial x} = \int_{-\infty}^{\infty} nx^{n-1} \exp[-ax^2]dx - 2a \int_{-\infty}^{\infty} x^{n+1} \exp[-ax^2]dx$$

$$= nI(n-1) - 2aI(n+1), \tag{A.17}$$

or, rearranging,

$$I(n+1) = \frac{n}{2a}I(n-1). \tag{A.18}$$

Equation A.15 is $I(0) = \sqrt{\pi/a}$, and $I(1) = 0$ by inspection, so equation A.18 gives $I(n)$ for all other n by recursion. For example,

$$\int_{-\infty}^{\infty} x^2 \exp[-ax^2]dx = I(2) = \frac{1}{2a}I(0) = \frac{1}{2a}\sqrt{\frac{\pi}{a}}. \tag{A.19}$$

Another approach to calculating the integrals is to use the moment generating function for the Gaussian probability distribution function. The moment generating function for the normalized Gaussian is (equation 2.58):

$$M(\zeta) = \exp\left[\frac{1}{2}\sigma^2\zeta^2\right]. \tag{A.20}$$

The function $\exp[-ax^2]$ becomes a normalized Gaussian if we set $\sigma^2 = 1/2a$ and normalize by $\sqrt{a/\pi}$. Expand $M(\zeta)$ as a Taylor series:

$$M(\zeta) = 1 + \frac{1}{1!}\left(\frac{1}{2}\sigma^2\zeta^2\right) + \frac{1}{2!}\left(\frac{1}{2}\sigma^2\zeta^2\right)^2 + \frac{1}{3!}\left(\frac{1}{2}\sigma^2\zeta^2\right)^3 + \cdots$$

$$= 1 + \frac{\sigma^2}{1!2^1}\zeta^2 + \frac{\sigma^4}{2!2^2}\zeta^4 + \frac{\sigma^6}{3!2^3}\zeta^6 + \cdots. \tag{A.21}$$

The mth moment is generated from the moment generating function by

$$M_m = \frac{\partial^m M(\zeta)}{\partial \zeta^m}\bigg|_{\zeta=0}. \tag{A.22}$$

For example, M_6 is given by

$$M_6 = \frac{\partial^6 M(\zeta)}{\partial \zeta^6}\bigg|_{\zeta=0} = \frac{\sigma^6}{3!2^3}\frac{\partial^6 \zeta^6}{\partial \zeta^6} = \frac{6!}{3!}\frac{\sigma^6}{2^3}. \tag{A.23}$$

The mth moment is therefore

$$M_m = \frac{m!}{(m/2)!}\frac{\sigma^m}{2^{m/2}}, \tag{A.24}$$

where m is a positive even integer. Thus we find

$$\int_{-\infty}^{\infty} x^m \exp[-ax^2]dx = \frac{m!}{2^m(m/2)!}\frac{1}{a^{m/2}}\sqrt{\frac{\pi}{a}}. \tag{A.25}$$

A.3 Gamma Function for Integers and Half Integers

The gamma function is defined to be

$$\Gamma(x) = \int_0^\infty \exp[-t] \, t^{x-1} dt. \tag{A.26}$$

Equation A.26 can be integrated by parts to give

$$\Gamma(x) = (x-1) \int_0^\infty \exp[-t] t^{x-2} dt - \left[\exp[-t] t^{x-1} \Big|_{-\infty}^\infty \right]$$

$$= (x-1)\Gamma(x-1). \tag{A.27}$$

If x is an integer or half integer, the gamma function becomes a factorial:

$$\Gamma(n) = (n-1)!. \tag{A.28}$$

To complete the factorial, one eventually needs an explicit expression for either 0! or $(1/2)!$. Evaluation of 0! is trivial:

$$0! = \Gamma(1) = \int_0^\infty \exp[-t] \, dt = 1. \tag{A.29}$$

Evaluate $(1/2)!$ from

$$\left(\frac{1}{2} \right)! = \Gamma\left(\frac{3}{2} \right) = \int_0^\infty \exp[-t] \, t^{1/2} dt. \tag{A.30}$$

Perform a change of variables, setting $z^2 = t$, and equation A.30 becomes

$$\left(\frac{1}{2} \right)! = 2 \int_0^\infty \exp\left[-z^2\right] z^2 dz = \int_{-\infty}^\infty \exp\left[-z^2\right] z^2 dz$$

$$= \frac{1}{2}\sqrt{\pi}, \tag{A.31}$$

where the last step uses equation A.19.

A.4 Beta Function

The beta function is defined to be

$$B(m,n) = \int_0^1 x^{m-1}(1-x)^{n-1} dx. \tag{A.32}$$

There are two useful alternate forms for the beta function. The first comes from setting $x = \sin^2\theta$. Since $dx = 2\sin\theta\cos\theta\,d\theta$, the beta function becomes

$$B(m,n) = \int_0^{\pi/2} (\sin^2\theta)^{m-1}(1-\sin^2\theta)^{n-1} 2\sin\theta\sin\theta\,d\theta$$

$$= 2 \int_0^{\pi/2} \sin^{2m-1}\theta \cos^{2n-1}\theta\,d\theta. \tag{A.33}$$

The second comes from setting $x = (1+y)^{-1}$. With $dx = -(1+y)^{-2}dy$, we have

$$B(m,n) = -\int_\infty^0 (1+y)^{-(m-1)} \left(\frac{y}{1+y}\right)^{n-1} (1+y)^{-2} dy$$

$$= \int_0^\infty y^{n-1}(1+y)^{-(m+n)} dy. \tag{A.34}$$

To evaluate the integral in equation A.32, we first evaluate the product of two gamma functions:

$$\Gamma(m)\Gamma(n) = \left(\int_0^\infty \exp[-x] x^{n-1} dx\right) \left(\int_0^\infty \exp[-y] y^{m-1} dy\right). \tag{A.35}$$

Substituting $x = u^2$ and $y = v^2$ and collecting terms, we have

$$\Gamma(m)\Gamma(n) = 4\int_0^\infty \int_0^\infty \exp\left[-(u^2 + v^2)\right] u^{2m-1} v^{2n-1} du dv. \tag{A.36}$$

Transforming to polar coordinates, $u = r\cos\theta$ and $v = r\sin\theta$, we obtain

$$\Gamma(m)\Gamma(n) = 4\int_{r=0}^\infty \int_{\theta=0}^{\pi/2} \exp[-r^2] r^{(2m+2n-2)} \cos^{2m-1}\theta \sin^{2n-1}\theta \, r dr d\theta, \tag{A.37}$$

and now setting $t = r^2$, we find

$$\Gamma(m)\Gamma(n) = \left[\int_{t=0}^\infty \exp[-t] t^{(m+n-1)} dt\right]\left[2\int_{\theta=0}^{\pi/2} \cos^{2m-1}\theta \sin^{2n-1}\theta \, r dr d\theta\right]. \tag{A.38}$$

The first integral in this equation is a gamma function, and from equation A.33, the second integral is the beta function, so we have

$$\Gamma(m)\Gamma(n) = \Gamma(m+n)B(m,n). \tag{A.39}$$

Rearranging this equation, we arrive at the desired result:

$$B(m,n) = \frac{\Gamma(m)\Gamma(n)}{\Gamma(m+n)}. \tag{A.40}$$

If m and n are integers, then

$$B(m,n) = \frac{(m-1)!\,(n-1)!}{(m+n-1)!}. \tag{A.41}$$

Method of Lagrange Multipliers

Suppose one wants to find the minimum or maximum of a function of n variables $f(x_1, x_2, \ldots, x_n)$. The differential of f is

$$df = \sum_{i=1}^{n} \frac{\partial f}{\partial x_i} dx_i = \nabla f \cdot d\mathbf{x}, \tag{B.1}$$

where ∇f is the gradient of f, and $d\mathbf{x}$ is the differential vector $(dx_1, dx_2, \ldots, dx_n)$. Extrema of f occur where f remains unchanged for arbitrary small changes in the x_i or, equivalently, where $df = 0$ for arbitrary $d\mathbf{x}$. Since $d\mathbf{x}$ is arbitrary in both magnitude and direction, df can be 0 only if all components of its gradient are 0:

$$\nabla f = 0, \tag{B.2}$$

or, equivalently,

$$\frac{\partial f}{\partial x_i} = 0, \quad i = 1, \ldots, n. \tag{B.3}$$

These are the standard equations for the extrema of f.

Suppose now that the x_i cannot vary independently but are limited by m constraints of the form

$$g_j(x_1, x_2, \ldots, x_n) = \text{constant}, \quad j = 1, \ldots, m. \tag{B.4}$$

Small changes in the x_i must satisfy the m equations

$$\sum_{i=1}^{n} \frac{\partial g_j}{\partial x_i} dx_i = 0. \tag{B.5}$$

The dx_i can no longer vary independently, so the derivatives of f in equation B.1 need not be 0, and equation B.3 no longer holds.

The direct way to solve this problem is to use the m constraint equations to eliminate m of the x_i in f and then set the derivatives of f with respect to the remaining x_i equal to 0. The method of Lagrange multipliers provides an alternate way to solve the problem if the direct approach is cumbersome or otherwise ineffective. First note that the g_j are a set of surfaces in the n-dimensional space of the x_i and that the normals to the surfaces are given by the gradients ∇g_j. Likewise, ∇f is the normal to the surfaces of constant f. Figure B.1 shows the

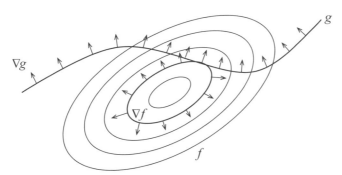

Figure B.1: Lagrange multipliers in two dimensions. We wish to minimize the function $f(x_1,x_2)$ subject to the constraint $g(x_1,x_2) = \text{constant}$. The ellipses are contours of constant f; the curved line is the function g. The arrows attached to one of the ellipses show the gradient of f, which is perpendicular to the contours of constant f. The arrows attached to g are the perpendiculars to g, which are given by ∇g. The constrained minimum of f occurs where one of its contours touches g at just one point and is, therefore, tangent to g. At that point, ∇f is parallel to ∇g, or $\nabla f = \lambda \nabla g$.

geometry for two dimensions. If there is just one constraint equation, g_1, the constrained minimum of f occurs where one of its contours touches the surface g_1 at just one point and so is tangent to g_1. At that point, ∇f is parallel to ∇g_1, or $\nabla f = \lambda_1 \nabla g_1$. If there are two constraint equations, the constrained minimum of f occurs where one of its contours just touches the intersection of surfaces g_1 and g_2. The normals to the intersection of g_1 and g_2 are any linear combination of the normals to the individual surfaces, $\lambda_1 \nabla g_1 + \lambda_2 \nabla g_2$, so $\nabla f = \lambda_1 \nabla g_1 + \lambda_2 \nabla g_2$. Generalizing to m constraint equations, we have

$$\nabla f + \sum_{j=1}^{m} \lambda_j \nabla g_j = 0. \tag{B.6}$$

The λ_j are called *Lagrange multipliers* (or sometimes *Lagrange undetermined multipliers*).

Equation B.6 is an n-dimensional vector equation. The individual components are

$$\frac{\partial f}{\partial x_i} + \sum_{j=1}^{m} \lambda_j \frac{\partial g_j}{\partial x_i} = 0. \tag{B.7}$$

These n equations plus the m constraint equations give $n + m$ equations for the n values of x_i and the m values of λ_j. The procedure for finding the location of the constrained minimum is:

1. Solve the n equations B.7 for x_i. The x_i will be expressed in terms of the λ_j, which are not yet known: $x_i = x_i(\lambda_1, \lambda_2, \ldots, \lambda_m)$.
2. Substitute these expressions for the x_i into the m constraint equations B.4, and solve the equations for the m values of λ_j.
3. Now that the λ_j are known, substitute them back into the equations $x_i = x_i(\lambda_1, \lambda_2, \ldots, \lambda_m)$. The result is the values of x_i at an extremum of f.

The method of Lagrange multipliers is sometimes presented in terms of a generating function F, given by

$$F = f + \sum_{j} \lambda_j g_j. \tag{B.8}$$

Setting the derivatives of F equal to 0, we find

$$\frac{\partial F}{\partial x_i} = 0 = \frac{\partial f}{\partial x_i} + \sum_{j=1}^{m} \lambda_j \frac{\partial g_j}{\partial x_i} \tag{B.9}$$

$$\frac{\partial F}{\partial \lambda_j} = 0 = g_j, \tag{B.10}$$

which are the same as equations B.7 and B.4, respectively. While the generating function is notationally clever and a useful mnemonic device, it has no computational and little conceptual advantage.

Example: Minimize $f = (x - x_0)^2 + (y - y_0)^2$, subject to the constraint $y = ax$, or equivalently, $g = y - ax = 0$. One can think of g as a line that cuts across f. The problem is to find the point along g where f reaches its lowest value.

The necessary derivatives are

$$\frac{\partial f}{\partial x} = 2(x - x_0), \qquad \frac{\partial g}{\partial x} = -a$$

$$\frac{\partial f}{\partial y} = 2(y - x_0), \qquad \frac{\partial g}{\partial y} = 1.$$

Equation B.7 becomes

$$2(x - x_0) - a\lambda = 0$$

$$2(y - y_0) + \lambda = 0,$$

which have the solutions

$$x = x_0 + \frac{a}{2}\lambda$$

$$y = y_0 - \frac{1}{2}\lambda.$$

Plugging these solutions back into the constraint equation, we find

$$y_0 - \frac{1}{2}\lambda = a\left[x_0 + \frac{a}{2}\lambda\right],$$

so λ is

$$\lambda = \frac{2(y_0 - ax_0)}{a^2 + 1}.$$

The values of x and y at which f is a minimum are then

$$x = x_0 + \frac{a}{2}\frac{2(y_0 - ax_0)}{a^2 + 1} = \frac{x_0 + ay_0}{a^2 + 1}$$

$$y = y_0 - \frac{1}{2}\frac{2(y_0 - ax_0)}{a^2 + 1} = a\frac{x_0 + ay_0}{a^2 + 1}.$$

For this simple case we can, of course, find the same result simply by substituting $y = ax$ into f, giving

$$f = (x - x_0)^2 + (ax - y_0)^2,$$

and the constrained minimum is the solution to

$$0 = \frac{df}{dx} = 2(x - x_0) + 2a(ax - y_0),$$

which is

$$x = \frac{x_0 + ay_0}{a^2 + 1}.$$

The constraint equation then yields, as before,

$$y = ax = a\frac{x_0 + ay_0}{a^2 + 1}.$$

Additional Properties of the Gaussian Probability Distribution

C.1 Other Derivations of the Gaussian Distribution

The Gaussian probability distribution function was derived in Section 2.4 as the end result of the central limit theorem. It can be derived in many other ways. The derivation here shows that it is the limit of the Poisson distribution as μ becomes large, and of the binomial distribution when the number of samples becomes so large that the Laplace approximation (see Section 7.3.2) becomes valid. Herschel's interesting derivation is also given here.

C.1.1 Gaussian Distribution as a Limit of the Poisson Distribution

The Poisson distribution is

$$P(k) = \frac{\mu^k}{k!} \exp[-\mu], \tag{C.1}$$

where k is an integer, and both $\langle k \rangle = \mu$ and $\sigma^2 = \mu$ (equations 2.29 and 2.31). The Gaussian distribution can be derived by taking the limit of the Poisson distribution as μ becomes large. Define $\epsilon = k - \mu$, so that

$$\begin{aligned} P(\epsilon) &= \frac{\mu^{\mu+\epsilon}}{(\mu+\epsilon)!} \exp[-\mu] \\ &= \frac{\mu^\mu \exp[-\mu]}{\mu!} \left\{ \frac{\mu}{\mu+1} \times \frac{\mu}{\mu+2} \times \cdots \times \frac{\mu}{\mu+\epsilon} \right\}. \end{aligned} \tag{C.2}$$

Evaluate $\mu!$ using Sterling's formula:

$$n! = (2\pi n)^{1/2} n^n \exp[-n] \qquad \text{for } n \gg 1, \tag{C.3}$$

and then as μ becomes large, the first term in equation C.2 becomes

$$\lim_{\mu \gg 1} \frac{\mu^\mu \exp[-\mu]}{\mu!} = \frac{\mu^\mu \exp[-\mu]}{(2\pi\mu)^{1/2} \mu^\mu \exp[-\mu]} = \frac{1}{(2\pi\mu)^{1/2}}. \tag{C.4}$$

To evaluate the series in equation C.2, take

$$\left\{ \frac{\mu}{\mu+1} \times \frac{\mu}{\mu+2} \times \cdots \times \frac{\mu}{\mu+\epsilon} \right\} = \exp\left[-\ln\left\{ \frac{\mu+1}{\mu} \times \frac{\mu+2}{\mu} \times \cdots \times \frac{\mu+\epsilon}{\mu} \right\} \right]. \tag{C.5}$$

To first order, the logs can be expanded as

$$\ln\left\{\frac{\mu+x}{\mu}\right\} = \ln\left\{1+\frac{x}{\mu}\right\} = \frac{x}{\mu}+\cdots, \tag{C.6}$$

and the series becomes

$$\left\{\frac{\mu}{\mu+1}\times\frac{\mu}{\mu+2}\times\cdots\times\frac{\mu}{\mu+\epsilon}\right\} = \exp\left[-\frac{1}{\mu}(1+2+\cdots+\epsilon)\right]. \tag{C.7}$$

Finally, as ϵ becomes large, the sum becomes

$$1+2+\cdots+\epsilon = \frac{\epsilon(\epsilon+1)}{2} \approx \frac{1}{2}\epsilon^2, \tag{C.8}$$

and we arrive at the Gaussian distribution:

$$P(\epsilon) = \frac{1}{(2\pi\mu)^{1/2}}\exp\left[-\frac{\epsilon^2}{2\mu}\right]. \tag{C.9}$$

Remembering that $\mu=\sigma^2$, we can put the Gaussian distribution in standard form:

$$P(\epsilon) = \frac{1}{(2\pi\sigma^2)^{1/2}}\exp\left[-\frac{\epsilon^2}{2\sigma^2}\right]. \tag{C.10}$$

C.1.2 Gaussian Distribution as a Limit of the Binomial Distribution

The binomial distribution is (see Section 2.2)

$$P(k) = \frac{n!}{k!\,(n-k)!}p^k q^{n-k} = \binom{n}{k}p^k q^{n-k}, \tag{C.11}$$

where k and n are integers and $q=1-p$. The mean value and variance of k are $\langle k\rangle = np$ and $\sigma^2 = npq$. In Section 2.3 we derived the Poisson distribution from the binomial distribution. Since we have just derived the Gaussian distribution from the Poisson distribution, we have already derived the Gaussian distribution from the binomial distribution. The Gaussian distribution can also be derived directly from the binomial distribution, albeit somewhat awkwardly. We give a derivation here.

Assume that k and np are so large that $P(k)$ is is essentially continuous in k, and then consider a region of the binomial distribution near its maximum—actually near the maximum of its log:

$$\left.\frac{\partial \ln P(k)}{\partial k}\right|_{k_{max}} = 0. \tag{C.12}$$

Expand $\ln P(k)$ about its maximum and, setting $k = k_{max} + \epsilon$, we have

$$\ln P(\epsilon) = \ln P(k_{max}) + \left.\frac{\partial \ln P(k)}{\partial k}\right|_{k_{max}}\epsilon + \frac{1}{2}\left.\frac{\partial^2 \ln P(k)}{\partial k^2}\right|_{k_{max}}\epsilon^2 + \cdots$$

$$= \ln P(k_{max}) + \frac{1}{2}\left.\frac{\partial^2 \ln P(k)}{\partial k^2}\right|_{k_{max}}\epsilon^2 + \cdots. \tag{C.13}$$

Exponentiating equation C.13 and retaining the low-order terms, we get

$$P(\epsilon) = \exp\left[\ln P(k_{max}) + \frac{1}{2}\frac{\partial^2 \ln P(k)}{\partial k^2}\bigg|_{k_{max}}\epsilon^2\right]$$

$$= P(k_{max})\exp\left[-\frac{1}{2}\left|\frac{\partial^2 \ln P(k)}{\partial k^2}\bigg|_{k_{max}}\right|\epsilon^2\right], \quad \text{(C.14)}$$

which takes advantage of the fact that the second derivative must be negative, because we are expanding about a maximum of $P(k)$. Equation C.14 is the Laplace approximation to the binomial distribution.

We now need to evaluate the derivatives of $P(k)$ explicitly. From equation C.11, we have

$$\ln P(\epsilon) = \ln n! - \ln k! - \ln(n-k)! + k\ln p + (n-k)\ln q. \quad \text{(C.15)}$$

For large n, k, and $n-k$, the logarithms are essentially continuous, so we have, for example,

$$\frac{\partial \ln k!}{\partial k} \approx \frac{\ln(k+\Delta k)! - \ln k!}{\Delta k} = \frac{\ln(k+1)! - \ln k!}{1} = \ln\left(\frac{(k+1)!}{k!}\right)$$

$$\approx \ln k. \quad \text{(C.16)}$$

Equation C.15 becomes

$$\frac{\partial \ln P(\epsilon)}{\partial k} = -\ln k + \ln(n-k) + \ln p - \ln q = \ln\left[\frac{(n-k)}{k}\frac{p}{q}\right]. \quad \text{(C.17)}$$

We can now find k_{max} by setting equation C.17 equal to 0:

$$\ln\left[\frac{(n-k_{max})}{k_{max}}\frac{p}{q}\right] = 0, \quad \text{(C.18)}$$

or

$$\frac{(n-k_{max})}{k_{max}}\frac{p}{q} = 1. \quad \text{(C.19)}$$

Recognizing that $q = 1 - p$ and then solving for k_{max}, we find

$$(n-k_{max})p = (1-p)k_{max} \quad \text{(C.20)}$$

$$k_{max} = np. \quad \text{(C.21)}$$

Differentiating equation C.17 a second time yields

$$\frac{\partial^2 \ln P(\epsilon)}{\partial k^2} = -\frac{1}{k} - \frac{1}{n-k} = -\frac{n}{k(n-k)}. \quad \text{(C.22)}$$

Evaluating the second derivative at $k = k_{max} = np$, we have

$$\frac{\partial^2 \ln P(\epsilon)}{\partial k^2}\bigg|_{k_{max}} = -\frac{n}{np(n-np)} = -\frac{1}{np(1-p)} = -\frac{1}{npq}. \quad \text{(C.23)}$$

Inserting this result into equation C.14, we get

$$P(\epsilon) = P(k_{max}) \exp\left[-\frac{1}{2}\frac{\epsilon^2}{npq}\right]. \tag{C.24}$$

Remembering that $\sigma^2 = npq$, we arrive at the (unnormalized) Gaussian distribution

$$P(\epsilon) = P(k_{max}) \exp\left[-\frac{1}{2}\frac{\epsilon^2}{\sigma^2}\right]. \tag{C.25}$$

C.1.3 Herschel's Derivation of the Gaussian Distribution

One of the most interesting derivations of the normal distribution is from John F. W. Herschel, who is better known for his astronomical than his mathematical contributions.

Consider the errors involved in measuring the position of a star as shown in Figure C.1. Assume that the errors in x and y are independent. The distribution of errors will then be given by

$$f_1(x)f_2(y)\,dxdy, \tag{C.26}$$

that is, the distribution is separable. Also assume that the errors in position are independent of angle. The distribution function must then be a function only of r, the distance from the center of the distribution:

$$g(r)\,dxdy. \tag{C.27}$$

Setting distributions C.26 and C.27 equal, we have

$$g(r) = f_1(x)f_2(y). \tag{C.28}$$

Since $g(r)$ is independent of θ, differentiating equation C.28 by θ yields

$$\frac{\partial g(r)}{\partial \theta} = f_1(x)\frac{\partial f_2(y)}{\partial \theta} + f_2(y)\frac{\partial f_1(x)}{\partial \theta} = 0. \tag{C.29}$$

To evaluate the derivatives, we change to (r,θ) coordinates, where

$$x = r\cos\theta \tag{C.30}$$
$$y = r\sin\theta, \tag{C.31}$$

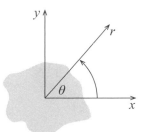

Figure C.1: Geometry for Herschel's derivation of the normal distribution.

and find

$$\frac{\partial f_1(x)}{\partial \theta} = \frac{\partial f_1(x)}{\partial x}\frac{\partial x}{\partial \theta} = -y\frac{\partial f_1(x)}{\partial x} \tag{C.32}$$

$$\frac{\partial f_2(y)}{\partial \theta} = \frac{\partial f_2(y)}{\partial y}\frac{\partial y}{\partial \theta} = x\frac{\partial f_2(y)}{\partial y}. \tag{C.33}$$

Plugging these results into equation C.29, we find

$$0 = xf_1(x)\frac{\partial f_2(y)}{\partial y} - yf_2(y)\frac{\partial f(_1 x)}{\partial x}, \tag{C.34}$$

or, rearranging,

$$\frac{1}{xf_1(x)}\frac{\partial f_1(x)}{\partial x} = K = \frac{1}{yf_2(y)}\frac{\partial f_2(y)}{\partial y}, \tag{C.35}$$

where K is a constant, because x and y can vary independently of each other. The solution to the x-component of the equation is

$$\frac{df_1(x)}{f_1(x)} = Kx\,dx$$

$$\ln f_1(x) = \frac{1}{2}Kx^2 + C$$

$$f_1(x) \propto \exp\left[\frac{1}{2}Kx^2\right]. \tag{C.36}$$

Since $f_1(x)$ must be normalizable, K must be negative. Therefore we can set $K = -1/\sigma^2$, where σ^2 is a positive constant, and write $f_1(x)$ as

$$f_1(x) = \exp\left[-\frac{1}{2}\frac{x^2}{\sigma^2}\right]. \tag{C.37}$$

In the same way we find

$$f_2(y) \propto \exp\left[-\frac{1}{2}\frac{y^2}{\sigma^2}\right], \tag{C.38}$$

so we arrive at the desired result. Physically, this derivation means that any cut whatsoever through a cylindrically symmetric, two-dimensional Gaussian yields the same function except for a normalization factor, and a Gaussian is the only two-dimensional function for which this is true. The crosscuts are always Gaussians with the same variance.

C.2 Product of Gaussian Probability Distributions

C.2.1 Product of Two Gaussian Probability Distributions

Let two Gaussian probability distributions be

$$f_1(x) = \frac{1}{\sqrt{2\pi}\sigma_1}\exp\left[-\frac{1}{2}\frac{(x-\mu_1)^2}{\sigma_1^2}\right] \tag{C.39}$$

$$f_2(x) = \frac{1}{\sqrt{2\pi}\sigma_2}\exp\left[-\frac{1}{2}\frac{(x-\mu_2)^2}{\sigma_2^2}\right], \tag{C.40}$$

where all symbols have their usual meanings. The product of the Gaussians is

$$f(x) = f_1(x)f_2(x) = \frac{1}{2\pi\sigma_1\sigma_2}\exp\left[-\frac{1}{2}\frac{(x-\mu_1)^2}{\sigma_1^2} - \frac{1}{2}\frac{(x-\mu_2)^2}{\sigma_1^2}\right]. \tag{C.41}$$

Set $w_1 = 1/\sigma_1^2$ and $w_2 = 1/\sigma_2^2$, and expand the exponent of $f(x)$:

$$\begin{aligned}
e(x) &= -\frac{1}{2}\frac{(x-\mu_1)^2}{\sigma_1^2} - \frac{1}{2}\frac{(x-\mu_2)^2}{\sigma_1^2} \\
&= -\frac{1}{2}\left[w_1 x^2 - 2w_1 x\mu_1 + w_1\mu_1^2 + w_2 x^2 - 2w_2 x\mu_2 + w_2\mu_2^2\right] \\
&= -\frac{w_1 + w_2}{2}\left[x^2 - 2\frac{w_1\mu_1 + w_2\mu_2}{w_1 + w_2}x + \frac{w_1\mu_1^2 + w_2\mu_2^2}{w_1 + w_2}\right].
\end{aligned} \tag{C.42}$$

Complete the square:

$$\begin{aligned}
e(x) &= -\frac{w_1 + w_2}{2}\left[x^2 - 2\frac{w_1\mu_1 + w_2\mu_2}{w_1 + w_2}x + \left(\frac{w_1\mu_1 + w_2\mu_2}{w_1 + w_2}\right)^2\right. \\
&\qquad\qquad \left. -\left(\frac{w_1\mu_1 + w_2\mu_2}{w_1 + w_2}\right)^2 + \frac{w_1\mu_1^2 + w_2\mu_2^2}{w_1 + w_2}\right] \\
&= -\frac{w_1 + w_2}{2}\left(x - \frac{w_1\mu_1 + w_2\mu_2}{w_1 + w_2}\right)^2 + \text{constant}.
\end{aligned} \tag{C.43}$$

Now set

$$\sigma^2 = \frac{1}{w_1 + w_2} = \frac{\sigma_1^2\sigma_2^2}{\sigma_1^2 + \sigma_2^2} \tag{C.44}$$

$$\mu = \frac{w_1\mu_1 + w_2\mu_2}{w_1 + w_2} = \frac{\sigma_2^2\mu_1 + \sigma_1^2\mu_2}{\sigma_1^2 + \sigma_2^2}. \tag{C.45}$$

The exponent becomes

$$e(x) = -\frac{1}{2}\frac{(x-\mu)^2}{\sigma^2} + c, \tag{C.46}$$

and the product of the two Gaussians becomes

$$f(x) \propto \exp\left[-\frac{1}{2}\frac{(x-\mu)^2}{\sigma^2}\right], \tag{C.47}$$

where the constant has been absorbed into the proportionality constant. From Section 2.4, we can write down the normalization constant by inspection, arriving at

$$f(x) = \frac{1}{\sqrt{2\pi}\sigma}\exp\left[-\frac{1}{2}\frac{(x-\mu)^2}{\sigma^2}\right], \tag{C.48}$$

where μ and σ^2 are given by equations C.44 and C.45. Thus, the product of two Gaussians is another Gaussian.

C.2.2 Product of n Gaussians

Now let us take the product of three Gaussians,

$$f(x) = f_1(x)f_2(x)f_3(x), \tag{C.49}$$

where $f_1(x)$ and $f_2(x)$ are given by equations C.39 and C.40, and $f_3(x)$ is given by

$$f_3(x) = \frac{1}{\sqrt{2\pi}\sigma_3} \exp\left[-\frac{1}{2}\frac{(x-\mu_3)^2}{\sigma_3^2}\right]. \tag{C.50}$$

We have just shown that the product of $f_1(x)$ and $f_2(x)$ is a Gaussian. Rewriting equation C.48 as

$$y(x) = \frac{1}{\sqrt{2\pi}\sigma_y} \exp\left[-\frac{1}{2}\frac{(x-\mu_y)^2}{\sigma_y^2}\right], \tag{C.51}$$

where

$$\sigma_y^2 = \frac{1}{w_1 + w_2} \tag{C.52}$$

$$\mu_y = \frac{w_1\mu_1 + w_2\mu_2}{w_1 + w_2}, \tag{C.53}$$

we have

$$f(x) = y(x)f_3(x). \tag{C.54}$$

Thus $f(x)$ is the product of two Gaussians and must also be a Gaussian. We can write it as

$$f(x) = \frac{1}{\sqrt{2\pi}\sigma} \exp\left[-\frac{1}{2}\frac{(x-\mu)^2}{\sigma^2}\right], \tag{C.55}$$

where now

$$\sigma^2 = \frac{1}{w_y + w_3} \tag{C.56}$$

$$\mu = \frac{w_y\mu_y + w_3\mu_3}{w_y + w_3}, \tag{C.57}$$

and $w_y = 1/\sigma_y^2 = w_1 + w_2$. Expanding σ^2 and μ, we get

$$\sigma^2 = \frac{1}{w_1 + w_2 + w_3} \tag{C.58}$$

$$\mu = \frac{w_y\mu_y + w_3\mu_3}{w_y + w_3} = \frac{w_1\mu_1 + w_2\mu_2 + w_3\mu_3}{w_1 + w_2 + w_3}. \tag{C.59}$$

By induction, we can now write down the product of n Gaussians without further effort:

$$\prod_{i=1}^{n} \frac{1}{\sqrt{2\pi}\sigma_i} \exp\left[-\frac{1}{2}\frac{(x-\mu_i)^2}{\sigma_i^2}\right] = \frac{1}{\sqrt{2\pi}\sigma} \exp\left[-\frac{1}{2}\frac{(x-\mu)^2}{\sigma^2}\right], \tag{C.60}$$

where

$$\sigma^2 = \frac{1}{\sum_{i=1}^{n} w_i} \tag{C.61}$$

$$\mu = \frac{\sum_{i=1}^{n} w_i \mu_i}{\sum_{i=1}^{n} w_i}, \tag{C.62}$$

and $w_i = 1/\sigma_i^2$. Thus the product of n Gaussians is a Gaussian with mean and variance given by equations C.61 and C.62.

C.3 Convolution of Gaussian Probability Distributions

We wish to convolve two normalized Gaussians probability distributions,

$$f_1(t) = \frac{1}{\sqrt{2\pi}\,\sigma_1} \exp\left[-\frac{1}{2}\frac{t^2}{\sigma_1^2}\right] \tag{C.63}$$

$$f_2(t) = \frac{1}{\sqrt{2\pi}\,\sigma_2} \exp\left[-\frac{1}{2}\frac{t^2}{\sigma_2^2}\right]. \tag{C.64}$$

Their convolution is given by

$$y(\tau) = \int_{-\infty}^{\infty} f_1(t) f_2(\tau - t) dt = \frac{1}{2\pi\sigma_1\sigma_2} \int_{-\infty}^{\infty} \exp\left[-\frac{1}{2}\frac{t^2}{\sigma_1^2}\right] \exp\left[-\frac{1}{2}\frac{(\tau - t)^2}{\sigma_2^2}\right] dt. \tag{C.65}$$

For convenience, set $w_1 = 1/\sigma_1^2$ and $w_2 = 1/\sigma_2^2$, and then expand the exponential to get

$$y(\tau) = \frac{1}{2\pi\sigma_1\sigma_2} \int_{-\infty}^{\infty} \exp\left[-\frac{1}{2}(w_1 t^2 + w_2 \tau^2 - 2w_2 \tau t + w_2 t^2)\right] dt$$

$$= \frac{1}{2\pi\sigma_1\sigma_2} \int_{-\infty}^{\infty} \exp\left[-\frac{w_1 + w_2}{2}\left(t^2 - 2\frac{w_2}{w_1 + w_2}\tau t + \frac{w_2}{w_1 + w_2}\tau^2\right)\right] dt \tag{C.66}$$

Now complete the square to get

$$y(\tau) = \frac{1}{2\pi\sigma_1\sigma_2} \int_{-\infty}^{\infty} \exp\left[-\frac{w_1 + w_2}{2}\left(t^2 - 2\frac{w_2}{w_1 + w_2}\tau t + \frac{w_2^2}{(w_1 + w_2)^2}\tau^2\right.\right.$$

$$\left.\left. - \frac{w_2^2}{(w_1 + w_2)^2}\tau^2 + \frac{w_2}{w_1 + w_2}\tau^2\right)\right] dt. \tag{C.67}$$

Note that

$$-\frac{w_1 + w_2}{2}\left(-\frac{w_2^2}{(w_1 + w_2)^2}\tau^2 + \frac{w_2}{w_1 + w_2}\tau^2\right) = -\frac{1}{2}\frac{w_1 w_2}{w_1 + w_2}\tau^2 = -\frac{1}{2}\frac{\tau^2}{\sigma_1^2 + \sigma_2^2}, \tag{C.68}$$

so the convolution becomes

$$y(\tau) = \frac{1}{2\pi\sigma_1\sigma_2} \exp\left[-\frac{1}{2}\frac{\tau^2}{\sigma_1^2 + \sigma_2^2}\right] \int_{-\infty}^{\infty} \exp\left[-\frac{w_1 + w_2}{2}\left(t - \frac{w_2}{w_1 + w_2}\tau\right)^2\right] dt. \tag{C.69}$$

We now recognize that the integral is that of an unnormalized Gaussian with variance

$$\sigma^2 = \frac{1}{w_1 + w_2} = \frac{\sigma_1^2 \sigma_2^2}{\sigma_1^2 + \sigma_2^2}. \tag{C.70}$$

Using equation A.15, we can evaluate the integral by inspection:

$$\int_{-\infty}^{\infty} \exp\left[-\frac{w_1 + w_2}{2}\left(t - \frac{w_2}{w_1 + w_2}\tau \right)^2 \right] dt = \sqrt{2\pi}\frac{\sigma_1 \sigma_2}{\sqrt{\sigma_1^2 + \sigma_2^2}}, \tag{C.71}$$

and the convolution becomes

$$\begin{aligned}
y(\tau) &= \frac{1}{2\pi\sigma_1\sigma_2}\sqrt{2\pi}\frac{\sigma_1\sigma_2}{\sqrt{\sigma_1^2 + \sigma_2^2}}\exp\left[-\frac{1}{2}\frac{\tau^2}{\sigma_1^2 + \sigma_2^2} \right] \\
&= \frac{1}{\sqrt{2\pi}(\sigma_1^2 + \sigma_2^2)^{1/2}}\exp\left[-\frac{1}{2}\frac{\tau^2}{\sigma_1^2 + \sigma_2^2} \right]. \tag{C.72}
\end{aligned}$$

This is a normalized Gaussian distribution with variance $\sigma_1^2 + \sigma_2^2$. Thus the convolution of two Gaussians is a Gaussian with variance equal to the sum of the variances of the two original Gaussians.

C.4 Probability Distribution of the Sum of Random Variables with Gaussian Distributions

Let z be the sum of n independent random variables x_i, $i = 1, \ldots, n$:

$$z = \sum_{i=1}^{n} x_i, \tag{C.73}$$

where the x_i are drawn from Gaussian distributions $f_i(x_i)$, whose means are all 0 and variances are σ_i^2. To find the probability distribution for z, we first derive the distribution for $n = 2$ and then extend the results to arbitrary n. Since the variables are independent, the joint distribution function for x_1 and x_2 is

$$f(x_1, x_2)dx_1 dx_2 = f_1(x_1)f_2(x_2)dx_1 dx_2. \tag{C.74}$$

Set $z = x_1 + x_2$ and then change variables from (x_1, x_2) to (x_1, z), which results in

$$f(x_1, z)dx_1 dz = f_1(x_1)f_2(z - x_1)dx_1 dz. \tag{C.75}$$

To find $f(z)$, integrate over x_1:

$$f(z)dz = \left[\int f_1(x_1)f_2(z - x_1)dx_1 \right] dz. \tag{C.76}$$

Thus we find

$$f(z) = \int f_1(x_1)f_2(z - x_1)dx_1. \tag{C.77}$$

This is the convolution of $f_1(x_1)$ with $f(x_2)$, so we can use the results of Section C.3 to write down the answer by inspection:

$$f(z) = \frac{1}{\sqrt{2\pi}\,(\sigma_1^2 + \sigma_2^2)^{1/2}} \exp\left[-\frac{1}{2}\frac{z^2}{\sigma_1^2 + \sigma_2^2}\right]. \tag{C.78}$$

Generalizing to n variables, we find

$$f(z) = \frac{1}{\sqrt{2\pi}\,\sigma} \exp\left[-\frac{1}{2}\frac{z^2}{\sigma^2}\right], \tag{C.79}$$

where

$$\sigma^2 = \sum_{i=1}^{n} \sigma_i^2. \tag{C.80}$$

The *n*-Dimensional Sphere

We will evaluate the integral

$$I = \int_{-\infty}^{\infty} \cdots \int_{-\infty}^{\infty} \exp\left[-\left(x_1^2 + x_2^2 \cdots x_n^2\right)\right] dx_1 dx_2 \cdots dx_n \tag{D.1}$$

twice. Comparison of the two results will yield the volume and surface area of the *n*-dimensional sphere.

Integral D.1 can be rewritten as

$$I = \left\{ \int_{-\infty}^{\infty} \exp\left[-x^2\right] dx \right\}^n, \tag{D.2}$$

which, from equation A.15, is

$$I = \left(\sqrt{\pi}\right)^n = \pi^{n/2}. \tag{D.3}$$

The volume of an *n*-dimensional sphere must be proportional to r^n, where r is its radius. The volume of a spherical shell of thickness dr must, then, be given by

$$dV_n = S_n r^{n-1} dr, \tag{D.4}$$

where S_n is a constant that depends on n. For example, $S_2 = 2\pi$ and $S_3 = 4\pi$. After it has been integrated over over $n - 1$ angles, integral D.1 can be written as

$$I = \int_0^{\infty} \exp\left[-r^2\right] S_n r^{n-1} dr. \tag{D.5}$$

With a change of variables to $t = r^2$, the integral becomes

$$I = \frac{1}{2} S_n \int_0^{\infty} \exp[-t] \, t^{(n/2-1)} dt. \tag{D.6}$$

The integral is the gamma function (see Section A.3 in Appendix A), so equation D.6 becomes

$$I = \frac{1}{2} S_n \Gamma\left(\frac{n}{2}\right) = \frac{1}{2} S_n \left(\frac{n}{2} - 1\right)!. \tag{D.7}$$

Since the two results D.3 and D.7 must be the same, we have

$$\frac{1}{2}S_n\left(\frac{n}{2}-1\right)! = \pi^{n/2} \tag{D.8}$$

$$S_n = \frac{2\pi^{n/2}}{(n/2-1)!}. \tag{D.9}$$

The volume element is therefore

$$dV_n = \frac{2\pi^{n/2}}{(n/2-1)!}\,r^{n-1}dr, \tag{D.10}$$

and the volume of the *n*-dimensional sphere is

$$V_n = \frac{S_n}{n}r^n = \frac{2\pi^{n/2}}{n\,(n/2-1)!}r^n. \tag{D.11}$$

Since the volume elements is also $dV_n = A_n dr$, where A_n is the surface area of the sphere, the surface area of an *n*-dimensional sphere is

$$A_n = \frac{2\pi^{n/2}}{(n/2-1)!}r^{n-1}. \tag{D.12}$$

For a three-dimensional sphere, equations D.11 and D.12 reduce to the familiar

$$V_3 = \frac{2\pi^{3/2}}{3\,(3/2-1)!}r^3 = \frac{4\pi}{3}r^3 \tag{D.13}$$

$$A_3 = \frac{2\pi^{3/2}}{(3/2-1)!}r^2 = 4\pi r^2. \tag{D.14}$$

For a four-dimensional sphere, they reduce to the less familiar

$$V_4 = \frac{2\pi^{4/2}}{4\,(4/2-1)!}r^4 = \frac{1}{2}\pi^2 r^4 \tag{D.15}$$

$$A_4 = \frac{2\pi^{4/2}}{(4/2-1)!}r^4 = 2\pi^2 r^3. \tag{D.16}$$

Review of Linear Algebra and Matrices

E.1 Vectors, Basis Vectors, and Dot Products

The term "vector" originally meant a geometric object with magnitude and direction—an arrow. Vectors became important in part because many of the quantities of classical physics could be described by arrows: velocities, accelerations, forces, and electric and magnetic fields. But it was soon realized that other, less-obvious things could also be described by vectors: linear equations; polynomials; images; and perhaps startlingly, wave functions in quantum mechanics. The three-dimensional vectors of classical physics became four dimensional in relativity and infinite dimensional in quantum mechanics. The concept of vector had to be generalized to include these other kinds of vector. Any quantity $\mathbf{a}$ that satisfies the following rules is now called a *vector*.

Vectors can be added together. If $\mathbf{a}$, $\mathbf{b}$, and $\mathbf{c}$ are vectors, the sum of $\mathbf{a}$ and $\mathbf{b}$ is denoted by

$$\mathbf{c} = \mathbf{a} + \mathbf{b}. \tag{E.1}$$

Addition of vectors is commutative and associative:

$$\mathbf{a} + \mathbf{b} = \mathbf{b} + \mathbf{a} \tag{E.2}$$

$$\mathbf{a} + (\mathbf{b} + \mathbf{c}) = (\mathbf{a} + \mathbf{b}) + \mathbf{c}. \tag{E.3}$$

For every vector $\mathbf{a}$ there is a negative vector $\mathbf{b}$, denoted by

$$\mathbf{b} = -\mathbf{a}, \tag{E.4}$$

and subtraction of two vectors is defined by $\mathbf{a} - \mathbf{b} = \mathbf{a} + (-\mathbf{b})$. There is a zero vector defined by

$$\mathbf{a} = \mathbf{z} + \mathbf{a}. \tag{E.5}$$

A vector $\mathbf{a}$ can be multiplied by a scalar number λ to give a new vector $\mathbf{b}$:

$$\mathbf{b} = \lambda\mathbf{a}. \tag{E.6}$$

Multiplying a vector by $\lambda = 1$ returns the same vector

$$\mathbf{a} = 1\mathbf{a}. \tag{E.7}$$

Multiplication by scalars is distributive and associative:

$$\lambda(\mathbf{a} + \mathbf{b}) = \lambda\mathbf{a} + \lambda\mathbf{b} \tag{E.8}$$

$$(\lambda_1 + \lambda_2)\mathbf{a} = \lambda_1\mathbf{a} + \lambda_2\mathbf{a} \tag{E.9}$$

$$\lambda_1(\lambda_2\mathbf{a}) = (\lambda_1\lambda_2)\mathbf{a}. \tag{E.10}$$

A collection of vectors that can be created by these operations is called a *vector space*.

An arrow is not a useful way to represent a vector in n-dimensional vector spaces. In this book, the vector $\mathbf{a}$ is represented by a vertical column of numbers a_i:

$$\begin{pmatrix} a_1 \\ a_2 \\ \vdots \\ a_m \end{pmatrix}$$

called the *elements of the vector*. Logically, a special symbol should be used to describe the relation between a vector and its representation, perhaps

$$\mathbf{a} \Rightarrow \begin{pmatrix} a_1 \\ a_2 \\ \vdots \\ a_m \end{pmatrix},$$

because the values of the a_i are different in different coordinate systems; but once a set of basis vectors has been chosen, there is a unique, one-to-one correspondence between the a_i and the vectors they represent. We can, therefore, denote the relation without confusion by

$$\mathbf{a} = \begin{pmatrix} a_1 \\ a_2 \\ \vdots \\ a_m \end{pmatrix}, \tag{E.11}$$

although one should keep in mind that this is actually a representation, not an equality. In this representation, vector addition is

$$\begin{pmatrix} c_1 \\ c_2 \\ \vdots \\ c_m \end{pmatrix} = \begin{pmatrix} a_1 \\ a_2 \\ \vdots \\ a_m \end{pmatrix} + \begin{pmatrix} b_1 \\ b_2 \\ \vdots \\ b_m \end{pmatrix} = \begin{pmatrix} a_1 + b_1 \\ a_1 + b_2 \\ \vdots \\ a_m + b_m \end{pmatrix}. \tag{E.12}$$

The negative vector and the zero vector are

$$-\mathbf{a} = \begin{pmatrix} -a_1 \\ -a_2 \\ \vdots \\ -a_m \end{pmatrix} \quad \text{and} \quad \mathbf{z} = \begin{pmatrix} 0 \\ 0 \\ \vdots \\ 0 \end{pmatrix}. \tag{E.13}$$

Finally, the product of a scalar and a vector is

$$\lambda \mathbf{a} = \begin{pmatrix} \lambda a_1 \\ \lambda a_2 \\ \vdots \\ \lambda a_m \end{pmatrix}. \tag{E.14}$$

A vector can be linear combinations of other vectors $\mathbf{b_i}$:

$$\mathbf{a} = a_1\mathbf{b_1} + a_2\mathbf{b_2} + a_3\mathbf{b_3} + \cdots, \tag{E.15}$$

where the a_i are scalars. The minimum number, n, of vectors from which all other vectors in a vector space can be constructed is the *dimension* of the vector space. Any such group of n vectors is called a *basis* for the vector space. We denote the members of a basis by $\mathbf{e_i}$. Any vector $\mathbf{a}$ can be constructed from a linear combination of the $\mathbf{e_i}$:

$$\mathbf{a} = a_1\mathbf{e_1} + a_2\mathbf{e_2} + \cdots + a_n\mathbf{e_n} = \sum_{i=1}^{n} a_i\mathbf{e_i}. \tag{E.16}$$

The elements in the vector,

$$\mathbf{a} = \begin{pmatrix} a_1 \\ a_2 \\ \vdots \\ a_n \end{pmatrix}, \tag{E.17}$$

are the coefficients in equation E.16. Equations E.16 and E.17 define the relation between a vector and its representation.

The inner product, or dot product, of two vectors is denoted by $\mathbf{a} \cdot \mathbf{b}$. If $\mathbf{a} \cdot \mathbf{b} = 0$, the vectors are said to be orthogonal to each other. The square root of the dot product of a vector with itself is called the norm of the vector:

$$|\mathbf{a}| = (\mathbf{a} \cdot \mathbf{a})^{1/2}. \tag{E.18}$$

Basis vectors for which

$$\mathbf{e_i} \cdot \mathbf{e_j} = \delta_{ij}, \tag{E.19}$$

where δ_{ij} is the Kronecker delta function, make up an orthonormal basis. As an example, the unit vectors $\mathbf{i}, \mathbf{j}, \mathbf{k}$ in the three-dimensional Cartesian coordinate system are an orthonormal basis, because they are defined to be 1 unit in length and are aligned along the x-, y-, and z-axes, respectively, and are therefore orthogonal to one another. Any vector in a Cartesian coordinate system can be written as

$$\mathbf{a} = a_x\mathbf{i} + a_y\mathbf{j} + a_z\mathbf{k} \quad \text{or} \quad \mathbf{a} = \begin{pmatrix} a_x \\ a_y \\ a_z \end{pmatrix}. \tag{E.20}$$

Orthonormal basis sets are useful because the dot product of any vector with basis vector $\mathbf{e_k}$ yields

$$\mathbf{a} \cdot \mathbf{e_k} = \left(\sum_{i=1}^{n} a_i\mathbf{e_i} \right) \cdot \mathbf{e_k} = \sum_{i=1}^{n} a_i\mathbf{e_i} \cdot \mathbf{e_k} = \sum_{i=1}^{n} a_i\delta_{ik} = a_k, \tag{E.21}$$

and the dot product of two vectors is simply

$$\mathbf{a} \cdot \mathbf{b} = \left(\sum_{i=1}^{n} a_i \mathbf{e_i} \right) \cdot \left(\sum_{k=1}^{n} b_k \mathbf{e_k} \right) = \sum_{i=1}^{n} \sum_{k=1}^{n} a_i b_k \mathbf{e_i} \cdot \mathbf{e_k} = \sum_{i=1}^{n} \sum_{k=1}^{n} a_i b_k \delta_{ik} = \sum_{i=1}^{n} a_i b_i. \quad \text{(E.22)}$$

The dot product of a vector with itself becomes

$$\mathbf{a} \cdot \mathbf{a} = \sum_{i=1}^{n} a_i^2. \quad \text{(E.23)}$$

This is the n-dimensional equivalent of the Pythagorean theorem, so the norm $|\mathbf{a}| = (\mathbf{a} \cdot \mathbf{a})^{1/2}$ is the length of the vector. Calculation of the dot product is substantially more complicated for nonorthonormal bases.

E.2 Linear Operators and Matrices

One can operate on a vector to produce another vector. Operators are thus the vector equivalent of functions of scalars. If the symbol $\mathbf{A}$ represents an operator, the operation of $\mathbf{A}$ on vector $\mathbf{a}$ to produce vector $\mathbf{b}$ is represented by the equation

$$\mathbf{b} = \mathbf{A}\mathbf{a}. \quad \text{(E.24)}$$

Linear algebra is concerned with the subset of operators called *linear operators*, which have the property

$$\mathbf{A}(\lambda_1 \mathbf{a} + \lambda_2 \mathbf{b}) = \lambda_1 \mathbf{A}\mathbf{a} + \lambda_2 \mathbf{A}\mathbf{b}, \quad \text{(E.25)}$$

where λ_1 and λ_2 are arbitrary scalars. Unless explicitly stated otherwise, all operators we discuss from now on will be linear operators.

Suppose that a set of basis vectors $\mathbf{e_i}$ have been chosen. The result of operating on any of the basis vectors is another vector, and this resulting vector must be expressible as a linear combination of the basis vectors. Thus,

$$\mathbf{A}\mathbf{e_j} = \sum_{i=1}^{n} A_{ij} \mathbf{e_i}, \quad \text{(E.26)}$$

where A_{ij} is double subscripted because the operation of $\mathbf{A}$ on each $\mathbf{e_j}$ produces a different resulting vector. The quantities A_{ij} are called the components of $\mathbf{A}$. The components depend explicitly on the basis vectors and have different values for different sets of basis vectors. Making use of equations E.16, E.25, and E.26, we can write equation E.24 as

$$\sum_{i=1}^{n} b_i \mathbf{e_i} = \mathbf{A} \sum_{j=1}^{n} a_j \mathbf{e_j} = \sum_{j=1}^{n} a_j \mathbf{A}\mathbf{e_j} = \sum_{j=1}^{n} a_j \left(\sum_{i=1}^{n} A_{ij} \mathbf{e_i} \right) = \sum_{i=1}^{n} \left(\sum_{j=1}^{n} A_{ij} a_j \right) \mathbf{e_i}. \quad \text{(E.27)}$$

Thus, the elements of vector $\mathbf{b}$ are given by

$$b_i = \sum_{j=1}^{n} A_{ij} a_j. \quad \text{(E.28)}$$

The quantities A_{ij} can be written as a rectangular array called a *matrix*:

$$\begin{pmatrix} A_{11} & A_{12} & \cdots & A_{1n} \\ A_{21} & A_{22} & \cdots & A_{2n} \\ \vdots & \vdots & & \vdots \\ A_{m1} & A_{m2} & \cdots & A_{mn} \end{pmatrix}, \tag{E.29}$$

and this matrix is a representation of operator $\mathbf{A}$. A matrix with the same number of rows and columns is called a *square matrix*. The distinction made in the previous section between a vector and its representation holds for operators and matrices, but here also there is a unique one-to-one correspondence between a linear operator and the matrix that represents it once a set of basis vectors has been chosen. As for vectors, we can safely write

$$\mathbf{A} = \begin{pmatrix} A_{11} & A_{12} & \cdots & A_{1n} \\ A_{21} & A_{22} & \cdots & A_{2n} \\ \vdots & \vdots & & \vdots \\ A_{m1} & A_{m2} & \cdots & A_{mn} \end{pmatrix}. \tag{E.30}$$

The A_{ij} are called the *elements of the matrix*. One can also use the notation $(\mathbf{A})_{ij}$ for the elements of a matrix, $(\mathbf{a})_i$ for the elements of a vector, or even $(\mathbf{a_j})_i$ for element i of vector j. Equation E.24 can now be written

$$\begin{pmatrix} b_1 \\ b_2 \\ \vdots \\ b_n \end{pmatrix} = \begin{pmatrix} A_{11} & A_{12} & \cdots & A_{1n} \\ A_{21} & A_{22} & \cdots & A_{2n} \\ \vdots & \vdots & & \vdots \\ A_{m1} & A_{m2} & \cdots & A_{mn} \end{pmatrix} \begin{pmatrix} a_1 \\ a_2 \\ \vdots \\ a_n \end{pmatrix}, \tag{E.31}$$

where the meaning of this equation is given by equation E.28. To be explicit, the value of, say, b_2 is given by

$$b_2 = A_{21}a_1 + A_{22}a_2 + \cdots + A_{2n}a_n. \tag{E.32}$$

The connection between linear operators and matrices is so close that the algebra of linear operators is often called *matrix algebra*.

E.3 Matrix Algebra

The rules of matrix algebra follow directly from the definition of linear operators. Two matrices $\mathbf{A}$ and $\mathbf{B}$ can be added to give a third matrix $\mathbf{C}$, but the sum has meaning if and only if all three matrices have the same number of rows and columns. The sum is denoted by

$$\mathbf{C} = \mathbf{A} + \mathbf{B}, \tag{E.33}$$

and the elements of $\mathbf{C}$ are given by

$$C_{ij} = A_{ij} + B_{ij}. \tag{E.34}$$

For every matrix $\mathbf{A}$ there is a negative matrix denoted by

$$\mathbf{B} = -\mathbf{A} \tag{E.35}$$

whose elements are given by $B_{ij} = -A_{ij}$. Subtraction of two matrices is defined by $\mathbf{C} - \mathbf{D} = \mathbf{C} + (-\mathbf{D})$. A matrix $\mathbf{A}$ can be multiplied by a scalar λ to give matrix $\mathbf{C}$,

$$\mathbf{C} = \lambda\mathbf{A}, \tag{E.36}$$

whose elements are $C_{ij} = \lambda A_{ij}$; and there is a zero matrix $\mathbf{Z}$ whose elements are $Z_{ij} = 0$.

These properties mean that matrices satisfy equations E.2–E.10, which are the defining properties of vectors. Matrices are therefore vectors. It is entirely consistent to think of an ordinary vector $\mathbf{c}$ as a matrix with a single column of quantities c_j:

$$\mathbf{c} = \begin{pmatrix} c_1 \\ c_2 \\ \vdots \\ c_m \end{pmatrix}. \tag{E.37}$$

One can also have a vector that is a row of quantities r_j:

$$\mathbf{r} = (\begin{array}{cccc} r_1 & r_2 & \cdots & r_n \end{array}), \tag{E.38}$$

and this can be thought of as a matrix with a single row. When necessary, the two types of vectors can be distinguished by calling them "row vectors" and "column vectors."

Matrix multiplication is defined to mean successive operation by two linear operators. Suppose a linear operator $\mathbf{B}$ operates on a vector $\mathbf{a}$ to produce another vector $\mathbf{b}$:

$$\mathbf{b} = \mathbf{Ba}. \tag{E.39}$$

Now operate on $\mathbf{b}$ with a second linear operator $\mathbf{A}$ to give vector $\mathbf{c}$:

$$\mathbf{c} = \mathbf{Ab} = \mathbf{A}(\mathbf{Ba}). \tag{E.40}$$

This is equivalent to the equation

$$\mathbf{c} = \mathbf{Ca} = (\mathbf{AB})\mathbf{a}, \tag{E.41}$$

where $\mathbf{C}$ is the single operator that produces the same result as successive operation first by $\mathbf{B}$ and then by $\mathbf{A}$, denoted by

$$\mathbf{C} = \mathbf{AB}. \tag{E.42}$$

From equation E.28, the elements of $\mathbf{c}$ are given by

$$c_i = \sum_k A_{ik} b_k = \sum_k A_{ik} \left[\sum_j B_{kj} a_j \right] = \sum_j \left[\sum_k A_{ik} B_{kj} \right] a_j = \sum_j C_{ij} a_j, \tag{E.43}$$

so the elements of $\mathbf{C}$ are

$$C_{ij} = \sum_k A_{ik} B_{kj}. \tag{E.44}$$

Combining two matrices in this way to make a third called *matrix multiplication*. Note that the number of columns in **A** must equal the number of rows in **B**. As examples,

$$
\begin{pmatrix} C_{11} & C_{12} \\ C_{21} & C_{22} \end{pmatrix} = \begin{pmatrix} A_{11} & A_{12} & A_{13} \\ A_{21} & A_{22} & A_{23} \end{pmatrix} \begin{pmatrix} B_{11} & B_{12} \\ B_{21} & B_{22} \\ B_{31} & B_{32} \end{pmatrix}
$$

$$
= \begin{pmatrix} A_{11}B_{11} + A_{12}B_{21} + A_{13}B_{31} & A_{11}B_{12} + A_{12}B_{22} + A_{13}B_{31} \\ A_{21}B_{11} + A_{22}B_{21} + A_{23}B_{31} & A_{21}B_{12} + A_{22}B_{22} + A23B_{32} \end{pmatrix} \quad \text{(E.45)}
$$

or

$$
\begin{pmatrix} C_{11} & C_{12} & C_{13} \\ C_{21} & C_{22} & C_{23} \end{pmatrix} = \begin{pmatrix} A_{11} \\ A_{21} \end{pmatrix} \begin{pmatrix} B_{11} & B_{12} & B_{13} \end{pmatrix}
$$

$$
= \begin{pmatrix} A_{11}B_{11} & A_{11}B_{12} & A_{11}B_{13} \\ A_{21}B_{11} & A_{21}B_{12} & A_{21}B_{13} \end{pmatrix}. \quad \text{(E.46)}
$$

Matrix multiplication is associative and distributive over addition, but it is not commutative:

$$
\mathbf{A(BC)} = \mathbf{(AB)C} \quad \text{(E.47)}
$$

$$
\mathbf{C(A + B)} = \mathbf{CA + CB} \quad \text{(E.48)}
$$

$$
\mathbf{(A + B)C} = \mathbf{AC + BC} \quad \text{(E.49)}
$$

$$
\mathbf{AB \neq BA}. \quad \text{(E.50)}
$$

To see that matrix multiplication is not commutative, note that

$$
\begin{pmatrix} A_{11} \\ A_{21} \end{pmatrix} \begin{pmatrix} B_{11} & B_{12} & B_{13} \end{pmatrix} \neq \begin{pmatrix} B_{11} & B_{12} & B_{13} \end{pmatrix} \begin{pmatrix} A_{11} \\ A_{21} \end{pmatrix}. \quad \text{(E.51)}
$$

The right-hand side of this equation is not even defined.

It should now be apparent that row and column vectors are, in fact, different kinds of objects. For orthonormal basis sets, a dot product is a matrix multiplication of a row vector times a column vector: The row vector operates on a column vector to produce a scalar. A row vector is then an operator—sometimes called a *linear functional*—and it resides in a different vector space, the *dual space*, from the vector space of the column vectors. The distinction between a vector space and its dual space is not a major issue in this book, but it becomes crucial in curvilinear coordinate systems and in such subjects as the general theory of relativity.

The unit matrix **I** of order n is defined to be the matrix such that

$$
\mathbf{IA} = \mathbf{AI} = \mathbf{A}, \quad \text{(E.52)}
$$

where **A** be an arbitrary square matrix with n rows and n columns. The unit matrix is the diagonal matrix

$$
\mathbf{I} = \begin{pmatrix} 1 & 0 & \cdots & 0 \\ 0 & 1 & \cdots & 0 \\ \vdots & \vdots & & \vdots \\ 0 & 0 & \cdots & 1 \end{pmatrix} \quad \text{(E.53)}
$$

also with n rows and columns. The elements of the unit matrix can be written compactly as

$$I_{ij} = \delta_{ij}, \quad i, j = 1, \ldots, n, \tag{E.54}$$

where δ_{ij} is the Kronecker delta. The unit matrix can multiply nonsquare matrices as long as the matrix has n rows if multiplied on the left by $\mathbf{I}$, or has n columns if multiplied on the right. Thus, if $\mathbf{a}$ is a column vector, the operation

$$\mathbf{Ia} = \mathbf{a} \tag{E.55}$$

is sensible, but only if $\mathbf{a}$ has n elements and $\mathbf{I}$ is order n.

E.4 Transpose Matrices

The transpose of $\mathbf{A}$, denoted by $\mathbf{A}^{\mathrm{T}}$, is the matrix produced by exchanging the rows and columns of $\mathbf{A}$. The elements of $\mathbf{A}^{\mathrm{T}}$ are given by

$$(\mathbf{A}^{\mathrm{T}})_{ij} = (\mathbf{A})_{ji}. \tag{E.56}$$

If $\mathbf{C}$ is the sum of two matrices

$$\mathbf{C} = \mathbf{A} + \mathbf{B}, \tag{E.57}$$

the transpose of $\mathbf{C}$ is the sum of their transposes:

$$\mathbf{C}^{\mathrm{T}} = \mathbf{A}^{\mathrm{T}} + \mathbf{B}^{\mathrm{T}}. \tag{E.58}$$

The transpose of a product is slightly more complicated. Let matrix $\mathbf{C}$ be the product of matrices $\mathbf{A}$ and $\mathbf{B}$:

$$\mathbf{C} = \mathbf{AB}. \tag{E.59}$$

Rewriting equation E.44 in the alternative notation for the elements of a matrix, we have

$$(\mathbf{C})_{ij} = \sum_k (\mathbf{A})_{ik} (\mathbf{B})_{kj}. \tag{E.60}$$

The elements of the transpose of $\mathbf{C}$ are

$$(\mathbf{C}^{\mathrm{T}})_{ij} = (\mathbf{C})_{ji} = \sum_k (\mathbf{A})_{jk} (\mathbf{B})_{ki} = \sum_k (\mathbf{A}^{\mathrm{T}})_{kj} (\mathbf{B}^{\mathrm{T}})_{ik} = \sum_k (\mathbf{B}^{\mathrm{T}})_{ik} (\mathbf{A}^{\mathrm{T}})_{kj}. \tag{E.61}$$

Thus,

$$\mathbf{C}^{\mathrm{T}} = \mathbf{B}^{\mathrm{T}} \mathbf{A}^{\mathrm{T}}. \tag{E.62}$$

This result generalizes to products of more than one matrix. If

$$\mathbf{D} = \mathbf{ABC}, \tag{E.63}$$

then

$$\mathbf{D}^{\mathrm{T}} = \mathbf{C}^{\mathrm{T}} \mathbf{B}^{\mathrm{T}} \mathbf{A}^{\mathrm{T}}. \tag{E.64}$$

Transpose matrices provide an alternative way to think about dot products. From equation E.22, the dot product of two column vectors expanded on an orthonormal basis set is

$$\mathbf{a} \cdot \mathbf{b} = \begin{pmatrix} a_1 \\ a_2 \\ \vdots \\ a_n \end{pmatrix} \begin{pmatrix} b_1 \\ b_2 \\ \vdots \\ b_n \end{pmatrix} = \sum_{i=1}^{n} a_i b_i. \qquad (E.65)$$

Making use of transposes, this can be written instead as a matrix multiplication

$$\mathbf{a} \cdot \mathbf{b} = \mathbf{a}^{\mathrm{T}} \mathbf{b} = \mathbf{b}^{\mathrm{T}} \mathbf{a}, \qquad (E.66)$$

and the dot product becomes a special example of matrix multiplication. However, it should be remembered that row vectors and column vectors are not really the same thing; the transpose of a vector is actually an operator.

If a matrix is square and if $A_{ij} = A_{ji}$, the matrix is symmetric. It is antisymmetric if $A_{ij} = -A_{ji}$ and $A_{ii} = 0$. The following are examples of symmetric and antisymmetric matrices:

$$\begin{pmatrix} A_{11} & A_{12} & A_{13} \\ A_{12} & A_{22} & A_{23} \\ A_{13} & A_{23} & A_{33} \end{pmatrix}, \quad \begin{pmatrix} 0 & A_{12} & A_{13} \\ -A_{12} & 0 & A_{23} \\ -A_{13} & -A_{23} & 0 \end{pmatrix}.$$

The definitions of symmetric and antisymmetric matrices can be written compactly using transpose matrices. A matrix is symmetric if $\mathbf{A} = \mathbf{A}^{\mathrm{T}}$ and antisymmetric if $\mathbf{A} = -\mathbf{A}^{\mathrm{T}}$. Every square matrix can be decomposed into the sum of a symmetric and an antisymmetric matrix. The symmetric component is equal to $(\mathbf{A} + \mathbf{A}^{\mathrm{T}})/2$, and the antisymmetric component is equal to $(\mathbf{A} - \mathbf{A}^{\mathrm{T}})/2$.

E.5 Functions of Matrices

A matrix can be multiplied by itself if it is a square matrix. If $\mathbf{A}$ is a square matrix, the nth power of a matrix is the matrix multiplied by itself n times

$$\mathbf{A}^n = \mathbf{A}\mathbf{A} \cdots \mathbf{A}, \quad n \text{ times}. \qquad (E.67)$$

Also, since

$$\mathbf{A}^0 \mathbf{A}^n = \mathbf{A}^n \mathbf{A}^0 = \mathbf{A}^n, \qquad (E.68)$$

we have

$$\mathbf{A}^0 = \mathbf{I}. \qquad (E.69)$$

One can construct polynomial functions of $\mathbf{A}$ by taking

$$f(\mathbf{A}) = \sum_{n=0}^{\infty} a_n \mathbf{A}^n. \qquad (E.70)$$

Other elementary functions of $\mathbf{A}$ can be constructed self-consistently by expanding the functions of ordinary scalar variables in a Taylor series and then replacing the scalar variable with $\mathbf{A}$. For example,

$$\cos(\mathbf{A}) = \mathbf{I} - \frac{1}{2!}\mathbf{A}^2 + \frac{1}{4!}\mathbf{A}^4 - \cdots \tag{E.71}$$

$$\sin(\mathbf{A}) = \mathbf{A} - \frac{1}{3!}\mathbf{A}^3 + \cdots \tag{E.72}$$

$$\exp[\mathbf{A}] = \mathbf{I} + \mathbf{A} + \frac{1}{2!}\mathbf{A}^2 + \cdots \tag{E.73}$$

and then

$$\exp[i\mathbf{A}] = \cos(\mathbf{A}) + i\sin(\mathbf{A}), \tag{E.74}$$

where in this case $i = \sqrt{-1}$.

E.6 Determinant

The determinant of a matrix is a unique scalar that can be calculated from the elements of the matrix. Meaningful only for square matrices, the determinant of the $n \times n$ matrix $\mathbf{A}$ is denoted by

$$|\mathbf{A}| = \begin{vmatrix} A_{11} & A_{12} & \cdots & A_{1n} \\ A_{21} & A_{22} & \cdots & A_{2n} \\ \vdots & \vdots & & \vdots \\ A_{n1} & A_{n2} & \cdots & A_{nn} \end{vmatrix}. \tag{E.75}$$

There are a variety of ways to define the determinant. It is convenient here to define it in terms of cofactors. The *minor* $|M_{ij}|$ of element A_{ij} in matrix $\mathbf{A}$ is the determinant of the $(n-1) \times (n-1)$ matrix produced by removing the row i and column j from the matrix. For example, let $\mathbf{A}$ be a 4×4 matrix. Its determinant is

$$|\mathbf{A}| = \begin{vmatrix} A_{11} & A_{12} & A_{13} & A_{14} \\ A_{21} & A_{22} & A_{23} & A_{24} \\ A_{31} & A_{32} & A_{33} & A_{34} \\ A_{41} & A_{42} & A_{43} & A_{44} \end{vmatrix}, \tag{E.76}$$

and its $|M_{23}|$ minor is

$$|M_{23}| = \begin{vmatrix} A_{11} & A_{12} & A_{14} \\ A_{31} & A_{32} & A_{34} \\ A_{41} & A_{42} & A_{44} \end{vmatrix}. \tag{E.77}$$

The *cofactor* $|C_{ij}|$ of element A_{ij} is the minor times $(-1)^{i+j}$:

$$|C_{ij}| = (-1)^{i+j}|M_{ij}|. \tag{E.78}$$

Thus, for the previous example,

$$|C_{23}| = (-1)^{2+3}|M_{23}| = -\begin{vmatrix} A_{11} & A_{12} & A_{14} \\ A_{31} & A_{32} & A_{34} \\ A_{41} & A_{42} & A_{44} \end{vmatrix}. \tag{E.79}$$

The determinant of the $n \times n$ matrix $\mathbf{A}$ is defined in terms of cofactors by the Laplace expansion

$$|\mathbf{A}| = \sum_{k=1}^{n} A_{ki}|C_{ki}|, \qquad (E.80)$$

where the sum can be carried out over any i. This is to be understood as a recursive relation. The determinant of the $n \times n$ matrix is first reduced to a sum over the determinants of $(n-1) \times (n-1)$ matrices (the cofactors). The determinants of the $(n-1) \times (n-1)$ matrices are then calculated from sums over the determinants of $(n-2) \times (n-2)$ matrices, and so on, ending in the sum over the determinants of 2×2 matrices, which are given by, for example,

$$\begin{vmatrix} A_{11} & A_{12} \\ A_{21} & A_{22} \end{vmatrix} = A_{11}A_{22} - A_{12}A_{21}. \qquad (E.81)$$

If the determinant of a matrix is equal to 0, the matrix is called a *singular matrix*.

The following useful properties of determinants are given without proof.

- The determinant of a matrix is equal to the determinant of its transpose:

$$|\mathbf{A}^{\mathrm{T}}| = |\mathbf{A}|. \qquad (E.82)$$

Equation E.80 could, therefore, be replaced by

$$|\mathbf{A}| = \sum_{k=1}^{n} A_{jk}|C_{jk}|, \qquad (E.83)$$

where the sum can be carried out over any j.

- A square matrix is called an *upper triangular matrix* if $A_{ij} = 0$ when $j < i$, and a *lower triangular matrix* if $A_{ij} = 0$ when $j > i$. A square matrix is a *diagonal matrix* if $A_{ij} = 0$ when $i \neq j$. The following are upper triangular, lower triangular, and diagonal 3×3 matrices:

$$\begin{pmatrix} A_{11} & A_{12} & A_{13} \\ 0 & A_{22} & A_{23} \\ 0 & 0 & A_{33} \end{pmatrix}, \quad \begin{pmatrix} A_{11} & 0 & 0 \\ A_{21} & A_{22} & 0 \\ A_{31} & A_{32} & A_{33} \end{pmatrix}, \quad \begin{pmatrix} A_{11} & 0 & 0 \\ 0 & A_{22} & 0 \\ 0 & 0 & A_{33} \end{pmatrix}.$$

If a matrix is a diagonal or a triangular matrix, its determinant is just the product of its diagonal elements:

$$|\mathbf{A}| = \prod A_{ii}. \qquad (E.84)$$

- The determinant of the product of two matrices is the product of their individual determinants:

$$|\mathbf{AB}| = |\mathbf{A}||\mathbf{B}|. \qquad (E.85)$$

The following properties of determinants can easily be derived from equation E.85.

- If all elements in a row or column of a matrix have a common factor λ, this factor may be divided out from the elements, and the determinant is equal to the determinant of the remaining matrix times λ.
- It follows that if all elements in a row or column of a matrix are 0, the determinant of the matrix is equal to 0.
- If two rows or two columns of a matrix are exchanged, the magnitude of the determinant remains the same, but its sign changes.

- A constant multiple of one row of a matrix can be added to another row without changing the determinant of the matrix. This surprising property provides an easy way to calculate the determinants of matrices numerically. Simply add or subtract multiples of rows to turn a matrix into a triangular matrix. The determinant is then the product of diagonal elements of the triangular matrix.
- If two rows or columns of a matrix are identical, the determinant of the matrix is equal to 0.

The Laplace expansion for the determinant (equation E.80) is the sum over terms like $A_{ki}|C_{ki}|$. One might guess that the sum over terms like $A_{kj}|C_{ki}|$ is a meaningless random number unless $j = i$. Perhaps surprisingly, this is not true. To see this, construct a new matrix $\mathbf{A}'$ by replacing all the elements in column i by the elements in column j, so that $(\mathbf{A}')_{ki} = A_{kj}$. The cofactors of elements in column i of matrix $\mathbf{A}'$ remain the same as the cofactors of column i in matrix $\mathbf{A}$: $|(\mathbf{C}')_{ki}| = |C_{ki}|$. Because two of the columns in $\mathbf{A}'$ are the same, though, its determinant is equal to 0, so

$$0 = |\mathbf{A}'| = \sum_{k=1}^{n} (\mathbf{A}')_{ki}|(\mathbf{C}')_{ki}|. = \sum_{k=1}^{n} A_{kj}|C_{ki}|, \quad i \neq j. \tag{E.86}$$

Equations E.80 and E.86 can be combined into a single equation using the Kronecker delta function:

$$|\mathbf{A}| = \sum_{k=1}^{n} A_{kj}|C_{ki}|\delta_{ij}. \tag{E.87}$$

While the Laplace expansion is useful for the analysis of determinants, the expansion is a terrible way to calculate the actual numerical value of the determinant if the matrix is large, because the number of operations increases as $n!$. Methods for calculating determinants in which the number of operations increase as n^3 are available, among them simple Gaussian elimination. Computer programs for calculating determinants are widely available. Issues such as roundoff error, memory size, and computation time soon become important as the size of the matrix increases, so it is wise to test "canned" programs on matrices with known determinants before relying on them.

E.7 Inverse of a Matrix

While matrices can be multiplied, they cannot be divided. The equivalent of division for matrices is the "undoing" of an operation. Let matrix $\mathbf{A}$ operate on vector $\mathbf{a}$ to give vector $\mathbf{b}$:

$$\mathbf{b} = \mathbf{Aa}. \tag{E.88}$$

The matrix that undoes this operation and returns the original vector is called the *inverse* of $\mathbf{A}$. It is denoted by $\mathbf{A}^{-1}$ and has the property

$$\mathbf{A}^{-1}\mathbf{b} = \mathbf{A}^{-1}\mathbf{Aa} = \mathbf{Ia} = \mathbf{a}. \tag{E.89}$$

Similarly, undoing the operation on a vector from the left, $\mathbf{b} = \mathbf{aA}$, yields

$$\mathbf{bA}^{-1} = \mathbf{aAA}^{-1} = \mathbf{aI} = \mathbf{a}, \tag{E.90}$$

and together these two equations yield the definition of the inverse

$$\mathbf{A}^{-1}\mathbf{A} = \mathbf{A}\mathbf{A}^{-1} = \mathbf{I}. \tag{E.91}$$

The inverse of a matrix is unique. The components of the inverse of $\mathbf{A}$ can be written in terms of the determinant of $\mathbf{A}$ and its cofactors:

$$(\mathbf{A}^{-1})_{ij} = \frac{\left(|C_{ij}|\right)^{\mathrm{T}}}{|\mathbf{A}|} = \frac{|C_{ji}|}{|\mathbf{A}|}. \tag{E.92}$$

This expression can be verified by taking

$$(\mathbf{A}\mathbf{A}^{-1})_{ij} = \sum_{k}(\mathbf{A}^{-1})_{ik}(\mathbf{A})_{kj} = \sum_{k}\frac{|C_{ki}|}{|\mathbf{A}|}(\mathbf{A})_{kj} = \frac{|\mathbf{A}|}{|\mathbf{A}|}\delta_{ij} = \delta_{ij}, \tag{E.93}$$

where the penultimate step uses equation E.87. Equation E.92 shows that if $|\mathbf{A}| = 0$, the matrix cannot be inverted. Some other properties of inverses are

$$(\mathbf{A}^{-1})^{-1} = \mathbf{A} \tag{E.94}$$

$$(\mathbf{A}\mathbf{B})^{-1} = \mathbf{B}^{-1}\mathbf{A}^{-1} \tag{E.95}$$

$$(\mathbf{A}^{\mathrm{T}})^{-1} = (\mathbf{A}^{-1})^{\mathrm{T}}. \tag{E.96}$$

Some matrices are easy to invert. The inverse of a diagonal matrix with components A_{ii} has components $(\mathbf{A}^{-1})_{ii} = 1/A_{ii}$:

$$\mathbf{A} = \begin{pmatrix} A_{11} & 0 & 0 & 0 \\ 0 & A_{22} & 0 & 0 \\ 0 & 0 & A_{33} & 0 \\ 0 & 0 & 0 & \ddots \end{pmatrix} \Rightarrow \mathbf{A}^{-1} = \begin{pmatrix} 1/A_{11} & 0 & 0 & 0 \\ 0 & 1/A_{22} & 0 & 0 \\ 0 & 0 & 1/A_{33} & 0 \\ 0 & 0 & 0 & \ddots \end{pmatrix}. \tag{E.97}$$

Triangular matrices are almost as easy to invert. In general, though, calculating the inverses of any but the very smallest matrices is cumbersome and is generally done numerically. Computer programs for calculating inverses are widely available and are typically based on Gauss-Jordan elimination, lower upper (LU) decomposition, or singular value decomposition.[1] As for determinants, roundoff error, memory size, and computation time can become issues for large matrices. It is wise to check that a program gives accurate results by multiplying the inverse back into the original matrix to verify that the result is the unit matrix.

---

[1] See, for example, Press et al. (2007).

E.8 Solutions to Simultaneous Linear Equations

One of the most common uses of matrices is to solve systems of simultaneous linear equations. Suppose one has a set of simultaneous linear equations

$$
\begin{aligned}
A_{11}x_1 + A_{12}x_2 + \cdots + A_{1n}x_n &= b_1 \\
A_{21}x_1 + A_{22}x_2 + \cdots + A_{2n}x_n &= b_2 \\
&\ \ \vdots \\
A_{m1}x_1 + A_{m2}x_2 + \cdots + A_{mn}x_n &= b_m,
\end{aligned}
\tag{E.98}
$$

where the values of the A_{ij} and b_i are given, and one wishes to solve for the x_i. Equation E.98 is equivalent to the matrix equation

$$
\begin{pmatrix}
A_{11} & A_{12} & \cdots & A_{1n} \\
A_{21} & A_{22} & \cdots & A_{2n} \\
\vdots & \vdots & & \vdots \\
A_{m1} & A_{m2} & \cdots & A_{mn}
\end{pmatrix}
\begin{pmatrix}
x_1 \\ x_2 \\ \vdots \\ x_m
\end{pmatrix}
=
\begin{pmatrix}
b_1 \\ b_2 \\ \vdots \\ b_m
\end{pmatrix},
\tag{E.99}
$$

or

$$
\mathbf{Ax} = \mathbf{b}.
\tag{E.100}
$$

To solve for $\mathbf{x}$, calculate the inverse of $\mathbf{A}$ and then multiply both sides of the equation by the inverse to give

$$
\mathbf{x} = \mathbf{A}^{-1}\mathbf{b}.
\tag{E.101}
$$

A unique solution exists if and only if $\mathbf{A}^{-1}$ exists, that is, if and only if $\mathbf{A}$ is nonsingular.

E.9 Change of Basis

It is often useful to change the set of basis vectors for a vector space. Suppose a vector has been expanded in terms of basis vectors $\mathbf{e_i}$

$$
\mathbf{a} = \sum_{i=1}^{n} a_i \mathbf{e_i},
\tag{E.102}
$$

and one wishes to change to a new set of basis vectors $\mathbf{e_i'}$:

$$
\mathbf{a} = \sum_{i=1}^{n} a_i' \mathbf{e_i'}.
\tag{E.103}
$$

The new basis vectors must be linear combinations of the old basis vectors:

$$
\mathbf{e_j'} = \sum_{i=1}^{n} S_{ij} \mathbf{e_i},
\tag{E.104}
$$

where S_{ij} is double subscripted because each $\mathbf{e_i'}$ is a different linear combination of the $\mathbf{e_j}$. We have then

$$\mathbf{a} = \sum_{j=1}^{n} a_j' \mathbf{e_j'} = \sum_{j=1}^{n} a_j' \left(\sum_{i=1}^{n} S_{ij} \mathbf{e_i} \right) = \sum_{i=1}^{n} \left(\sum_{j=1}^{n} S_{ij} a_j' \right) \mathbf{e_i} = \sum_{i=1}^{n} a_i \mathbf{e_i}, \quad (E.105)$$

where

$$a_i = \sum_{j=1}^{n} S_{ij} a_j'. \quad (E.106)$$

Note that the elements a_i transform inversely to the way the basis vectors transform. This must be true if the vector itself is to remain the same. Equation E.106 is a matrix equation

$$\mathbf{a} = \mathbf{S}\mathbf{a}', \quad (E.107)$$

where the elements of matrix $\mathbf{S}$ are S_{ij}. Because the $\mathbf{e_i'}$ are a basis set, it is possible to expand every $\mathbf{e_i}$ in terms of them and return $\mathbf{a}$ back to its expansion in terms of $\mathbf{e_i}$. Thus the inverse of $\mathbf{S}$ must exist, and

$$\mathbf{a}' = \mathbf{S}^{-1}\mathbf{a}. \quad (E.108)$$

Suppose the matrix and the vectors in an arbitrary matrix equation

$$\mathbf{b} = \mathbf{A}\mathbf{a} \quad (E.109)$$

have been expanded in terms of basis vectors $\mathbf{e_i}$. We want the equation to have the same form and same meaning after changing the basis to $\mathbf{e_i'}$:

$$\mathbf{b}' = \mathbf{A}'\mathbf{a}'. \quad (E.110)$$

Multiply equation E.109 by $\mathbf{S}^{-1}$

$$\mathbf{S}^{-1}\mathbf{b} = \mathbf{S}^{-1}\mathbf{A}\mathbf{a}. \quad (E.111)$$

Now insert $\mathbf{I} = \mathbf{S}\mathbf{S}^{-1}$ between $\mathbf{A}$ and $\mathbf{a}$ in equation E.111:

$$\mathbf{S}^{-1}\mathbf{b} = \mathbf{S}^{-1}\mathbf{A}\mathbf{S}\mathbf{S}^{-1}\mathbf{a}. \quad (E.112)$$

Recognizing that $\mathbf{b}' = \mathbf{S}^{-1}\mathbf{b}$ and $\mathbf{a}' = \mathbf{S}^{-1}\mathbf{a}$, we achieve our goal by setting

$$\mathbf{A}' = \mathbf{S}^{-1}\mathbf{A}\mathbf{S}. \quad (E.113)$$

Transformations like equation E.113 are called *similarity transformations*. Note that the unit matrix has the same form for all vector bases, since

$$\mathbf{I}' = \mathbf{S}^{-1}\mathbf{I}\mathbf{S} = \mathbf{S}^{-1}\mathbf{S} = \mathbf{I}. \quad (E.114)$$

E.10 Eigenvectors and Eigenvalues

If $\mathbf{A}$ is an $n \times n$ square matrix, there is at least one nonzero vector $\mathbf{x}$ and scalar λ such that

$$\mathbf{A}\mathbf{x} = \lambda\mathbf{x}. \quad (E.115)$$

The vectors that satisfy equation E.115 are called *eigenvectors* of $\mathbf{A}$, and the corresponding λs are called *eigenvalues*. Equation E.115 can be converted to

$$(\mathbf{A} - \lambda \mathbf{I})\mathbf{x} = 0, \tag{E.116}$$

where $\mathbf{I}$ is the unit matrix. Set $\mathbf{B} = \mathbf{A} - \lambda \mathbf{I}$, so that equation E.116 becomes $\mathbf{Bx} = 0$. When λ is an eigenvalue and $\mathbf{x}$ is an eigenvector, matrix $\mathbf{B}$ cannot have an inverse, because there is no matrix $\mathbf{B}^{-1}$ such that $\mathbf{B}^{-1}\mathbf{Bx} = \mathbf{B}^{-1}0$ retrieves vector $\mathbf{x}$. The inverse does not exist. Thus, when λ is an eigenvalue, the determinant of $\mathbf{B}$ must be 0, or equivalently,

$$|\mathbf{A} - \lambda \mathbf{I}| = 0. \tag{E.117}$$

Equation E.117 yields a polynomial in λ of degree n called the *characteristic equation* for $\mathbf{A}$. The n roots of the polynomial, some of which may be repeated, are the eigenvalues of $\mathbf{A}$. Because they are roots of the characteristic equation, the eigenvalues and eigenvectors are often called *characteristic values* and *characteristic vectors*, respectively.

The values of the eigenvalues are independent of the set of basis vectors on which $\mathbf{A}$ is expanded. To see this, convert $\mathbf{A}$ to $\mathbf{A}'$ by a similarity transformation:

$$\mathbf{A}' = \mathbf{S}^{-1}\mathbf{A}\mathbf{S}. \tag{E.118}$$

The eigenvalues of $\mathbf{A}'$ are given by

$$|\mathbf{A}' - \lambda \mathbf{I}| = |\mathbf{S}^{-1}\mathbf{A}\mathbf{S} - \lambda \mathbf{I}|. \tag{E.119}$$

Because $\mathbf{I} = \mathbf{S}^{-1}\mathbf{I}\mathbf{S}$, this equation becomes

$$\begin{aligned}
|\mathbf{A}' - \lambda \mathbf{I}| &= |\mathbf{S}^{-1}\mathbf{A}\mathbf{S} - \lambda \mathbf{S}^{-1}\mathbf{I}\mathbf{S}| \\
&= |\mathbf{S}^{-1}(\mathbf{A} - \lambda \mathbf{I})\mathbf{S}| = |\mathbf{S}^{-1}||\mathbf{S}||\mathbf{A} - \lambda \mathbf{I}| \\
&= |\mathbf{A} - \lambda \mathbf{I}|.
\end{aligned} \tag{E.120}$$

Thus the eigenvalue equation and the resultant eigenvalues are the same for all basis sets.

If a matrix is symmetric and its elements are real numbers (not complex numbers), the eigenvalues and eigenvectors have two important properties. First, the eigenvectors corresponding to two different eigenvalues are orthogonal. Take any two different eigenvectors $\mathbf{x}_i$ and $\mathbf{x}_j$, and form the products $\mathbf{x}_i^{\mathrm{T}}\mathbf{A}\mathbf{x}_j$ and $\mathbf{x}_j^{\mathrm{T}}\mathbf{A}\mathbf{x}_i$. Because they are eigenvectors, the products are

$$\mathbf{x}_i^{\mathrm{T}}\mathbf{A}\mathbf{x}_j = \lambda_j \mathbf{x}_i^{\mathrm{T}}\mathbf{x}_j \tag{E.121}$$

$$\mathbf{x}_j^{\mathrm{T}}\mathbf{A}\mathbf{x}_i = \lambda_i \mathbf{x}_j^{\mathrm{T}}\mathbf{x}_i. \tag{E.122}$$

From equation E.66, $\mathbf{x}_i^{\mathrm{T}}\mathbf{x}_j$ is the dot product of the two vectors. The transpose of equation E.122 is

$$\mathbf{x}_i^{\mathrm{T}}\mathbf{A}\mathbf{x}_j = \lambda_i \mathbf{x}_i^{\mathrm{T}}\mathbf{x}_j, \tag{E.123}$$

where we have explicitly assumed that $\mathbf{A}^{\mathrm{T}} = \mathbf{A}$. Now subtract equation E.123 from equation E.121 to get

$$(\lambda_j - \lambda_i)\mathbf{x}_i^{\mathrm{T}}\mathbf{x}_j = 0. \tag{E.124}$$

If the two eigenvalues are different, then $\lambda_j - \lambda_i \neq 0$, and the only way that equation E.124 can be satisfied is for $\mathbf{x}_i^{\mathrm{T}}\mathbf{x}_j = 0$, which means the corresponding eigenvectors are orthogonal. If λ_i is a repeated root of the characteristic equation, repeating m times, it is possible to construct m mutually orthogonal eigenvectors all with eigenvalue λ_i using, for example, the Gram-Schmidt orthogonalization procedure.[2]

If an eigenvector satisfies equation E.116, any multiple of the eigenvector also satisfies the equation. Thus eigenvectors are specified only up to a multiplicative constant. This allows the eigenvectors to be normalized by adding the additional requirement:

$$\mathbf{x}^\mathrm{T}\mathbf{x} = 1. \tag{E.125}$$

The normalized eigenvectors form an orthonormal basis set for the vector space in which $\mathbf{A}$ resides.

Second, the eigenvalues of real, symmetric matrices are real numbers. To see this, rewrite equation E.115, now explicitly allowing λ and the elements x_i of $\mathbf{x}$ to be complex:

$$\mathbf{A}\mathbf{x} = \lambda\mathbf{x}. \tag{E.126}$$

This equality still holds for the complex conjugates of λ and $\mathbf{x}$:

$$\mathbf{A}\mathbf{x}^* = \lambda^*\mathbf{x}^*, \tag{E.127}$$

where λ^* is the complex conjugate of λ, and $\mathbf{x}^*$ represents the vector whose individual elements are x_i^*. Now multiply equation E.126 on the left by $(\mathbf{x}^*)^\mathrm{T}$ and equation E.127 on the left by $\mathbf{x}^\mathrm{T}$ and subtract the two equations to get

$$(\mathbf{x}^*)^\mathrm{T}\mathbf{A}\mathbf{x} - \mathbf{x}^\mathrm{T}\mathbf{A}\mathbf{x}^* = \lambda(\mathbf{x}^*)^\mathrm{T}\mathbf{x} - \lambda^*\mathbf{x}^\mathrm{T}\mathbf{x}^*. \tag{E.128}$$

The two terms on the left-hand side are equal to each other because

$$(\mathbf{x}^*)^\mathrm{T}\mathbf{A}\mathbf{x} = \sum_i \sum_j x_i^* A_{ij} x_j = \sum_i \sum_j x_i^* A_{ji} x_j = \sum_i \sum_j x_j A_{ji} x_i^* = \mathbf{x}^\mathrm{T}\mathbf{A}\mathbf{x}^*, \tag{E.129}$$

where the second step explicitly uses the symmetry of $\mathbf{A}$. By similar logic, $(\mathbf{x}^*)^\mathrm{T}\mathbf{x} = \mathbf{x}^\mathrm{T}\mathbf{x}^*$, so equation E.128 becomes

$$0 = (\lambda - \lambda^*)(\mathbf{x}^*)^\mathrm{T}\mathbf{x}. \tag{E.130}$$

Except for the trivial case $\mathbf{x} = 0$,

$$(\mathbf{x}^*)^\mathrm{T}\mathbf{x} = \sum_i x_i^* x_i > 0 \tag{E.131}$$

and therefore

$$\lambda - \lambda^* = 0. \tag{E.132}$$

This can only be true if λ is real. This result is remarkable, because the roots of polynomials can generally be complex.

As a simple example, let $\mathbf{A}$ be the symmetric matrix

$$\mathbf{A} = \begin{pmatrix} 3 & -1 \\ -1 & 3 \end{pmatrix}. \tag{E.133}$$

---

[2] See, for example, W. Cheney and D. R. Kincaid. 2010. *Linera Alegebra: Theory and Applications*, second edition. Subury, MA: Jones and Bartlett Publishers.

Equation E.117 becomes

$$0 = |\mathbf{A} - \lambda\mathbf{I}| = \begin{vmatrix} (3-\lambda) & -1 \\ -1 & (3-\lambda) \end{vmatrix} = (3-\lambda)^2 - 1, \qquad (\text{E.134})$$

so the eigenvalues are $\lambda_1 = 2$ and $\lambda_2 = 4$. The eigenvector corresponding to λ_1 is given by

$$\begin{pmatrix} 3 & -1 \\ -1 & 3 \end{pmatrix} \begin{pmatrix} x_1 \\ x_2 \end{pmatrix} = 2 \begin{pmatrix} x_1 \\ x_2 \end{pmatrix}, \qquad (\text{E.135})$$

which yields $3x_1 - x_2 = 2x_1$, or $x_2 = x_1$. For the eigenvector $\lambda_2 = 4$, we find $x_2 = -x_1$. The normalized eigenvectors are therefore

$$\mathbf{x_1} = \begin{pmatrix} \sqrt{2}/2 \\ \sqrt{2}/2 \end{pmatrix} \quad \text{and} \quad \mathbf{x_2} = \begin{pmatrix} \sqrt{2}/2 \\ -\sqrt{2}/2 \end{pmatrix} \qquad (\text{E.136})$$

The various dot products of these eigenvectors are $\mathbf{x_i}^{\mathrm{T}}\mathbf{x_j} = \delta_{ij}$, so they form a set of orthonormal basis vectors for a two-dimensional vector space.

Orthogonal matrices are square matrices for which the inverse is equal to the transpose:

$$\mathbf{A}^{\mathrm{T}} = \mathbf{A}^{-1}. \qquad (\text{E.137})$$

or, equivalently,

$$\mathbf{A}^{\mathrm{T}}\mathbf{A} = \mathbf{I}. \qquad (\text{E.138})$$

Since the determinant of the transpose of a matrix is equal to the determinant of the matrix itself, we have

$$1 = |\mathbf{I}| = |\mathbf{A}^{\mathrm{T}}||\mathbf{A}| = |\mathbf{A}||\mathbf{A}| = |\mathbf{A}|^2, \qquad (\text{E.139})$$

and thus $|\mathbf{A}| = \pm 1$. Orthogonal matrices often have simple geometric interpretations. For a 2×2 orthogonal matrix, the possible orthogonal matrices and their interpretations are

$$\underset{\text{Identity}}{\begin{pmatrix} 1 & 0 \\ 0 & 1 \end{pmatrix}} \quad \underset{\text{Exchange of axes}}{\begin{pmatrix} 0 & 1 \\ 1 & 0 \end{pmatrix}} \quad \underset{\text{Rotation}}{\begin{pmatrix} \cos\theta & \sin\theta \\ -\sin\theta & \cos\theta \end{pmatrix}} \quad \underset{\text{Reflection}}{\begin{pmatrix} \cos\theta & \sin\theta \\ \sin\theta & -\cos\theta \end{pmatrix}}$$
$$(\text{E.140})$$

Suppose that $\mathbf{A}$ is an orthogonal matrix. Like any other matrix, its eigenvalues and eigenvectors satisfy the equation

$$\mathbf{A}\mathbf{x} = \lambda\mathbf{x}. \qquad (\text{E.141})$$

Multiply this equation on the left by $\mathbf{A}^{\mathrm{T}}$ to get

$$\mathbf{A}^{\mathrm{T}}\mathbf{A}\mathbf{x} = \lambda\mathbf{A}^{\mathrm{T}}\mathbf{x}. \qquad (\text{E.142})$$

Because the matrix is orthogonal, the left-hand side of this equation is $\mathbf{A}^{\mathrm{T}}\mathbf{A}\mathbf{x} = \mathbf{I}\mathbf{x}$, and the right-hand side is $\lambda\mathbf{A}^{\mathrm{T}}\mathbf{x} = \lambda\mathbf{A}\mathbf{x}$, so equation E.142 becomes

$$\mathbf{x} = \lambda\mathbf{A}\mathbf{x} = \lambda^2\mathbf{x}. \qquad (\text{E.143})$$

Thus $\lambda^2 = 1$ and $\lambda = \pm 1$. The geometric interpretation of this result is that the operations listed in E.140 do not change the lengths of vectors.

E.11 Matrix Diagonalization

Assume that a matrix $\mathbf{A}$ is expanded on a set of orthonormal basis vectors $\mathbf{e_i}$, and further that the eigenvalues and eigenvectors of the matrix are λ_i and $\mathbf{x_i}$, respectively. The $\mathbf{x_i}$ can serve as a new set of basis vectors. Matrix $\mathbf{A}$ is transformed to the new basis by the similarity transformation

$$\mathbf{A}' = \mathbf{S}^{-1}\mathbf{A}\mathbf{S}, \tag{E.144}$$

where, from the prescription of Section E.9, the elements of the matrix $\mathbf{S}$ of the similarity transformation are given by

$$\mathbf{x_j} = \sum_{i=1}^{n} S_{ij}\mathbf{e_i}. \tag{E.145}$$

In this case the components of $\mathbf{S}$ are simply the components of the eigenvectors, each eigenvector corresponding to one column:

$$\mathbf{S} = \begin{pmatrix} (\mathbf{x_1})_1 & (\mathbf{x_2})_1 & \cdots & (\mathbf{x_n})_1 \\ (\mathbf{x_1})_2 & (\mathbf{x_2})_2 & \cdots & (\mathbf{x_n})_2 \\ \vdots & \vdots & & \vdots \\ (\mathbf{x_1})_n & (\mathbf{x_2})_n & \cdots & (\mathbf{x_n})_n \end{pmatrix}, \tag{E.146}$$

or

$$(\mathbf{S})_{ij} = (\mathbf{x_j})_i. \tag{E.147}$$

Note the reversal of the indices on the two sides of the equation. If the eigenvectors have been normalized, calculation of the inverse of $\mathbf{S}$ is trivial, because $\mathbf{S}^{-1} = \mathbf{S}^{\mathrm{T}}$. To see this, calculate the individual elements of $\mathbf{S}^{\mathrm{T}}\mathbf{S}$:

$$(\mathbf{S}^{\mathrm{T}}\mathbf{S})_{ij} = \sum_{k=1}^{n}(\mathbf{S}^{\mathrm{T}})_{ik}(\mathbf{S})_{kj}\sum_{k=1}^{n}(\mathbf{x_i})_k(\mathbf{x_j})_k = \mathbf{x_i}\cdot\mathbf{x_j} = \delta_{ij}. \tag{E.148}$$

Thus $\mathbf{S}^{\mathrm{T}}\mathbf{S} = \mathbf{I}$, and the similarity matrix for the transformation is an orthogonal matrix.

Now consider the components of the transformed matrix. They are

$$(\mathbf{A}')_{ij} = (\mathbf{S}^{-1}\mathbf{A}\mathbf{S})_{ij}$$

$$= \sum_{k=1}^{n}(\mathbf{S}^{-1})_{ik}\left(\sum_{r=1}^{n}A_{kr}(\mathbf{S})_{rj}\right) = \sum_{k=1}^{n}(\mathbf{S}^{-1})_{ik}\left(\sum_{r=1}^{n}A_{kr}(\mathbf{x_j})_r\right)$$

$$= \sum_{k=1}^{n}(\mathbf{S}^{-1})_{ik}\lambda_j(\mathbf{x_j})_k = \sum_{k=1}^{n}\lambda_j(\mathbf{S}^{-1})_{ik}(\mathbf{S})_{kj}$$

$$= \lambda_j\delta_{ij}. \tag{E.149}$$

Thus the transformed matrix has the form

$$\mathbf{A}' = \begin{pmatrix} \lambda_1 & 0 & \cdots & 0 \\ 0 & \lambda_2 & \cdots & 0 \\ \vdots & \vdots & & \vdots \\ 0 & 0 & \cdots & \lambda_n \end{pmatrix}. \tag{E.150}$$

The matrix is diagonal, and the diagonal elements are the eigenvalues. The order of the eigenvalues is arbitrary and results from the choice for the order of the eigenvectors. Thus all real symmetric matrices can be converted to a diagonal matrix by a similarity transformation. Geometrically, this means that all real symmetric matrices merely change the lengths of (or the sign of) vectors by amounts λ_i in the directions of the corresponding $\mathbf{x_i}$.

E.12 Vectors and Matrices with Complex Elements

Up to here in this appendix, it has been assumed that the elements of all vectors and matrices are real. This is appropriate, because all vectors and matrices of data analysis are real. The extension to complex vectors and matrices is trivial for most of the discussion. The most important exception is the discussion of eigenvalues and eigenvectors. The role played by symmetric real matrices is taken over by Hermitian matrices. A *Hermitian matrix* is a matrix that equals the complex conjugate of its transpose:

$$\mathbf{H}^\dagger \equiv (\mathbf{H}^T)^* = \mathbf{H}. \tag{E.151}$$

The eigenvalues of a Hermitian matrix are real. The role played by orthogonal matrices is taken over by unitary matrices. A *unitary* matrix is a matrix for which

$$\mathbf{H}^{-1} = \mathbf{H}^\dagger \tag{E.152}$$

or equivalently, $\mathbf{H}^\dagger \mathbf{H} = \mathbf{I}$. While less relevant for data analysis, complex vectors and matrices play major roles in other subjects, notably in quantum mechanics.

Limit of $[1+f(x)/n]^n$ for Large n

We wish to find the limit of $[1+f(x)/n]^n$ as n becomes large. Set

$$y = \left[1+\frac{f(x)}{n}\right]^n. \tag{F.1}$$

Take the derivative

$$dy = n\left[1+\frac{f(x)}{n}\right]^{n-1}\frac{1}{n}d\{f(x)\}, \tag{F.2}$$

and then divide both sides by y to get

$$\frac{dy}{y} = \left[1+\frac{f(x)}{n}\right]^{-1}d\{f(x)\}. \tag{F.3}$$

In the limit as $n \to \infty$ this becomes

$$\frac{dy}{y} = d\{f(x)\}. \tag{F.4}$$

Equation F.4 integrates immediately to

$$\ln y = f(x) + \text{constant}, \tag{F.5}$$

or

$$y = \exp[f(x)], \tag{F.6}$$

where the integration constant has been set to 0 to retrieve equation F.1. Some notable applications are

$$\lim_{n\to\infty}\left[1+\frac{1}{n}\right]^n = e \tag{F.7}$$

$$\lim_{n\to\infty}\left[1-\frac{x}{n}\right]^n = \exp[-x] \tag{F.8}$$

$$\lim_{n\to\infty}\left[1-\frac{ax^2}{n}\right]^n = \exp[-ax^2]. \tag{F.9}$$

Green's Function Solutions for Impulse Response Functions

Suppose that a system is described by the ordinary differential equation

$$z_n(t)\frac{d^n g}{dt^n} + z_{n-1}(t)\frac{d^{n-1}g}{dt^{n-1}} + \cdots + z_1(t)\frac{dg}{dt} + z_0(t)g = 0, \tag{G.1}$$

where the $z_j(t)$ are continuous functions of t. Each of the terms in the differential equation must be finite if they are to sum to 0. We will assume that the solution $g(t)$ is known, remembering that $g(t)$ has n integration constants that are determined by boundary conditions. The Green's function $G(t, t_0)$ for the system is the solution to the inhomogeneous differential equation

$$z_n(t)\frac{d^n G}{dt^n} + z_{n-1}(t)\frac{d^{n-1}G}{dt^{n-1}} + \cdots + z_1(t)\frac{dG}{dt} + z_0(t)G = \delta(t - t_0), \tag{G.2}$$

where $\delta(t - t_0)$ is a Dirac delta function located at $t = t_0$.

G.1 General Green's Function

We wish to solve equation G.2 for $G(t, t_0)$. For $t \neq t_0$, the delta function is equal to 0, and equation G.2 reduces to equation G.1. So, for $t \neq t_0$, we must have $G(t, t_0) = g(t)$. The constants of integration in $G(t, t_0)$ can, however, change abruptly at t_0, because $G(t, t_0)$ or its derivatives must respond to the delta function. We will absorb the response to the delta function into the highest-order derivative and attempt to make $G(t, t_0)$ and its other derivatives continuous. We will find, however, that the $n - 1$ derivative must be discontinuous at t_0. The justification for this traces back to the physical properties of the system described by the differential equation. If the system is a simple harmonic oscillator, for example, a sudden impulse will cause a large instantaneous jump in acceleration, a discontinuous change in velocity, but no discontinuity in position.

We proceed by integrating the differential equation over an infinitesimal interval from $t_0 - \epsilon$ to $t_0 + \epsilon$:

$$\int_{t_0-\epsilon}^{t_0+\epsilon} z_n(t)\frac{d^n G}{dt^n}\,dt + \int_{t_0-\epsilon}^{t_0+\epsilon}\left[z_{n-1}(t)\frac{d^{n-1}G}{dt^{n-1}} + \cdots + z_1(t)\frac{dG}{dt} + z_0(t)G\right]dt = \int_{t_0-\epsilon}^{t_0+\epsilon}\delta(t)\,dt. \tag{G.3}$$

The integral over the delta function is equal to 1. All the terms inside the square brackets are finite, so, as ϵ goes to 0, the integral over the terms in the square bracket goes to 0. Therefore, Equation G.3 reduces to

$$\int_{t_0-\epsilon}^{t_0+\epsilon} z_n(t) \frac{d^n G}{dt^n} dt = 1. \tag{G.4}$$

Integrate this by parts:

$$\int_{t_0-\epsilon}^{t_0+\epsilon} z_n(t) \frac{d^n G}{dt^n} dt = \left[z_n(t) \frac{d^{n-1}G}{dt^{n-1}} \right]_{t_0-\epsilon}^{t_0+\epsilon} - \int_{t_0-\epsilon}^{t_0+\epsilon} \frac{dz_n(t)}{dt} \frac{d^{n-1}G}{dt^{n-1}} dt. \tag{G.5}$$

The integral on the right-hand side goes to 0 as ϵ goes to 0, because the integrand is finite, so equation G.3 reduces yet further to

$$z_n(t_0+\epsilon) \left. \frac{d^{n-1}G}{dt^{n-1}} \right|_{t_0+\epsilon} - z_n(t_0-\epsilon) \left. \frac{d^{n-1}G}{dt^{n-1}} \right|_{t_0-\epsilon} = 1, \tag{G.6}$$

or, since we have required that $z_n(t)$ be continuous, to

$$\left. \frac{d^{n-1}G}{dt^{n-1}} \right|_{t_0+\epsilon} - \left. \frac{d^{n-1}G}{dt^{n-1}} \right|_{t_0-\epsilon} = \frac{1}{z_n(t_0)}. \tag{G.7}$$

We must also now specify that the condition only holds for $z_n(t_0) \neq 0$. Letting ϵ go to 0, equation G.7 specifies that there is a discontinuous jump in the $n-1$ derivative at $t = t_0$.

In summary, the Green's function is $G(t, t_0) = g(t)$, but the constants of integration change at t_0. The relations between the constants above and below t_0 are fully determined by the n boundary conditions:

- $G(t, t_0)$ and all its derivatives up to $n-2$ are continuous at t_0.
- The boundary condition on the $n-1$ derivative is given by equation G.7.

G.2 Application to Impulse Response Functions

The impulse response function $h(t)$ is the response of a system that is initially at rest to a sudden impulse at $t = 0$. We recognize that the impulse response function is the Green's function with $t_0 = 0$. Therefore, $h(t) = G(t, 0) = g(t)$ with the following additional conditions:

- For $t < 0$, $h(t)$ and all its derivatives are equal to 0.
- For $t \geq 0$,

$$h(0) = 0 \tag{G.8}$$

$$\left. \frac{d^k h}{dt^k} \right|_{t=0} = 0, \ k < n-1 \tag{G.9}$$

$$\left. \frac{d^{n-1} h}{dt^{n-1}} \right|_{t=0} = \frac{1}{z_n(0)}. \tag{G.10}$$

G.2.1 Damped Harmonic Oscillator

The differential equation for the damped harmonic oscillator is

$$\frac{d^2g}{dt^2} + 2\alpha\frac{dg}{dt} + \omega_0^2 g = 0, \tag{G.11}$$

for which the solution is

$$g(t) = A\exp[-\alpha t]\sin(\omega_1 t + \phi), \tag{G.12}$$

and $\omega_1^2 = \omega_0^2 - \alpha^2$. The impulse response function is then

$$h(t) = \begin{cases} 0, & t < 0 \\ A\exp[-\alpha t]\sin(\omega_1 t + \phi), & t \geq 0 \end{cases}. \tag{G.13}$$

The constants are determined from the boundary conditions:

$$h(0) = A\sin(\phi) = 0 \tag{G.14}$$

$$\left.\frac{dh}{dt}\right|_{t=0} = -A\alpha\sin(\phi) + A\omega_1\cos(\phi) = 1. \tag{G.15}$$

The solution to this pair of equations is $\phi = 0$ and $A = 1/\omega_1$, so the impulse response function is

$$h(t) = \begin{cases} 0, & t < 0 \\ \dfrac{1}{\omega_1}\exp[-\alpha t]\sin(\omega_1 t), & t \geq 0 \end{cases} \tag{G.16}$$

This is sometimes written more compactly—but less transparently—as

$$h(t) = H(t)\frac{1}{\omega_1}\exp[-\alpha t]\sin(\omega_1 t), \tag{G.17}$$

where $H(t)$ is the Heaviside function.

G.2.2 Object Slowed by Damping

The differential equation for an object whose velocity is slowed by a damping force proportional to its velocity is

$$\frac{dv}{dt} + \alpha v = 0, \tag{G.18}$$

where α is the damping constant. The impulse response function is the solution to the differential equation

$$\frac{dh}{dt} + \alpha h = \delta(t). \tag{G.19}$$

For $t \neq 0$, this yields

$$h(t) = A\exp[-\alpha t]. \tag{G.20}$$

We impose $h(t) = 0$ for $t < 0$ and use the boundary conditions to determine A for $t \geq 0$. For this case, there could be some confusion about the meaning of the derivatives in equations G.9 and G.10, so is best to begin afresh at equation G.3. At $t_0 = 0$ that equation becomes

$$\int_{-\epsilon}^{\epsilon} \frac{dh}{dt} dt + \int_{-\epsilon}^{\epsilon} \alpha h dt = \int_{-\epsilon}^{\epsilon} \delta(t) dt. \tag{G.21}$$

Following the same logic that led from equation G.3 to equation G.7, we find

$$h(\epsilon) - h(-\epsilon) = 1, \tag{G.22}$$

so h jumps from 0 to 1 at $t = 0$. The impulse response function is now fully determined:

$$h(t) = \begin{cases} 0, & t < 0 \\ \exp[-\alpha t], & t \geq 0. \end{cases} \tag{G.23}$$

Second-Order Autoregressive Process

A second-order autoregressive process is a discrete sequence y_r produced by the recurrence relation

$$y_r = a_1 y_{r-1} + a_2 y_{r-2} + \epsilon_r, \tag{H.1}$$

where a_1 and a_2 are constants and ϵ_r is uncorrelated zero-mean random noise. It is the discrete equivalent of a damped simple harmonic oscillator driven by noise. To see this, begin with the differential equation for a driven damped oscillator (see equation 10.118)

$$m\frac{d^2 y}{dt^2} + b\frac{dy}{dt} + ky = \epsilon(t). \tag{H.2}$$

Substituting

$$\frac{dy}{dt} \approx \frac{y_r - y_{r-1}}{\Delta t} \tag{H.3}$$

$$\frac{d^2 y}{dt^2} \approx \frac{y_r - 2y_{r-1} + y_{r-2}}{\Delta t^2} \tag{H.4}$$

and setting $\Delta t = t_r - t_{r-1} = 1$, one soon arrives at the recurrence relation.

H.1 Solution of the Homogeneous Recurrence Relation

Since the homogeneous recurrence relation is the discrete equivalent of a damped simple harmonic oscillator, search for a solution of the form

$$y_r = A\exp[\alpha r \Delta t]\exp[i\omega_1 r \Delta t] = A\exp[\alpha r]\exp[i\omega_1 r], \tag{H.5}$$

where ω_1 is the frequency of the oscillation, and α is the damping coefficient. Substituting this trial solution into the homogeneous recurrence relation, we have

$$A\exp[\alpha r]\exp[i\omega_1 r] = a_1 A\exp[\alpha(r-1)]\exp[i\omega_1(r-1)]$$
$$+ a_2 A\exp[\alpha(r-2)]\exp[i\omega_1(r-2)]$$
$$1 = a_1\exp[-\alpha]\exp[-i\omega_1] + a_2\exp[-2\alpha]\exp[-i2\omega_1]. \tag{H.6}$$

The real and imaginary parts of equation H.6 determine a pair of simultaneous equations:

$$0 = a_1 \exp[-\alpha] \sin(\omega_1) + a_2 \exp[-2\alpha] \sin(2\omega_1) \qquad \text{(H.7)}$$

$$1 = a_1 \exp[-\alpha] \cos(\omega_1) + a_2 \exp[-2\alpha] \cos(2\omega_1). \qquad \text{(H.8)}$$

After some uninteresting algebra, these two equations yield the relations between (a_1, a_2) and (α, ω_1):

$$a_1 = 2 \exp[\alpha] \cos(\omega_1) \qquad \text{(H.9)}$$

$$a_2 = -\exp[2\alpha] \qquad \text{(H.10)}$$

$$\alpha = \frac{1}{2} \ln(-a_2) \qquad \text{(H.11)}$$

$$\cos^2(\omega_1) = -\frac{a_1^2}{4a_2}. \qquad \text{(H.12)}$$

The conditions for a stable underdamped oscillator are $\alpha < 0$ and $0 < \cos^2(\omega_1) \le 1$, which lead to the constraints

$$-1 < a_2 < 0 \qquad \text{(H.13)}$$

$$a_1^2 < -4a_2. \qquad \text{(H.14)}$$

H.2 Frequency Response and Gain Functions

The frequency response function is found by driving the recurrence relation with a sine curve. A discretely sampled sinusoidal driving force can be written in complex form as

$$f_r = \exp[i\omega r \Delta t] = \exp[i\omega r], \qquad \text{(H.15)}$$

where ω is the frequency of the sine curve. Replace ϵ_r by f_r on the right-hand side of equation H.1, and look for a solution of the form

$$y_r = H(\omega) \exp[i\omega r], \qquad \text{(H.16)}$$

where $H(\omega)$ is the frequency response function. We have

$$H(\omega) \exp[i\omega r] = a_1 H(\omega) \exp[i\omega(r-1)] + a_2 H(\omega) \exp[i\omega(r-2)] + \exp[i\omega r], \quad \text{(H.17)}$$

or

$$H(\omega) = a_1 H(\omega) \exp[-i\omega] + a_2 H(\omega) \exp[-2i\omega] + 1. \qquad \text{(H.18)}$$

The frequency response function is therefore

$$H(\omega) = \frac{1}{1 - a_1 \exp[-i\omega] - a_2 \exp[-2i\omega]}. \qquad \text{(H.19)}$$

The gain function is

$$
\begin{aligned}
Z(\omega) = |H(\omega)| &= \left[H^*(\omega)H(\omega)\right]^{1/2} \\
&= \big\{1 + a_1^2 + a_2^2 - a_1\left(\exp[-i\omega] + \exp[i\omega]\right) - a_2\left(\exp[-i2\omega] + \exp[i2\omega]\right) \\
&\quad + a_1 a_2\left(\exp[-i\omega] + \exp[i\omega]\right)\big\}^{-1/2} \\
&= \left\{1 + a_1^2 + a_2^2 - 2a_1\cos(\omega) - 2a_2\cos(2\omega) + 2a_1 a_2\cos(\omega)\right\}^{-1/2} \\
&= \left\{1 + a_1^2 + a_2^2 + [2a_1 a_2 - 2a_1]\cos(\omega) - 2a_2\cos(2\omega)\right\}^{-1/2}.
\end{aligned}
\tag{H.20}
$$

Equations H.19 and H.20 should be compared to equations 10.120 and 10.121, which are the frequency response function and gain function for the continuous driven damped harmonic oscillator. If desired, one can calculate the phase shift from (see equation 10.31)

$$
\tan[\phi(\omega)] = \frac{\operatorname{Im}[H(\omega)]}{\operatorname{Re}[H(\omega)]}.
\tag{H.21}
$$

H.3 Resonant Frequency

The *resonant frequency* is the frequency at which the gain is greatest, which occurs at

$$
\frac{d}{d\omega}Z(\omega) = 0,
\tag{H.22}
$$

or

$$
\begin{aligned}
\frac{d}{d\omega}\left\{1 + a_1^2 + a_2^2 + [2a_1 a_2 - 2a_1]\cos(\omega) - 2a_2\cos(2\omega)\right\} &= 0 \\
-[2a_1 a_2 - 2a_1]\sin(\omega) + 4a_2\sin(2\omega) &= 0 \\
-[2a_1 a_2 - 2a_1]\sin(\omega) + 8a_2\sin(\omega)\cos(\omega) &= 0.
\end{aligned}
\tag{H.23}
$$

Calling the resonant frequency ω_0, we find

$$
\cos(\omega_0) = \frac{a_1(a_2 - 1)}{4a_2}.
\tag{H.24}
$$

From equations H.9 and H.10, the relation between ω_0 and ω_1 is

$$
\cos(\omega_0) = \frac{1}{2}\left(\exp[\alpha] + \exp[-\alpha]\right)\cos(\omega_1).
\tag{H.25}
$$

H.4 Mean Autocovariance Function

The *mean autocovariance function* of a sequence y_r of length n is

$$
\langle\gamma_{yy}(k)\rangle = \frac{1}{n}\sum_{r=1}^{n-k}\langle y_r y_{r+k}\rangle.
\tag{H.26}
$$

If $n \gg k$ and if the y_r were produced by a second-order autoregressive process, then

$$\langle \gamma_{yy}(k) \rangle = \frac{1}{n} \sum_{r=1}^{n} \langle y_r(a_1 y_{r+k-1} + a_2 y_{r+k-2} + \epsilon_{r+k}) \rangle.$$

$$= a_1 \left(\frac{1}{n} \sum_{r=1}^{n} \langle y_r y_{r+k-1} \rangle \right) + a_2 \left(\frac{1}{n} \sum_{r=1}^{n} \langle y_r y_{r+k-2} \rangle \right) + \frac{1}{n} \sum_{j=1}^{n} \langle y_r \epsilon_{r+k} \rangle. \quad \text{(H.27)}$$

The first term on the right-hand side if this equation is $\langle \gamma_{yy}(k-1) \rangle$, and the second term is $\langle \gamma_{yy}(k-2) \rangle$. The third term on the right is zero if $k \geq 1$, because ϵ_{r+k} is generated later than y_r and is independent of and uncorrelated with y_r. For $k = 0$ the third term becomes

$$\frac{1}{n} \sum_{r=1}^{n} \langle y_r \epsilon_r \rangle = \frac{1}{n} \sum_{r=1}^{n} \langle a_1 y_{r-1} \epsilon_r + a_2 y_{r-2} \epsilon_r + \epsilon_r^2 \rangle. \quad \text{(H.28)}$$

Once again, ϵ_r is generated later than y_{r-1} and y_{r-2}, so it is independent and uncorrelated with them. Since $\langle \epsilon_r \rangle = 0$, the third term becomes

$$\frac{1}{n} \sum_{r=1}^{n} \langle y_r \epsilon_r \rangle = \frac{1}{n} \sum_{r=1}^{n} \langle \epsilon_r^2 \rangle = \sigma_\epsilon^2. \quad \text{(H.29)}$$

Thus we find

$$\langle \gamma_{yy}(k) \rangle = a_1 \langle \gamma_{yy}(k-1) \rangle + a_2 \langle \gamma_{yy}(k-2) \rangle + \delta_{k0} \sigma_\epsilon^2. \quad \text{(H.30)}$$

This can be understood as a recurrence relation for the autocovariance function. The sequence is driven by a single impulse at $k = 0$ with amplitude σ_ϵ^2. For $k = 1$ and later, the recurrence relation is identical to the homogeneous form of the recurrence relation for the original autoregressive process! We can therefore write down the mean autocovariance function by inspection:

$$\langle \gamma_{yy}(k) \rangle \propto \exp[\alpha k] \exp[i \omega_1 k], \quad \text{(H.31)}$$

with α and ω_1 given by equations H.11 and H.12, respectively.

Bibliography

Basic Mathematical Methods
  - Arfken, G. B., and H. J. Weber. 2000. *Mathematical Methods for Physicists*, fifth edition. Waltham, MA: Harcourt Academic Press.
  - Courant, R., and D. Hilbert. 1989. *Methods of Mathematical Physics*, New York: Wiley-Interscience.
  - Mathews, J., and R. L. Walker. 1970. *Mathematical Methods of Physics*, second edition. Boston: Addison-Wesley.
  - Riley, K. F., M. P. Hobson, and S. J. Bence. 2006. *Mathematical Methods for Physics and Engineering*, third edition. Cambridge: Cambridge University Press.
Probability
  - Feller, W. 1968. *An Introduction to Probability Theory and Its Applications*, third edition. Hoboken, NJ: John Wiley and Sons.
  - Forbes, C., M. Evans, N., Hastings, and B. Peacock. 2010. *Statistical Distributions*, fourth edition. Hoboken, NJ: John Wiley and Sons.
  - Grinstead, C. M., and J. L. Snell. 1997. *Introduction to Probability*, second revised edition. Providence, RI: American Mathematical Society.
  - Tijms, H. 2012. *Understanding Probability*, third edition. Cambridge: Cambridge University Press.
Statistics and Data Analysis
  - Bernardo, J. M., and A. F. M. Smith. 2000. *Bayesian Theory*. New York: John Wiley.
  - Bevington, P. R., and D. K. Robinson. 2002. *Data Reduction and Error Analysis for the Physical Sciences*, third edition. New York: McGraw-Hill.
  - Cowan, G. 1998. *Statistical Data Analysis*. Oxford: Oxford University Press.
  - Edwards, A. W. F. 1992. *Likelihood*, Expanded edition. Baltimore: Johns Hopkins University Press.
  - Gelman, A., J. B. Carlin, H. S. Stern, D. B. Dunson, A. Vehtari, and D. B. Rubin. 2013. *Bayesian Data Analysis*, third edition. London/Boca Raton, FL: Chapman and Hall/CRC Press.
  - Gregory, P. 2010. *Bayesian Logical Data Analysis for the Physical Sciences*. Cambridge: Cambridge University Press.
  - Hoel, P. G. 1947. *Introduction to Mathematical Statistics*. New York: John Wiley.
  - Jaynes, E. T. 2003. *Probability Theory, The Logic of Science*. Cambridge: Cambridge University Press.

– Kendall, M. G. 1969. *The Advanced Theory of Statistics*, third edition. New York: Hafner Publishing.
– Meyer, S. L. 1975. *Data Analysis for Scientists and Engineers*. Hoboken, NJ: John Wiley and Sons.
– Sivia, D. S. 1996. *Data Analysis: A Bayesian Tutorial*. Oxford: Oxford University Press.

Spectral Analysis

– Blackman, R. B., and J. W. Tukey. 1959. *The Measurement of Power Spectra*. New York: Dover.
– Box, G. E. P., G. M. Jenkins, and G. C. Reinsel. 1995. *Time Series Analysis: Forecasting and Control*, third edition. Prentice Hall.
– Bracewell, R. N. 1965. *The Fourier Transform & Its Applications*. New York: McGraw-Hill.
– Jenkins, G. M., and D. G. Watts. 1968. *Spectral Analysis and Its Applications*. San Francisco: Holden-Day.
– Papoulis, A. 1962. *The Fourier Integral and Its Applications*. New York: McGraw-Hill.
– Proakis, J. G., and D. K. Manolakis. 2006. *Digital Signal Processing*, fourth edition. Upper Saddle River, NJ: Prentice Hall.
– Stoica, P., and R. L. Moses. 2005. *Spectral Analysis of Signals*. Upper Saddle River, NJ: Prentice Hall.

Numerical Techniques and Tables

– Press, W. H., S. A. Teukolsky, W. T. Vetterling, and B. P. Flannery. 2007. *Numerical Recipes*, third edition. Cambridge: Cambridge University Press.
– Zwillinger, D., and S. Kokoska. 1999. *CRC Standard Probability and Statistics Tables and Formulae*. Boca Raton, FL: CRC Press.

Index